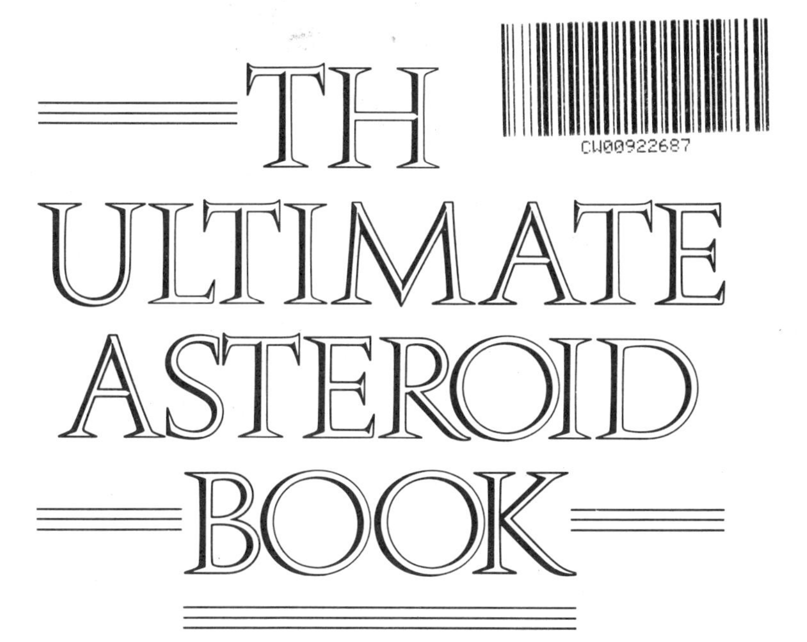

BY J. LEE LEHMAN, PH.D.

1469 Morstein Road
West Chester, Pennsylvania 19380 USA

Cover design by Clare Ultimo
Edited by Skye Alexander
Charts by Astro Computing Services

International Standard Book Number: 0-914918-78-8
Library of Congress Catalog Card Number: 88-50479

Published by Whitford Press
a division of Schiffer Publishing Ltd.
1469 Morstein Road
West Chester, Pennsylvania 19380

Manufactured in the United States of America

This book may be purchased from the publisher.
Please include $2.00 postage
Try your bookstore first.

Contents

Part Four: Mundane, Theme and Variations

Introduction

In autumn of 1979, Al H. Morrison convinced me to work with him in producing ephemerides of a number of asteroids which had not been studied previously. When I began calculating the ephemerides, I had little expectation that they would be especially interesting or useful. In fact, I almost did it as a joke.

I was wrong. Three of the first ten we published, Sappho, Eros and Amor, provide many insights into relationship analysis. When I began speaking about these bodies, it became clear that astrologers were very receptive to new methods of relationship analysis. Accordingly, we present here ephemerides for these and other asteroids, along with delineations of the meanings of these bodies.

I would like to thank Tee Corinne and the Southern Oregon Women Writers' Group for their comments on the manuscript. I also would like to thank Margaret Meister and Barbara May for their many contributions to the concept of this manuscript. Debbie Kempton-Smith deserves kudos for coming up with the delightful nickname "'roids." I also want to thank the following people for discussions and help: Mark Pottenger, Zipporah Dobyns, Marilyn Muir, Martha Wheelock, and Frances McEvoy.

1

Introducing the Asteroids

The asteroids are a series of small, heavenly bodies whose orbits fall mainly between those of Mars and Jupiter. The largest asteroid is 620 miles in diameter, the smallest, less than one mile.

Asteroids are classified according to their orbital characteristics: Those in the Apollo group have orbits which extend inside the orbits of Venus or Mercury; the Trojan group asteroids move in the same orbit as Jupiter and at a distance of 60 degrees (seven of these asteroids are in front of Jupiter; five are behind). Other groups include the Hilda Group, the Minerva Group, the Hestia Group and the Flora Group. The asteroids which are located between Mars and Jupiter typically have sidereal periods (i.e., the time it takes to make one orbit around the Sun) of three to five years. However, the asteroids that spend part of their orbit within that of the Earth may have sidereal periods of much less. For instance, the sidereal period of Icarus is 1.12 years, and Icarus has passed within only four million miles of the Earth. By contrast, the mean distance between the Moon and the Earth is 239,000 miles.

The first asteroid, Ceres, was discovered on January 1, 1801 by M. Piazzi. Since then, over 2200 asteroids have been confirmed. Traditionally, the person who discovers the asteroid names it, although today naming is delayed pending confirmation of the siting by other astronomers. Possibly because this confirmation was not always required, or because the asteroids in question were unstable, a few of the asteroids are considered to be "lost."

Although Alan Leo attempted to generate interest in the asteroids, the first astrologer to successfully create a lasting interest in them was Eleanor Bach. In 1973 she published an ephemeris of the first four asteroids, Ceres, Pallas, Juno and Vesta and her interpretation of their effects. Bach has continued to expand her observations, and has been joined subsequently by a large number of astrologers. A more accurate ephemeris for the first four asteroids, along with interpretations by Zipporah Dobyns, was published by TIA Publications in 1977. Another relatively early player was Emma Belle Donath, who published a series of books on the first four asteroids.

Such was the state of affairs in 1979 when Al Morrison asked me to use my computer to calculate the zodaical positions of some additional asteroids. We first looked at the asteroids Sappho, Eros, Lilith, Hidalgo, Toro and Icarus. At that time, two principal criteria were used for selecting asteroids: the names and the orbits. Sappho, Eros and Lilith were selected for their names. Toro was selected because it has passed so close to the Earth that when it was discovered there was some speculation about whether it was a second moon. Hidalgo's orbit extends outside the orbit of Jupiter and is highly eccentric. (Eccentricity refers to the angle between the plane of the orbit of the body in question and the Earth.) Icarus was the only asteroid known to pass within the orbit of Mercury.

These six ephemerides were published early in 1980. Subsequent ephemerides have been published, both by CAO Times and by the National Council of Geocosmic Research. The work Al and I started has been confirmed by a number of people who have worked with these ephemerides. In addition, the Asteroid Subcommittee of the New York Chapter of N.C.G.R. extended the study into common name asteroids. One member of that subcommittee, Diana Rosenberg, encouraged me to study the asteroid nodes, and that has been another fruitful area of research. These different study groups have reached the following consensus:

1. The asteroids have astrological effects which may be studied.
2. The name of an asteroid has astrological significance.

The use of the asteroids brings an influx of new symbols, and the addition of new symbols means the number of ideas which may be communicated is increased. It is as if you were working with a language which contained a vocabulary of ten words. You could create many more meanings than ten by combining these words in different ways. But consider how much richer the possible communication becomes if you suddenly have more than twice as many words to play with.

Several legitimate questions should be considered concerning the use of the asteroids astrologically. These questions fit into two major groups. First, some astrologers tell me that the asteroids are an advanced technique, and that it is important to stick to basics. This is probably true.

I certainly would not want to go around discussing Hidalgo if I didn't already have some understanding of Saturn. This objection, however, merely postpones delving into the asteroids; it has nothing to do with their validity. Another objection stems from the need to deal with over two thousand asteroid bodies and analyze them in someone's chart! This objection seems to be primarily a quantitative one. If there are that many bodies, how can any one be that significant? And since they are so small, are they less significant than the planets?

The objection that there are too many asteroids is a practical one, and I will deal with it shortly. The size issue, being a physical (i.e., quantitative) one, may be considered by looking at the quantitative difference in a known force: gravity. Using the formula for gravitational attraction (which varies according to mass and the distance squared), one may calculate how large a body which is three astronomical units from the Sun (midway between Mars and Jupiter) would have to be in order to equal the gravitational effect of the planet Pluto on the Earth. Substituting the appropriate values, the result is a body with a radius of 240 miles—which is slightly less than the radius of Ceres.

Since most astrologers would admit that Pluto is presumably not the outer limit (except in interpretation!), this means that many more asteroids should be large enough to have an effect. I do not mean to imply that gravity in fact has anything to do with how astrology works; I simply point out that those who attempt to make a quantitative argument when we don't know what is the "cause" (if, indeed, there is one) of astrological phenomena must consider known forces in their modeling.

The dreaded question which follows is: Must I have 2200 ephemerides sitting around my house, can I get my computer to do the calculations, and how am I going to read the chart with all those extra critters crawling all over it? The answer lies in going back to the conclusions that research has reached so far.

The name of an asteroid is significant. This suggests that not all asteroids properly belong in a reading, which is necessarily a summary of the important things in a person's life as seen by the astrologer. I have no desire to put the position of the asteroid Rumpelstilz (#1773) or Aida (#861) in the chart of everyone for whom I read. However, if the client is a soprano who is a Verdi freak and has had a lifelong obsession with singing Aida, it might be interesting to find out the position of Aida, or at least, the nodal position of Aida in her chart.

This example suggests one of the real values of using asteroids, whether we are talking about only the first four, or the ones I will be discussing in this book. The interpretation of a planet is a bit tricky, since a planet can mean many things. Venus, for example, could refer to love, harmony, magnetic attraction, the veins, diabetes, erotica, potatoes, or a host of other things. The asteroid Eros refers to passionate attachment. The asteroids have fewer concepts allocated to them. The fact that they are small and numerous may allow for many very exact meanings.

In this book I will discuss a number of the asteroids, some of which have been published already, and some additional ones. I also will show how the asteroids may be used. This includes both classic delineations and the use of the asteroid nodes. As will be seen, the methods used are exactly the same as the methods used for the planets. Therefore, using any particular asteroid which may appeal to you is easy; it is simply necessary to understand what the asteroid means, and then use it the same way you use any other celestial body.

There are four major ways to approach the astrology of an unknown asteroid:

1. Use the "common name" aspect of the asteroid, which means that you will be looking for a literal effect of the name. For example, the asteroid Apollo may appear prominently in the chart of someone involved in the Apollo space program, or in the chart of a company named Apollo regardless of whether the company has anything to do with other aspects of the meaning of the asteroid Apollo.
2. Use the mythology of a name as a guide to the meaning of the asteroid. The difficulty here lies in the fact that different times and generations interpret mythological symbols differently. For example, there is a renaissance of interest in and reinterpretation of the mythology of goddesses, both in astrology and in society. This is part of the living nature of myths: they change over time. Thus, while the mythology is significant in the interpretation of an asteroid, you must be careful to be aware not only of the layers created by multiple reinterpretations, but also the essence of the myth itself.
3. Use historical information that relates to the name. This helps you to get an idea of the essence of a symbol. However, like the

use of mythology, the interpretation of an asteroid (or of a planet, for that matter) may change some over time, but the essence should remain recognizable.

4. Use the asteroids in a variety of charts to see what effect they have. This work, however, will always be dependent on the first three methods mentioned above, because something has to prompt the question before it is researched.

In this book, I will use all four approaches to illustrate the meanings of various asteroids. Usually, my approach is a combination. Occasionally, I will refer to the approach used by name, such as the "common name" method.

Several ephemerides for the asteroids and a listing of asteroid node positions are included in the back of this book. For further information about asteroid positions, two additional sources are available. Mark Pottenger has written a computer program which calculates the positions of more than four hundred asteroids. This program requires a PC compatible, and preferably a hard disk. He has also printed out a number of ephemerides for these asteroids, including the Sappho, Eros and Amor positions contained in this volume. The second source is Astro Computing Services, Box 16430, San Diego, CA 92116-0430, which uses programs written by Mark to either produce an ephemeris or print out the natal positions of asteroids.

Before proceeding, a word is in order about asteroid symbols. Some asteroids were given their symbols by astronomers, others are "assigned" when an ephemeris publication occurs. This practice follows the conventions set up in various scientific disciplines concerning precedence in naming, discussed earlier. To take this a step further, we suggest that asteroids only be given symbols at the time when they are introduced to the astrological community in a usable way, which means as an ephemeris, preferably of positions for at least a century. These protocols keep the number of different symbols low. There are certainly enough asteroids to go around, without multiplying the number of glyphs.

A few years ago, Eleanor Bach was kind enough to provide me with some xeroxed pages of E. Otis Kendall's *Uranography; or A Description of the Heavens,* which was published in 1844. Kendall's work was designed as a textbook in astronomy. There were two curious points to be gleaned from this. One is the title, which will surely give astrologers who have connected astrology with Uranus or Urania something to consider! The second is that glyphs had been assigned to the first four asteroids Ceres, Pallas, Juno and Vesta. The symbols which Eleanor published for Ceres and Pallas are identical to those used in Kendall's book. The two that he gives for Vesta and Juno are different, and fortunately they were not adopted astrologocially because they would have proved difficult to draw.

In the years since I started working on the asteroids, I have been awed repeatedly, not only because the asteroids work, but because this is just one indication of the degree of interconnection between our alleged separate selves and the Universe. Perhaps it is the very large number of asteroids which finally drives this point home.

Part One
Motivations

2

The Mind Asteroids: Psyche, Pallas, Athene, Minerva

> *"Mathematics may be defined as the subject in which we never know what we are talking about, nor whether what we are saying is true." (Bertrand Russell)*

For many thousands of years astrologers have been interested in the functioning of the human being, as shown through the planets. More recently, this realm, which previously was the province of religion, has been claimed by the "science" of psychology. As in many fields of human endeavor, the latest advocates often fail to acknowledge their debt to those who came before them. One psychologist, however, did not fall prey to this tendency. Carl Jung very clearly set out the spiritual dimensions of the mind while elaborating on Freud's concepts of the unconscious and subconscious.

Since the seventeenth century astrologers often have worked in a kind of intellectual vacuum, absorbing quotes from socially-acceptable fields to buttress their opinions. More recently, humanistic astrology has become quite conversant in humanistic psychology as astrological emphasis has shifted from horary and predictive astrology to individual chart delineation.

In this shift to natal astrology, one of the more difficult concepts we have had to come to grips with is how to study those features unique to humans, those things which don't show up in questions or events. Perhaps the core issue to this astrology of human beings, therefore, is the concept of the mind.

The mind: a term that has been interpreted so many ways that it has almost lost all meaning, and is often entwined with the brain, the presumptive organic house or seat of the mind. We have to distinguish between such concepts as lower mind and higher mind (based on the medical/biological research which shows that portions of our brains survive from previous stages in our evolution), mind versus Mind (the mind that the self believes is all there is or the Mind of the transcendant Self), the ego, the id, unconsciousness, subconsciousness and the rest. The mind is a mine field.

In this chapter we will consider four asteroids that have to do with the mind in a general sense: Psyche, Pallas, Athene and Minerva. All these names come down to us from mythology. I will discuss these asteroids in two groups, first Psyche, then the triumvirate of Pallas, Athene and Minerva. These two groups represent the two sides of mind being considered: mind in its abstract form and mind in its applied form.

Psyche

Let's begin the study of Psyche with the myth of Psyche. This myth is presented in detail in Erich Neumann's *Amor and Psyche.* To summarize briefly, Psyche was a beautiful mortal. Many mortal men forsook the adoration of Aphrodite and worshipped the beauty of Psyche. Psyche herself was indifferent to this worship. Aphrodite, on the other hand, was furious, and sent her son Eros to dispose of Psyche. Instead, Eros fell in love with her and married her, though he would not let Psyche see his face (i.e., know who he was). Psyche's jealous sisters convinced her that she had married a monster and that she must kill him. One night she bared a lamp and saw that she was married to a beautiful god. However, a drop of oil from the lamp fell on Eros and he awoke and departed from her. In despair, she traveled the Earth, and eventually threw herself at the mercy of Aphrodite. Aphrodite sent Psyche on a series of "impossible" tasks which she was able to accomplish with the help of various creatures who came to her aid. During the final quest, Eros rescued her after he "escaped" from "imprisonment" in his mother's house. Psyche was taken to Mount Olympus where she was granted immortality and she resumed her position as the wife of Eros.

Psychologists have had a field day with this myth. Some, such as Linda Schiese Leonard, interpret Psyche as the quintessential female heroine, emphasizing the "passivity" of Psyche (others came to her aid during all of her quests). Others, such as Erich Neumann, see Psyche as the first human to become divine, emphasizing that what Psyche did was take charge of her fate (by looking at her sleeping husband, etc.). The essential point, according to these psychologists, is that Psyche's love worked to transform her *and* her husband. When the tale begins, Psyche is essentially unconscious, living in the garden of a husband she doesn't know. She ends up the mistress of her fate. To get from Point A to Point B she had to transcend not only her own unconsciousness but also her husband's tendency to keep her in ignorance. Thus, she could be considered his savior as well as her own.

Jung defined Psyche as the intermediary between "the physiological sphere of instinct and the spiritual."[1] When we first published the ephemerides of Psyche, I called this asteroid "the interface between the conscious, the subconscious and the unconscious." This would be in accord with the myth of Psyche. But by understanding the meaning of her transformation and that of Eros from Mama's boy to husband, one may with some justification consider Psyche the ruler of the individual mind. By *individual mind,* I refer to that sense which most people have of separate existence. Psyche represents the stage of development in which the separate self emerges. Psyche stands between the unconscious pre-self and the superconscious beyond-self. Most of the Eastern religions have stated to one degree or another that this individual self is either ultimately nonexistent or subordinate to the higher Self (God). Yet it seems that this egoic individuation is nonetheless necessary before the ego can be transcended.

Another way we may examine the workings of the individual mind is by making a model of it, a sort of semi-black box.

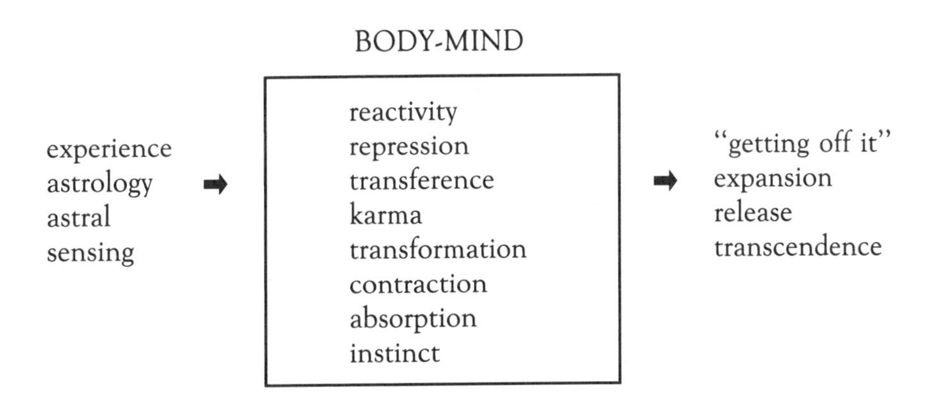

Here we see a process. Our individual body-mind exists within a matrix. (The term body-mind is one representation of the concept which Ken Wilber calls the centaur; others have called it by other names, but the essential idea is that we can achieve a level of awareness beyond simply that of the body or that of the mind.[2]) We have experiences, we sense things and people (through our sense organs or extra-sensory perception), there are influences on nonmaterial planes (represented by the word astral, though there are myriad definitions for that term which need not concern us here). Of course, as astrologers we should not omit the astrological influences going on, primarily the transits if we are consider these influences to be "outside" the self.

All this "stuff" is going on: people, events, possessions, ideas. We could ignore it, of course. However, what most of us do is suck it up: we identify with certain pieces of stuff, reject others. In this way, each of us can imagine we are different from other people, who pick up and reject different pieces than we do.

Once we have picked up this stuff, we process it. This is the domain of classical psychology: we can sublimate it, transfer the emotional impact to something other than the original source, transform one emotion into another, react to it (consciously or unconsciously, the latter often being instinctual), project it onto someone else, etc.

This very processing is what we may consider to be the Psyche, the mind. In the words of the Zen teacher Shunryu Suzuki

> " . . . usually our mind is very busy and complicated, and it is difficult to be concentrated on what we are doing. This is because before we act we think, and this thinking leaves some trace. Our activity is shadowed by some preconceived idea. The thinking not only leaves some trace or shadow, but also gives us many other notions about other activities and things. These traces and notions make our minds very complicated. When we do something with a quite simple, clear mind, we have no notion of shadows, and our activity is strong and straightforward. But when we do something with a complicated mind, in relation to other things or other people, our activity becomes very complex."[3]

There is another possibility, too, and that is release. Release is profoundly difficult for most people because it involves completely letting go. This is, in fact, analogous to the difficulty of the mourning process: the unwillingness to let go of our attachments to the dead person.

From an astrological standpoint, we may view this threefold process as being somewhat analogous to the cardinal, fixed and mutable positions. Most people do not act out the true potentials of the triplicity: for example, most mutables either avoid situations in the first place (so there is nothing to release) or simply sublimate what is going on rather than release it.

How does all this relate to the asteroid Psyche? Psyche expresses how this function we call mind operates. For example, you may observe that someone with Psyche positioned in a fire sign may allow lots of things to glance off without letting them "in," or this person may vent whatever does get in explosively. When Psyche is in a water sign, there is more of a flow. People with air-sign Psyches may try to get things out by talking, while those with earth-sign Psyches tend to bury things.

I would suggest that it is the people with air-sign Psyches who are most predisposed toward traditional psychoanalysis, since this technique primarily involves talking. In fact, one could theorize about what kinds of therapies would appeal to which signs: rebirthing to Cancer, primal to Scorpio, Gestalt to Sagittarius, etc.

Psyche in aspect to the planets may further illustrate the workings of this asteroid. For instance, Psyche-Moon connections between two people often show that one person is psyching out the other. In an individual birth chart, a Psyche-Mercury connection may mean that mental processes operate principally on an intuitive level.

Psyche-Neptune aspects may represent the interface to the collective unconsciousness.

> " . . . instincts are impersonal, universally distributed, hereditary factors of a dynamic or motivating character, which very often fail so completely to reach consciousness that modern psychotherapy is faced with the task of helping the patient to become conscious of them. Moreover, the instincts are not vague and indefinite by nature, but are specifically formed motive forces which, long before there is any consciousness, and in spite of any degree of consciousness later on, pursue their inherent goals. Consequently they form very close analogies to the archetypes, so close, in fact, that there is good reason for supposing that the archetypes are the unconscious images of the instincts themselves, in other words, they are *patterns of instinctive behavior.*"[4]

Note two levels of "instinct": the biological (i.e., material) and the spiritual. Psyche may be thought of as the modulator between the two. The spiritual may be thought of as acting through the psychic sense, or synchronicity:

> "Synchronicity . . . consists of two factors: a) An unconscious image comes into consciousness either directly (i.e., literally) or indirectly (symbolized or suggested) in the form of a dream, idea, or premonition. b) An objective situation coincides with this content."[5]

Here we are talking about a system that is not at all like the Western classical conception. We are into the quantum mechanics of the mind.

> " . . . the relation between the properties of a physical system on the one hand, and the projections (wave function) on the other, makes possible a sort of logical calculus with these. However, in contrast to the concepts of ordinary logic, this system is extended by the concepts of 'simultaneous decidability' (the uncertainty principle) which is characteristic for quantum mechanics."[6]

This model I am presenting for Psyche requires viewing the human being from a different perspective than that usually adopted by the astrologer. My original characterization of Psyche as the personal Neptune may be understood in this light: "Personalizing" Neptune is the process of getting beyond the usual self conceptualization of this body, this ego, this self. This is the sense of dissolution astrologers notice with regard to Neptune. "Chaos" replaces order as the nice neat boxes we put ourselves into are removed. As personal order dissolves we become aware of an entirely different mode of existence, in which the boundaries are more fluid than we previously understood.

As many have noted before, this is not the ultimate, for after Neptune comes Pluto. The climax of the dissolution process may be analogous to what in Eastern systems is called *Nirvikalpa Samadhi*: a state of bliss in which the person is oblivious to outer surroundings. Most of the Eastern scriptures note that this is a conditional, not an ultimate state. For example, in the *Tripura Rahasya,* the lesson learned by Hemachuda from his enlightened wife Hemalekha is that the ultimate state is available at any moment, regardless of one's activity or inactivity.

Understanding your Psyche may help you come to grips with the a-rational side of your nature, that part of the mind which is *not* ruled by Mercury.

Pallas, Athene and Minerva

The three names Pallas, Athene and Minerva all apply to the same mythological goddess, although each connotes something slightly different. Zipporah Dobyns summarizes the relationship this way:

> "Pallas, or Pallas Athene, or Minerva (alternate names) was born full-grown and armed from the forehead of Zeus (Jupiter). Like Zeus, she hurled thunderbolts, but in addition to her war-like nature, she was associate with wisdom and handicrafts. She was credited with inventing weaving, the potter's wheel, spinning, needlework, and was also associated with navigation and agriculture, especially the olive tree. One myth described her taming of the winged horse, Pegasus. She was the special patron of Athens, as her name suggests, defeating Neptune in a contest when her gift of the olive tree was judged more helpful to humans than Neptune's gift of the horse. Her name, Minerva, testified to her wisdom in her role as counselor to her father, Zeus, and her sacred bird was the owl. She was given the name Pallas because she killed a famous giant with that name. She was also called 'Hippia' because she tamed horses. Although frequently involved in combat, Pallas was considered less aggressive and bloodthirsty than Mars. She usually fought for justice."[7]

Dobyns considers Pallas and Juno to be Libra asteroids, using the key words justice, equality, fair play, both sides of the issue for both of them. Donath lists a number of key words for Pallas: agile, allocate, connect, counsel, defend, discipliner, earnest, guardian, intuitive, inventive, liaison, mathematical, modest, organized, pattern, peace maker, perceptive, practical wisdom, protects, shielding, stern, tamer, valiant, works.[8]

Pallas has been thoroughly studied astrologically since the time of

Eleanor Bach's work, but how can we distinguish these three facets of the goddesshead? In the first place, Pallas Athene was the Greek name of the goddess, and Minerva the Roman one. When the Romans adopted other countries' pantheons, their gods and goddesses displayed similarity to the earlier manifestations, but included some changes that were associated with Roman culture. Pallas/Athene/Minerva was perhaps one of the most "liberated" of the goddesses: she was always assumed to have a mind. She was unique nexus of intellect and craft, and astrologically she remains so today.

Creativity of an artistic sort does not flow from this asteroid grouping. However, it does depict interest/ability in areas which combine the functioning of the mind and the body. For example, a scientist I know who has worked extensively in electron microscopy, has Athene rising, twenty-five minutes off the Ascendant. She also has a T-square, with Moon (21 degrees Leo) opposite Jupiter (20 degrees Aquarius)/Mercury (22 degrees Aquarius) square Pallas (20 Scorpio). A physician has Pallas quincunx Athene. An artist who does technical photography has Athene in Gemini trine Neptune in Libra. All of these fields represent examples of work that involves the application of intellect to manual dexterity.

The quintessential example was the association of Pallas/Athene/Minerva with weaving. Arachne, a mortal, made the mistake of attempting to compete with the Goddess and ended up a spider. Though weaving may not be the valued craft it once was, museums contain examples of numerous original and intricate patterns worked out by weavers all over the world performing their mundane and essential craft. Yet many weavers were not content to do only the functional; they added their own ideas and inventions to the pattern.

In the birth chart of a person who is truly talented in eye-hand coordination, more than one of these asteroids may be prominent. Each could be said to represent the application of skill and dexterity to the kind of application suggested by the associated planet or asteroid.

There are some subtle differences between the interpretations of the three asteroids, however. Pallas represents righteousness more than either Minerva or Athene. This agrees with Donath's and Dobyn's interpretations. Her inventiveness is inventiveness with a purpose, or a vengeance. For example, the birth chart of a former business associate of mine has a Yod pattern with Pallas retrograde in Pisces in the eleventh house at the apex, quincunx Sun in Libra in the sixth and Pluto in Leo conjunct the

I.C. She always was obsessed with fairness and used her articulation of the *right* way as a means to control the outcome of the situation—as would be expected from the Sun sextile Pluto. One might be tempted to attribute the concern with righteousness to the Sun in Libra, but the Pisces overtones were evident from her tendency to favor the underdog and her compassion for the people involved. Her willingness to fight for the rights of the downtrodden (or those she perceived as being downtrodden) could override her desire for balance, which would have been suggested by her Sun.

Pallas people are concerned about being right, Athene people are more interested in being competent, and Minerva people with being accomplished. Thomas Merton had Athene in Cancer quincunx his Sun. Mother Teresa also has Athene in Cancer, but it is conjunct her Neptune. Christopher Isherwood had Minerva in Taurus trine his Moon. Merton and Mother Teresa both were able to bring their religious visions into the world, but they did not seem to be doing it in order to be known; instead, they did so almost reluctantly. Their service came first. Isherwood conducted his spiritual quest more publicly, and published a book on his Guru's Guru. I am not in any way denigrating Isherwood for this more public stance; I merely want to point out that Minerva is more publicly-oriented and Athene is more private.

Notes

1. Carl G. Jung, *On the Nature of the Psyche* (Princeton University Press, 1960), p. 92.

2. Ken Wilber, *The Atman Project* (Wheaton, IL: Quest Books, 1980), p. 45.

3. Shunryu Suzuki, *Zen Mind, Beginner's Mind* (Weatherhill, NY), p. 62.

4. Carl G. Jung, *The Archetypes and the Collective Unconscious* (Princeton University Press, 1959), pp. 45-45.

5. Carl G. Jung, *Synchronicity* (Princeton University Press, 1960), p.31.

6. John von Neumann, *The Mathematical Foundations of Quantum Mechanics* (Princeton University Press, 1955).

7. Zipporah Pottenger Dobyns, Rique Pottenger and Neil Michelsen, *The Asteroid Ephemeris* (Los Angeles: TIA Publications, 1977).

8. Emme Belle Donath, *Asteroids in the U.S.A.* (Dayton, OH: Geminian Institute, 1979).

3

Asteroid Power Plays

"Energy comes from exerting energy."

Stephen Levine
A Gradual Awakening

"Power alone, like knowledge alone, is only less dangerous than Love alone."

A.R. Orage
On Love

Power is a complex subject. First of all, there are so many different kinds of power. Funk & Wagnalls' *Standard College Dictionary* lists thirteen of them. With all these possibilities, one must be cautious in using the term without some allusion to the kind of power in question. For instance, what does the statement "Pluto rules power" mean? Surely the statement is not referring to the order of angels, but which other definition is appropriate?

The common denominator of the definitions is the concept of available energy, whether it is actually used or potential. Then the question of power is a dynamic one: the distribution of energy, the disparity in its allocation, and so forth.

When Jung discussed the differences between his theories and those of Adler and Freud, the key concept was energy:

> " . . . life can flow forward only along the path of the gradient. But there is no energy unless there is a tension of opposites; hence it is necessary to discover the opposite to the attitude of the conscious mind . . . . (T)his compensation of opposites also

> plays its part in the historical theories of neurosis: Freud's theory espoused Eros, Adler's the will to power. Logically the opposite to love is hate, . . . but psychologically it is the will to power."[1]

I suggest that power is not exclusive to any one astrological body. In fact, all bodies have a power aspect because all bodies represent a type of energy. The living condition could be characterized simply as an intricate series of energy (power) transfers. Accordingly, we could talk of solar power, lunar power, intellectual power, aesthetic power, sexual power, or any other kind of power for that matter.

Given this overview, it is also apparent that most people refer to a particular kind of energy exchange as power: the relative quantities of energy (power) "possessed" by different individuals *vis-a-vis* each other. This form of power will be referred to here as power dynamics, and is the basis for this chapter.

Even given this definition, there are still multiple kinds of power dynamics. Some of these are summarized in Table 3.1.

Table 3.1: Power Dynamics and Associated Celestial Bodies

Type	Rulership
Brute Force	Mars
Macho control	Hidalgo
Conquest	Attila
Individual	Psyche
Institutional	Jupiter, Saturn
Intellectual	Mercury
Knowledge	Pallas, Minerva, Athene
Magnetic	Venus
Revolutionary	Uranus
Sexual	Sappho, Eros
Shamanistic	Icarus
Transformative	Pluto
Charismatic	Neptune
Violence	Amor

As may be seen from the table, all the planets rule some form of power dynamics, as do most of the asteroids. In this chapter I will discuss the forms of power dynamics expressed by the three asteroids Hidalgo, Icarus and Attila.

Hildago

Unlike most of the other asteroids discussed in this volume, there is little mythology to relate concerning the name. Hidalgo was a Spanish designation for the lowest rank of nobility entitled to use the appellation "don." There was one person worth noting who used the name: Miguel Hidalgo y Costilla, a Mexican priest turned revolutionary during the early nineteenth century. After leading an unsuccessful rebellion, he was defrocked and then shot. Even so, he remains dear to the Mexican people as one of their founding fathers.

The asteroid Hidalgo was discovered by astronomer Walter Baade at the Hamburg Observatory in 1920. Astrologically, Hidalgo represents an assertion of will *over* others. Specifically, Hidalgo *expects* to be in control, to be the general in all situations. When Hidalgo is prominent in a synastry comparison, it can show two people who try to dominate each other, even over petty details.

Hidalgo's energy can be used positively in fighting for other people's rights. However, it is almost impossible to play this energy out without an ego component. For example, Hidalgo was exactly rising in Washington, D.C. when Neil Armstrong first stepped on the Moon. This represented a considerable propaganda triumph for the United States. Although there was no violent component, there was a considerable macho component. I suspect that this macho component in the thinking around the Space Program was partially to blame for the extreme delay in admitting women astronauts.

The configuration at the time of Ayatollah Khomeini's return to Iran included a T-square, with Chiron in the eleventh house opposite Hidalgo in the fifth, square to Jupiter in the third. Clearly the involvement of Hidalgo shows what we now know: the revolution was enunciated in terms of an extreme form of orthodox religion (Jupiter), with Chiron providing the vision for the reorganization of all of society along Medieval rules and strictures. Jupiter represents the legalistic side of religion, not the compassionate.

Though I am using the term "macho" for Hidalgo, women are just as capable of being macho as men are, however, one could argue that fewer women exercise this tendency. But when they do, it can be in spades. One woman I know, who has Hidalgo at the Ascendant in Scorpio, is one of the most controlling people I have ever met. She is excellent at playing political games at work (she also has a Capricorn Sun and Scorpio rising), where she always seeks to be in control. She also insists on playing it out in her relationships with men. With Hidalgo in Scorpio rising, she is not the least inhibited about using sexuality to get what she wants in any field. A few years ago when she was divorced, she seemed at least as upset about the fact that her husband had initiated the procedure as she was about the divorce itself!

Hidalgo's orbit is interesting. The most distant of the conventional asteroids, Hidalgo has an orbital period of 3.96 years. (By comparison,

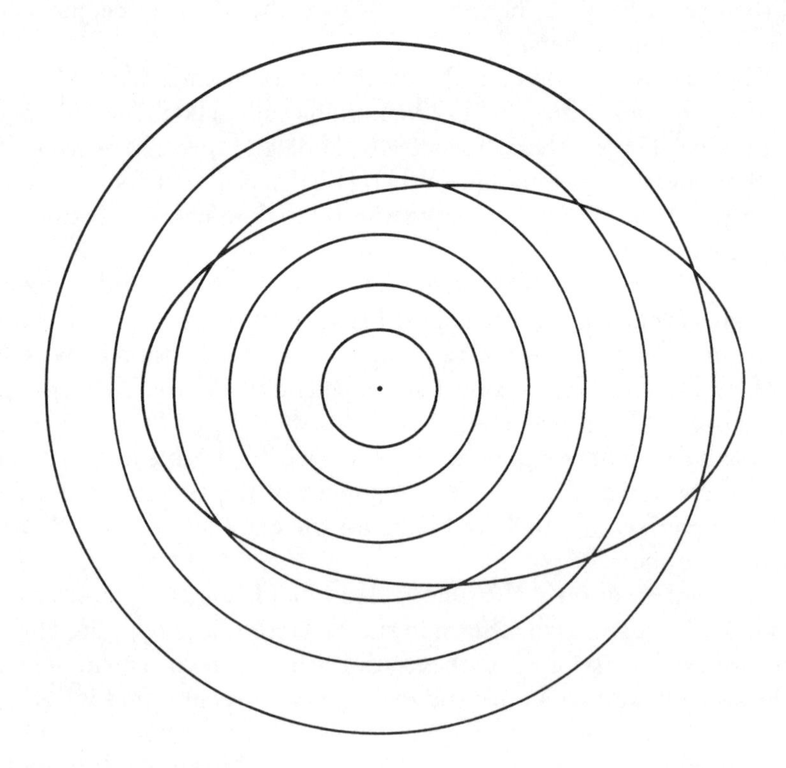

Figure 3.1: Orbit of Hidalgo

Mars has a period of 1.88 years, Jupiter 11.87 years.) At times, Hidalgo passes outside the orbit of Jupiter. This orbit is shown schematically in Figure 3.1. (The diagram has been simplified by showing the planetary orbits as circular.) Its orbit has a high eccentricity, which means that the orbital plane of Hidalgo is considerably different than the Earth's orbital plane. The two orbital planes of Hidalgo and the Earth differ by 42.5 degrees, which is one of the greatest eccentricities known. Consequently, the ecliptic latitude (or declination, if you want the other set of coordinates) is a significant factor in determining the effect of Hidalgo. For example, Hidalgo was in the south latitude of Berlin in declination when Hitler crossed into the Rhineland.

Because domination can be either given or received, Hidalgo may show up prominently in the chart at a time when a person is "overpowered," or dominated in some way. Gloria C. Leggett (personal communication) reported a few years back that Hidalgo was conjunct a client's daughter's Sun/M.C. when the daughter was attacked at knife point by an intruder who tried to rape her. She was successful in repelling the attacker forcefully, but not without getting cut up in the process.

Because domination often involves force, the association with violence in the above examples should not be surprising. I do not believe that Hidalgo is *intrinsically* violent. However, the need to be dominant, indicated by a strong Hidalgo placement, is clearly consistent with the maxim "the ends justify the means."

Icarus

The asteroid Icarus was the first body discovered which passes inside the orbit of Mercury. Like Hidalgo, it was discovered by Walter Baade, in this case on June 26, 1949 at the Mount Wilson Observatory. A schematic of the orbit is shown in Figure 3.2.

Because of the extreme nature of its orbit, going inside that of Mercury (from the standpoint of the Earth), Icarus can appear to move very quickly. It may be that there is a difference in the effect of Icarus depending on whether it is moving quickly or slowly with respect to the Earth. Specifically, a quicker Icarus is more precipitate, extreme or crisis-oriented.

In Table 3.1 above I related Icarus to shamanistic power. This term is not in common usage, so let me define it. The shaman state refers to

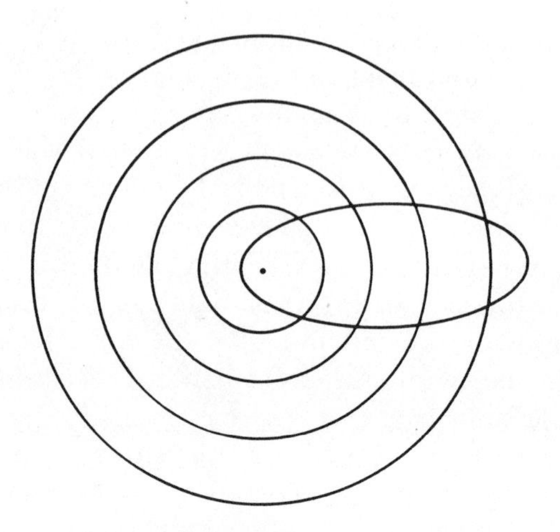

Figure 3.2: Orbit of Icarus

a particular stage of spiritual development. Ken Wilber, who has synthesized Eastern mysticism and Western psychology, has come up with a series of states of consciousness, or as he puts it, the spectrum of consciousness. To understand the shamanistic level, we must first consider more primitive states, or levels of consciousness. In Wilber's scheme, the earliest state is the uroboric, a state which could be characterized as pre-egoic. The uroboric individual is not aware of her/himself as an individual. This stage is not much different from the animal state: the "I" is absent, or very primitive. While it is possible for "I" to be hungry, the idea of "I" extending in time—backwards and forwards—has not yet developed.

Later, as consciousness emerges, the typical person is in the typhonic stage. With the typhonic stage comes a primitive sense of individual. "I" am, and further, "I" am mortal. "I" am interested, therefore, in doing anything which prolongs "my" life, or at least, makes "me" less aware of "my" death. This concept will be addressed more fully later, but for now, suffice it to say that this stage is characterized by the evolution of magic and rituals. As language develops, a more sophisticated ego emerges based upon mental concepts, including the concept of time and the tendency to view self as mind. As higher stages are reached, the mortality avoidance construct (or Eros-Thanatos archetype) becomes more sophisticated, but is not eliminated until the ego is transcended and the issue becomes meaningless.

It is within this context of evolution that the concepts of translation, transformation and transcendence become significant. In this process, the world of the person breaks down, whether through physical exhaustion, mental exhaustion or breakdown, or a crisis of a so-called external event. In Wilber's words:

> "One of the best ways of describing translation . . . is to say that it seeks to preserve the life of the separate self-sense and hold it against those forces . . . which threaten its present form of existence . . . .
>
> "However, should Thanatos exceed Eros, then the present form of translation tends to fail and even break down . . . . At that point, translation tends to fail miserably—thought processes become disoriented . . . . I must emphasize that the 'breakdown' or surrender of a mode of translation and the subsequent transformation is not necessarily . . . a bad thing. Growth and evolution . . . require transformation—the replacement of old translations by newer ones . . . .
>
> " . . . transformation can go in any number of different directions. There can be *regressive* transformation back into archaic structures . . . . There can be *progressive* transformations to higher and more organized structures of consciousness . . . . There also can be truly transcendent transformations into realms of the superconscious—a giant leap upward . . . "[2]

The shaman was that person who, rather than developing psychopathic symptoms following such a breakdown, emerged with a greater level of consciousness. From paleolithic times, the shaman has been known in many cultures: the Eskimo, the Ainu of Japan, the Buryats of Siberia, most Native American tribes, the Bon-po religion of Tibet, the Northern Aranda of Australia, the Mongolians, and pre-Roman Europe. One of the better treatments is Joan Halifax's *Shaman: The Wounded Healer.* Two other good sources, of a more experiential nature, are the books by Lynn Andrews and Carlos Castenada. The saga of the shaman through the many cultures remains remarkably similar. The shaman goes through a quest or journey, experiences a "breakdown" during which "death" is encountered, then emerges at the other end with a deeper understanding of life and the inevitable intertwining of death. Interestingly, one of the universal symbols of the shaman dates back to cave paintings in Europe: the bird, and with it, the description of the ability to fly.

Flight is an essential part of the symbolism of Icarus. The myth of Icarus is that Icarus and his father Daedalus were imprisoned in Crete in the labyrinth made famous by the Minotaur, a half-human monster. To escape, Daedalus fashioned wings of bird feathers and wax for himself and his son. They escaped by flying away, but Icarus, ignoring his father's instructions, flew too near the Sun. The wax melted, and he plunged to his death in the sea.

Icarus died quite literally during his quest or test. The person who becomes a shaman survives bodily, but often experiences either a kind of ego death, or a major transformation of awareness. Death in this context can be either physical, psychological or spiritual. In any case, the aspirant ends up on the other "side" of his/her prior awareness.

Of course, not everyone goes off on such a quest, much less invests sufficient energy in it to have results. The person who became a shaman was often an outcast before, and certainly the shaman, while respected, had a reputation for being strange.

For the person who is not on the shaman journey, what is the effect of Icarus? There are several possibilities. One is quite literal: flight. This is especially easy in this day of air travel. An astrologer who has Sun conjunct Icarus natally told me that as a child she jumped out of second story windows, completely convinced she could fly. Unfortunately, not everyone who encounters Icarus in this literal way survives.

There is one caution with this literal interpretation of Icarus: When Icarus flew, flight was dangerous. The danger—and almost certain death—was as necessary to the myth as the fantasy of flying around with bird's feathers. Today, most people who fly routinely see it as is a business and quite unlike the experience of pilots in the first half of the century. Though Icarus might be more significant in the charts of earlier pilots, flight today, with its high degree of organization and redundancy, seems more Saturnian. In examining the charts of contemporary pilots and flight attendants, I have not found consistent Icarus connections.

However, Antoine de St.Exupery, an early French pilot who wrote about flight in *Vol de Nuit, Le Petit Prince* and other works, had Icarus opposite Mercury, trine Mars and quincunx to his Ascendant. He apparently suffered the Icarus fate, disappearing during a flight. Charles Lindbergh had Icarus in a wide (2 degree) conjunction to the Sun, trine Pluto (geocentric and heliocentric) and sextile helio Uranus. Amelia Earhart had natal Icarus retrograde in Pisces, quincunx her Sun and sextile her Ascendant. Today, however, people who fly for a living or for fun need not have Icarus prominent in their charts, unless there is also an element of danger—real or perceived—in the *way* that they fly.

For example, consider the chart of the Smuggling Pilot (Figure 3.3). He has Icarus 4 minutes off of exactly quincunx helio Neptune, and a degree away from quincunx geo Neptune. It is also interesting that his Hidalgo is conjunct geo Uranus and quincunx helio Mercury. Here is a man who fantasizes about the dangers of his work, and also gets his macho kicks from doing something different.

Charles Emerson stated that there was a relation between the position of Icarus and the plane crashes examined in the New York National Council for Geocosmic Research study. However, a re-evaluation of these charts by Margaret Meister revealed no such relationship (personal communication).

There are other ways of interpreting Icarus, too. Icarus shows that area of life in which we are willing to risk all, where we have no patience. Because of the inherent risk-taking associated with this asteroid, the area is often one in which we may be hurt, because our actions are hasty. Icarus was not mentally prepared for his escape. He had the courage, but not the discipline; Daedalus had both. The impatience of risk-taking often reveals something about which we are uncomfortable for some reason, something which, if explored, could serve as a catalyst for transformation. When it doesn't, accidents may result from hastiness or ill-planning.

To use the model of transference, when something is experienced as uncomfortable, it generally means it's "hitting too close for comfort." Seen in this light, Icarus is providing the opportunity for growth. Growth, however, is not compulsory. Instead, most of us develop the most elaborate mechanisms to *avoid* growth, thus we become impatient with anything that even suggests the possibility of growth.

Growth is not easy, because if we are to grow, we must confront all those nasty unconscious drives and desires which are the core of our "problem." In the words of Maurice Nicoll:

> "Through inner growth man finds the real solution of his difficulties. It is necessary to understand that the direction of this growth is not outwards . . . but inwards, in the direction of knowl-

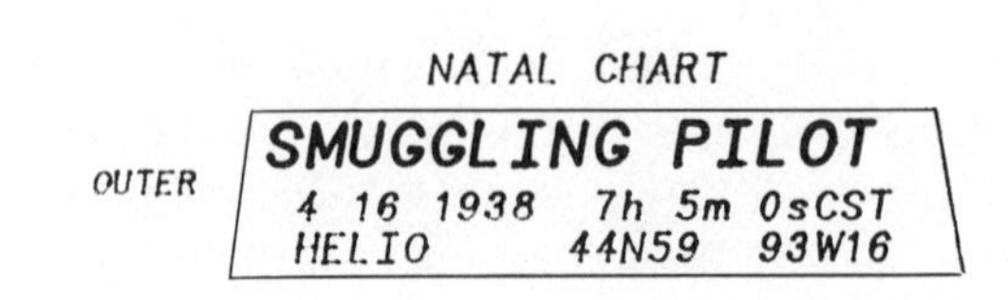

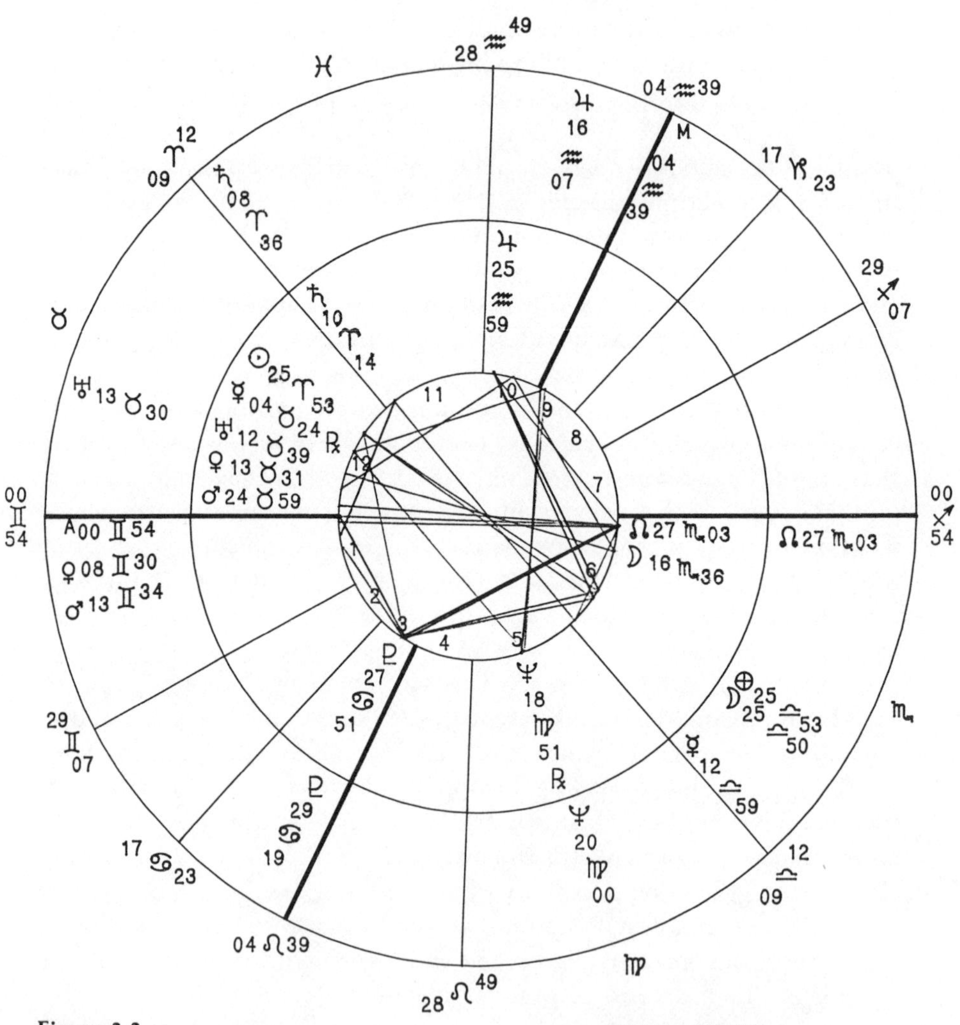

NATAL CHART
INNER CUSPS
SMUGGLING PILOT
4 16 1938 7h 5m 0sCST
REGMNTNS 44N59 93W16

Figure 3.3
Asteroids

Attila	7	CN	41
Hidalgo	12	SC	31R
Icarus	19	AQ	54

> edge of himself, through which there comes a *change of consciousness.* As long a man is turned only outwards, as long as his beliefs turn him toward sense as the only criterion of the 'real', as long as he believes only in appearances, he cannot change himself.
>
> " . . . He must relate himself to the 'world of ideas' before he can begin to grow. That is, he must feel that there is *more* in the universe than is apparent to the senses."[3]

Icarus can be understood as the spark of imagination which sometimes allows us to grow. It represents the "flight of fantasy" that can allow us to escape from our mundane existence. But there also are other ways to shake free of the outmoded status quo.

One rather obvious way that Icarus may be played out is through accidents. "Accidents" are often a momentary lapse into unconsciousness. They frequently occur when an individual is unable to cope with the circumstances fully, or when s/he needs to move on to a different plane of meaning. In this sense, the accident becomes a substitution for a vision quest.

I experienced one example of this kind of accident when I was in a car accident in January 1986. The transits for the accident were interesting. All transiting aspects that were within one degree of exact are shown in Table 3.2.

Table 3.2: Lee's Accident—Transits

Transiting Icarus quincunx natal Mercury
Transiting helio Uranus square natal Mercury
Transiting North Node trioctile natal Mercury
Transiting helio Venus quincunx natal Sun
Transiting geo Venus opposite natal Venus
Transiting helio Jupiter quincunx natal Moon
Transiting geo Jupiter trine natal Jupiter
Transiting helio Mars conjunct natal Neptune
Transiting geo Mars sextile Moon

Transiting Icarus formed part of a group of transits to my natal Mercury, which is my final dispositor. During this accident I had a vision of life and death in the split second before the other driver crashed his car into mine. That suspended moment showed me vividly what my attachments in life were and were not. Therefore, the accident could be seen as a transcendental experience. While not all accidents may be viewed that way, each can serve as a direct learning experience if the individual is open to the lesson—and if s/he survives the accident!

The association of Icarus with "accidents" (Al Morrison also finds some association with political assassinations) suggests something worth pondering.

It is patently clear that in one lifetime an individual does not achieve mastery or even grapple seriously with many issues or planetary pictures or archetypes. Most astrologers see clients who seem to deal over and over again with the same "problems." Most of us manage to reach our level of incompetence—as Dr. Peter would say—in life and then settle down into a rut. Often this repetition is accompanied by an impatience concerning that issue, or an impatience about the idea that there might be a rut at all! To a large degree, risk-taking is, after all, an attempt to transcend. As Pauline de Dampierre has indicated, the reason for climbing a mountain is the feeling one experiences at the top. The danger experienced on the climb up is an integral preparation for the exhilaration of looking over the edge. The question is whether the risk-taker actually transcends, or instead becomes addicted to the process, and thus doomed to perpetually repeat the danger as a kind of "high."

The impatience we see in today's "fast track" may be symptomatic of an inability to cope with issues in consciousness which were previously dealt with only metaphorically. The fact is that in our current era we "fly" all the time: literally, or through telecommunication networks. No longer do we just live in our home town, or a state, or a nation. We have transformed our material matrix without fully developing its concomitant psychological/astrological/spiritual implications.

By most scales of "spiritual evolution," after the development of the ego comes the point of realizing that the ego is not the absolute. There is something beyond the ego. Previously, at least in a Western context, any intimation of this level would have been considered a flight of fancy or the imagination. And it is with this flight that we return to Icarus.

The risk-taking and impatience may be symptomatic of a primal resistance on the part of the ego to allow its own dissolution. If you took a poll, I doubt that you would find "transcending the ego" very high up on the list for "what is the meaning of life."

Without true transcendence, Icarus can show areas of thrill-seeking or risk-taking. For example, with Sun-Icarus, the person either may fantasize that virtually every circumstance in life has an element of danger, and may make a game out of seeing how much s/he can get away with before being shot down. With Saturn-Icarus, the individual may risk or gamble with an authority figure (such as a guru) or rebel against authority figures, or even men in general, as is the case of some feminists. The Mars-Icarus person attempts to run his/her own quest. Considerable energy can be thrown into dangerous enterprises. The Venus-Icarus individual may be looking for danger/transcendence in relationships.

Because Icarus deals with the whole question of ego death/transcendence, it may either act in a literal or a metaphorical way. For those who are not ready, able or willing to tread the path of the shaman, it tends to appear in a way which is directly related to flight. For those who tackle the metaphorical side, the position of Icarus may show the key issue which will be the catalyst for going over the edge. From there, whether the person will be able to fly successfully remains to be seen.

Attila

Attila's historical reputation is not altogether deserved. Known as the Scourge of the West, Attila's Huns (Mongols) attacked and overran Roman provinces circa 433. He was neither the first nor the last Hun to do so, however, his reputation remains, and has considerable bearing on the interpretation of the asteroid Attila.

The interpretation of the asteroid Attila is in some ways similar to Hidalgo's. Both involve different facets of the power/dominance definition. The person who has Hidalgo prominent in the birth chart assumes that s/he will get his/her way. The Attila person will do whatever is necessary to get it. One could almost say that Hidalgo represents passive dominance and Attila active dominance.

Involvement of either asteroid—but especially Attila—with Pluto increases the desire for dominance by an order of magnitude. Pluto has quite a reputation as a manipulator, but this is not entirely correct. Pluto is an obsessor, and it does whatever is necessary to get the object of its obsession. Attila and Hidalgo signify dominance for dominance's sake, rather than for an object or person.

To understand these asteroids further, let us look at the charts of two people with Machiavelli complexes in recent times: Richard Nixon and Henry Kissinger.

Nixon natally has the Sun in Capricorn, Moon in Aquarius and Virgo rising. Normally, we would expect someone with a Capricorn Sun to be a good administrator who delegates authority fairly well, and who is possibly amoral in a business sense. He has Mars opposite Pluto, which would make him a formidable opponent, since Mars-Pluto types have a tendency to get what they want, by any means necessary.

Consider, however, the additional information provided by the asteroids. His Attila is conjunct his Moon! This suggests that he had an emotional need to be dominant. Interestingly, some of the ways he chose to express this dominance are shown through other asteroids. The asteroid Karl Marx is at 28 Aquarius in his chart, trine Pluto, sextile Mars and square Saturn. Karl Marx is a rather literal indicator of Communism, and Nixon broke into national prominence with an anti-Communist crusade. The presence of helio Jupiter in the same degree as Mars suggests that Nixon would achieve public recognition for this course. (T. Pat Davis has noted that heliocentric placements often express more publicly than geocentric ones.) Nixon's asteroid America is conjunct his Neptune, so I suspect that he alternates between a vision for what America could/should be and confusion about what America really is.

Henry Kissinger has a Mars-Pluto conjunction in helio, but not in geo. A double Gemini with a Libra Moon, his verbal expressiveness has always been his forte. His strong Gemini certainly was useful in the kind of "shuttle diplomacy" for which he was famous, but it is hard to see the kind of manipulation and power-brokering he performed. When we examine his asteroids, however, we see that his Attila is quincunx his Neptune and Hidalgo is conjunct his Moon. It is also interesting to note that the asteroid America is conjunct his Mars and both are trine his Europa.

Consider the difference between Nixon's Moon-Attila conjunction and Kissinger's Moon-Hidalgo conjunction. Nixon had to claw his way

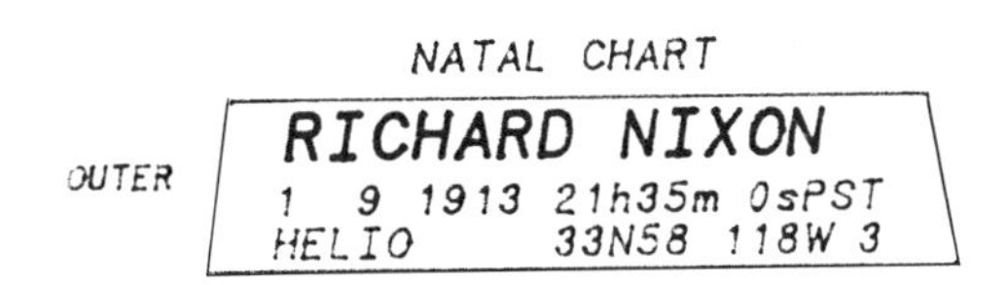

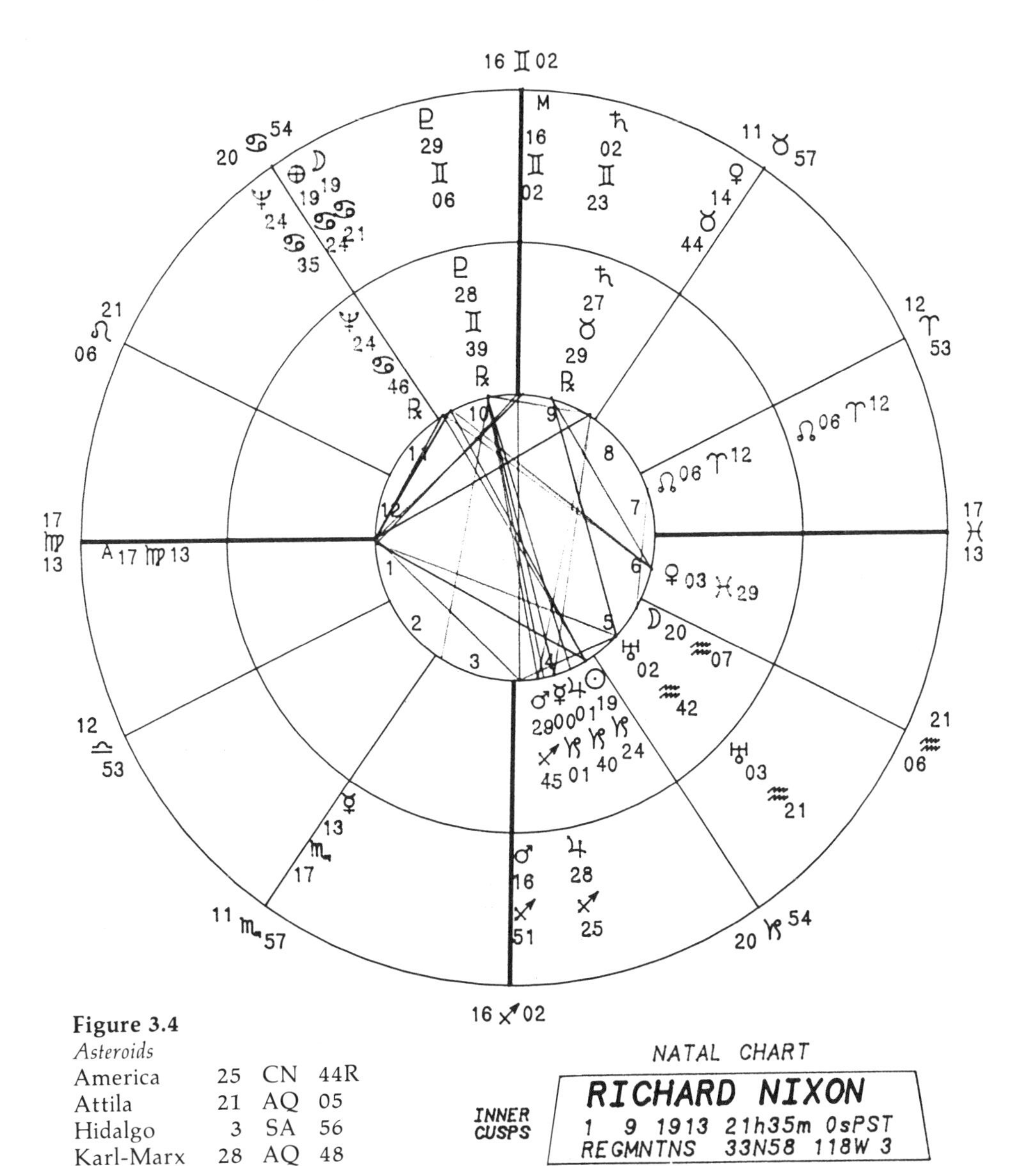

Figure 3.4
Asteroids

America	25	CN	44R
Attila	21	AQ	05
Hidalgo	3	SA	56
Karl-Marx	28	AQ	48

NATAL CHART

OUTER

HENRY KISSINGER
5 27 1923 5h30m 0sMET
HELIO 49N39 11E 0

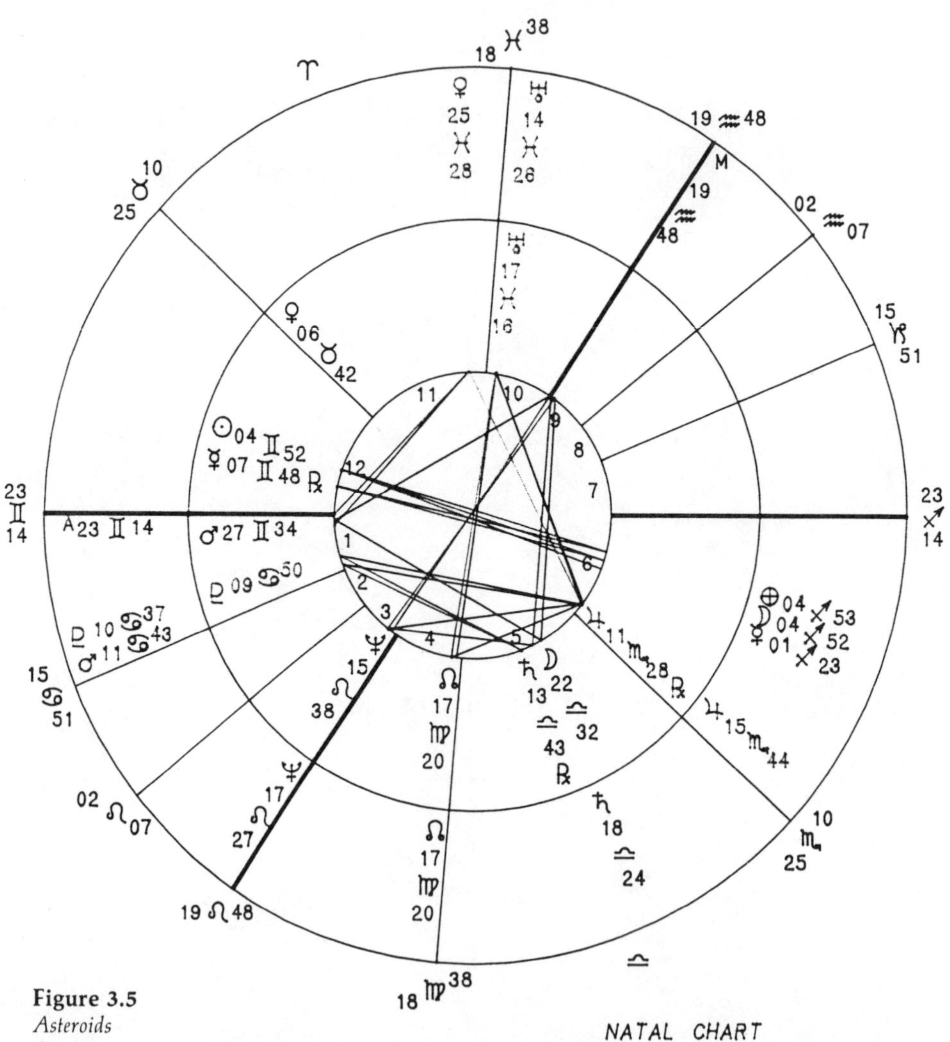

Figure 3.5

Asteroids

America	22	TA	48
Attila	24	CP	37R
Europa	5	LE	39
Hidalgo	23	LI	26R

NATAL CHART

INNER CUSPS

HENRY KISSINGER
5 27 1923 5h30m 0sMET
REGMNTNS 49N39 11E 0

to the top of the heap, first in Congress, then in the Vice Presidency, then the governorship of California, and finally the Presidency. By contrast, Kissinger was appointed to his position of power. This shows the more active nature of the Attila dominance pattern compared to the Hidalgo one.

Attila signifies the fighter. The Attila type does not retire gracefully. And unfortunately, the Attila person fights even when fighting is unnecessary. This fighting need not be physical—although it may be, especially if the asteroid is in aspect to Mars—but it is fighting nonetheless.

Notes

1. Carl Jung, *The Collected Works, Vol 7* (Princeton University Press, 1970).

2. Ken Wilber, *Up From Eden* (Boulder, CO: Shambhala, 1983).

3. Maurice Nicoll, *Living Time* (London: Watkins, 1952).

Sources for Data

Smuggling Pilot: Personal contact.

Richard Nixon: T. Pat Davis, photocopy of birth certificate, communicated by Lois Rodden.

Henry Kissinger: Holliday quotes Edmund H. Troinski from birth certificate, from Lois Rodden's *The American Book of Charts* (San Diego, CA: Astro Computing Services).

4

Concepts: Pandora, Pax, Armisticia, Utopia, Fanatica, Harmonia, Prudentia

Concepts apply to ideas rather than to concrete things. Unlike mythological names, concepts do not require much explanation. This chapter will approach the concepts from two directions: personal and mundane. The view is that certain concepts may apply more strongly to aggregates (such as countries) than to most individuals. However, each concept still may express itself in an individual chart, though perhaps in a different way than it would in society at large.

Rather than deal with each of these concepts separately, I will consider them together. This task is actually quite easy because each of the names—except one—are common nouns, nouns without personal histories. The one exception is Pandora.

Pandora

In mythology, Pandora is supposed to have released evil into this world when she opened "Pandora's box." When Al H. Morrison and I first published a Pandora ephemeris in 1981, I stated that the effects of the asteroid Pandora in an individual's chart were twofold: to stir the person into doing something, and to produce unintended options for the person, what I have called the "cascade effect."

In some ways Pandora is similar to Uranus, however, Pandora represents a process, while Uranus may not. With Pandora, one "isolated" action (often a signature with Uranus) may open up a whole series of unintended effects. During Uranus transits, the individual might experience sudden changes or insights, but after the transit has passed s/he may "spring" back to "normal." With Pandora, that one anomaly opens up a whole new pathway. Pandora may be thought to have certain analogies to the "many worlds" hypothesis in physics, which states that one moment opens up a decision tree.

Suppose I win a trip to the Bahamas. As a Uranus transit, this is certainly an unexpected event. But this opens a series of crucial moments. First, I have to decide whether I can take the trip. In the many worlds hypothesis, one world splits off in which I take the trip, a second in which I don't. I decide to take the trip, but then I can either go in February or in March. In the February path, I meet somebody there who can help me vocationally. In March I meet someone with whom I can become romantically involved. The next step is that the romantic entanglement may be just a vacation romance, or perhaps my lover lives in my own city—now a casual contact could become a major involvement.

Pandora represents the ways and moments when we are open to a chain of events which alters our course, for good or bad. Zip Dobyns calls Pandora "an insatiable intellect willing to investigate anything."[1] Demetra George describes the asteroid as follows:

> "Pandora serves as an intermediary in the octave progression between Mercury and Uranus. Through her curiosity, Pandora acts as an agent of change, inviting the unexpected and opening up new possibilities. By rising from the earth, she brings to light what had previously been hidden. Pandora may signify a process of deep cleansing, purification and healing. Pandora is a Gemini-Scorpio blend who through her unorthodox actions raises the rational intellect of Mercury to the intuitive and revolutionary vibration of Uranus."[2]

The Concept Asteroids

The concept asteroids will be examined first with regard to significant events of war and peace, and then in the charts of people. Four events associated with World War I are presented in Table 4.1. The first two are declarations of war. The declaration of war by Germany against Russia represents and unsuccessful declaration: Neither Germany nor Russia won the war since Russia pulled out after the Bolsheviks came to power, and the German government also fell. The second declaration was successful from the viewpoint that the United States was on the winning side, while Germany lost, at least in terms of the peace treaty, the Treaty of Versailles, the third chart of this group.

After World War I was concluded, there was a tremendous pacifistic response, a natural result of the savagery of the War. That "Great War" was to be the war to end all wars, certainly a contradiction in terms. The fourth chart in this batch is for a Naval Disarmament Treaty, which was negotiated from this perspective and signed in a highly significant year, 1923. The year began with the January 11th occupation of the Ruhr by French and Belgian troops, and ended with the November 8-11 "Beer Hall Putch," which landed Hitler in jail where he wrote *Mein Kampf* and planned his eventual rise to power.

In the chart for the declaration of war by Germany against Russia, the principal asteroid activity—geo and helio—was in Pandora and Utopia. Helio Armisticia was sextile Mars, certainly an appropriate position for the outbreak of war. Geo Pandora was semi-sextile the Sun, and quincunx both geo and helio Uranus. Helio Pandora was conjunct Saturn and semi-sextile geo and helio Neptune. At the time of the declaration, there was a considerable grouping of planets in the seventh house of open enemies: Sun, Neptune, Mars, Pluto, Mercury, South Node, and asteroids Armisticia and Fanatica. This did not bode well for the declaration.

Table 4.1: Events Related to World War I

A. Data

Germany Declares War on Russia
08 01 1914 17:29 UT
St. Petersburg, Russia 59 N 55 30 E 15

U.S. Declares War on Germany
04 06 1917 13:00 EST
Washington, DC 38 N 54 77 W 02

Treaty of Versailles
06 28 1919 14:12 UT
Versailles, France 48 N 48 2 E 08

Limitation of Naval Armaments Treaty
08 17 1923 12:00 EST
Washington, DC 38 N 54 77 W 02

B. Asteroid Positions

		Germany-Russia	U.S.-Germany	Versailles Treaty	Naval Disarmament
Armisticia	Geo	22 CN 23	18 AQ 19	2 CN 00	5 TA 12
	Helio	26 CN 36	2 AQ 35	0 CN 40	16 AR 41
Fanatica	Geo	22 LE 19	27 AR 41	6 SC 37	5 SA 45
	Helio	18 LE 23	2 TA 50	26 SC 24	29 SA 24
Harmonia	Geo	13 SC 37	9 LE 23	17 TA 21	27 CN 21
	Helio	19 LI 26	2 VI 26	26 AR 47	16 CN 01
Pandora	Geo	28 GE 31	18 CP 28	19 CN 39	4 CN 53
	Helio	9 CN 28	27 SA 26	24 CN 42	17 GE 19
Pax	Geo	3 SC 55	18 CN 56	8 SA 24R	25 SC 01
	Helio	16 LI 02	11 LE 01	18 SA 21	16 SA 46
Prudentia	Geo	25 PI 27	7 SC 27R	7 CN 31	4 LE 22
	Helio	24 AR 33	29 LI 06	8 CN 09	27 CN 43
Utopia	Geo	28 VI 53	12 PI 12	13 LE 15	2 GE 06
	Helio	17 VI 43	1 PI 18	24 LE 39	9 TA 27

When The United States declared war on Germany, the only two of these asteroids making aspects geocentrically were Harmonia and Prudentia. When we look at them heliocentrically, helio Armisticia was opposite Neptune, quincunx Pluto and conjunct helio Mars (helio positions tend to demonstrate themselves more obviously in the public sphere). Like the other declaration, an Armisticia-Mars aspect was present. However, this time the accumulation of planets—Mars, Venus, Sun and Mercury—was in the ninth house of distant journeys. The ruler of the seventh house of open enemies was Uranus (positioned in the seventh) and the co-ruler Saturn (in the twelfth). Harmonia was quincunx Armisticia, and trine Fanatica.

The Versailles Treaty was probably one of the most important events of the twentieth century, since the vindictiveness of the reparations directly influenced the factors which led to the Second World War. At the time of that treaty, geo Armisticia was semi-sextile the Vertex, quincunx the Mean North Node and trine Uranus. Geocentric Pax was trine both geo and helio Neptune. Helio Prudentia was semi-sextile the two Neptunes. Geo Utopia was square the East Point, while helio Utopia was semi-sextile Mercury and helio Jupiter. This was not a successful treaty. This was a desperate attempt to justify the expense in lives, money and suffering of the war to such a ludicrous degree that another war was inevitable. The geo Fanatica was trine Pluto, while the helio one was trine helio Jupiter. The Plutonian forces were not far from emergence.

The Treaty of Versailles also represented the birth of the League of the Nations, which, like the American Articles of Confederation, was a premature attempt at a previously untried form of government.

The Sun of the Limitation of Naval Armaments Treaty was the Saturn of the Versailles Treaty, and the Ascendent of the Naval Treaty was the East Point at Versailles. For this second treaty, there were no geo aspects of these asteroids except Pax opposite the East Point. This perhaps expresses the wishes of the treaty—peace. Helio Armisticia was trine Venus, opposite Saturn and semi-sextile Uranus. Helio Harmonia was square helio Armisticia, and so made aspects with the same planets, as did helio Pax, trine helio Armisticia.

Pax and Armisticia are not always in aspect. Peace and treaties are not always the same thing!

The second group of charts, given in Table 4.2, concerns the Second World War. One of them is worth mentioning in some detail, because in this case the observer was Ivy Goldstein-Jacobson! The chart is for the intended attack on Los Angeles by the Japanese, which was covered up to maintain morale. Goldstein-Jacobson described it as follows:

> "This is the chart set for the Japanese air invasion over Los Angeles that, to avoid national panic, was played-down at the time by official disavowal of an attack. This writer, driving home from the desert, was on the scene at the time & questioned the later denial, thus setting this chart for the correct PST. Since then, a secret and confidential document having come to light gave an official account covering what the military expected from the

> Japanese during the tense weeks following their Pearl Harbor attack. On Oct. 29th, 1945, the Los Angeles TIMES published an actual photograph (in this writer's possession), showing searchlights centered on a target with anti-aircraft shells seen bursting nearby. In the text, Lt. Gen. John L. DeWitt, Commanding General of the Army and Western Defense Command, said 'It has been definitely ascertained that the early-morning blackout & anti-aircraft firings in the Los Angeles area were caused by the invading presence of one to five 'then publicly unidentified' enemy airplanes.' "[3]

The Pearl Harbor attack was a preemptive strike which was designed to disable the United States fleet. The attack on Los Angeles also was supposed to have great symbolic significance; in both cases the Japanese military command wanted to demonstrate its technical capabilities in an effort to dishearten the Americans. Needless to say, that represented a poor read-

---

Table 4.2: Events Related to World War II

A. Data

Event	Coordinates
Attack on Pearl Harbor 12 07 1941 7:55 HST Pearl Harbor Base, HI	21 N 21 157 W 56
Attack on Los Angeles 02 25 1942 3:12 PWT Los Angeles, CA	34 N 03 118 W 15
Germany Surrenders 05 07 1945 0:41 UT Reims, France	49 N 15 4 E 02
Japan Surrenders 09 02 1945 9:03 Zone -9.00 Tokyo, Japan	35 N 42 139 E 46

B. Asteroid Positions

		Pearl Harbor	Los Angeles	Germany Surrenders	Japan Surrenders
Armisticia	Geo	12 SC 51	7 SA 30	10 GE 10	25 CN 14
	Helio	2 SC 23	17 SC 55	18 GE 08	11 CN 15
Fanatica	Geo	25 SC 32	27 SA 35	5 LI 58R	29 LI 05
	Helio	18 SC 11	6 SA 24	20 LI 16	16 SC 19
Harmonia	Geo	25 SC 43	2 CP 14	18 SC 28R	24 SC 45
	Helio	17 SC 36	9 SA 30	17 SC 27	20 AR 03
Pandora	Geo	13 GE 24R	11 GE 33	7 PI 55	24 PI 09R
	Helio	14 GE 08	3 CN 41	15 AQ 51	17 PI 52
Pax	Geo	6 LE 55R	19 CN 25R	13 AR 53	27 GE 51
	Helio	17 CN 26	5 LE 40	25 PI 14	24 TA 24
Prudentia	Geo	25 AR 21	13 TA 36	9 AR 46	29 TA 13
	Helio	14 TA 26	5 GE 14	22 PI 23	2 TA 30
Utopia	Geo	6 LI 21	12 LI 05R	24 AR 17	6 GE 51
	Helio	20 VI 11	2 LI 04	16 AR 14	14 TA 08

ing of the character of the Americans! It also represented a lot of wishful thinking, since many of the Japanese command had lived in the States and were perfectly aware of the foolishness of engaging such a larger country in a war. The Void of Course planet at this time was Neptune, at 29 Virgo.

A brief digression about the Void of Course idea is in order here. Because when we deal with the asteroids, there are perhaps three thousand positions that we could use (if we were masochistic enough), there will be conjunctions of asteroids to any position, and further, there will be asteroids with positions at 29 degrees plus (i.e., higher than the Void of

Course planet) 99.93 percent of the time. This is one case where it would be a mistake to tamper with the Void of Course concept by inserting the asteroids. Some asteroid workers attempt to insert the asteroids into *all* astrological ideas, including rulerships, and, conceivably, something like Void of Course. This is a mistake, because the rulerships were a highly technical system developed over thousands of years, and only make sense if their original purpose is remembered. The rulerships were part of the system of essential dignities, and were a method to evaluate the strength of a planet independently of its aspect patterns. The Void of Course concept was designed to show when the energy of a planet was waning. Either of these concepts would be completely diluted if so many additional asteroid bodies were inserted. I realize that one attempt to circumvent this is to declare the first four asteroids as the major asteroids, and the rest as minor ones, but this is a completely artificial system. Are the outer planets "minor?"

At the time of Pearl Harbor, neither geo nor helio Armisticia made any aspects to the planets. Fanatica is a bit more telling: geo Fanatica was opposite helio Saturn, while helio Fanatica was sextile the Mean Node. The opposition to helio Saturn suggests the very real conflict between the realists and the extremists in the Japanese command. Harmonia was conjunct Fanatica in both geo and helio, suggesting the degree to which the two sides pulled together once the decision was made. Pandora, like Armisticia, was quiescent. The Pandora unexpected twist of fate was absent. What followed did so in logical sequence, not through an unpredictable sequence. Geo Pax was trine Mercury; helio was sextile the Mean North Node and semi-sextile Jupiter, again reminding us that Jupiter doesn't always act like the Greater Benefic. Prudentia was semi-sextile helio Saturn and quincunx the Sun. The attack was not prudent!! Geo Utopia was sextile Pluto, helio was trine helio Mars and sextile helio Mercury. The utopian ideas of the Japanese extremists ran into the Mars-Pluto power realities.

The attack on Los Angeles was a somewhat different circumstance. The commitment had already been made to the war. The Japanese were interested in demonstrating their ability to bring the war to the North American continent. It failed because of the decision of the American military to squelch the publicity. Again Fanatica was active, with geo square helio Neptune, and helio trine the Vertex and square the Sun. It was certainly a bold move, but one with a weakness. Geo Pandora was trine Mercury and conjunct Jupiter, while Pandora was semi- sextile Pluto. The Pluto coverup effectively cut off any impact of the attack. Like Pearl Harbor, Utopia had a Mars component, in this case helio Utopia square helio Mars, which focused the "plan" or ideal on the martial issue at hand.

When Germany surrendered, geo Armisticia was sextile helio Pluto, while helio Pluto was sextile Venus. Geo Pandora was trine the chart co-ruler Saturn and helio Pandora was square the Sun. At the surrender of Japan, geo Armisticia was square the Ascendant and square helio Mercury. Geo Fanatica was square the Midheaven, and helio Fanatica was opposite helio Mars and quincunx Uranus. Geo Utopia was conjunct the Vertex, trine helio Neptune and conjunct helio Venus; helio Utopia was sextile the Moon and semi-sextile Uranus. This was not a bad surrender. Japan came out of the war in much better shape than Germany did after World War I.

The Concept Asteroids in Natal Chart Comparison

Though it may be that the easiest way to use these asteroids is in event work, it is also possible to use them in natal astrology. For the first example, I present below a study of twenty-eight devotees of a spiritual master known by many names: Da Free John, Franklin Jones, or more recently, Heart Master Da. Since the theory here is that a spiritual commitment is a private commitment, and none of these devotees is well known, I have used the devotees' geo placements exclusively, and have simplified the study by eliminating the alternate ascendants East Point and Vertex. The definition of "devotee" is a person who formally joined his church, which means, among other things, a financial commitment to tithing. This requires a more serious commitment than joining a mainstream church (although this doesn't mean that these devotees are more "spiritual" than more conventional churchgoers). The results are shown in Table 4.3. The devotees have been separated into three groups. Normally, I would not divide a group by gender, but the general experience of male and female devotees has been different, because in the context of the institution, men and women are often separated, and different roles are assigned to each. The children were either born into the church or brought in when their parents joined, so their involvement was less a matter of choice.

This study is really too small for good statistical analysis, and is meant to be suggestive of possible trends, not a final statement. Within this context, I look for total aspects by asteroid which are either 50 percent above

Table 4.3: Aspects between geo planets, Asc & M.C. of devotees of Franklin Jones (aka Da Free John, Heart Master Da) to Jones' natal asteroids. Expected number in total column is approximately 20.

		Women	Men	Children	Total
Armisticia	Geo	7	3	4	14
	Helio	14	9	7	30
Fanatica	Geo	10	4	4	19
	Helio	10	5	5	20
Harmonia	Geo	16	2	2	20
	Helio	6	2	6	13
Pandora	Geo	14	7	6	27
	Helio	12	5	3	20
Pax	Geo	12	5	4	21
	Helio	15	2	2	19
Prudentia	Geo	10	4	5	19
	Helio	18	4	2	24
Utopia	Geo	9	5	5	19
	Helio	5	3	8	16

or 50 percent below the average. There is only one match: there were 50 percent more aspects between the devotees' charts and Jones' helio Armisticia than the average. What does this suggest? Perhaps the devotees are achieving an armistice—a peace-making—in themselves because of this spiritual involvement.

If the women—who comprise half the sample—are considered separately, helio Prudentia and geo Harmonia are higher than anticipated, and helio Utopia is lower. The men show helio Armisticia and geo Harmonia high, consistent with the observation that the experience of the men is different from the experience of the women. The emphasis on Prudentia and Harmonia in the women may reflect an interesting psychological aspect of the women's involvement. There has been a heavy emphasis on marriage within the group, and the women often find that to leave the group would be tantamount to leaving their husbands, a condition that does not seem to affect the men to the same extent. Thus, the women may be taking the prudent or the harmonious course, rather than the one in which they truly believe. The low number of hits to helio Utopia certainly indicates that these women do not necessarily believe that this community is superior to the "outside."

Armisticia has already been considered in the group as a whole. The other high aspect in the charts of the men is Pandora, and this could represent the flip side of the women's dilemma. Men in this community have much more status than men in mainstream American society, on a par with "traditional" (read sexist) patriarchal society. Most devotees describe an emotional outpouring with regard to Free John (often through his books) as the basis for their commitment, but the increase in their relative status may be an unanticipated bonus to their continued involvement.

Interpreting the Concept Asteroids in a Natal Chart

We also may look at positions in natal charts directly. The first example I present is Karl E. Krafft, who had the reputation (incorrectly) of being Hitler's astrologer. Krafft was an interesting case: a Swiss citizen who became enamored of Nazism and made the mistake of moving to Germany. This mistake eventually resulted in his incarceration by the Nazis and to his premature death. Before this chapter in his life, Krafft was involved in attempting to use the statistical method to verify certain features of Cosmobiology. His heart was in the right place, but his effectiveness was somewhat spotty. An excellent account of his life is given by Ellic Howe. Krafft's asteroids are given in Table 4.4.

What is the anatomy of an obsession? Krafft's Taurus Sun and Virgo Ascendant suggest someone who was interested in practical issues and the Libra Moon indicates an emotional need for approval by others. With both geo and helio Mars in Aries, Krafft was ready to do what needed to be done in his research, although his working style was likely to be a series of spurts. With Saturn opposite Venus in a T-square with the Moon, he

NATAL CHART

OUTER

KARL KRAFFT

5 10 1900 12h45m 0sMET
HELIO 46N32 6E30

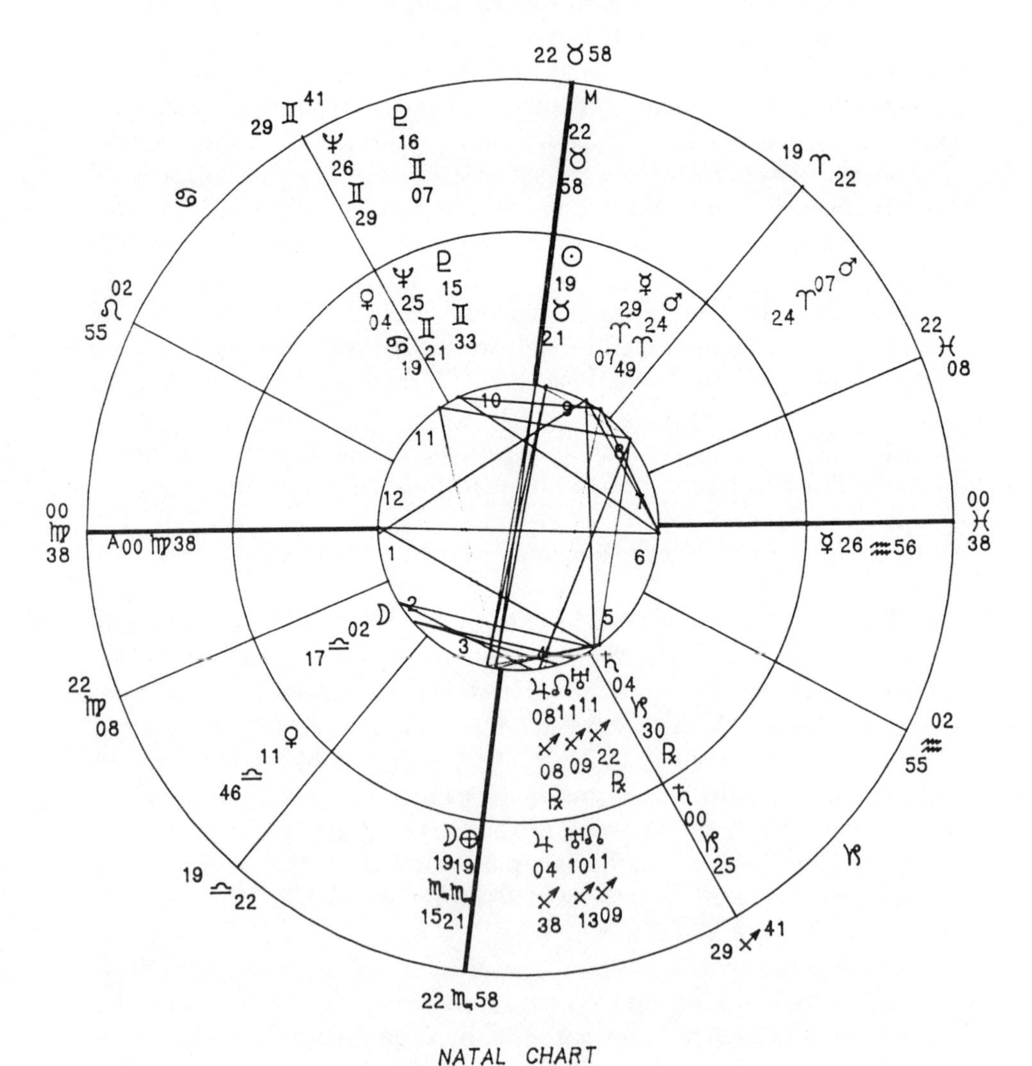

NATAL CHART

INNER CUSPS

KARL KRAFFT

5 10 1900 12h45m 0sMET
PLACIDUS 46N32 6E30

had difficulty believing that others thought well of him, or approved of him, regardless of whether that was true. (By the way, patterns like T-squares are another area of astrology where asteroids may not apply. Though I might look at what asteroids fall opposite the apex planet in the T, I would not assign such asteroids to fill in a Grand Cross. By that reasoning, a T-square could never exist because there will *always* be at least one asteroid in that position.)

In Krafft's chart, geo Fanatica was trine geo and helio Pluto, and helio Fanatica was opposite Mercury. Pluto-Fanatica indicates that Krafft would be a fanatic over a Plutonic issue—and Nazism was pretty Plutonic! With Mercury-helio Fanatica, Krafft couldn't wait to spout off about his more extreme enthusiasms. His geo Pandora was also trine his Plutos, so what to him might have seemed a straightforward path developed some unexpected turns. Further, with all the aspects between both geo and helio Utopia, Krafft was operating out of what he believed was the blueprint for a better, or utopian society.

Table 4.4: Asteroid Positions of Karl E. Krafft, Janis Joplin, H. R. Haldeman and Anita Bryant

A. Data:

Janis Joplin
1 19 1943 9:45 CWT
Port Arthur, Texas 29 N 54 93 W 56

H. R. Haldeman
10 27 1926 3:30 PST
Los Angeles, CA 34 N 04 118 W 15

Anita Bryant
3 25 1940 15:10 CST
Barnsdall, OK 36 N 34 96 W 10

B. Asteroid Positions

		Krafft	Joplin	Haldeman	Bryant
Armisticia	Geo	1 SC 05R	19 CP 57	0 SA 00	7 GE 14
	Helio	7 SC 17	17 CP 13	8 SA 27	24 GE 45
Fanatica	Geo	16 LI 28R	23 AQ 05	9 SC 08	29 TA 08
	Helio	28 LI 30	3 PI 33	11 SC 22	19 GE 45
Harmonia	Geo	22 LE 58	3 PI 01	18 CN 09	10 TA 50
	Helio	18 VI 12	17 PI 55	22 GM 52	26 TA 13
Pandora	Geo	15 TA 12	24 VI 22R	8 CP 31	20 CP 56
	Helio	13 TA 28	8 VI 57	29 CP 01	0 CP 41
Pax	Geo	2 CN 16	18 LI 26	24 LI 14	5 CP 55
	Helio	19 CN 15	0 LI 02	21 LI 21	14 SA 43
Prudentia	Geo	1 CN 00	15 LE 27R	29 GE 13R	27 LI 43R
	Helio	14 CN 42	9 LE 55	10 GE 15	19 LI 11
Utopia	Geo	26 AQ 52	4 SA 12	28 SC 11	8 TA 07
	Helio	8 AQ 09	20 SC 16	5 SA 28	20 TA 11

I have included three other people in Table 4.4 to illustrate particular asteroids. In Joplin's case, geo Harmonia was very active: conjunct the East Point, square helio Uranus, trine helio Mercury and square helio Mars. Her geo Utopia was near to square this Harmonia position, so it made aspects to most of the same bodies. Harmonia is certainly an important asteroid for a singer!

H. R. Haldeman presents another side. With his geo Prudentia quincunx geo Utopia, the prudent path and the path dictated by the greater vision were in conflict, the severe conflict of the quincunx which is so hard to resolve. Both asteroids make aspects to his helio Jupiter and East Point. In addition, his Prudentia is square the Ascendant and conjunct the Mid-

Table 4.5: Aspect Summaries for Four Groups

A. People in Each Group

Peace	Women's Issues	Political	Spiritual
Joan Baez	Anita Bryant	A. Eickmann	Franklin Jones
Amy Carter	Betty Friedan	Larry Flynt	Jim Jones
Jane Fonda	Germaine Greer	H.R. Haldeman	Komar
Mary Travers	Gloria Steinem	Patty Hearst	Gurudev Muktananda
		H. Himmler	Sathya Sai Baba
		John Mitchell	Jose Silva
		Madalyn O'Hair	Pakh Subud

B. Results (underlined items are 50% above or below an average of one aspect/person):

		Peace	Women's	Political	1-3 Total	Spiritual
Armisticia	Geo	6	3	7	16	12
	Helio	6	6	8	20	8
Fanatica	Geo	5	3	3	11	6
	Helio	6	7	5	18	7
Harmonia	Geo	4	5	8	17	12
	Helio	4	6	10	20	11
Pandora	Geo	4	4	10	18	6
	Helio	6	4	9	19	14
Pax	Geo	9	4	14	27	7
	Helio	8	1	9	18	10
Prudentia	Geo	7	1	8	16	7
	Helio	2	3	9	14	9
Utopia	Geo	5	2	11	18	8
	Helio	3	9	9	21	13

heaven, while Utopia is trine Uranus and conjunct helio Saturn. In helio, the major factor among these asteroids is Pandora, which is trine the Ascendant-East Point, quincunx the Midheaven and semi-sextile helio Jupiter. Haldeman was faced with the choice between ethics and his belief in and commitment to Richard Nixon. His morality (personified by Prudentia) came into conflict with his ideals (Utopia).

Anita Bryant has a number of aspects to both geo and helio Pandora. Geo is square the Mean North Node and helio Jupiter, and trine helio Uranus. Helio Pandora is trine Saturn and quincunx Pluto. In interviews, it is apparent that it never occurred to Bryant that her commitment to conservative Christian causes—and especially the Dade County homosexuality issue—would generate so much controversy and criticism.

The effects of these asteroids in groups of people is shown in Table 4.5. The people selected were divided into four categories, three of which were political, the fourth being spiritual/human potential. If the three political groups are combined, the only high total is for geo Pax, which may summarize the desires of most people who become involved politically for reasons other than personal gain. Among the people in the Spiritual/growth group, geo Armisticia, geo Harmonia, helio Pandora and helio Utopia are high. This is not surprising since these asteroids represent aspirations. The Pandora emphasis may show up because the spiritual quester is simply incapable of anticipating the details of the "path."

Among the Peace group, both geo and helio Pax are high. Among the Women's Issues group (pro and con), geo Prudentia and helio Pax are low. This has not been the most cautious of movements, militancy often being something feminists had to counter.

Notes

1. Zipporah Dobyns, *Expanding Astrology's Universe,* (San Diego: ACS Publications, 1983).

2. Demetra George, *Asteroid Goddessess* (San Diego: ACS Publications, 1986), p. 194.

3. Ivy M. Goldstein-Jacobson, *Astrological Essays* (Tempe, AZ: AFA, and Los Angeles: First Temple & College of Astrology, 1979).

Sources for Data

Major Peace Treaties of Modern History from 1648 to 1967. New York: Chelsea House Publishers, McGraw-Hill, 1967.
Miami Herald
New York Times
San Francisco Chronicle
Goldstein-Jacobson, Ivy M. *Astrological Essays.* Tempe, AZ: AFA, & Los Angeles: First Temple & College of Astrology, 1979.

Sources for Charts

Rodden, Lois M., *The American Book of Charts* (San Diego: ACS Publications, 1980).
Rodden, Lois M., *Profiles of Women* (Tempe, AZ: AFA, 1979).
Rodden, Lois M., *Astro Data III* (Tempe, AZ: AFA, 1986).

5

Escapism: Bacchus, Paradise, Ganymed, Beer

Perhaps one of the most powerful temptations is to be able to escape at will from the present set of circumstances. We all do it, and frequently. Mystics remind us to "be here now" and to keep our minds on what we are doing. In the negative, escapism is what we do in the moments between moving forward—or backward. Escapism represents the inertia, the delusion that this moment will last forever. In the positive, escapism is the way we free ourselves from the bonds of our existence, consciously and ecstatically.

Often the routine methods of escapism, such as daydreaming, are not enough to satisfy us. When stronger distractions are desired, drugs provide some of the most potent and successful avenues of escape. Other distractions may include relationships, sex, gurus, savior figures, religion, money and politics.

In this chapter we will consider the temptations to distraction by examining a series of asteroids which are related to escape. The extreme case of distraction is dependency on it—also known as addiction.

The nature of our biology is such that repeated use of chemical or other methods to produce altered states (another term for distraction) may result in addiction to the substance or method used to produce the altered state. It is usually true that in addiction, the chemical used becomes less effective in producing the desired effect, while other side-effects become more prevalent. While physiologists have studied the physical nature of this problem, many groups, such as Alcoholics Anonymous, have spoken to the spiritual dimension of this issue. It has become increasingly clear that addicts, rather than being criminal/nasty types as they have often been portrayed, are in fact engaged in a legitimate search for meaning beyond the mundane bounds of existence. In their quest, they may not be that dissimilar to religious querents, although they have gotten themselves mired in the physical side-effects of the quest. However, other forms of allegedly spiritual practice, such as flagellation, or tapas (ascetic practices in the East), or celibacy, or prolonged meditation, may produce altered states that are not so different from those which are drug-induced.

In a sense, the various asteroids in this chapter may represent the ways that we seek a greater whole, a union with some level of reality which transcends our small selves. Whether this need to reach beyond results in enlightenment, or addiction, or indifference, is an individual decision.

Bacchus

Bacchus was an early fertility/nature god, presented in later Greek and Roman mythology as an ecstatic drunkard. Bacchus' drunkenness was a metaphor for divine drunkenness, or madness, or ecstasy. This state was not necessarily drug-induced, despite the wine heraldry.

The Bacchanalia was the culmination of a religious rite, a festival, like many of the early fertility festivals, in which all polite pretext was dropped, and drunkenness, promiscuity, and madness were the order of the day. Through dropping the usual restrictions of social existence, the participant (hopefully) was thrown into an altered state as a result of the disorientation. The direct experience of the ecstasy of divine union which could result was an important religious experience. The actual events—drunkenness or sex—were merely a method for stripping off the social veneer and relaxing the self-imposed restrictions which ordinarily kept civilized humans from experiencing their emotions fully. Directed properly, the resulting emotional rush could be interpreted as a transcendental experience.

The fertility cults represented a curious coupling. By stripping off civilized behavior through the encouragement of normally taboo or lower passions, the emotional reaction could be channeled to produce a higher result.

This is the basis of the so-called tantric approach to the spiritual path: The direct experience of ecstasy can be used to propel the being to higher levels of contemplation. Unlike the majority of Eastern spiritual methods, in which the direct experience of the stronger passions is discouraged, tantric schools use these passions as a direct channel to the godhead. In these schools, the lower passions are harnessed as a direct intuition of the higher passions. Both Hindu and Buddhism—especially Tibetan Buddhism—had practitioners of this tradition. In Tibet, it was called the Crazy tradition. Individual lamas were able to speed the enlightenment of many people this way. These traditions were the basis of the *Kama Sutra,* a whole yoga of sexuality. In some schools, both the men and women practiced these techniques. In others, the men practiced the control of the orgasmic process while encouraging their partners to have as many orgasms as possible under the theory that the men could benefit from the energy being released by their partners.

In all cases, these practices were a minority method. One of the principal reasons for this was the unenlightened could be taken advantage of by corrupt gurus or alleged gurus of these schools. There has never been an entirely satisfactory way to distinguish the truly enlightened or "egoless" masters from the unscrupulous charlatans.

In the West these practices were dealt with differently, because, at least until the success of the followers of Jesus Christ, the West did not have the kind of emphasis on the guru-disciple relationship found in the East. In the practices of the Bacchants, or the followers of the Greek god Dionysus, whole groups of people together practiced the setting aside of convention. One person was not in control in the same sense as in the East.

The asteroid Bacchus represents the way that a person seeks ecstasy through direct experience or passion (if s/he chooses to do this at all). If the individual lacks aspects of Bacchus to the planets in the natal chart, s/he still may end up exercising this asteroid through synastry with another person.

Mother Teresa, for example, has Bacchus in Sagittarius in a tight square to her Sun in Virgo. The Sun in Virgo emphasizes the need for service to others: This becomes her Passion, the fire through which she purifies herself and approaches God. Thomas Merton had Bacchus in Capricorn closely conjunct helio Mars. The Trappist Monastery that he entered emphasizes heavy work, silence and a vegetarian diet—a combination I certainly would consider conducive to the exercise of Mars, since it blocks other kinds of energies such as those of Mercury and Venus.

Aspects of Bacchus to the Moon may indicate that the person attempts to bridge the emotions to the Higher Self. Every response becomes a crucible for transformation. Bacchus-Mercury aspects suggest that the intellect becomes the instrument for enlightenment: In this case, a passion for knowledge may be literally true. Venus aspects to Bacchus can denote several possibilities: women in general, the anima, or the magnetic aspect of the body system could be utilized as steps to a higher level. With Jupiter, the individual tries to connect with an institution, whether religious, political or philanthropic in nature. The tendency is to expect that this institution will "trickle down" its understanding to the members or celebrants.

The Saturn-Bacchus person believes that hard work is the basis of the transformational experience. This may occur either through the Protestant Ethic that one *should* work hard, or because his/her work becomes the *tapas* or heat through which passion is burned. Aspects between Bacchus and Uranus indicate an emphasis on the peak experience, those elusive, powerful moments that overwhelm the mundane mind. If this aspect is strong, the individual may be a bit of a thrill-seeker, climbing mountains just to get the rush at the top. Neptune-Bacchus denotes a person who experiences the higher order through psychic experiences, chemical induction, or through helping others. In aspect with Pluto, the Bacchus passion is felt in crisis, in those moments when the breakdown in the old order takes place.

The Bacchus mode is a means of direct involvement. It is not passive. Nor is it necessarily easy—primal experiences never are.

Paradise

Doesn't Paradise sound wonderful? The root question asked by the asteroid Paradise is: Can perfection or Paradise be found in this existence? If the asteroid Paradise is well-aspected in an individual's birth chart, then s/he believes or expects that this is possible or a least a good goal to be pursued. If the asteroid is poorly aspected, then the person is less than optimistic that Paradise exists outside of the movies.

Paradise in aspect to a planet suggests the area of life in which the person believes perfection is possible (if well-aspected), or impossible (if poorly aspected). The energies represented by those planets which are not in aspect to Paradise show areas in which the issue may be simply irrelevant.

Ganymed

Ganymed was the cup-bearer to Jupiter/Zeus. Originally, he was a Trojan, from before the Trojan War. He was deified because Zeus was attracted to him for his beauty. (Zeus had male as well as female lovers).

Ganymed was a popular figure in Renaissance painting. With Saint Sebastian, he shared the distinction of being portrayed most frequently in a homosexual context, which at the time, meant allegorically.

Ganymed, like Psyche, represents a mortal who became deified. As such, both show something of the human interface to the divine—the human who becomes divine. It is interesting that neither achieved this status because of what they did but because of who they were.

As cup-bearer to Zeus, Ganymed, though immortal, was destined always to wait on the Higher God. Here the meaning suggests divinity by association, the ascension to higher levels through service to those already there. When the asteroid Ganymed is stressfully aspected in an individual's chart it may indicate that the person believes such service is unnecessary.

In religious context, the word submission denotes a way of surrendering to the divine that is considered a positive virtue. Whether Jesus' submitting to the Roman authorities, or Gautama Buddha in a previous incarnation giving himself up to a hungry tiger and her cubs, or the aspirant bowing before the guru or Sensei, the act symbolizes the awareness of the submittee that the smaller self is indeed a lesser vehicle to the Higher Self.

In positive manifestation, the asteroid Ganymed shows how we are able to submit ourselves to that which is beyond our personal power. In negative form, it is the way we evade even the awareness that there *is* anything beyond our own powers.

Christopher Isherwood had Ganymed conjunct the Sun in Virgo. In addition to being a successful author (the original of the play and movie "Cabaret" was one of his many works), Isherwood was deeply committed to Vedanta, in which he submitted to one of the spiritual heirs of the great Ramakrishna. Mother Teresa's Libran Ganymed is conjunct Jupiter, further confirming that the way she chooses to express submission is through a traditional religious hierarchy.

I have seen the Ganymed-Mars aspect play itself out in several different ways. One less-than-inspiring example is a man, a heavy drug consumer, who uses drugs as sexual and other bait. He has Ganymed in Virgo conjunct Mars, and believes that he is all-powerful, that he alone can vanquish whatever comes his way. Another, a woman who has been married for over thirty-five years, has a trine of Ganymed in Aries to Mars in Leo. She submits to her husband with few complaints or qualms to his face, and rarely challenges him overtly even when she thinks he is wrong. Another woman, who has a quincunx of a Virgo Ganymed to an Aquarius Mars, tilts at windmills constantly. She began her "career" in Leftist politics with Marxism, proceeding to the Civil Rights Movement, Feminism, and then finally Lesbian Feminism. Another woman with the trine of Ganymed in Gemini to Aquarius Mars, has worshipped the Goddess of Lesbian Feminism by writing about it.

It should be clear from these examples that the Ganymed submission issue is not restricted to conventional dogmas of golden calves. It is truly amazing how many different things humans will put up on pedestals and worship!

The different expressions of Ganymed will be tinted by the planets aspecting it. Mars may suggest an active form of submission or rebellion against same, Venus might represent a submission to Beauty or Harmony. Saturn-Ganymed may show a person who is forced by the School of Hard Knocks to submit. Uranus aspects could reveal someone who has sudden, intense experiences which overwhelm the defenses momentarily. Neptune involvement with Ganymed suggests the awareness of the Higher through visions. Pluto aspects are perhaps the hardest, because here the individual is forced to submit to circumstances analogous to rape, although not necessarily in a physical sense.

Beer

Beer, Bacchus and Vinifera, are the asteroids of addiction. Beer, however, may be prominent in the charts of heavy beer drinkers, brewers or beer alcoholics.

Beer is a noun representing an alcoholic substance brewed in a particular way. As such, there is little mythology associated with this asteroid's name, though there are many tales engendered under the influence of the substance. As such, the interpretation of the asteroid Beer is quite literal.

Discussion

An interesting dichotomy is presented by the interpretation of the asteroids discussed in this chapter. The exercise of these asteroids can represent the gamut from the highest spiritual endeavor to the abyss of chemical (or other) dependency. Yet these two extremes are really two sides of the same issue, the balance of the different components of the person: the mind, emotions, spirit and body. When these components are not in balance, inertia may result. This inertia then becomes a way of forgetting even the possibility of balance. Taking an active role in any of the endeavors listed above tends to break down the inertia. In the words of Pauline de Dampierre,

> "We all have had experiences of a nonpassive state—perhaps after a shock or after a great effort or a deep impression of nature. Why do some people like to climb mountains? There comes a moment when they feel that everything is in accordance. The body has worked and worked and is absolutely relaxed; it feels free and light. The air is light, and the impressions are light, and there is a sense of reaching a part of oneself that one seldom reaches, where there is that vividness. That kind of harmony is something very special. Afterwards one has the memory of it and will decide to go on and on climbing mountains. But will that help me to have more freedom in my ordinary life? That is much more important than climbing mountains. Will it help me in the midst of the responsibilities in my life, in society? Probably not."[1]

The kicker is that either the position of movement or the position of inertia can become addicting. The same experience that is liberating for one person can be oppressive to another. The experience of these asteroids is truly a double-edged sword. One can go to the edge and awake or go to the edge and fall asleep.

Note

1. Pauline de Dampierre, "The Center of Our Need," *Parabola* Vol. 12, No. 2, 1987, pp. 24-31.

Part Two
Mythology

6

Moon and Sky Goddesses: Artemis, Diana, Lilith, Walkure

In the early days of asteroid astronomy, it was considered appropriate to name the asteroids after women, especially goddesses. Initially, this was quite easy because there are thousands of such names from which to choose. Each culture had multiple goddesses and demi-goddesses, many of whom merged with each other as cultures and peoples were thrown together by time and necessity.

The history of myth is tortuous at best. The earliest myths predate writing by thousands of years, and early entities are overlaid with later accretions as circumstances changed. In the case of female goddesses, there is the further complication pointed out by many feminists that male historians, anthropologists and archaeologists have generally trivialized the feminine figures, while emphasizing the masculine ones. As astrologers, we are forced to delve into a realm which can never be fully known. Yet these archetypes gave birth to the meaning of astrologically significant names.

In the beginning was the Sky, the Sun, the Moon and the Earth. Cultures have differed over which of these (if any) came first. We will consider the asteroids of these realms in the next three chapters, but in this chapter, we will focus on the Sky and Moon goddesses.

To primitive people viewing the sky, the Sun and the Moon appeared to be almost identical in size. In our post-Copernican society, this point may be difficult to truly understand. The Moon was not the mere satellite of the Earth, but one of the lights and equal in importance, if not in candlepower, to the Sun. The lights and the planets moved through a belt in the sky known as the zodiac. The stars themselves were generally believed to be fixed, although the more sophisticated societies worked out the niceties of precession. But the Sky was generally conceived to be the ground of activity, the Primum Mobile, the realm of the Gods, or of Heaven.

Different cultures had different gods or goddesses associated with the lights and the sky. At times, the gender of the Moon and Sun divinities reversed or changed. Today, we are generally acquainted with the idea that the Moon is feminine and the Sun masculine, but that was not always the interpretation. And what of the Sky itself? Being of the Judeo-Christian culture, we often think only of Yahweh ruling both Heaven and Earth, but there have been other sky deities as well. To explore these questions, let us look at four asteroids as representations of these goddesses: Artemis, Diana, Lilith, and Walkure.

Artemis

Artemis is usually thought of as the Greek Moon goddess, who although chaste ruled childbirth. Her origins are actually much older, and she was known also as a woodlands goddess, with connections to agricultural, animal husbandry and fishing. In Greek mythology, she was the twin of the Sun God Apollo, and the two of them shared many duties and abilities, so much so that it is tempting to suggest that originally they may have been one entity.

Artemis was a hard task master. In myths she punished individuals or entire peoples, often in far greater measure than their "crimes" seemingly deserved.

In mythology, Artemis is a hunter, always hunting that which is in her path. Astrologically, this is also true, and the aspects the asteroid Artemis makes show what we hunt. Hunting is an old ritual and the quarry is deeply respected, almost divine. The quarry is sacrificed, and then the magical aspect of the quarry is passed to those who partake of the meal. From this came the belief that one becomes more brave by killing a brave man.

When the asteroid Artemis is in contact with a planet, the individual's approach is to hunt down the energy represented by the planet, "kill" it and then absorb it. For example, if Artemis aspects the Moon, the person does not merely feel emotions. They are tracked. They are killed. They are eviscerated. Then they are eaten. Finally, they become a part of the person. The individual with this contact in his/her chart may be seen by others as not very emotional, or not one who values emotions. Actually, emotions for this person are the divine food.

Artemis-Venus aspects indicate that attractiveness is the target. This person may "play" with another's love or admiration. It might seem to the other person that s/he is being used or trifled with, when in fact, the Artemis-Venus person is attempting to transform that bond magically into the divine.

Transformation in the Artemis context means transmutation and absorption. The emotion or bond is brought into the person and then becomes part of him/her. Someone who has Artemis contacts does not experience the energy of the contacted planet directly. This individual must go through an elaborate process before that energy becomes available. Accordingly, it is difficult to react quickly to that planetary energy. Artemis does not operate on the level of the conscious mind. And it is difficult for others to understand that process in the Artemis person.

Diana

Diana is the Roman manifestation of the goddess Artemis. She was, in fact, a local goddess who was grafted to Artemis. Whenever the Romans did this, the god or goddess theoretically absorbed all the older characteristics, and took on some new ones as well. But the flavor always changed somewhat, since Roman society was different from that of the Greeks. According to Patricia Monaghan, the original Roman Diana was

> ". . . queen of the open sky, worshipped only out of doors, where her domain stretched overhead. Possibly she was ruler of the sun as well as the moon, for the early Italians had no sun-god and had to adopt Apollo for that role. Diana's name comes from the word for light; probably she was the original Italian ruler of the sun."[1]

Artemis was a much more independent and important goddess than the Diana that existed after the Roman merging of the Greek pantheon. Diana maintained one aspect of Artemis' personality very clearly, though: she took no truck from mere mortals. Diana never forgot who she was and she never forgot the difference between the divine and the mortal.

The position of Diana in the chart shows the place and area of life in which a person expects absolute respect and obedience, as if s/he were divine. For example, someone with Diana-Sun or Diana-Moon aspects may truly see him/herself as absolutely superior to other mere mortals.

On a more mundane note, Zipporah Dobyns has noted some correlation between Diana positions and the desire to have children independently, often without the need for marriage. The emphasis was on having the child, not relating to the child's father. Demetra George sees Diana as a protector of whatever it contacts.

Intolerance of lesser types is this asteroid's astrological hallmark. The Mercury-Diana person believes her/himself endowed with divine vision or intellect. The Venus-Diana person believes s/he possesses divine beauty. The Mars-Diana person believes s/he has divine courage, strength, or follow-through.

Lilith

Lilith hovers between the mortal and the immortal, depending on the source. On the Hebrew side, there are two versions of the creation myth in Genesis. In one version, "male and female he created them" (Genesis 1:27); in the other, God formed man first (Genesis 2:7), and then created woman later. Lilith is held to be the first wife of Adam (Genesis 1), but she would not acknowledge Adam's superiority, since they were created at the same time. Lilith was expelled from Eden, and then the more pliant Eve was created.

Lilith appears to be associated with the Babylonian "maid of desolation," who was a demon of waste places, thus somewhat resembling thc Indian goddess Kali. Her reputation in Hebraic circles was universally bad and that idea carried over into the Christian context. Her name seems to be derived from the Hebrew word for night, hence her reputation as a night demon, often a mother of demons. The Hebrews called the fixed star Algol by the name Lilith.

Recently, some feminists have called attention to the fact that Lilith's principle crime was to assert equality with Adam, something that the patriarchal Hebrews did not want to support. Hence, Lilith is enjoying some rehabilitation as a womanifestation of the feminist principle of gender equality.

Lilith appears to represent what in Jungian terms might be called the feminine shadow: that aspect of the feminine which could be called negative, immoral, or evil. The difficulty is that the dominant cultural response to shadows—male or female—is to try to suppress them. The challenge is to recognize that this repression is harder on the self than the acceptance of the shadow.

In astrology, Lilith is better known as the Dark Moon, mainly (in America) through the work of Ivy Goldstein-Jacobson. This Lilith is a hypothetical planet, not the asteroid number 1181 to which we refer here. However, there are certain difficulties with the hypothetical planet, whether one considers that any hypothetical planet represents a difficulty. First, while the orbits of the trans-Neptunian hypothetical planets are at least theoretically possible, the orbit of hypothetical Lilith is not. According to Rob Hand (oral communication), if this Lilith existed it would have crashed into one of the known planets ages ago, not to mention the fact that there are no known instances of perfectly circular orbits. The second difficulty with hypothetical planet Lilith is that completely different (and incompatible) ephemerides are used on the two sides of the Atlantic.

Goldstein-Jacobson is completely unambiguous about her opinion of the operation of the hypothetical planet Lilith:

> "Lilith is always SINISTER and MALEVOLENT in her intent and ultimate effect so that the matters and people represented by the house she is in will not be granted full measure of the good that otherwise could develop in the native's life in that department. She is DENYING, FRUSTRATING & CATASTROPHIC, bringing CHAOS to the affairs ruled by that house."[2]

When I first starting working with Lilith I said that Lilith shows the area of life where the individual is forced to bring to light that which has previously been unconscious. Al Morrison felt that Lilith and Eros represent contradictory principles. (See chapter 14 for more about Eros.)

> "Where Eros seems to open the floodgates of sexual expression, Lilith often seems to shut them, at least temporarily. Thus, Eros and Lilith are like planetary pairs in their somewhat opposed and somewhat quincunx-like thrust against each other."[3]

Zip Dobyns finds asteroid Lilith to be Pluto-like, agreeing with the connection to the unconscious mind. Demetra George believes that Lilith "represents the principle of repressed anger and conflict resolution." She emphasizes the independent nature of Lilith and the refusal to submit to unequal relationships. While these themes exist, I don't believe that they represent the essence of the Lilith issue.

Although the original issue for Lilith was her awareness of her own autonomy and independence, that Lilith was repressed for millennia. Through repression, the Lilith version of female sexuality and desire was suppressed, and through suppression emerged sporadically with a life of its own, bubbling up to the surface in moments of crisis or low resistance. Lilith represents the wild woman in us, the unrestrained aspect of sexuality that, when restrained, leads to violence and death. In men, the suppression may lead to misogyny. In women, a self-hatred may develop which is centered around sexuality.

We may consider two examples of women with Sun-Lilith aspects. One, who has the Sun and Ascendant in Scorpio quincunx Lilith in Gemini, is an erotic artist. The other, with Lilith in Capricorn sextile Sun in Pisces, believes that no woman likes sex. An interesting sidelight of this comparison is that, where Lilith is concerned, the softer aspects may be more difficult than the harder ones. The harder aspects tend to bring the issues of Lilith to the surface, and once conscious, they are more easily integrated.

Because repression often results in a change of subject if not in intensity, the Lilith wildness can be transferred to other areas of life. Thus, while a Lilith conflict may not appear to be sexual, the root issue is. The major issue with Lilith, then, is whether the individual chooses to deal with the Lilith nature consciously or unconsciously. Dealing openly with Lilith does not mean giving in to extreme sexual depravity. Rather, it means openness to one's irrational or a-rational side, of which Lilith is an ambassador.

Our society, which has emphasized logical or rational thinking as the ideal, has tended to repress this "less" than rational side. Actually, other modes are simply that: other modes. However, when repression has not worked, the alternative has been to project the irrational/a-rational onto less empowered members of society, for example, women ("sinister") and people of color ("dark").

Since Lilith represents the issue of bringing this a-rational side to conscious awareness, Lilith transits can be devastating. But this is only true if the individual is unwilling to allow this process to take place. Regardless of one's attitude, Lilith transits represent the potential destruction of rationalizations.

Walkure

The Walkure are best known through the intercession of Wagner and the Ring Cycle. The more usual English spelling—outside of Wagner and the Asteroid Listing—is Valkyries. These are the warrior maidens of Germanic legend, the attendants of Odin (Woden or Votan), the Norse king of the gods. On the battlefield, they wore armor, carried weapons and rode white horses. Among their duties was to carry slain warriors to Valhalla, where they served mead to the dead heroes. (And that was supposed to be a reward!) The most famous of the Walkure was Brundhilde, also the name of an asteroid.

The Walkure represent an interesting dichotomy. They dressed for battle, rode to the scene of the battle, witnessed the battle, but did not actually fight themselves. Thus, we find a key to the meaning of the asteroid Walkure: enjoining the battle, without actually engaging in the fighting. One feminist I know who has Walkure in Scorpio conjunct the Sun and sextile Neptune in Virgo, wrote a column in a lesbian publication for years, back before the official beginnings of the gay and lesbian movements. Her column was on book reviews! While encouraging others to get involved in the political issues, she was not actually on the barricades!

An academic with Walkure in Taurus trine helio Jupiter in Capricorn spent many years berating what he saw as the lack of standards at his university, yet he was unable to do what was necessary to affect a change.

Three women clients who have Walkure square Neptune have an interesting connection. One, with the square to helio Neptune, has been involved in a number of relationships with alcoholics. The next, with Walkure in Aquarius square Neptune in Scorpio, is a drug addict, and uses sexuality to get her supply. The third is a child of an alcoholic. An important thread with these three is that none of them actually got involved directly in "fighting" the addiction that affected them.

If Walkure is to be effective, Mars must be strong in the chart. Walkure alone will not do what needs to be done. While the idea of handmaidens may not be popular in some circles, it is appropriate here. Walkure may deliver the message that the battle is to begin, but Walkure then steps back and lets the rest of the nativity fight it.

Notes

1. Patricia Monaghan, *The Book of Goddesses and Heroines* (NY: E. P. Dutton, 1981), p. 84.

2. Ivy M. Goldstein-Jacobson, *The Dark Moon Lilith in Astrology* (Alhambra, CA: Frank Severy Publishing, 1961).

3. J. Lee Lehman and Al H. Morrison, *Lilith* (NY: CAO Times, 1980).

7

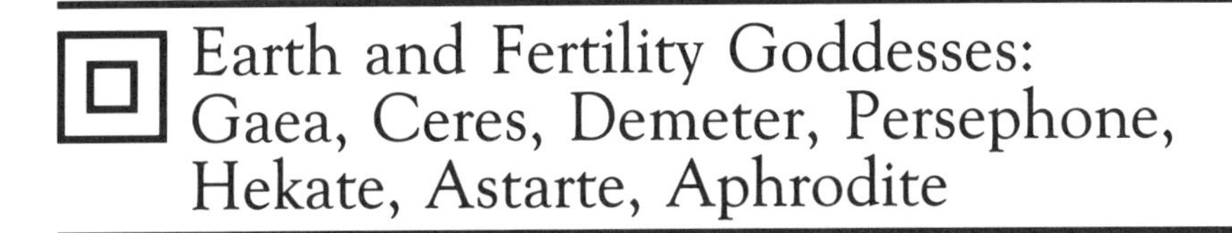

Earth and Fertility Goddesses: Gaea, Ceres, Demeter, Persephone, Hekate, Astarte, Aphrodite

The oldest known human-like gods or goddesses are fertility goddesses. These are represented in their most early form by crude figures with oversized breasts. The art of many cultures has favored presentation of people with oversized genitals and breasts; this is associated with the awesome powers of fertility and reproduction. Fertility in an agricultural society means the growth of plants, especially cereal grains. In a society dependent on livestock, fertility means the growth of the herd. In any society, fertility includes human fertility.

In mythology there is often a linkage between something and the place where that something occurs. Accordingly, since the Earth is the substrate for plant growth, the Earth is linked to the fertility process, often in the form of a goddess. Here we consider several manifestations of this ancient reverence for fertility and abundance: the asteroids representing Earth, fertility and love.

Gaea

Gaea comes through history to us as a personification of Mother Earth. That would give her domain over just about everything we experience. And yet, with such supreme influence, Gaea remains relatively impersonal. She is the *place* where almost everything in our lives unfolds, rather than being an active player.

The asteroid Gaea represents the ground of being. As such, it is not easy to interpret the meaning of this asteroid in a person's life. This may be shown by considering Gaea through the signs.

Gaea in Aries shows someone whose life focuses on newness and trail-lazing, on being different or unique. Example: Judy Garland.

Gaea in Taurus emphasizes the Taurean side of the Earth element:

the slow, patient farmer, going through the required steps in the appropriate order, with the hope of a good harvest at the end.

Gaea in Gemini means that the ground of being is in the mind and in using the mind to communicate with others. Examples: Jill St. John, Greta Garbo (hers is trine Saturn), Elizabeth Taylor.

Gaea in Cancer shows someone whose basis is the primacy of emotions, of feeling. Examples: Thomas Merton, Raquel Welch (trine Moon and quincunx Pluto), Joan Baez.

Gaea in Leo means that the individual's *raison d'etre* is playing out the hero/ine role, and presenting him/herself or priorities as larger than life. Examples: Mother Teresa (square Saturn), Marlene Dietrich, Vanessa Redgrave (opposite Sun), Mary Travers (quincunx Saturn).

Gaea in Virgo, the second sign of the Earth element, whose zodiacal symbol shows the Earth Goddess in another form, indicates a person who is perhaps closest to the archetype of Mother Earth. S/he may see the Earth in a spiritual way, and be conservative in an environmental, not political sense. There is also the sexual side of Earth. Examples: Katherine Hepburn, Cher, Lauren Bacall, Bette Midler (sextile Moon), Liza Minelli, Yul Brynner.

Gaea in Libra shows that this person's life is based on the Other, through cooperation or competition. Examples: Barbara Stanwyck, Chris Evert.

Gaea in Scorpio represents the person whose ground is that plumbing depth for which Scorpio is noted. Sexuality also may become the ground of being. Examples: Jane Fonda, Anais Nin, Sally Ride.

Gaea in Sagittarius emphasizes the underlying importance of travel or foreign influences to the individual. Examples: Grace Kelly, Simone de Beauvoir.

Gaea in Capricorn, the last sign of the Earth element, is involved with the structure of personal reality. Political concepts are important. Examples: Gloria Steinem, Brigitte Bardot, Germaine Greer, Shirley MacLaine, Jane Russell.

Gaea in Aquarius would represent someone who operates within the context of a greater humanity—or that person's vision of humanity. Example: Billie Jean King.

Gaea in Pisces emphasizes compassion and renunciation as the ground of being. Examples: Lily Tomlin, Christopher Isherwood, Doris Day, Marilyn Monroe.

Ceres, Demeter and Persephone

Ceres has been written about extensively by most of the earlier authors on asteroids, since Ceres was the first asteroid to be discovered. In mythology, Ceres was the ancient Italian goddess of grain. She was connected to the Greek goddess Demeter. As usual, the Roman goddess is far less interesting than the Greek one, because the Greeks embellished their pantheon with many stories and subplots, while the Romans looked on their deities from a legalistic and duty standpoint.

Demeter was an Earth goddess, specifically concerned with the fertility of the Earth. Demeter formed a pair with her daughter Persephone, who was the goddess of new grain. The Demeter-Persephone connection, and Persephone's abduction/rape by Hades, was used to explain the changing seasons.

Not surprisingly, considering the practical nature of the Romans, Ceres the asteroid is connected to the industry of agriculture, to the business end of things. Both Emma Belle Donath and Demetra George have given extensive lists of keywords for Ceres, and both authors basically follow the lines set down by Eleanor Bach when she first published on this asteroid. Zipporah Dobyns has argued that Ceres should be assigned rulership of the sign Virgo. The primary word Donath uses for Ceres is domestication.

These authors have covered the field well, and I have little else to say, except that I reject the idea of asteroid rulerships. The reason that I do is because the original meaning of rulerships is considerably more than simple affinity. The affinity of Ceres—and Gaea, Demeter and Persephone as well—with Virgo is hardly in question.

As might be expected, there appears to be some overlap between the meanings of Demeter and Ceres. Both were goddesses of agriculture and fertility. Demeter has a more spiritual connotation, while Ceres is connected more with the mundane. The Demeter-Persephone pair was wrapped up in the Eleusian Mysteries. Demeter and Persephone, along with Hekate, formed the goddess triumvirate of Mother, Daughter and Post-menopausal Woman. While their activities were "down to Earth," they were being done for their own sake (at least theoretically), not for profit. Ceres, with her Roman overlays, has no problem with profit.

To illustrate these differences, I will use some examples of female movie stars. For better or worse, film actresses, more often than not, have been latter-day representations of fertility goddesses, hence the connection.

Brigitte Bardot, certainly one of the more "earthy" stars, has both Ceres and Demeter in Cancer. However, her Ceres is sextile her Venus in Virgo, while Demeter is quincunx the Sagittarius Ascendant. Because the Venus aspect is the easier, Bardot's Ceres side may be said to predominate. With the Demeter quincunx, Bardot was not going to give it away for free!

Cher has Ceres in Aquarius opposite Mars, and Demeter in Cancer without close aspect to any planets. The Ceres presentation of the fertility goddess is definitely different, and Cher always appears in striking costume! Away from her profession, she is undoubtedly a warm person, but that is not the persona she projects on the screen—at least when she is playing "herself."

Raquel Welch has Ceres in Libra conjunct the Part of Fortune! Meanwhile, her Demeter is in Scorpio quincunx Moon in Gemini. Her sex appeal is almost too distracting; it has been so over-emphasized that it has detracted from her considerable skill as an actress.

Greta Garbo has Ceres in Pisces opposite Mercury, and Demeter in Virgo square Mars. On screen, Garbo often projected the Piscean compassion and vulnerability. Off the screen, Garbo has a very strong idea of herself and how she wants to live her life.

Vanessa Redgrave has both Ceres and Demeter in Capricorn. However, her Ceres is semi-sextile her North Node, while Demeter is trine her Midheaven. Redgrave is more notorious for her political work, which she clearly felt necessary. And yet, Janus-like with the Capricorn Ceres, she has been able to look both toward her acting and toward her politics.

Liza Minelli also has Ceres in Capricorn; in this case Ceres is quincunx her North Node. Demeter is in Gemini, sextile Pluto. Away from her professional life, she has had to overcome drugs and the powerful shadow of her mother.

All of the agriculture goddesses were working goddesses. With the exception of Hephaestus/Vulcan, these are almost the only ancient deities shown working. The difference is that Ceres-work is for profit, while Demeter-work is not. Ceres represents vocational work; Demeter is avocational work.

Ceres and Demeter are the Working Women archetypes in the fertile portion of their lives. Persephone the Daughter signifies the promise of things to come. When a person has Persephone aspecting a planet, the indication is that the skills associated with that planet may not be well-developed yet, but the potential is there.

The less comfortable side of Persephone is as the innocent, yet willing, victim. Persephone was abducted by Hades, and forced by the gods to remain with him—despite Ceres' strike—because she ate pomegranate seeds while in Hades' world. She was forced to stay with Hades only half the year, but it was her own willing participation—she chose to eat the seeds—that made it necessary for her to stay at all.

A person with Persephone conjunct the Sun may show promise in many things, and yet continually be involved with people who bilk or use him/her. Persephone aspecting the Moon suggests that the individual's emotions are used by other people.

Persephone-Pluto is perhaps one of the most interesting aspects, because that pair represents abductor-abductee. These people seem to alternate between roles, sometimes playing one side of the pair, sometimes the other.

Hekate

Hekate, as mentioned above, represented the third aspect of the goddess trinity: the older woman, the crone, the wise woman, the woman beyond reproductive years. According to Barbara Myerhoff, these women "are specialists in those critical moments when the designs of culture are threatened by a breakthrough of nature—birth, illness and death—moments when we are reminded of our animal origins and human limits." These women are granted privileges in many societies—Amer-Indian, Asian, European and African—and not necessarily those which hold the most enlightened views of women. Myerhoff explains that this is because women during the reproductive years are useful to society and hence under the most restrictions. Post-menopause, women can assert their roles as humans, not reproductive vessels.

Hekate was the destroyer aspect of the goddess group. Just as Persephone represents new growth, and Demeter mature fruition, Hekate represents the wheat stalks after harvest, the death that is necessary before the new cycle can begin.

Hekate was one of two witnesses to the rape of Persephone, and she became Persephone's companion in Hades. A goddess of the night, she was the ruler of magic. In more primitive European societies, the crone or wise woman was a more useful healer than the male physician. She was the midwife, and her association with the old religion was later repressed through the witch-hunts, which were, according to Szasz, at least partially a process of consolidating power with the establishment (i.e., the male church and later, the male medical establishment).

People who have Hekate strong in their charts tend to break down other people's images of themselves, as well as their own. They may see and criticize that which is going on in society and attempt to change it. The destruction they encourage is not wanton, but part of the cycle. The meaning of Hekate is somewhat analogous to the meaning that we ascribe to a Balsamic Moon: the end of the old that precedes the beginning of the new.

Lily Tomlin has Hekate in Pisces opposite Venus in Virgo. Through her characterizations over the years, she has poked fun at many foibles, both personal and political. And yet, in true Piscean form, she shows considerable compassion for her characters and their plights.

Gloria Steinem has Hekate in Pisces as well, conjunct her Part of Fortune. Simone de Beauvoir has Hekate in Taurus, trine Mercury in Capricorn and sextile the North Node in Cancer. Both feminists are/were involved in the breakdown of the old order.

Astarte and Aphrodite

Astarte and Aphrodite represent a different side of the Earth Goddess image: the fertility goddess. Astarte was Phoenician in origin, goddess of sex, fertility, motherhood and war. She was identified with the Grecian goddess Aphrodite. In Greece, Aphrodite represented sex and fertility, but also beauty. The Roman representation was Venus. Aphrodite was married to Hephaestus, but she also was the lover of Ares. She was the mother of Eros.

It may seem strange to our eyes that Astarte would be associated with war. Aphrodite was not completely separated from war: her bribe of Paris was the cause of the Trojan War. Fertility and war—population increase and population decrease—were functions of the same entity. The only difference between them was that Aphrodite was more refined about her war interests. Astarte was out there in the thick of it.

In chart comparisons, Astarte and Aphrodite contacts between people are much like Venus contacts, with a slightly different twist. Venus (a Roman goddess, like Ceres) is the more mercenary of the two. For example, people with Venus in Taurus seem to mentally put price tags on everything they see. Aphrodite in Taurus people may have expensive taste, but sensuality is what appeals to them, not the price associated with it. People who have Astarte in Taurus are interested in comfort.

When it comes to sexuality, all three are very similar. For instance, Venus in Capricorn is often a very randy placement; likewise Aphrodite in Capricorn and Astarte in Capricorn.

Astarte, Aphrodite and Venus are basically three aspects of the same entity, more so than most related goddesses. The difference is subtle: Astarte is the most primal, Aphrodite the most refined, Venus the most manipulative.

Note

1. Barbara Myerhoff, "The Older Woman as Androgyne," *Parabola* Vol. 3, No. 4, 1978, p.74-89.

8

Sky and Heaven Gods: Apollo, Lucifer, Helio, Quezalcoatl, Tezcatlipoca

In many societies, the original duality was Earth and Sky, however, this duality came much later than the appearance of the Earth and Fertility Goddesses discussed in the previous chapter. Ken Wilber has called the earlier period the *typhonic;* the latter process he refers to as *solarization.*

Wilber looks at human development as an evolution of consciousness. The earliest period in this scheme is the *uroboric.* During this stage there was little differentiation between the self and the environment. The typhonic stage, which corresponds to the period when the fertility/earth goddesses were worshipped, represented the dawn of awareness of the difference between self and environment, which included awareness of personal mortality. During the typhonic period body consciousness developed.

The solar period represents the transition from body consciousness to mind consciousness. It also corresponds to the time of the influx of male deities into the Western belief system.

Wilber relates this state of affairs to the Oedipus stage in human development. First there is no sense of separateness between the baby and her/his environment. The initial separation is from Mother, who is not so much a female being as a bisexual all-powerful entity. As the infant develops, this super-Mom eventually becomes personalized, a distinct human female who is generally involved with a distinct human male, i.e., the father. At this stage, sexual differentiation occurs in the mind if not in the body, and the infant is faced with the uncomfortable reality that there are three people of two types—and s/he is losing out in the quest for sexual unification with the opposite sex parent.

The solar period represents the emergence of super-Dad, the sky or heaven god, to correspond to super-Mom, the earth or fertility goddess. There can be more than one god/goddess, as is the case with parents. The existence of solar gods did not necessarily preclude the existence of earth goddesses. Wilber expressed this transition in mythology as follows:

> " . . . the body is Earth, the mind is Heaven. Now this particular 'Heaven' is not to be confused with a truly Transcendent Heaven . . . , any more than the Great Mother is to be confused with the Great Goddess. This heaven is not anything so lofty as a Dharmakaya or Buddha Realm or Christian Paradise. Rather, this particular heaven simply represents the ascendance of the mind . . . over the body . . . —it is precisely the *heaven of Apollonian rationality* (not ultimate transcendence). Thus, . . . the mind was Heaven, the body was Earth, and the transcendence of the latter by the former was everywhere celebrated in the Hero Myths of this period."[1]

The hero and heroine myths will be discussed in chapters 10 and 11. Several of the heroines—Circe, Medusa and Medea—came into conflict with the heroes. Their defeat was a representation of the defeat of the body by the mind—and the defeat of women by men.

The asteroids discussed in this chapter are named for gods that represented the solar principle: the rational mind.

Apollo

The asteroid Apollo gave its name to the Apollo group, those asteroids whose orbits take them inside the orbit of Mercury. As such, Apollo is one of the faster-moving bodies in the geocentric system, though not quite as speedy as Icarus (another asteroid in the Apollo group). I have observed that at periods when an asteroid is traveling faster than its average motion, its effect is accentuated, provided that the asteroid makes aspects to the natal chart (either geocentrically or heliocentrically). Thus, when Apollo is moving faster than one degree per day, its impact on an individual's chart is heightened.

In mythology, Apollo was the Sun God of both the Greeks and the Romans; since the Romans did not have an indigenous Sun God, they adopted Apollo wholesale, including his name. Apollo, however, was never limited to being merely a sun god. He also was the god of healing, plagues, sudden death, oracles, male beauty, poetry, music and mice. His facet as Phoebus Apollo was later identified with Helios, as we shall see. Like his sister Artemis/Diana, Apollo was a hunter.

Apollo was nothing if not complex! There are three essential sides to his character:

1. The personification of the Sun;
2. The personification of the Ideal Man (male) in Greek society;
3. The giver/healer of disease.

When interpreting the asteroid Apollo astrologically one is never sure which of these faces is going to be prominent. Thus, we see Leonard Bernstein, with Apollo in Leo semi-sextile Sun in Virgo, clearly playing out the musical side of the Apollo myth. (Remember that, especially in Athens, poetry and music had higher social standing than in our own society, and adeptness in the arts was an attribute of the ideal man.) The beauty side (in this case female) is evident in Elizabeth Taylor, who has Apollo in Scorpio trine Sun in Pisces. The athletic side (again an "ideal" trait) is manifested in Olympian Mark Spitz, whose Apollo is in Scorpio square Sun in Aquarius. The "giver of plagues" side is shown in Heinrich Himmler, who was known for conducting insane medical experiments and operating concentration camps that brought about swift death (another Apollonean trait). He had Apollo retrograde in Cancer trine Sun in Pisces. For astronauts of the Apollo space program, this asteroid has a literal interpretation. John H. Glenn has Apollo in Virgo conjunct his Midheaven; Sally Ride has Apollo in Leo sextile her Sun.

The use of an asteroid as the name for an entity will be discussed in greater detail later. However, we can see how the asteroid Apollo functions with regard to astronauts in the U. S. space Apollo Program.

The first chart set (geo and helio) given is for the unfortunate outcome of the testing of Apollo I: the deaths of Virgil I. Grissom, Edward H. White and Roger Chaffee in an oxygen fire of the command module. The second chart set shows the natal chart of Gus Grissom, and his natal positions progressed to the time of the accident. The positions of the asteroid Apollo are summarized below in Table 8.1.

In the transits of the event, heliocentric Apollo forms a grand trine with geocentric Venus and geo Neptune. While Venus and Neptune would not normally be considered fatal, note that Venus is the co-ruler of the gas oxygen, and Neptune would have some relationship to smothering. The mixing of geo and helio in this grand trine should not be surprising given T. Patrick Davis' interpretation of geo and helio as being private and

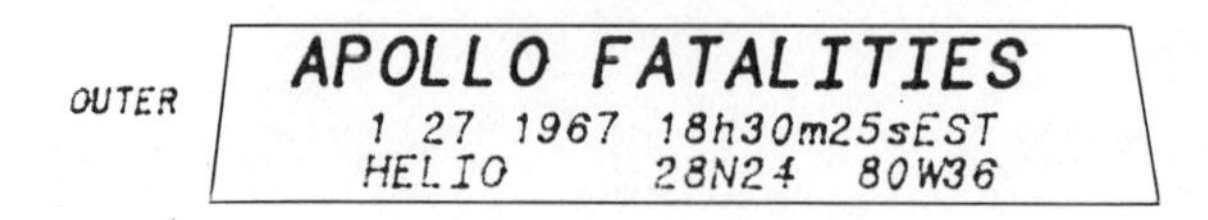

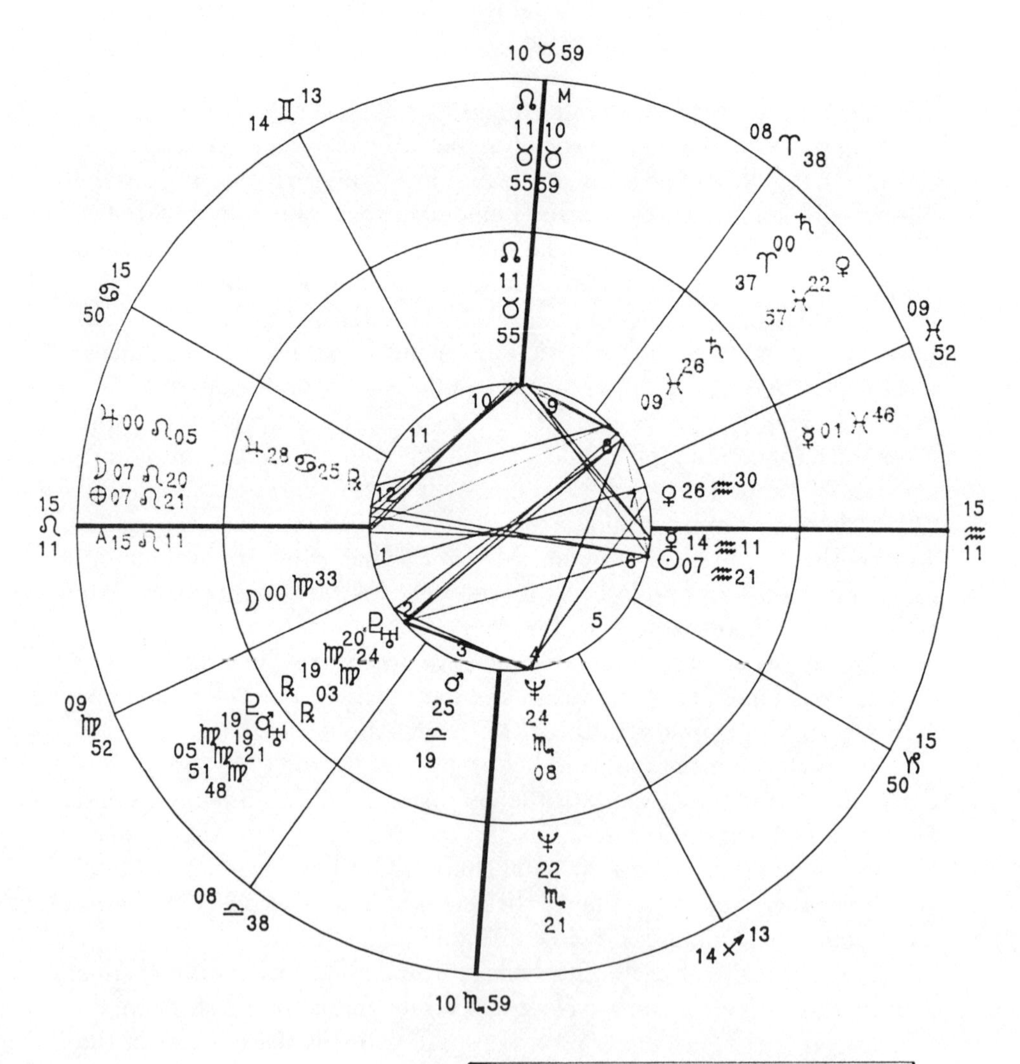

INNER CUSPS

APOLLO FATALITIES
1 27 1967 18h30m25sEST
PLACIDUS 28N24 80W36

Figure 8.1

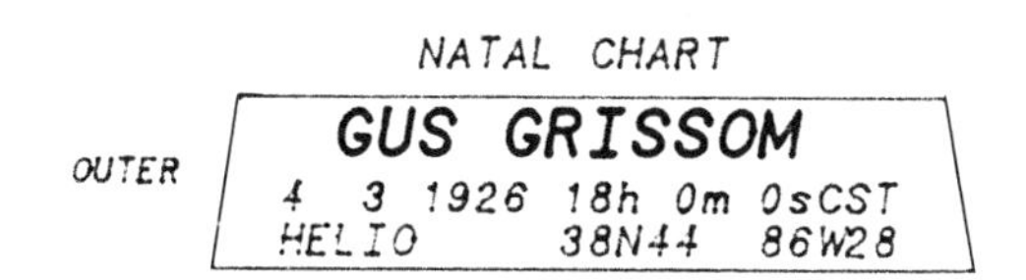

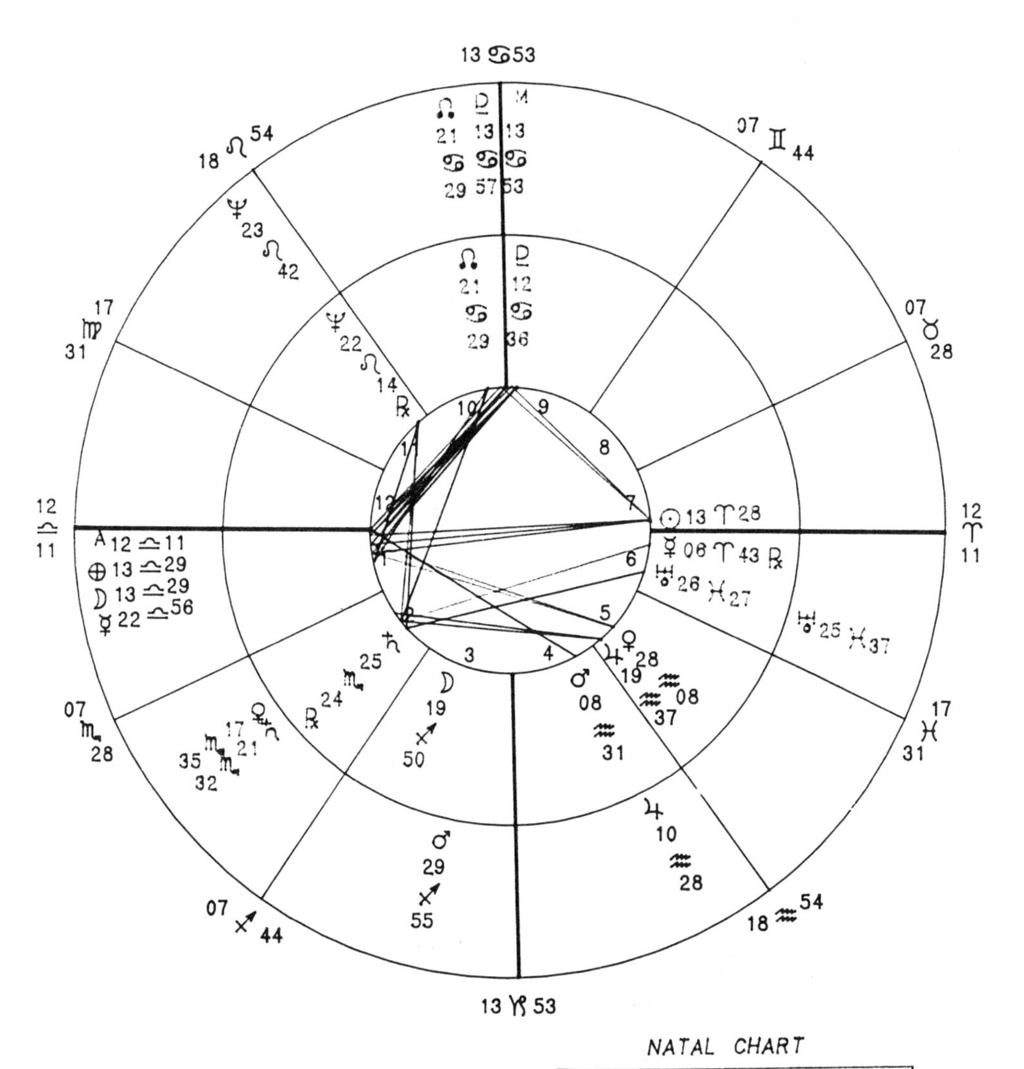

NATAL CHART

INNER CUSPS

GUS GRISSOM

4 3 1926 18h 0m 0sCST

REGMNTNS 38N44 86W28

Figure 8.2

public respectively. Here are three deaths—private matters for the individuals involved—in a very public setting.
The accident occurred when transiting helio Apollo was past Grissom's natal geo position and transiting geo Apollo was approaching that position. Heliocentric Apollo was within twenty minutes to sextile Grissom's Ascendant and within eleven minutes to quincunx his natal helio Mars. Geocentric Apollo was twenty minutes past trine to his geo progressed Saturn and within seven minutes past sextile his geo progressed Sun. The asteroid was within the same degree semi-sextile his helio natal and progressed Neptune.

Transiting geo Mars was within fourteen minutes of square Grissom's natal geo Apollo. There had been a heliocentric Mars-Pluto conjunction the day before, within thirty minutes of quincunx Grissom's natal Jupiter. This is far more indicative of the death than the Apollo aspects. The prominence of Apollo in these charts simply indicates that the name would be prominent. The Apollo positions at the Mars-Pluto conjunction are shown in Table 8.2.

Since the hypothesis taken here is that the name of the asteroid Apollo should indicate prominence in a situation involving either Apollo symbolism or the name Apollo, I have included the positions of the asteroid Soyuz-Apollo as well. This is a case where the asteroid was unequivocally named for the Apollo Program, instead of the god Apollo. At the time of the accident, heliocentric transiting Soyuz-Apollo is widely trine geo transiting Apollo, but in a very tight trine to Neptune.

Table 8.1: Apollo Positions in the Apollo Tragedy Charts

	Fatality	Gus Grissom	Prog. Gus Grissom
Geocentric	23 CN 05 R	25 CN 33	3 LE 41
Heliocentric	29 CN 43	20 LE 52	28 LE 37

Lori Efrein characterized the asteroid Apollo with the following keywords: "fixation, intensity, crisis-mastery, exaltation, commitment, challenge. The liabilities—especially with conjunctions—are tunnel vision, blind spots, irrational awe, imbalance, obsession."[2] I have not found any evidence for any of this. Apollo may manifest according to the three basic groupings given above, but in each case, the emphasis is on rather cold, or at least rational, detachment. Apollo was the embodiment of the rational point of view, and may be symbolic of the mechanistic materialistic metaparadigm that so infects the scientific *Weltanschauung* of today.

Table 8.2: Apollo Positions at the Mars-Pluto Conjunction

1 26 1967 5 1 0 UT (delta T is 38 seconds)

Geocentric Asteroids	
1862 APOLLO	23 CN 51R
2228 SOYUZ-APOLLO	12 PI 38
Heliocentric Asteroids	
1862 APOLLO	29 CN 20
2228 SOYUZ-APOLLO	24 PI 21

Apollo Examples

Apollo-Sun: Leonard Bernstein (semisextile), Rex Harrison (trine), Heinrich Himmler (sextile), Henry Kissinger (square), Mark Spitz (square), Billie Jean King (square), Sally Ride (sextile), Elizabeth Taylor (trine)

Apollo-Moon: Bob Dylan (semi-sextile), Tom Smothers (sextile), Katherine Hepburn (sextile), Anaïs Nin (trine), Raquel Welch (sextile)

Apollo-Mercury: Jack Anderson (semi-sextile), Mick Jagger (semi-sextile), Jane Fonda (semi-sextile)

Apollo-Venus: David Bowie (trine), Sean Connery (conjunct), Richie Valens (sextile), William Shatner (trine), Judy Garland (conjunct)

Apollo-Mars: Richie Valens (semi-sextile), Orson Welles (square), Doris Day (opposite), Sally Ride (sextile)

Apollo-Jupiter: Liberace (semi-sextile), Cat Stevens (square), Anaïs Nin (opposite), Placido Domingo (opposite-exact), Mother Teresa (semi-sextile)

Apollo-Saturn: Placido Domingo (opposite), Christopher Isherwood (opposite helio)

Apollo-Uranus: Orson Welles (quincunx), Jill St. John (square), Mother Teresa (trine)

Apollo-Neptune: James Dean (square), Jack Nicholson (semi-sextile), Sydney Poitier (opposite), Arnold Schwarzenegger (sextile), Joan Baez (semi-sextile), Barbara Stanwyck (semi-sextile), Elizabeth Taylor (sextile)

Apollo-Pluto: Jack Kerouac (semi-sextile), William Shatner (semi-sextile), Lauren Bacall (sextile), Doris Day (sextile), Gloria Steinem (quincunx)

Apollo-Ascendant: Leonard Bernstein (semi-sextile), Mary Travers (trine)

Apollo-Midheaven: John Glenn (conjunct), Doris Day (conjunct), Bette Midler (trine)

Lucifer

Two separate mythological beings are named Lucifer. Lucifer is the Latin rendering of the Greek Phosphorus, which was the designation of Venus as the Morning Star. Lucifer literally means light-bearer, and Venus in the morning is certainly brilliant. The Greeks, like many other cultures, ascribed different deities to describe the Morning and Evening Stars, the two different visible manifestations of Venus.

Much later, the name was applied to the Archangel who rose the highest and then sank the lowest: Satan. This interpretation was much in vogue in the early days of the Christian Church, in the period before the separation of Eastern and Western Orthodox traditions. According to Bultmann, the Gnostics, both Christian and Pagan, were very concerned with the fall of Man and his eventual return to the higher levels. Gnosticism in the West is the main progenitor of Demonology. The demons were said to have stolen sparks of light.

Jaroslav Pelikan notes that the Council of Nicea (A.D. 325) first canonized the existence of angels. This followed intense speculation about the nature of both the angels and demons who had been very prominent in the second and third centuries. That Lucifer had not been equated with Satan at this time is evident from the historical personage Lucifer of Calaris, who precipitated the Council of Milan in 355 which dealt with the Arian Heresy. While Lucifer was a perfectly logical term for a light-stealing super-demon of the Gnostic sort, one would hardly expect a prominent Christian to be named after the Devil if the term was then current. It is perhaps significant that Lucifer of Calaris was instrumental in driving the wedge that eventually resulted in the schism between the Eastern and Western Churches.

Lucifer: a sometimes disruptive bearer of light, who is not the source of the light, which is divine. One might call Lucifer the torch bearer. Shirley MacLaine, whose book *Dancing in the Light* is a partial chronicle of her spiritual quest, has Lucifer in Pisces sextile Mars in Taurus. Joan Baez, with Lucifer retrograde in Gemini quincunx Sun in Capricorn, has acted the role of the torch bearer for issues of world peace. Lucifer's dark side can be seen in Himmler, whose retrograde Lucifer in Taurus was quincunx his Libra Sun and square his Aquarius Ascendant; he carried a different kind of torch.

Lucifer Examples

Lucifer-Sun: Heinrich Himmler (quincunx), Joan Baez (quincunx)

Lucifer-Moon: Greta Garbo (conjunct), Katherine Hepburn (sextile), Billie Jean King (square), Jill St. John (square)

Lucifer-Mercury: David Bowie (trine), Antoine de St.Exupery (square), George Wallace (opposite), Anaïs Nin (trine)

Lucifer-Venus: Alan Watts (square), Joan Crawford (quincunx), Vanessa Redgrave (quincunx)

Lucifer-Mars: A. de St.Exupery (semi-sextile), Marlene Dietrich (opposite), Chris Evert (semi-sextile), Shirley MacLaine (sextile), Bette Midler (semi-sextile), Mary Travers (semi-sextile)

Lucifer-Jupiter: Jack Anderson (semi-sextile), A. de St.Exupery (quincunx), Peter Sellers (quincunx)

Lucifer-Saturn: A. de St.Exupery (trine), Jean Sartre (sextile), Doris Day (sextile)

Lucifer-Uranus: Sean Connery (sextile), Dustin Hoffman (trine), Henry Kissinger (quincunx), Raquel Welch (semi-sextile)

Lucifer-Neptune: David Bowie (semi-sextile), Henry Mancini (opposite), Paul Newman (quincunx), Jack Nicholson (semi-sextile), Vincent Price (semi-sextile), Judy Garland (conjunct), Grace Kelly (opposite), Elizabeth Taylor (conjunct), Raquel Welch (square)

Lucifer-Pluto: Cher (semi-sextile), Vanessa Redgrave (semi-sextile), Mother Teresa (semi-sextile)

Lucifer-Ascendant: David Bowie (semi-sextile), James Dean (trine), Heinrich Himmler (square), A. de St.Exupery (trine), Jean Sartre (opposite), Orson Welles (sextile), Cher (conjunct), Anaïs Nin (conjunct), Mother Teresa (quincunx)

Lucifer-Midheaven: Yul Brynner (semi-sextile), Marlene Dietrich (sextile)

Helio

Helio means Sun. The Greek god's form was Helios, who is depicted as the Charioteer of the Chariot of the Sun, a representation never given to Apollo. Helios was the son of Hyperion, a Titan, who in turn was the son of Uranus and Gaea. Helios was all-seeing. He was worshipped at Rhodes; the Colossus was his image. The Romans called him Sol.

A person who has the asteroid Helio prominent in his/her chart shines like the Sun God. Helios is remarkably uncomplicated compared to many of his fellow gods. Every day he did his job unfailingly. The Helio person does her/his job, and expects to shine as a result of it.

An amusing sidelight of the asteroid Helio is that the name corresponds exactly to the slang term in use for heliocentric astrology. Recently, a Heliocentric Special Interest Group (Helio S.I.G.) has formed within the National Council for Geocosmic Research. The four founding board members are Edith Custer, T. Patrick Davis, Margaret M. Meister and myself. Edith has geo Helio sextile her Midheaven, and helio Helio conjunct her Descendent and square both geo and helio Neptune. T. Pat has geo Helio opposite her Moon, and helio Helio sextile geo Mars and trine helio Venus. (T. Pat has published more on helio Mars than on any other planet!) Margaret's geo is sextile her Midheaven, and her helio Helio makes no aspect within one degree. My geo Helio is conjunct my North Node, and my helio Helio is quincunx my Sun.

Helio Examples

Helio-Sun: Heinrich Himmler (sextile), Henry Kissinger (opposite), Vincent Price (quincunx), Robert Redford (opposite), A. de St.Exupery (quincunx)

Helio-Moon: Liberace (square)

Helio-Mercury: Sandy Koufax (conjunct), Joan Crawford (sextile), Doris Day (opposite), Greta Garbo (sextile), Elizabeth Taylor (sextile), Sally Ride (quincunx)

Helio-Venus: Leonard Bernstein (quincunx), Jack Kerouac (opposite), Jane Fonda (quincunx)

Helio-Mars: Peter Sellers (quincunx), Tom Smothers (trine)

Helio-Jupiter: Leonard Bernstein (opposite)

Helio-Saturn: Sydney Poitier (quincunx), Richie Valens (trine), Alan Watts (semi-sextile), William Shatner (sextile), Marlene Dietrich (semi-sextile), Anaïs Nin (square)

Helio-Uranus: Mick Jagger (conjunct), Lauren Bacall (sextile)

Helio-Neptune: Leonard Bernstein (quincunx), James Dean (opposite), Jack Nicholson (quincunx), Joan Baez (square), Cher (sextile), Gloria Steinem (sextile)

Helio-Pluto: Rex Harrison (square), Mick Jagger (sextile), Robert Redford (quincunx), Cat Stevens (trine), Marlene Dietrich (trine), Katherine Hepburn (trine), Barbara Stanwyck (trine), Lily Tomlin (sextile)

Helio-Ascendant: Sean Connery (semi-sextile), Joan Crawford (opposite), Chris Evert (quincunx)

Helio-Midheaven: Bob Dylan (square), Brigitte Bardot (semi-sextile), Doris Day (square)

Quetzalcoatl

The Aztecs, also known as the Crane People, were one of the last tribes to enter the Central Valley of Mexico. They were not immediately of any note, but before the Spaniards arrived, they had achieved dominance of the region.

One of the differences between the Aztec religion and the religions of the Mediterranean is that in Europe, conquests resulted in the amalgamation of gods and goddesses of the conquered region into the pantheon of the victorious one. Thus, we see that Artemis becomes one face of Diana, and Demeter becomes one aspect of Ceres. When the Aztecs conquered different regions and different peoples, the number of their deities increased because similar entities were not merged. Thus, there are multiple representations of concepts and forces which are depicted as singular in Europe.

Quetzalcoatl, or the Plumed Serpent God, was a Wind God who was instrumental in the life of humankind. The Aztecs believed in multiple ages, somewhat as the Hindus do. At the end of each age, the god destroyed man. The last time around, Quetzalcoatl descended into the realm of the dead to gather up the remains of the past generations of humans, sprinkled them with his own blood, and thus created humans anew. Because Quetzalcoatl had used his own blood, the priests held that blood sacrifices were necessary.

Quetzalcoatl was the ruler of Venus, and had been a sun god in a previous cycle. He brought maize to the Indians, thus serving a similar function as Prometheus.

Quetzalcoatl is the Sun God as Warrior. One could think of the asteroid Quetzalcoatl as having a flavor much like the Sun and Mars combined. He is the Hero-God: more active than the Sun, more creative than Mars. Consider James Dean, with Quetzalcoatl and Sun conjunct in Aquarius. His role in the movie "Rebel without a Cause" forever froze him as an anti-hero, who then became a cult hero in absentia. Robert Redford, with Quezalcoatl in Taurus square his Mars in Leo, almost always plays a hero in his films, but in a more traditional way. Dean did not play a traditional Martian role, but his actions got him into trouble/glory. Redford has played traditional Martian roles such as a baseball player and a naval officer, yet his appeal comes more from his essence or being.

Antoine de St.Exupery, with Quetzalcoatl retrograde in Pisces square Pluto in Gemini, wrote extensively about the aviator as hero. However, his most compelling work, *The Little Prince,* was one of the most transcendent books of the century, with the hero/Prince undergoing transformations throughout the book, finally transforming himself through death back to his place among the stars.

Quetzalcoatl Examples

Quetzalcoatl-Sun: David Bowie (semi-sextile), James Dean (conjunct), Henry Mancini (semi-sextile), William Shatner (semi-sextile)

Quetzalcoatl-Moon: John Glenn (opposite), George Wallace (opposite), Yul Brynner (semi-sextile)

Quetzalcoatl-Mercury: Rex Harrison (conjunct), Jack Kerouac (semi-sextile), Peter Ustinov (semi-sextile), Yul Brynner (trine)

Quetzalcoatl-Venus: Leonard Bernstein (quincunx), Liberace (trine), Jack Nicholson (sextile), George Wallace (quincunx)

Quetzalcoatl-Mars: Robert Redford (square)

Quetzalcoatl-Jupiter: Leonard Bernstein (opposite), Rex Harrison (quincunx), Dustin Hoffman (square)

Quetzalcoatl-Saturn: Sandy Koufax (conjunct), Jean Sartre (sextile), Yul Brynner (quincunx)

Quetzalcoatl-Uranus: David Bowie (trine), Arlo Guthrie (square), Jean Sartre (trine)

Quetzalcoatl-Neptune: Jack Anderson (quincunx), Leonard Bernstein (quincunx), John Glenn (semi-sextile), Yul Brynner (trine), Mary Travers (quincunx)

Quetzalcoatl-Pluto: Henry Kissinger (trine), Liberace (trine), A. de St.Exupery (square), Cat Stevens (square), Alan Watts (quincunx)

Quetzalcoatl-Midheaven: Bob Dylan (sextile), Sally Ride (conjunct)

Tezcatlipoca

The Aztecs had two gods who alternately created humanity in the different ages: Quetzalcoatl and Tezcatlipoca. This pair of gods struggled perpetually for dominance; their ages alternated. Tezcatlipoca was the dark side of the creator god: god of the night, of the form of the jaguar, Ursa Major. Tezcatlipoca was a sorcerer, and to this day the jaguar is a very important animal in the shamanism of Mexico.

Tezcatlipoca is anti-Quetzalcoatl; Quetzalcoatl is anti-Tezcatlipoca. They are a polarity: the god of lightness and the god of darkness. This is not exactly a contest between good and evil, but a contest in the nature of power. Quetzalcoatl was associated with much of the important technology of the Aztecs. Their culture was based on the cultivation of maize, and he was associated with weaving, jade crafting and the calendar. Tezcatlipoca was associated with sorcery, thievery and obsidian. Quetzalcoatl gave, Tezcatlipoca took. If Quetzalcoatl is a Sun-Mars force, then so is Tezcatlipoca, with Tezcatlipoca presenting the darker, or more ominous side.

Hits to Tezcatlipoca were less frequent in my prominent persons survey, possibly because we tend to be a bit squeamish about the dark side of the force. One interesting example of a "hit" is the conjunction of Sun-Tezcatlipoca in Virgo experienced by Christopher Isherwood. Isherwood, as author of *Berlin Stories,* portrayed the seedy side of life as Germany prepared for Hitler. He also practiced in the Vedanta sect founded by the Indian Master Ramakrishna. His biography of that great man, *Ramakrishna and His Disciples,* is one of the clearest renditions of that very interesting period of spiritual development. It is interesting from our asteroid perspective that Ramakrishna throughout his life worshipped Kali, the Great Black Mother Goddess, who very much fulfilled the role in the Indian pantheon that Tezcatlipoca did in the Aztec one. Study of Ramakrishna's life makes it clear that the worship of the dark side is not necessarily any less holy, unless it is a corrupt version. There is great power on the dark side, including the unlocking of the serpent—or Kundalini—power, which I will discuss in more detail in chapter 13.

James Dean, with the Moon in Scorpio quincunx Tezcatlipoca in Gemini, played the role of hero (Quetzalcoatl) and anti-hero (Tezcatlipoca). To a great extent, his role depended on one's point of view.

Tezcatlipoca Examples

Tezcatlipoca-Sun: Sandy Koufax (square), Christopher Isherwood (conjunct)

Tezcatlipoca-Moon: James Dean (quincunx), Mark Spitz (conjunct), Peter Ustinov (quincunx)

Tezcatlipoca-Mercury: Henry Kissinger (quincunx), Mother Teresa (square)

Tezcatlipoca-Venus: Leonard Bernstein (sextile), Henry Kissinger (opposite), Orson Welles (sextile)

Tezcatlipoca-Jupiter: George Wallace (semi-sextile)

Tezcatlipoca-Saturn: Jack Nicholson (conjunct)

Tezcatlipoca-Uranus: Sean Connery (sextile), James Dean (sextile), Arnold Schwarzenegger (square), Peter Ustinov (sextile), Yul Brynner (opposite)

Tezcatlipoca-Neptune: Leonard Bernstein (sextile)

Tezcatlipoca-Pluto: Rex Harrison (trine)

Tezcatlipoca-Ascendant: Alan Watts (sextile)

Tezcatlipoca-Midheaven: Dustin Hoffman (trine)

Notes

1. Ken Wilber, *Up from Eden* (Boulder, CO: Shambhala, 1983), p. 224.

2. Laurie Efrein, "Apollo in the Horoscope," *Ephemeris of Apollo 1890-1999* (New York: CAO Times, 1982).

Source of Data Used

Apollo Fatalities: National Aeronautics and Space Administration.

Gus Grissom: Cited in Lois Rodden, *The American Book of Charts.* AFA, November 1967, "from his mother."

9

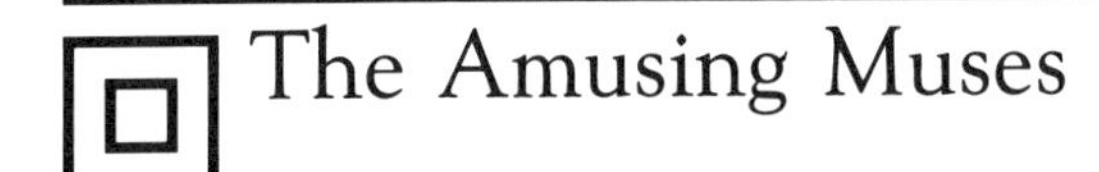

The Amusing Muses

"Never, where lovers of the Muses dwell,
Should dirges sound, for us that were not well."

Sappho

The nine Muses were the daughters of Mnemosyne and Zeus. Mnemosyne was the goddess of memory. The nine Muses were Calliope (eloquence and epic poetry), Clio (history), Erato (lyric and love poetry), Euterpe (lyric song and music), Melpomene (tragedy), Polyhymnia (sacred song), Terpsichore (dancing), Thalia (comedy) and Urania (astronomy). The muse is a common theme in Western literature, usually female, always inspirational. The Muses were universally considered beneficial and benign. Seven of the Muses have been commemorated as asteroid names. The asteroid numbers and node positions are shown below.

Table 9.1: The Muses

Muse	Asteroid Number	Node Position
Calliope		
Clio		
Erato	62	5 LE 18
Euterpe	27	4 CN 23
Meipomene	18	0 VI 07
Polyhymnia	33	8 AR 17
Terpsichore	81	1 AR 29
Thalia	23	6 GE 42
Urania	30	7 AQ 22

The most interesting thing immediately obvious in this list is that all seven Muses have nodal positions in the first decanate. As a result, there are an extraordinary number of aspects between these nodal positions.

What is unique about the Muses and the astrological energy they may symbolize? The Muses represent knowledge and creativity—art and what has since become known as science. In fact, in Grecian times the Muses ruled over all designated fields of knowledge. As such, they represent abstractions, more so than other Grecian mythological figures do.

The rest of the Greek pantheon is full of strong personalities and idiosyncrasies. One sees Zeus attempting to hide his dalliances from Hera; Ceres weeping for the loss of her daughter; Aphrodite, Hera and Athena quarreling over who is most beautiful, etc. It is practically impossible to discover *anything* about the personalities of the Muses, however. We thus have goddesses with oversight but no attributes; substance but no form.

Having no preconceived notions, we are free to ascribe whatever attributes we like to these figures, and therein lies their mystery and appeal. The appeal is that of new love, for how much sweeter is love when we are totally free to paint in the characteristics of our ideal lover with as broad a brush as we choose?

Undoubtedly, this impersonality has heightened vastly the popularity of the Muses through the ages, as countless poets and artists have portrayed their often tempestuous relationships to their personal versions. Each person may redefine the Muses in a relevant way. In impersonality the seeds of genius were planted. In order to explore this amusing connection, I will discuss two asteroid Muses, Erato and Urania.

Erato

In mythology, Erato was the muse of lyric and love poetry. The asteroid Erato represents emotional inspiration. If one person's Erato contacts another person's chart, the Erato-person is likely to be inspiring to other person in the area of life relating to the other person's planet or asteroid.

Erato may be thought of as the Muse in its purest form—the inspiration of one person by another person, place or thing. Erato's connection with lyric and love poetry is merely the expression in its most graphic form of the inspiration felt by the poet. In life we are all poets, spouting our verses over that which touches us most, and it is only a question of whether these verses fit a literary pattern or some other kind.

Erato also may bring insight. Within the context of relating to another, the inspiration may not be from love, but from that other person's loving assistance—consciously or otherwise. In other words, one of the distinguishing characteristics of the Erato insight is the emotional component, the genuine feeling for the other which frees the experiencer from inhibitions, thereby leaving him/her more free to experience in a different way. The Erato person *feels* in that way known as "touching the heart strings."

It may be a little difficult to distinguish the actions of Erato from those of Eros (passion) or Sappho (sexuality). Erato, however, represents the *inspiration* of love, not the consummation of it. The heart of the inspiration is that the love is *not* consummated. The "lover" remains always distant, always pure, always ideal. The "lover" never has a chance to have morning mouth.

Unfortunately, we all have a tendency to try to make our Muses flesh-and-blood, or else to make our flesh-and-blood lovers Muses. It seems perfectly natural to expect to get our inspiration from the same source as we get other goodies. Therefore, we confuse Erato, Eros (passion) and Sappho (sex), and when one is present, we think we have all three. Perhaps one of the greatest lessons of all these asteroids is that it is possible to differentiate the multitude of interactions we have—at least until true enlightenment, when presumably the matter is completely clear, or at any rate, irrelevant.

Short of enlightenment, we live in a world in which boundaries occur. The boundary that Erato suggests is that emotional inspiration and love do not necessarily require a physical component.

If we are talking about emotional inspiration, how does Erato differ from Venus? Venus has been spoken of by many astrologers as denoting harmony, magnetism, affinity, attraction. It is possible to go for a walk in the woods on a sunny autumn day, observing the brilliant hues of the changing leaves, and see it all as beautiful without being inspired by the scene. One person may see it as beautiful, and simultaneously reflect on the depressed property values in rural America. Another may go home and write sixteen stanzas about the vista.

Similarly, a person may be attractive, but not inspiring. Venus gives no assurance of inspiration. In fact, ease and inspiration—if we are to reflect on many writers' lives—have little in common.

Erato contacts in synastry suggest that one person is capable of being the Muse for the other. The Erato-person is the Muse to the planet/asteroid-person in the ways suggested by the planet(s)/asteroid(s) involved. Cross Erato contacts indicate that each individual can play that role for the other. The Muse function in the case of Erato shows the inspiration, the hope of love, not its conclusion or eventual course.

Erato Delineations

Erato-Sun: The inspiration of love—the concept of love in the abstract—can be more interesting than the practice of love to the individual with this combination in his/her chart. This person may make a better part-time lover than a full-time one. For him/her, familiarity breeds dissatisfaction.

Erato-Moon: This person invests more emotion into the idea or concept of love than into the physical side of it. To him/her, love can be a creative inspiration. This person may find difficulty in accomplishing anything—or at least valuing the accomplishment—unless a beloved can share that accomplishment with him/her.

Erato-Mercury: For this person, speaking and writing—though not necessarily about love—are a means of expressing love for another. A good discussion may be an ecstatic experience.

Erato-Venus: This person may be very attractive to others who find themselves inspired in his/her presence. This individual performs the role of the Muse for others.

Erato-Mars: This person works at understanding inspiration, at experimenting with inspiration. Possibly s/he experiments with different circumstances where inspiration, especially through love, can be experienced.

Erato-Jupiter: This individual is intrigued by things foreign to her/his experience. Ideas and inspiration may flow much more freely when traveling.

Erato-Saturn: Classification and systemization may be very important to this person. Without an outline, s/he might find it hard to feel inspired.

Erato-Uranus: This person may feel most inspired or creative during a crisis. It might be entertaining to encounter roadblocks just to see what new course events will take.

Erato-Neptune: This person is not noted for choosing Muses who are particularly appropriate. Nor are these Muses available for anything but a fantasy relationship. The beloved is imbued with qualities that are just "wishful thinking."

Erato-Pluto: This person may attempt to control the Muse, or to impose his/her will on the Muse. The Muse is not supposed to have free will or be capricious; the Muse is supposed to be under control. This individual is terrified that the Muse will have the upper hand, and this doesn't work very well in practice!

Erato-Ascendant: This individual likes to be seen as being loving, however, the skills for loving lie elsewhere.

Erato-Amor: This person may vacillate between wanting to be close to his/her Muse and distancing him/herself from this person. This is a classic approach/avoidance situation. Such an individual is attempting to experience intimacy with the Muse, which is a contradiction in terms. The Muse may feel very used.

Erato-Eros: Passion is the Muse.

Erato-Vesta: This person may have difficulty finding inspiration outside of work.

Urania

When we published the Urania ephemeris in 1981, the late John Addey had sent us part of the piece which appears there before he ever saw the ephemeris. He was very interested in the idea of an asteroid which applied specifically to astrology. I gather that he found the planet Uranus to be inadequate.

Allow me to digress and briefly mention one of the main problems, in my opinion, with the use of Uranus as an indicator of astrology. The Gauquelins' study on the prominence of Saturn in the charts of successful scientists is, in my opinion, quite pertinent. Much of science heretofore had been connected to Uranus, presumably because of the assumed connection to new ideas, discovery, electricity, and so forth. However, the Gauquelins clearly demonstrated something that most of us in the scientific establishment have known, but consistently ignored: that "successful" does not necessarily mean "good."

In Europe in the nineteenth and early twentieth centuries, successful meant being a member of one of the national academies, and being a member of an academy meant jumping through the political hoops necessary to be nominated and approved, much like fraternity hazing today. Thus, the scientific *establishment* became rather rigid. Likewise, so long as astrology could be described as different, nonconformist, heretical, or revolutionary, the rulership of Uranus might be justified. But when we see the astrological "establishment" setting up testing criteria, standards and so forth, we see not Uranus, but Saturn prevailing.

Addey considered Urania as a better candidate for indicator of astrology, and from his post-receipt article, Urania was strong in his own chart and in other charts of astrologers he examined.

Much as he was interested in Urania and astrology, he also recognized that Urania was probably not restricted to astrology. After all, mythological Urania was said to rule astronomy (remember that in Grecian times astronomy and astrology were not clearly delimited as they are now), and one could make a case for the idea that Urania ruled mathematics, the "science" of harmony (as in the Pythagorean school), and especially geometry, that stepchild of astronomy. As such, in our modern classification, one could say that Urania rules scientific inspiration.

Now why, you may ask, would we need a muse for science, that denizen of rationality? The answer lies in the difference between real and stereotyped science. Inspiration is part and parcel of the scientific process, in spite of the trappings of "objective" behavior. Inspiration is what led to Kekule's hypothesis of the structure of benzene, as noted in practically every introductory organic chemistry textbook, and to a host of other situations and discoveries. The inspiration in science lies not in the scientific method (the observations, the repetition), but in asking the right questions, or in the explanation based only partially on the data collected.

This impinges on the heart of the inspiration expressed by Urania: synthesis. The connection to astrology thus becomes obvious. The astrologer is sitting there with ten planets, umpteen asteroids, twelve houses and myriad aspects; Urania represents the process of taking that accumulation of "data" and producing a succinct reading.

In fact, Urania may represent a particular kind of talent: the ability to take a subject or an idea and measure it, or in other ways make its manifestation knowable in the world. Urania might symbolize the ability to translate the universe into an intelligible form.

Discussion

Perhaps we should reflect on what the Greeks meant by the arts and sciences being ruled over by the Muses. Divine inspiration implies the recognition that artistic or creative endeavor goes beyond the realm of the individual to the Divine, or universal. The skills presided over by the Muses are meant to be shared because they are not owned by the practitioner.

The intrinsically impersonal, or more properly, transpersonal nature of those fields represented by the Muses suggests that the interpretation of these asteroids in the natal horoscope (apart from their interactions with other charts or with the transits) may not only be difficult, but also misleading. Many have looked at Urania aspects as indicating astrological talents, but how does one interpret this outside the context of that person's ability to communicate with other people?

One may also comment on the Muses and learning in general. There are essentially two types of learning: the rote memorization type—a Mercury function—and the insight type—a Muse function. This insight may come through synthesis with Urania, or through dance if Terpsichore is involved, or ideal love in the case of Erato.

Alternately, one may consider the Muses to be conduits for that Divine spark which breaks through our usual veneer of ignorance. In fact, it is very difficult to speak for any length of time about inspiration without relating it to the Divine. Like most things in life, we cannot control the Muses.

10

Heros and Villains: Odysseus, Arthur, Lancelot, Galahad, Parsifal

The hero/ine motif is often a somewhat disguised allegory for the spiritual quest, or the duties which face "Everyman" in this life. (One may note ruefully that the *male* experience has too frequently been the standard for the *human* experience.) As such, the hero/ine may be a mythologized version of history, an exemplar of some valued trait, a teaching device, or an allegory. The stories of the heroes of this chapter and the heroines of the next are part of our oral tradition, the way that our tribes and cultures have remembered and cherished those who have gone before.

What is a hero/ine? A hero/ine is a person (a male in some contexts) who faces a challenge and meets it. Usually the hero/ine survives, but not always. In earlier times, the hero/ine may be a singular representation of an entire army. One example is Perseus slaying the Medusa; the Medusa may have been a symbol for a whole city-state, as we shall see in the next chapter. The hero/ine thus can be a device for glorifying the accomplishments of one's group. A hero/ine may be characterized by boldness, courage, cunning or intelligence. Often the gods and goddesses aided or impeded the hero/ine's way.

When people today complain about the absence of heros, they are usually referring to the fact that the swashbuckling style of hero/ine has pretty much disappeared from the modern scene, except in the guise of Rambo and other fictional modes.

Different cultures have valued different traits in their heros. The Japanese Samurai may be seen by us as the equivalent of our gunslinging Westerners, yet their virtues included attention to duty and the ability to face death with equanimity. Our Western sheriffs and Masked Men were better at cheating death, and when all else failed, facing death with bravado, not with a poem.

In Classical Greece man (almost always male) was seen as poised between his own actions, his duties, and the Fates and caprices of the divine pantheon. Thus, the Greek hero/ine required a more varied repertoire than the Medieval chivalrous knight or the Western marshall. And yet, the Greek hero/ine could get caught up in destiny as defined by the gods. Oedipus had no choice about marrying his mother; his only choice was how to deal with it. The Greek tragedy often would highlight the heroic individual caught in a web of duty and tradition, like Antigone, caught between her duty to her brother and her obedience to her uncle Creon.

Later heros would emphasize different virtues or challenges, depending on the prevailing culture. In the Middle Ages, a true knight combined chivalry, Christianity and fighting ability. A good example is Richard I of England, Coeur de Lion, who cost his country much in taxation and misery through his crusade and later, ransom, yet he still was hailed as the flower of his age.

In our modern West, the hero has become much more one-dimensional. Now we prefer military macho to the ability to write poetry. Occasionally, we revere those who have a more intellectual or political contribution, such as the signers of the Declaration of Independence.

Throughout the ages, we honor those who courageously stand by their principles in the face of death, or at least in the face of overwhelming odds—provided, that is, that the presumptive hero/ine doesn't disagree too terribly with our own cherished delusions about the truth and the nature of things.

The hero function appears in many guises. Joseph Campbell may have put it best in the title of his book *The Hero with a Thousand Faces* in which he shows that heros are much like the Greek or Hindu pantheons of god/desses: many aspects/faces for the human—or divine—condition. Ultimately, we are the hero. The hero may represent history, however garbled by time and translation. But the hero is a part of us, and we are all called to engage in heroic struggles through our existence on Earth.

As indicated previously, there has been the unfortunate tendency in the West to view the actions of women as special cases, while those of men are considered universal. While I may appear to be following this convention by separating this chapter from the next, in actually the heroines of the next chapter have at least as many interesting lessons as the heros of this one.

Before proceeding with certain of the asteroids, I should mention three asteroids I worked with that did not seem to pan out very well: Merlin, Gawain and Spartacus. I had expected Spartacus to show up prominently among fighters, and it did not. Gawain should have appeared in some heroic capacity, and again it was not convincing. Neither was Merlin, and I even worked with the chart of a friend who had named his cat Merlin. I mention this to point out something important about the asteroids: People do not tune in to all possible archetypes. The fact that I could not get these to work does not mean that everyone else will hit a brick wall with these asteroids. However, it does show that an asteroid will only work if enough people are relating to its meaning. Apparently, Merlin and Gawain have lost some of their luster through the years.

It is also likely that there are cases where I strike out because I simply cannot relate to an archetype personally. This is a phenomenon that the reader may expect to happen with the asteroids from time to time, because the asteroids do not necessarily represent a set of meanings as universal as those of the planets. Few people are truly unable to relate to the symbolism of Mercury, but Gawain may be another matter.

Odysseus

Odysseus was the King of Ithaca during the Trojan War. He was unenthusiastic about going to fight in the first place, but served well once he was there. On his way home, he encountered all sorts of calamities and adventures, as detailed in that classic by Homer, *The Odyssey.*

Odysseus was noted for his cleverness, not his overwhelming fighting ability. Thus, when we see the asteroid Odysseus prominent in someone's chart, it bespeaks a person who solves problems by guile of the best kind: the ability to view a situation from a fresh perspective, without the past history that most of us ascribe to each new moment.

William Shatner has Odysseus in Libra opposite his Sun in Aries. If one looks at his most prominent persona, that of Captain James T. Kirk, the Odysseus nature is evident: Kirk was a fighter in the sense that he never hesitated to use his fists. Yet, more often than not, he got out of danger by relying on a bluff, a ruse, or a totally original reading of the facts.

Odysseus Examples

Odysseus-Sun: Sally Ride (conjunct), Arnold Schwarzenegger (opposite), William Shatner (opposite)

Odysseus-Moon: Antoine de St. Exupery (opposite)

Odysseus-Mercury: David Bowie (semi-sextile helio), John Glenn (square), Arlo Guthrie (semi-sextile), Sidney Poitier (semi-sextile), Clint Eastwood (square), Robert McNamara (conjunct), Jackie Robinson (trine)

Odysseus-Venus: Yul Brynner (square), James Dean (conjunct), John Glenn (trine helio), Henry Kissinger (sextile helio), Jack Nicholson (semi-sextile), Vincent Price (opposite), Peter Sellers (quincunx), O. J. Simpson (trine helio)

Odysseus-Mars: John Glenn (trine), Arlo Guthrie (square helio), Sally Ride (conjunct), Jean-Paul Sartre (trine), William Shatner (semi-sextile helio), Mark Spitz (opposite), Lawrence Welk (opposite), O. J. Simpson (square helio)

Odysseus-Jupiter: David Bowie (sextile), Bob Dylan (sextile), Henry Mancini (sextile), Richie Valens (sextile), Larry Csonka (sextile)

Odysseus-Saturn: Arlo Guthrie (opposite), Dustin Hoffman (conjunct), Sandy Koufax (semi-sextile), Orson Welles (sextile helio), Jackie Robinson (semi-sextile), O. J. Simpson (opposite)

Odysseus-Uranus: Sidney Poitier (semi-sextile), Richie Valens (sextile), George Wallace (quincunx), Alan Watts (sextile), Larry Csonka (quincunx)

Odysseus-Neptune: Arlo Guthrie (trine), Dustin Hoffman (quincunx), Henry Mancini (opposite), Jean Sartre (conjunct), Cat Stevens (quincunx), Richie Valens (sextile), O. J. Simpson (trine)

Odysseus-Pluto: Sally Ride (quincunx)

Odysseus-Ascendant: Robert Redford (semi-sextile), Cat Stevens (quincunx)

Odysseus-Midheaven: John Glenn (sextile), Robert Redford (sextile), William Shatner (sextile)

Arthur

Arthur was a chieftain in Britain during the fifth or sixth century. He is known not for what he did, but for the adventures that were woven around him during the Age of Chivalry. Mentioned by the Welsh cleric Nennius, Arthur's deeds were even more inflated by the chronicler Geoffrey of Monmouth. Eventually, he became king over Celtic gods and local heroes, and was connected to totally unrelated legends like the Holy Grail. Next to nothing is known of Arthur personally. The important issues in the interpretation of the asteroid Arthur astrologically are his position as king and through it, his sponsorship of the Round Table.

The person who has Arthur the asteroid prominent in his/her chart is a hero who presides. The heroic nature of this asteroid comes from properly executing the duty of assigning someone else to the job of the quest, or alternately, from commanding the troops as general.

Presence is what makes a king. Usually, a king does not need not to command by brute force. He does not have to be able to defeat each of his knights in single combat. He is successful in battle, but he also is successful at delegating. The genius of a general is strategy. He instinctively knows how to defeat his opponent, how to take advantage of every weakness. Like Odysseus, he works with the combination of brawn and brain.

The difference between Arthur and Odysseus is that Odysseus was always thrown back on his own resources and those of his men. Arthur had the backing of the kingdom. We may contrast William Shatner and his Sun-Odysseus with Sean Connery, who has the Sun, Neptune (ruling movies) and Arthur conjunct in Virgo. Connery's persona was James Bond. Like Kirk, Bond was a wily, crafty type, who used his muscles, and who had not only his wits, but Her Majesty's resources behind him.

Arthur Examples

Arthur-Sun: Simone de Beauvoir (sextile), Greta Garbo (semi-sextile), Germaine Greer (semi-sextile), Lily Tomlin (square), Sean Connery (conjunct), John Glenn (square), Cat Stevens (quincunx)

Arthur-Moon: Lauren Bacall (square), Jack Nicholson (sextile), Sidney Poitier (sextile), Muhammad Ali (square)

Arthur-Mercury: Joan Baez (trine), Elizabeth Taylor (square helio), Raquel Welsh (semi-sextile helio), Arlo Guthrie (square), Henry Kissinger (trine helio), Henry Mancini (square helio), Sally Ride (square), William Shatner (opposite), Peter Ustinov (square helio)

Arthur-Venus: Brigitte Bardot (sextile helio), Chris Evert (trine), Lily Tomlin (square), Heinrich Himmler (quincunx helio), Jean Sartre (trine), Arnold Schwarzenegger (square)

Arthur-Mars: Anaïs Nin (quincunx helio), Robert Redford (sextile), Peter Sellers (conjunct), George Wallace (conjunct), Lawrence Welk (quincunx helio), Muhammad Ali (quincunx helio)

Arthur-Jupiter: Lily Tomlin (sextile), Dustin Hoffman (sextile), Paul Newman (quincunx), Mary Travers (sextile helio), Jackie Robinson (sextile helio)

Arthur-Saturn: Cher (semi-sextile), Jane Fonda (square), Liza Minelli (semi-sextile), Jane Russell (semi-sextile helio), John Glenn (semi-sextile helio), Mick Jagger (quincunx helio), Jack Kerouac (square), George Wallace (semi-sextile), Orson Welles (sextile), Joe Frazier (trine)

Arthur-Uranus: Doris Day (sextile helio), Anaïs Nin (trine helio), Arlo Guthrie (trine), Heinrich Himmler (semi-sextile), Sally Ride (semi-sextile helio), O. J. Simpson (trine)

Arthur-Neptune: Sean Connery (conjunct), Rex Harrison (square), Clint Eastwood (semi-sextile helio)

Arthur-Pluto: Jane Fonda (quincunx helio), Judy Garland (opposite), Yul Brynner (sextile), Antoine de St. Exupery (trine), Richie Valens (semi-sextile)

Arthur-Ascendant: Lauren Bacall (square), Jane Fonda (semi-sextile), Vanessa Redgrave (square), Barbara Stanwyck (opposite), Yul Brynner (semi-sextile), Dustin Hoffman (sextile), Sally Ride (semi-sextile), Mark Spitz (trine)

Galahad

Galahad was a son of Arthur and Elaine, and the only knight to successfully pass through the test of the Holy Grail on the first attempt (Parsifal succeeded on the second). He thus succeeded Gawain as the most perfect knight.

The Holy Grail was a complex image. Thought to originate in one of the Celtic magic cauldron myths, the Grail was alleged to contain the blood of Jesus from the crucifixion. It arrived in Britain—most likely in Glastonbury—by the action of Joseph of Arimathea. It was an overtly religious symbol, and by the time it made it into the Arthurian Cycle, it was thoroughly Christianized.

It was Galahad's purity that made him the only completely successful Grail aspirant. The chivalrous knight was a complex, and highly inconsistent type. The fighting part of a knight's work was easy; more difficult was balancing the Christian ideal of chastity with the reality of Medieval—or modern—life. The quest for the Grail was a purely spiritual quest, and the purity of motive required was in direct contradiction to the bawdy life of the knights.

What Galahad achieved is mystical vision, the union of Christ in Man, the ultimate achievement of mortal existence. Galahad, thus, is the model for the Knight-Mystic, the connection between the hu/man of action and the hu/man of God.

The challenge of Galahad is to merge the facets of action and contemplation, to achieve the position of Arjuna in the way of the warrior as spiritual path. Arjuna was taught by the Lord Krishna, disguised as his charioteer, to disconnect his ego from the results of his actions. The trick, as elucidated by Krishna, was not to leave the world, but to cease to be concerned about the fruits of the world.

Most people do not conceive of their movement through life as a spiritual process. Galahad is there to remind us of that prospect if we choose to seize it. Shirley MacLaine has the asteroid Galahad in Sagittarius square her Virgo Ascendant, and she certainly has focused on spiritual issues—not always without difficulty. Mother Teresa has Galahad retrograde in Aries, quincunx her Sun.

An interesting connection, which illustrates the "common name" facet of the asteroid, is shown in the case of Joan Baez. Baez has Galahad in Aries very close to her Ascendant. (Since her time is given as 10:45 am EST, it is possible that the time is not exact.) One of songs Baez has popularized is "Sweet Sir Galahad." Here the common name facet shows some literal connection between the individual and the name of the asteroid.

Galahad Examples

Galahad-Sun: Mother Teresa (quincunx), Chris Evert (semi-sextile), Germaine Greer (sextile)

Galahad-Moon: Judy Garland (opposite), Liza Minelli (square), Vanessa Redgrave (trine), Antoine de St. Exupery (trine)

Galahad-Mercury: Brigitte Bardot (sextile), Henry Mancini (trine), Paul Newman (semi-sextile), Jack Nicholson (square helio), Cat Stevens (sextile), Orson Welles (sextile)

Galahad-Venus: Christopher Isherwood (square)

Galahad-Mars: Joan Baez (trine), Cher (square), Elizabeth Taylor (square), Rex Harrison (sextile), Jack Kerouac (quincunx helio), Liberace (quincunx), Jack Nicholson (quincunx helio), Sidney Poitier (conjunct), Vincent Price (square helio), Peter Ustinov (semi-sextile helio), Clint Eastwood (conjunct helio)

Galahad-Jupiter: Germaine Greer (square), Liza Minelli (opposite helio), Placido Domingo (semi-sextile), Liberace (quincunx helio), Jack Nicholson (quincunx helio), William Shatner (conjunct), Robert McNamara (semi-sextile helio)

Galahad-Saturn: Christopher Isherwood (sextile), Katherine Hepburn (square helio), Jack Anderson (sextile), Placido Domingo (semi-sextile), Arlo Guthrie (semi-sextile helio), Paul Newman (square), Vincent Price (conjunct), William Shatner (square), O. J. Simpson (semi-sextile helio)

Galahad-Uranus: Jack Anderson (quincunx helio), Yul Brynner (semi-sextile), Henry Kissinger (quincunx), Sidney Poitier (sextile helio), Jean Sartre (square helio), Muhammad Ali (conjunct helio), Robert McNamara (square helio)

Galahad-Neptune: Lauren Bacall (trine), Barbara Stanwyck (conjunct), Lily Tomlin (square), Henry Kissinger (conjunct helio), Paul Newman (quincunx), Muhammad Ali (trine)

Galahad-Pluto: Joan Baez (trine), Doris Day (semi-sextile), Arlo Guthrie (semi-sextile), Walter Schirra (semi-sextile), O. J. Simpson (semi-sextile)

Galahad-Ascendant: Shirley MacLaine (square), Jane Russell (sextile), Barbara Stanwyck (sextile), Sean Connery (square), Paul Newman (semi-sextile), Antoine de St. Exupery (quincunx), Arnold Schwarzenegger (conjunct), Cat Stevens (semi-sextile)

Lancelot

Lancelot is today the most famous of the Arthurian knights. He is best known because he practiced the arts of chivalry with Guinevere as his object of desire. He rescued her (and other damsels in distress) from many scrapes. Eventually, when his adultery with her could not be ignored, Lancelot rescued her from death at the stake and returned to his homeland of France.

Lancelot (like Parsifal) was the beneficiary of the Gawain legend, because many of the adventures originally credited to Gawain were later passed to them. One name of the Irish sun god Lug was Lamfada, which was corrupted through translations into the French name Lancelot.

Lancelot is interesting to us not because of the uniqueness of the legends associated with him, but with what he became. The original story of Guinevere's cuckolding of Arthur did not involve Lancelot. The term "du Lac," meaning Lancelot of the Lake, was not originally applied, and we have already seen that many of the actions of Lancelot were inherited from Gawain.

This stresses the nature of Lancelot from the asteroid perspective: Lancelot—more than the other heroic asteroids—represents "glory" thrust on the individual, not due to any personal qualities or actions, but simply from being in the right place at the right time.

This can be seen in one interesting thread: the Lancelot positions in the astronauts of the pre-Moon landing period. In the listing of aspect positions below, all four of the pre-Shuttle astronauts—Walter Schirra, Edward White Jr., John Glenn and Gus Grissom—have Lancelot aspects. The other asteroids do not present this pattern. Consider Lancelot as the archetype for the astronaut. These men were considered heros, but not because of their actions: they were heros before they ever took off into space. They were heros because they were anointed to the position of astronaut.

To test this hypothesis further, I examined the takeoff times of all the Mercury flights for the U.S., and the first flight, Vostok I, of the Soviet Union. When space flight was new, all the astronauts were celebrities, much more so than with some of the interim Gemini and Apollo flights until the Moon shot. The results are shown in the following table.

Table 10.1: Lancelot Aspects to Charts

Flight/Astronaut:	Lancelot Geo Aspects	Lancelot Helio Aspects
Freedom 7		square true Node
		square mean Node
Alan Shepard	semi-sex. H. Mercury	semi-sex. Sun
Alan Shepard ➡ Freedom 7	trine East Point	sextile H. Uranus
	semi-sex. H. Venus	square H. Saturn
		square H. Jupiter
		semi-sex. Vertex
Liberty Bell 7	quincunx Jupiter	
	square H. Venus	
Friendship 7	quincunx Sun	Square H. Neptune
John H. Glenn, Jr.	square H. Uranus	trine Saturn
		quincunx East Point
		square H. Mercury
John Glenn ➡ Friend. 7	trine Mercury	trine M.C.
		square H. Jupiter
		quincunx H. Mercury

Flight/Astronaut:	Lancelot Geo Aspects	Lancelot Helio Aspects
Aurora 7	conjunct true Node	opposite H. Jupiter
	square Neptune	
	opposite Saturn	
	quincunx Jupiter	
Sigma 7	sextile Mars	
Walter Schirra	opposite Sun	opposite Sun
Sigma 7 ➡ Walter Schirra		opposite S. Node
Faith 7		sextile Mars
Vostok I	(no aspects)	(no aspects)
Yuri Gagarin	trine Moon	square H. Chiron
	quincunx Mars	
	trine Uranus	
Yuri Gagarin ➡ Vostok I		opposite mean Node
		quincunx Jupiter
		quincunx Mercury

The Freedom 7 flight by Alan Shepard took place on May 5, 1961. On April 16th, just previous to the flight, there had been a helio conjunction of Jupiter and Saturn, which squared Shepard's Lancelot position. As indicated, the Lancelot position shows glory thrust upon the person. The helio Jupiter-Saturn-Lancelot effect on Shepard was described as follows by Tim Furniss (who, of course, thought he was only describing history!):

> "Though Alan Shepard was the first American in space, most of the laurels went to John Glenn, the first into orbit. It is said that this piqued Shepard, making him doubly keen to fly again and achieve his share of recognition. But before he had a chance to go into space again he was stricken with an inner-ear infection called Meniere's disease which caused balance impairment and resulted in his being barred from all types of flying. Bitterly disappointed, Shepard took a desk job as chief of the astronaut corps, but he never gave up hope of flying again. His confidence was justified: by 1969 he was recovered and back on flight status. Much to the chagrin of many in the space team, and with no Apollo or Gemini experience behind him, he was able to use his powerful position to help win selection as the commander of Apollo 14, with two rookies making up the crew. Thus on January 31, 1971, the three men heading for what would be only the third Moon landing had just 15 minutes of space experience between them."[1]

On January 10, 1971 helio Jupiter was opposite helio Saturn at 21 degrees Scorpio-Taurus, square the geo Uranus of the Freedom 7 flight, and square Shepard's geo Neptune.

Lancelot Examples

Lancelot-Sun: Brigitte Bardot (conjunct), Simone de Beauvoir (semi-sextile), Greta Garbo (semi-sextile), Bette Midler (trine), Sandy Koufax (conjunct), Vincent Price (sextile), Peter Ustinov (conjunct), Muhammad Ali (semi-sextile), Walter Schirra

Lancelot-Moon: Bette Midler (quincunx), Vanessa Redgrave (trine), Clint Eastwood (trine)

Lancelot-Mercury: Joan Baez (sextile helio), Doris Day (quincunx helio), Sean Connery (sextile), Sidney Poitier (semi-sextile helio), Orson Welles (sextile helio)

Lancelot-Venus: Brigitte Bardot (semi-sextile helio), Greta Garbo (semi-sextile helio), Heinrich Himmler (semi-sextile), Mick Jagger (quincunx), Tom Smothers (sextile), Larry Csonka (trine helio)

Lancelot-Mars: Germaine Greer (trine), Liza Minelli (square helio), Lily Tomlin (opposite helio), Sean Connery (quincunx), Liberace (quincunx), Cat Stevens (square helio)

Lancelot-Jupiter: Joan Baez (opposite helio), Greta Garbo (square helio), Leonard Bernstein (trine), Henry Kissinger (sextile helio), Liberace (quincunx helio), Henry Mancini (semi-sextile), Arnold Schwarzenegger (conjunct helio), Mary Travers (conjunct helio)

Lancelot-Saturn: Doris Day (square), Jane Russell (trine helio), Raquel Welsh (quincunx helio), Jack Anderson (semi-sextile), Jack Nicholson (semi-sextile helio), Lawrence Welk (conjunct helio), Muhammad Ali (quincunx helio)

Lancelot-Uranus: Lauren Bacall (trine), Jane Fonda (sextile), Liza Minelli (trine helio), Anaïs Nin (semi-sextile), Gloria Steinem (trine), Jack Anderson (opposite helio), Yul Brynner (conjunct), John Glenn (square), Henry Mancini (trine)

Lancelot-Neptune: Billie Jean King (opposite), Heinrich Himmler (square), Henry Kissinger (semi-sextile), Henry Mancini (square helio), Sidney Poitier (trine), Robert Redford (square), Alan Watts (quincunx), Muhammad Ali (square helio), Clint Eastwood (square helio), Robert McNamara (semi-sextile)

Lancelot-Pluto: Shirley MacLaine (semi-sextile helio), Bette Midler (trine helio), Jack Anderson (sextile), Jack Nicholson (quincunx helio), Jean Sartre (semi-sextile), Robert McNamara (conjunct), Edward White, Jr. (quincunx helio)

Lancelot-Ascendant: Mick Jagger (sextile), Alan Watts (sextile)

Lancelot-Midheaven: Simone de Beauvoir (square), Grace Kelly (square), Mark Spitz (conjunct)

Parsifal

Parsifal, or Percival in the English rendition, was the star of Chretien de Troyes's novel of the same name. Parsifal was a country innocent brought up by his widowed mother, who went off to seek fame and fortune at Arthur's court. His adventures after knighthood were less than sterling, and he failed in his visit to the castle of the Holy Grail because he stayed silent and did not ask the correct question. Later, when he came into touch with God, he learned something of human sympathy.

Parsifal appears to be descended from the Welsh Peredur of the Mabinogion. Since, like Lancelot's, Parsifal's adventures outside of the Grail are mainly borrowings from Gawain, we will concentrate here on the Grail.

The Grail was kept by the Fisher King, who was wounded. In his first visit to the Grail Castle, Parsifal remained silent, and thus failed at the quest. Many years later, after human suffering had increased his compassion, or alternately, after religious austerities and prayer had achieved the same effect, Parsifal returned, asked the question, healed the King, and became King of the Grail himself.

When I looked at positions of the asteroid Parsifal in my "celebs" set, there were only occasional hints of what was going on. However, when I examined the position of the asteroid in the charts of people I knew, the meaning became quite clear. The meaning of Parsifal is spelled out very clearly by P. L. Travers in a retelling of the Grail story. In this version, the young Perceval (Parsifal) meets a hermit in the woods, who tells him that there are three initiatory stages to knighthood. When asked to elaborate, the hermit tells him that they are Induction, Action and Contemplation. Only when the last stage is reached can the Grail adventure truly begin.

A person with a Parsifal aspect in his/her chart can be at any one of these three stages, and that is the difficulty an astrologer faces when reading Parsifal in the chart of someone s/he does not know. The Parsifal individual senses the quest and longs for knighthood. But often knighthood is conceived of as extending no further than the second stage.

Heinrich Himmler, who had Parsifal retrograde in Aries opposite the Sun, got caught up in that twisted form of knighthood invented by Hitler, seemingly without ever realizing how limiting the perspective was. On the other hand, Christopher Isherwood, with Moon conjunct Parsifal in Pisces, spent a good deal of his life following an Eastern religious teaching.

Parsifal in the chart shows a key point in the lesson of compassion. Compassion often is learned through suffering. Unfortunately, the flip side is that if a person wants to break the cycle without learning, s/he can always try to inflict the suffering on someone else.

A Note About Fighting

One of the differences between male heros and female heroines is that virtually all male heros fight. Their heroism may extend to other areas, but there is nothing like a fight to separate the good guy from the bad. A number of the asteroid names we have considered in this chapter are associated with that ideal of Medieval knightly chivalry. We no longer have lances and armor for our knights, but we do have a sport which harkens to those days of single combat: boxing. Accordingly, shown below are the charts of two famous boxers who actually met in the ring: Muhammad Ali and Joe Frazier.

Muhammad Ali has geo Arthur in Scorpio square his Moon in Aquarius and conjunct his I.C. His helio Arthur is in Libra, square his Capricorn Sun and quincunx his geo Taurus Uranus. His geo Pisces Attila is sextile geo Taurus Saturn, semi-sextile geo Aquarius Venus, while helio Attila is in Aries, semi-sextile his Taurus geo Mars. Geo Galahad is conjunct helio Uranus and trine geo Neptune, while helio Galahad is trine geo Mercury and opposite geo Hidalgo. Geo Lancelot is quincunx helio Saturn and semi-sextile his Sun; helio Lancelot is trine his Ascendant. Geo Merlin is sextile his Moon and opposite Jupiter and helio Mars. This is far more than the usual number of hits with this set of asteroids—normally one would expect one or two of the asteroids to contact planets. This indicates a clear involvement of the motif represented by these asteroids, which would either suggest fighting or some other form of heroism.

Now let's consider the synastry of these asteroids with Joe Frazier's. JF's helio Arthur is sextile MA's geo Lancelot; JF's geo Arthur is sextile MA's helio Lancelot. JF's geo Lancelot is trine MA's geo Hidalgo. JF's geo Attila is square MA's geo Arthur; MA's geo Attila is semi-sextile JF's helio Parsifal. These contacts don't even include the contacts with the planets. Here are two fighters who were linked in people's minds for years, and were, if you will, "destined" to fight each other.

	NATAL CHART		NATAL CHART
OUTER	MOHAMMAD ALI 1 17 1942 18h35m 0sCST HELIO 38N15 85W46	INNER CUSPS	MOHAMMAD ALI 1 17 1942 18h35m 0sCST PLACIDUS 38N15 85W46

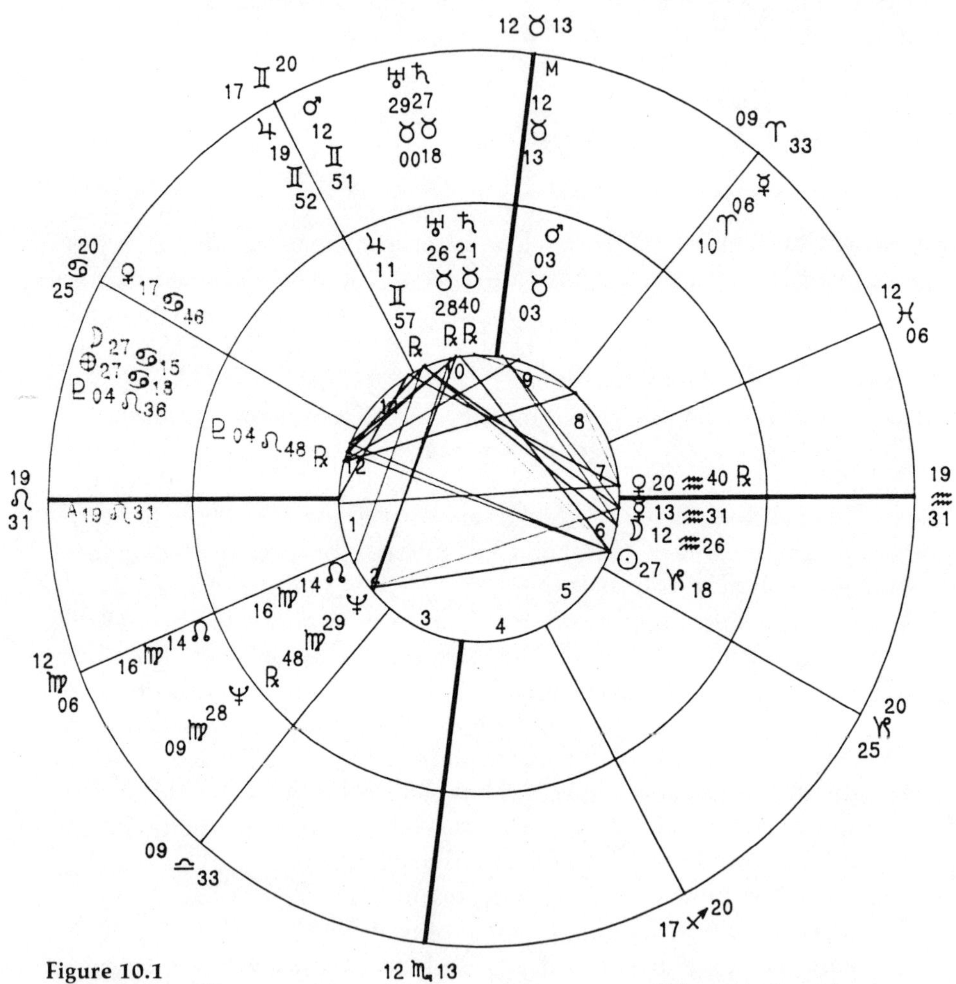

Figure 10.1

Geocentric Asteroids

Arthur	12	SC	28
Attila	21	PI	09
Galahad	29	TA	30R
Gawain	7	VI	19R
Hidalgo	14	SA	37
Lancelot	27	SA	34
Merlin	12	SA	37
Parsifal	9	SA	52

Heliocentric Asteroids

Arthur	26	LI	34
Attila	3	AR	40
Galahad	13	GE	53
Gawain	23	LE	47
Hidalgo	9	SA	46
Lancelot	19	SA	49
Merlin	0	SA	44
Parsifal	23	SC	57

	NATAL CHART		NATAL CHART
OUTER	**JOE FRAZIER** 1 17 1944 7h 0m 0sEWT HELIO 32N26 80W40	INNER CUSPS	**JOE FRAZIER** 1 17 1944 7h 0m 0sEWT PLACIDUS 32N26 80W40

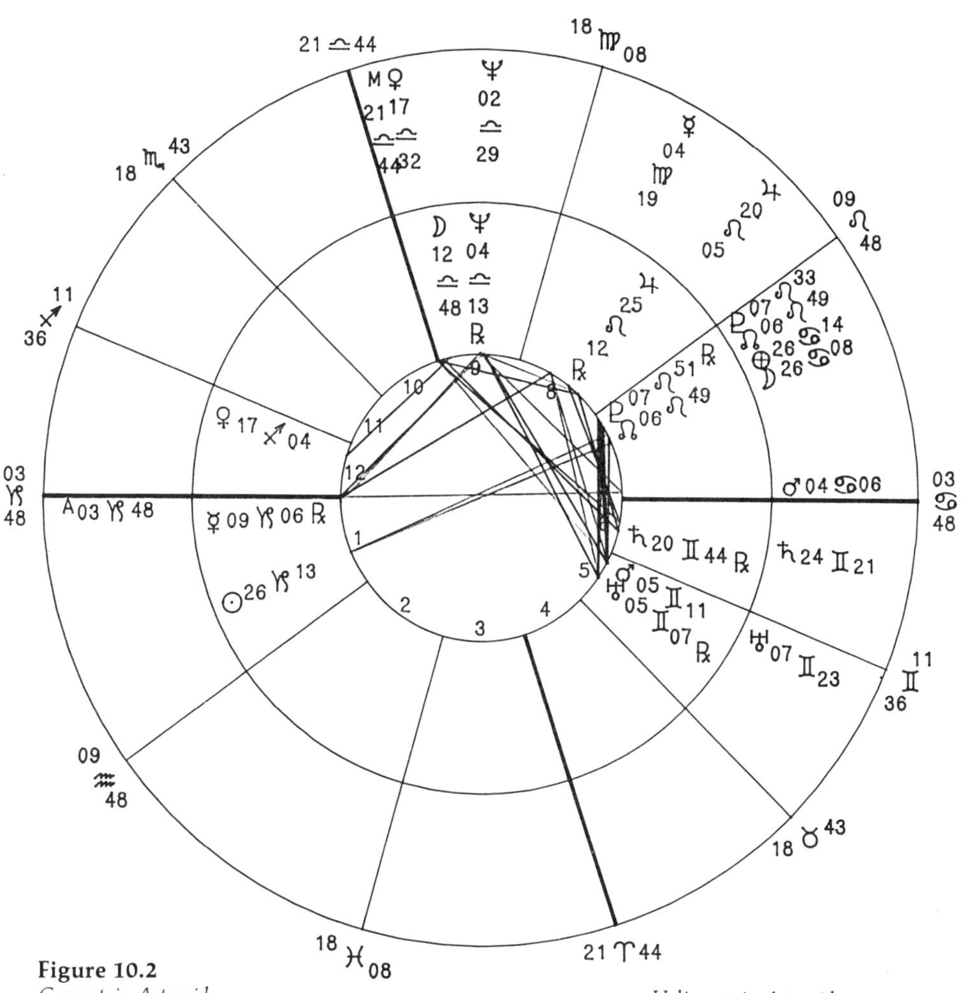

Figure 10.2

Geocentric Asteroids

Arthur	19	AQ	49
Attila	12	LE	48R
Galahad	1	SA	05
Gawain	15	CP	20
Hidalgo	0	CP	32
Lancelot	14	AR	16
Merlin	7	AR	33
Parsifal	21	AR	31

Heliocentric Asteroids

Arthur	28	AQ	04
Attila	6	LE	53
Galahad	11	SC	58
Gawain	12	CP	04
Hidalgo	27	SA	01
Lancelot	5	TA	52
Merlin	2	TA	33
Parsifal	13	TA	25

Note

1. Tim Furniss, *Manned Spaceflight Log* (New York: Jane's Publishing, Inc., 1986), p. 70.

Sources for Data

Tim Furniss, Mercury and Vostok Launch data.

Lois Rodden, Muhammad Ali: Edwin Steinbrecker states B.C. in *Mercury Hour,* July 1979.

Lois Rodden, *Joe Frazier: Current Biography,* 1971.

11

Heroines and Leading Ladies: Circe, Medusa, Medea, Kassandra, Godiva

When we deal with the accomplishments of mortals in mythology—just as when we consider those of the divinities—we are not necessarily looking at actual historical situations. Some of the women covered in this chapter really lived, others were representations of whole peoples, and others were pure invention.

Circe

Circe was an Oceanid, the daughter of Helios and Perse. She was a sorceress who murdered her husband. After being exiled to an island, she amused herself by turning men into beasts. She was not successful in the case of Odysseus, who was warned to take precautions, something his men did not do.

In mythology, Circe is a temptress who uses her power over people if she chooses. When the asteroid Circe and Uranus are in aspect in a person's chart, the combination suggests someone who may use this power over people who normally might not be under Circe's sway. Circe-Neptune aspects indicate a temptress who works either through spirituality or drugs. Other planetary combinations may be interpreted in much the same light.

Circe was anything but mild—in fact, she was a very strong-minded woman. And she was not fond of men. Circe is an archetype for the heterosexual woman who hates men, yet is dependent on them sexually. She may be enchanting, but she believes men are pigs.

Circe has two sides: the magical side and the temptress side. One example is Greta Garbo, who has Circe retrograde in Pisces quincunx Venus in Leo and sextile Moon in Taurus. Here is a woman who could act—i.e., create magic—when her back was turned to the audience. She could create an emotional moment out of thin air. With the quincunx to Venus, however, she ended up with roles where the romance was not right: Often the role was silly, or the romantic plot was contrived. And yet, Garbo could play the part!

Jack Kerouac, with his Sun in Pisces sextile Circe in Capricorn, was one of the creators of that bit of magic called the beat generation—he even coined the term. One of the features of that lifestyle was a sexual openness unknown otherwise in Fifties America.

Circe Examples

Circe-Sun: Jack Anderson, Bob Dylan, Jack Kerouac, Vanessa Redgrave, A. de St. Exupery

Circe-Moon: Lauren Bacall, Greta Garbo, Rex Harrison, Jane Russell, Orson Welles

Circe-Mercury: Grace Kelly, Liberace, Jean Sartre, Arnold Schwarzenegger, William Shatner

Circe-Venus: Marlene Dietrich, Greta Garbo, Ritchie Valens

Circe-Mars: Katherine Hepburn, Ritchie Valens

Circe-Jupiter: David Bowie, Henry Kissinger, Marilyn Monroe, Jane Russell, Arnold Schwarzenegger, George Wallace

Circe-Saturn: Arlo Guthrie, Sandy Koufax

Circe-Uranus: Sean Connery, Billie Jean King, Liberace, Bette Midler

Circe-Neptune: Shirley MacLaine, Cher, Liza Minelli, Paul Newman, Jane Russell, Gloria Steinem

Circe-Pluto: Elizabeth Taylor, Lily Tomlin, Cat Stevens, Lawrence Welk

Circe-Ascendant: John Glenn, Lawrence Welk

Medusa

The Gorgons—Stheno, Euryale and Medusa—were three sisters, sometimes depicted with snakes for hair, other times represented as beautiful. They were usually described as monsters. Originally, they lived in the West, but later they were relocated to Libya. Medusa alone was mortal. Depending on the source, she was slain either by Perseus or Athene, who put Medusa's head on her shield.

Medusa is a problem in mythology and astrology. Two schools of thought exist. One, advocated by Robert Graves, Demetra George and others, is that the slaying of Medusa by Perseus was an allegory for the triumph of the patriarchal forces of Greece over the matriarchal Gorgon Amazons of Lake Triton. This explanation is based on the relocation of the Gorgons to Africa, the supposed origin point of Athene. In another explanation, provided by Costa de Loverdo, Medusa and the Gorgons were representations of their people and land, much like Uncle Sam would be a personification of the United States. De Loverdo believes that the probable meaning of the three Gorgons living in the West is that the three sisters were representations of the three main islands of the Azores, and the slaying by Perseus represented a naval triumph.

Whatever its origin, the myth of Medusa comes down to us as a woman of deadly abilities: to look at her face was to turn to stone. She was the very model of an ancient major monster. De Loverdo feels that the reason for the snake hair was that the Azores were volcanic, and the snakes represented lava flows. Think of the Gorgon energy as volcanic: all quiet, then suddenly erupting violently and uncontrollably. The volcanic nature of the asteroid Medusa is that Medusa is dangerous, but not constantly. In small doses, she may even add spice.

Vincent Price has Moon and Medusa conjunct in Taurus. A mild-mannered man in private life, he is well-known for his starring roles in monster movies. Two other people with Moon-Medusa contacts are successful tennis players: Chris Evert, with Medusa in Sagittarius semi-sextile Moon and Saturn, and Billie Jean King, with Medusa in Leo sextile the Moon. Both are terrific competitors. The volcanic nature of Medusa may provide them with an extra dose of adrenaline when the going gets tough.

Medusa Examples

Medusa-Sun: Jack Anderson, Germaine Greer, Sandy Koufax, Jack Nicholson

Medusa-Moon: Chris Evert, Billie Jean King, Vincent Price, Orson Welles

Medusa-Mercury: David Bowie, Yul Brynner, Judy Garland

Medusa-Venus: Marlene Dietrich, Rex Harrison, Raquel Welch

Medusa-Mars: Liza Minelli, Peter Sellers, Ritchie Valens

Medusa-Jupiter: Joan Baez, Sean Connery, Marlene Dietrich, Jane Fonda

Medusa-Saturn: Doris Day, Chris Evert, John Glenn, Tom Smothers, Elizabeth Taylor

Medusa-Uranus: Lily Tomlin, Robert Redford, Arnold Schwarzenegger

Medusa-Neptune: David Bowie, Yul Brynner, Sydney Poitier, Vincent Price

Medusa-Pluto: Placido Domingo, Dustin Hoffman, Jack Kerouac, Mother Teresa

Medusa-Ascendant: David Bowie, Lawrence Welk

Medea

Another lady with a bad rep! In mythology, Medea was another sorceress. She was the niece of Circe and the daughter of Aetes, king of Colchis. Medea fell in love with Jason and helped him to obtain the prize of her country: the golden fleece. Later, Jason found other amorous interests.

Medea killed various and sundry people, either strategically, or more commonly, out of pique or revenge, and most of the killings were very messy. She killed her brother to prevent her father from successfully pursuing herself and Jason. She even killed her own children because of jealousy for Jason.

Medea was tempting, but fatal. A priestess of Hekate, she practiced her own private destruction. Both love and hate run strong, and Medea shows that love and hate are not truly opposites. Emotions do not come in opposite pairs. The important question is the presence or absence of *any* emotion. The absence of any emotion means noninvolvement in the issue. But because there are multiple emotional types, one can be uninvolved in hate, yet involved in love. To love and hate someone simultaneously is not vacillating between ends of a polarity, since both love and hate represent a type of bond between people.

Joan Baez has Medea in Libra square her Capricorn Sun. She has spent years working in politics. She has also shown an extremely practical side. When her personal or political finances needed replenishing, she released an album calculated to be very popular, more popular, in fact, than her beliefs.

Henry Kissinger, with Venus in Taurus sextile Medea in Cancer, was Nixon's master strategist. He also had a reputation for being quite interested in the ladies before his marriage. Coercion was not outside the realms of his diplomatic strategy.

Medea Examples

Medea-Sun: Joan Baez, Billie Jean King, Elizabeth Taylor, Lawrence Welk

Medea-Moon: David Bowie, Rex Harrison, Sandy Koufax, Tom Smothers

Medea-Mercury: Yul Brynner

Medea-Venus: Lauren Bacall, Germaine Greer, Henry Kissinger

Medea-Mars: Sydney Poitier, Raquel Welch

Medea-Jupiter: Mick Jagger, Jack Nicholson, Cat Stevens

Medea-Saturn: Doris Day, Arlo Guthrie, Jack Kerouac, Arnold Schwarzenegger, William Shatner

Medea-Uranus: Rex Harrison, Vanessa Redgrave

Medea-Neptune: Yul Brynner, Greta Garbo, Dustin Hoffman, Robert Redford, Jean Sartre

Medea-Pluto: Joan Crawford, Bette Midler, Jack Nicholson, Anais Nin

Medea-Ascendant: Judy Garland, Dustin Hoffman

Medea-Midheaven: Simone de Beauvoir

Kassandra

What a punishment Kassandra got! She was the daughter of Priam and Hecube. As a child, she had slept in the temple of Apollo, and he had granted her the gift of prophecy. Later, he became angry with her (the reason differs in different stories), and, unable to revoke the prophecy, Apollo simply ordained that her prophecies would not be believed. Time after time through the Trojan War, Kassandra foretold correctly, and no one believed her.

That is the meaning of the asteroid Kassandra, too: clear foresight, but no one believes you! Anaïs Nin had Kassandra in Aries sextile Mercury in Aquarius. Her writing, in both her diaries and her novels, was extremely insightful. Yet if one reads her descriptions of her conversations with friends, it seems that her friends either were unable to hear the brutally truthful things she told them, or they were unable to heed her words.

Bob Dylan, with Kassandra in Aquarius square the Moon in Taurus, produced all sorts of songs of protest, principally during the sixties. Much of what he said was true, many people sang his lyrics, but the message of peace and brotherhood was ignored in practice.

Kassandra Examples

Kassandra-Sun: Germaine Greer, Grace Kelly, Jack Kerouac
Kassandra-Moon: Bob Dylan, Chris Evert, A. de St. Exupery, Tom Smothers
Kassandra-Mercury: Anaïs Nin, Jill St. John
Kassandra-Venus: David Bowie, Henry Mancini, Marilyn Monroe, Mark Spitz
Kassandra-Mars: Lauren Bacall, Marlene Dietrich, Placido Domingo, Katherine Hepburn
Kassandra-Jupiter: James Dean, Dustin Hoffman, Henry Kissinger, Liberace, Elizabeth Taylor, Alan Watts
Kassandra-Saturn: Jack Anderson, Chris Evert, Jill St. John
Kassandra-Uranus: Billie Jean King, Marilyn Monroe, Sydney Poitier
Kassandra-Neptune: Leonard Bernstein, Robert Redford, Jean Sartre
Kassandra-Pluto: Grace Kelly, Billie Jean King, Sandy Koufax
Kassandra-Midheaven: Jack Anderson, Katherine Hepburn

Godiva

Lady Godiva was one of the flesh-and-blood characters. Her famous ride has become legendary, but what is sometimes forgotten is that her ride was done for political reasons. She wanted her husband, the Earl of Mercia, to remove certain taxes which she considered oppressive. He, in turn, was not so anxious for the entire population of Coventry to see her private parts.

There are several ways to approach this story which became myth. One is to explore the possible correlation between the asteroid Godiva and exhibitionism. Another is to emphasize the political component. After examining a number of charts, I have found no confirmation of the exhibitionism theory. However, there is some very good evidence to support the political theory.

Jack Anderson has Godiva in Leo sextile his Sun in Libra. For years, he has done what needed to be done: making many of the Washington plots public. However, because Kassandra is semi-sextile his M.C., he hasn't always been believed!

Simone de Beauvoir had Godiva in Cancer trine Moon-Mars in Pisces. She was involved with French politics, a mover and shaker of feminism, a major philosophical force, and someone who was willing to do what needed to be done. One example of this was her relationship with Jean-Paul Sartre.

Central to the political side of the story of Lady Godiva was that she was willing to do whatever was necessary to get what she wanted, regardless of the social cost. We have no way of knowing whether she enjoyed her famous ride, but one suspects that her life was never the same afterward. One would also suspect that she knew this was likely to be the case beforehand, and was willing to "bare" the consequences.

Godiva Examples

Godiva-Sun: Jack Anderson, Joan Baez, Doris Day, A. de St. Exupery

Godiva-Moon: Simone de Beauvoir, Sydney Poitier, Jane Russell, Tom Smothers

Godiva-Mercury: Jane Fonda, Sydney Poitier, Jean Sartre

Godiva-Venus: David Bowie, Liza Minelli, Elizabeth Taylor

Godiva-Mars: Simone de Beauvoir, Shirley MacLaine, Bette Midler

Godiva-Jupiter: Yul Brynner, Mick Jagger, Grace Kelly, Jane Russell, Peter Ustinov

Godiva-Saturn: Marilyn Monroe

Godiva-Uranus: Sean Connery, Vincent Price

Godiva-Neptune: Cher, Jane Russell, Jill St. John, Alan Watts

Godiva-Pluto: Lauren Bacall, Mick Jagger

Godiva-Ascendant: Greta Garbo, Jill St. John

Godiva-Midheaven: Germaine Greer, Mark Spitz

12

Egyptian Lore: Isis, Osiris, Anubis, Horus

Western occultism exudes Egyptian mythology. The romance of ancient Egypt, along with the hypothetical world of Atlantis, has captivated thousands of people for thousands of years. Part of the romance is that even though the Egyptian civilization existed thousands of years ago, it was more highly advanced than many modern societies.

Egyptian mythology was actually fragmented between multiple periods and different cities. According to Pierre Grimal,

> "Although the inhabitants of the Nile Valley were all faithful to the same fundamental concepts (hence their religion affords a measure of psychological unity) theologians did not think it necessary to coordinate their beliefs in a really rational system, and they laid even less claim to the foundation of a unified doctrine applicable to the country as a whole. Mythological concepts varied from place to place, and to a certain degree, from period to period. Egyptian religion actually consists of a powdering of local religions. In addition, the Egyptians, being conservative by nature, were reluctant to abandon ancient concepts, even when the concepts themselves evolved, or contact with neighboring cult-centres led to the introduction of new doctrines . . .
>
> "If a unifying principle exists in this diversity of beliefs, it is to be found in the unitary organization of the cult of local gods. For the Egyptians, the pharaoh, as son of the gods, was responsible for the standard practice of a cult. In the final analysis, the essence of Egyptian religion for its worshippers lay not in the acceptance of a dogma, but in the ritual practice of cult."[1]

Egyptian mythology developed in the different city-states, and over time, there were various accretions. From the asteroid standpoint being considered here, the most germane system was the Heliopolitan System, i.e., local cults of Heliopolis. Other systems or principal cults include the Hermopolitan System and the Memphite System. In this chapter, I will discuss four asteroids with Egyptian pedigrees: Isis, Osiris, Anubis and Horus.

Isis and Osiris

In the Heliopolitan System, Isis and Osiris were born of Geb (an earth god) and Nut (a sky goddess), born of Atum, the original god, "the Whole."

Isis was the Mother Goddess. Originally independent, she later was known as the sister/wife of Osiris. She was also the mother of Horus. Isis was both Earth and Fertility goddess, subsuming a number of functions served by separate goddesses in the Grecian pantheon.

Veronica Ions noted that "Osiris was an ancient corn-deity whose followers, coming probably from Syria, identified their god with a pastoral deity called Andjeti and established themselves in his city of Delta in predynastic times."[2]

Osiris was the death and resurrection god for the Egyptians. He was killed and dismembered by his brother Set, and afterward, Isis gathered up the pieces and brought him back to life. Because of the resurrection element, he was associated with the fertility cults more than was Anubis, who was more singly connected with the Cult of the Dead.

As is typical of the Egyptian system, Isis-Osiris represent a pair, a special kind of polarity. Like any true polarity, it is the opposite manifestation of the same function, expressed through the genders of female-male.

The whole Egyptian system was imbued with this gender polarity. Originally matrilineal (if not matriarchal), the system of sister-brother marriages of the pharaohs during some of the dynasties represented an extreme example of this system, although it had political ramifications as well.

In the gender polarity scheme, each being represents either an incomplete person (as in the myth of Plato) or an enantiomer—a mirror image. Plato's myth was that originally there were beings who were male-male, male-female or female-female. These archetypal folks were split, and now each half seeks its other half in order to feel complete. The alternate hypothesis is that any function can be performed in the male way or in the female way. In our updated cultural terminology, this would translate as left-brained and right-brained approaches, or yin and yang ways.

Isis and Osiris represent the archetypal pharaoh pair. Brother and sister, husband and wife, ruling and reigning, Osiris and Isis were inseparable. It was the jealousy and treachery of their brother Set who sundered the bond. When Osiris was killed, Isis did what was necessary to bring him back to life.

As the two sides of the same function, Isis and Osiris together represent androgyny. It is interesting that David Bowie, who has popularized the topic more than anyone else in his generation, has Isis and Osiris conjunct in Libra, sextile Saturn in Leo. Together Isis and Osiris represent the whole fertility cycle through to death and beyond.

They also represent this cycle from the regal or royal perspective. Here the activities of Isis have not created the seasons, as was the case with Demeter. Isis and Osiris are marshallers of resources: teachers to mortals, planners, doers.

As with all of the Egyptian deities, one only needs to scratch the surface of Osiris to run into the death motif. After his defeat by Set, Osiris became Lord of the Dead. Both Isis and Osiris show the combination of love and death, of fertility and abundance, juxtaposed with mortality. There is a poignant quality to the abundance of the asteroids Isis and Osiris. On the one hand, they symbolize the ephemeral beauty of the flower as it puts forth so much energy into its lovely, yet temporary blossoms. On the other hand, they represent the wantonness of the blood worshipper who is deluded into believing that the shedding of other's blood will improve the quality of his or her life.

The androgyny represented by these asteroids suggests that when the two qualities are merged, the distinctions of life and death fall away. If the gender polarity can be transcended, then the other aspects of separateness likewise fall like a rain of petals.

For the person with Isis and/or Osiris aspects in his/her chart, androgyny is an issue, whether this is expressed consciously or unconsciously. With Osiris contacts, the movement toward (or away from) androgyny comes from the "male" side, the objectivist mode. With Isis aspects, the movement is from the "female" or subjectivist perspective. When both asteroids make significant aspects, the person experiences both ends going to the middle—or escaping the middle.

The elemental modality of the planets and asteroids involved are important in the elucidation of the androgyny issue. For example, Paul Newman has Isis in Sagittarius trine geo Mars and helio Chiron in Aries. While Newman has played a number of roles which could be described as very masculine or even macho, he has always been appealing to women, and has never come across as insensitive to women or to feminine issues. With Isis and Mars both in "masculine" signs, he is going to appear masculine in any case, especially with Mars in its house (rulership) in Aries.

In movies "Butch Cassidy and the Sundance Kid" and "The Sting," Newman was paired with Robert Redford, who has Isis and Osiris conjunct in Scorpio. With Pisces rising, Redford is a true chameleon. Like Newman, Redford is sensitive to women and women's issues. However, with these asteroids placed in the arguably feminine sign Scorpio (I, for one, would argue for variation in the degree of masculinity and femininity of the signs), Redford is able to act by not acting—there is a passivity in some of his roles which alternates with the intensity of Scorpio. The effect is an androgyny which originates not from bisexuality, not from what he does, but from the way he can approach life from either the male or female perspective.

Whether in aspect or not, the sign placements of these two asteroids will tell something about the masculine-feminine or left-brain-right-brain balance of the person. This doesn't so much show whether the person would be classified as masculine or feminine in appearance or bearing, but in his/her mental approach to the world.

Like everything else in astrology, there is a polarity in how an individual can play out the energy. One way to express the issue is through androgyny avoidance. Heinrich Himmler, scarcely a feminist on anybody's list, had Isis in Virgo quincunx Moon in Aries, sextile Mercury in Scorpio, and sextile helio Mars in Cancer. He also had Osiris in Cancer at the midpoint of Venus in Leo and Neptune in Gemini. In Himmler's case, the extreme Nazi glorification of the "male" warrior mystique and the subjugation of women in the name of motherhood expressed the rebellion in his own psyche against the two sides of the coin. As Keith Critchlow reminds us, the androgyne is in our own souls.

Isis Examples

Isis-Sun: Arnold Schwartzenegger (sextile), Mark Spitz (quincunx)

Isis-Moon: Rex Harrison (square), Heinrich Himmler (quincunx), Robert Redford (sextile), Jean-Paul Sartre (semi-sextile)

Isis-Mercury: James Dean (trine helio), Heinrich Himmler (sextile), Mary Travers (sextile helio)

Isis-Venus: Liberace (conjunct), Peter Ustinov (trine), Orson Welles (trine helio)

Isis-Mars: Heinrich Himmler (sextile helio), Paul Newman (trine), Peter Sellers (quincunx), Mary Travers (trine)

Isis-Jupiter: Placido Domingo (sextile), Sandy Koufax (semi-sextile), Orson Welles (square)

Isis-Saturn: Jack Anderson (square), David Bowie (sextile), Robert Redford (trine), Tom Smothers (semi-sextile), Cat Stevens (quincunx)

Isis-Uranus: Jack Anderson (trine), Rex Harrison opposition), Henry Kissinger (trine), Jack Nicholson (semi-sextile), Sally Ride (square)

Isis-Neptune: Henry Kissinger (semi-sextile)

Isis-Pluto: Jack Anderson (conjunct), Sean Connery (conjunct), Jack Kerouac (square helio), Liberace (conjunct)

Isis-Ascendant: Sally Ride (square), Tom Smothers (opposite), Lawrence Welk (semi-sextile)

Isis-Midheaven: Sean Connery (trine), Jack Kerouac (sextile), Mary Travers (sextile)

Osiris Examples

Osiris-Sun: Sandy Koufax (sextile), Cat Stevens (opposite)

Osiris-Moon: James Dean (quincunx), Robert Redford (sextile), Sally Ride (trine)

Osiris-Venus: Heinrich Himmler (semi-sextile), Arnold Schwartzenegger (square helio), William Shatner (opposite), Tom Smothers (quincunx helio), Mary Travers (conjunct)

Osiris-Mars: Arlo Guthrie (trine), Mick Jagger (opposite helio), Liberace (conjunct helio exact)

Osiris-Jupiter: Sean Connery (semi-sextile), Dustin Hoffman (sextile), Henry Kissinger (quincunx)

Osiris-Saturn: David Bowie (sextile), Robert Redford (trine), Arnold Schwartzenegger (sextile), Mary Travers (square helio)

Osiris-Uranus: Yul Brynner (opposite), Mark Spitz (semi-sextile), Cat Stevens (quincunx)

Osiris-Neptune: Sean Connery (square), Bob Dylan (opposite), Heinrich Himmler (semi-sextile), George Wallace (semi-sextile)

Osiris-Pluto: Sydney Poitier (sextile), Vincent Price (sextile helio), Sally Ride (sextile helio), Arnold Schwartzenegger (sextile), Tom Smothers (opposite)

Osiris-Ascendant: Arlo Guthrie (semi-sextile), Rex Harrison (trine), Henry Mancini (trine), Jack Nicholson (quincunx)

Anubis

Anubis was the jackal-headed god of the dead. According to Ions, his cult apparently originated at Thinis near Abydos on the Nile, and then spread to most parts of Egypt. A desert animal, the jackal became associated with the Egyptian home of the dead, the western desert. He was the patron of embalming. Anubis was the judge of the afterworld, who weighed the heart of the deceased against a feather. In early times, he was the main god of the dead, a role that later was taken over by Osiris. When the transition to overlord Osiris occurred, Anubis was made the son of Osiris by Nephthys. Anubis was associated with the Grecian god Hermes. This may have been because he could foresee a mortal's destiny, and so he was associated with magic, much as Hermes was.

As God of the Dead, Anubis in his asteroid manifestation may represent someone for whom death is more than a passing issue. An example of this is a man who described himself to others as not being of this world, and who attempted suicide on numerous occasions from early childhood until he finally "succeeded" at the age of thirty-four. This man had helio Anubis in Pisces trine helio Saturn in Cancer. His geo Anubis, retrograde in Pisces, was square helio Mars in Gemini. At the time of his death, transiting geo Anubis was retrograde in Aries, trine his East Point (the East Point is just as prominent as the Ascendant at the time of death), while helio Anubis was square his helio Saturn and opposite geo Mercury, his chart ruler.

A less dramatic illustration comes again from Christopher Isherwood, who, the reader may recall from earlier chapters, wrote about the decadence and decay in Germany in the 1930s, and was the disciple of a disciple of Ramakrishna. Ramakrishna was a lifelong worshipper of the Goddess Kali, the Indian patron of death and graveyards. Isherwood had Anubis in Scorpio sextile helio Mercury in Capricorn. His published presentation of these matters meshed well with T. Pat Davis' description of the helio placements as representative of public matters.

James Dean had Anubis in Pisces conjunct his I.C., suggesting that his death, more than most, became the quintessential fact of his life. Dustin Hoffman, who has Anubis in Leo, square Uranus in Taurus, semi-sextile Mercury in Virgo and quincunx helio Mars in Capricorn, produced a vivid picture of death and decay on the screen in "Midnight Cowboy."

Anubis Examples

Anubis-Sun: Yul Brynner (quincunx), Mick Jagger (sextile), Jack Nicholson (sextile), Tom Smothers (trine)

Anubis-Moon: Arlo Guthrie (trine), Henry Kissinger (square), Sandy Koufax (trine), Liberace (semi-sextile)

Anubis-Mercury: Placido Domingo (semi-sextile), Dustin Hoffman (semi-sextile), Paul Newman (sextile helio), Orson Welles (trine helio)

Anubis-Venus: Mark Spitz (conjunct), George Wallace (square)

Anubis-Mars: Arlo Guthrie (sextile), Dustin Hoffman (quincunx helio), Sydney Poitier (semi-sextile helio) Vincent Price (semi-sextile), Tom Smothers (quincunx), Mary Travers (square)

Anubis-Jupiter: Sydney Poitier (square), Robert Redford (opposite)

Anubis-Saturn: Henry Mancini (sextile), Sydney Poitier (opposite)

Anubis-Uranus: Dustin Hoffman (square), Jean-Paul Sartre (semi-sextile)

Anubis-Neptune: Arlo Guthrie (sextile), Mick Jagger (conjunct), Robert Redford (square), Antoine de St. Exupery (quincunx)

Anubis-Pluto: Jack Kerouac (semi-sextile), Peter Sellers (quincunx), Mary Travers (semi-sextile)

Anubis-Ascendant: Arlo Guthrie (semi-sextile), Jack Nicholson (semi-sextile)

Anubis-Midheaven: James Dean (opposite), Sally Ride (sextile)

Horus

In the Major Trumps (Arcana) of the Tarot as interpreted through the Kabbalah, the first two cards represent thesis and antithesis, and the third card synthesis. Horus is the synthesis of the thesis of Osiris and the antithesis of Isis. Horus is a rather confusing god, because he actually represented a number of different—and sometimes conflicting—deities. He was originally a sky god represented as a falcon, also called the Egyptian Apollo. As such, he was the brother of Isis, Osiris and Set, possibly equivalent to Ra, represented as the rising or setting Sun. Later he was represented as the son who avenged the death of his father Osiris by killing Set, losing an eye in the process. However, the myth was not that simple, as Ions illustrates:

> "The god Horus, symbolized by a falcon, was in the beginning an exception to the general rule that the chief gods were associated with fertility or creation. As we have remarked, he was originally the god of a hunting people, or possibly a war-god, but after the unification of Egypt by Menes he began to be identified with the Sun cult of a king who had earlier unified the Two Lands. The falcon became the symbol of majesty and the archetype of the pharaohs, who were said to be seated on his throne. Attempts to incorporate worship of the sun into the Horus legend took the form of making the divine falcon a sky god, the sun being his right eye and the moon his left eye."[3]

The representation of Horus as a falcon seems to be the one common factor in the myriad incarnations of this god. The falcon is a bird of prey, whose major claims to fame are its prowess in hunting and the acuteness of its sight.

Thus, we see two aspects of the myth of Horus as it applies to the asteroid of the same name: far-sightedness and avenging nature. While avenge and revenge are synonyms, the meaning applied here is consonant with the ancient view of revenge, both in the West and in Samurai Japan. Revenge was an obligation imposed on the next of kin (male) or other interested parties. Revenge was to be swift and final. Not the extended plotting hatched as a result of petty slights, this is revenge in the tragic sense, a duty.

Far-sightedness as it applies to Horus need not be restricted to issues involving revenge. Horus aspects may indicate a person who is capable of foreseeing a whole sequence of necessary steps, each one a small movement toward the eventual goal. The awareness of the end which this asteroid brings can breed patience, because the end is never really lost.

Thomas Merton had Horus in Gemini trine Sun in Aquarius. His days in the monastery were filled with hard work, hard weather and silence. Yet through each "obstacle" he could see himself drawing nearer to his God and to God-awareness.

Alan Watts also had Horus in Gemini, but his was retrograde square the Moon in Virgo. Watts did as much as anyone to introduce Zen to the West. Yet his major accomplishment was to talk about Zen. Toward the end of his life, he realized that for all his talk (Gemini), his practice (Horus quincunx Mars) of Zen had never really deepened.

Horus Examples

Horus-Sun: Mick Jagger (semi-sextile), Thomas Merton(trine)

Horus-Moon: Tom Smothers (semi-sextile), Alan Watts (square)

Horus-Mercury: David Bowie (trine), Mary Travers (sextile)

Horus-Venus: Liberace (conjunct helio), Vincent Price (semi-sextile), William Shatner (sextile), Mary Travers (quincunx), Peter Ustinov (semi-sextile), George Wallace (trine)

Horus-Mars: Jack Kerouac (quincunx), Tom Smothers (conjunct), Alan Watts (quincunx), Lawrence Welk (quincunx)

Horus-Jupiter: Rex Harrison (semi-sextile), Jack Nicholson (trine), Tom Smothers (square)

Horus-Saturn: Jack Kerouac (trine), Sandy Koufax (semi-sextile), Jack Nicholson (opposite helio)

Horus-Uranus: Yul Brynner (semi-sextile), Sean Connery (quincunx), Tom Smothers (quincunx), Mark Spitz (square), Peter Ustinov (semi-sextile), Alan Watts (trine), Orson Welles (quincunx)

Horus-Neptune: David Bowie (quincunx), Mick Jagger (square), Liberace (semi-sextile), Arnold Schwartzenegger (sextile), Cat Stevens (semi-sextile)

Horus-Pluto: Robert Redford (conjunct), Peter Sellers (conjunct helio), Peter Ustinov (square), Mother Teresa (sextile)

Horus-Ascendant: David Bowie (quincunx), Robert Redford (trine), Mother Teresa (trine)

Horus-Midheaven: Robert Redford (quincunx), Cat Stevens (trine)

Notes

1. Pierre Grimal, ed. *Larousse World Mythology* (London: Paul Hamlyn, 1965), p. 25.
2. Veronica Ions, *Egyptian Mythology* (London: Paul Hamlyn, 1968), p. 50.
3. Ions, p. 21.

Part Three

Sex, Passion and Intimacy

13

Sappho: Sex

"It is to be a god, methinks, to sit before you and listen close by the sweet accents and winning laughter which have made the heart in my breast beat fast, I warrant you. When I look on you, Brocheo, my speech comes short or fails me quite, I am tongue-tied; in a moment a delicate fire has overrun my flesh, my eyes grow dim and my ears ring, the sweat runs down me and a trembling takes me altogether, till I am as green and pale as grass, and death itself seems not very far away . . . "

Sappho, Ode

Sappho, the poet of ancient Greece, was one of the more liberated ladies of her day. Considered one of the foremost lyric poets of all time, Sappho's subject was love, or sex to put it more bluntly. Her emotional expression is clear and compelling even in translation. Sappho knew what she wanted. Here is a succinct expression of primal energy. Sappho understood her own sexuality—this is the significance. She also made art from her experiences. The asteroid Sappho reflects both traits of its namesake: an interest in sex and a belief in the importance of its expression.

The asteroid Sappho (asteroid 80) was discovered May 16, 1866 by Norman Robert Pogson. Pogson discovered eight asteroids, including Isis and Asia. Asteroid 1830, Pogson, is named after him. Sappho's orbit is in the typical asteroid location between Mars and Jupiter. This asteroid has an estimated diameter of fifty-two miles and a sidereal period of 3.48 years.

Mythology/History

The life of the poet Sappho—what we know of it anyway—was described by Hans Licht:

> "(Sappho) was the daughter of Scamandronymus, born about 612 B.C. at Eresus in the island of Lesbos, or according to others, in Mitylene . . . . Since in all the fragments love for men is only mentioned once and then decidedly rejected, Clais is more likely to have been one of Sappho's girl friends than her daughter. Her amour with the handsome Phaeon is certainly to be banished to the realm of fable; and similarly, the famous leap to death into the sea, alleged to have been taken because Phaeon tired of her, is to be attributed to a misunderstanding of the metaphor current among the Greeks—'to spring from the rock of Leucadia into the sea', that is, to purify the soul of passions. Sappho's life and poetry are filled with the love of her own sex; she is in antiquity—perhaps in all time—the best known priestess of this type of love . . . ."[1]

Over the centuries writers have wasted much energy arguing over—or covering over—Sappho's sexual preference. While I have no evidence that the asteroid Sappho is any more prominent in the charts of lesbians, it would be wrong to completely ignore this facet of Sappho's life. Sappho's school for women on Lesbos gave the sexual preference known as lesbianism two of its names, the other being the less frequently used term sapphic or sapphism.

In tracing the evolution of the mythology of Sappho (i.e., the legends for which there is no historical evidence), it is worth noting that the term lesbian was not consistently applied to women who had a sexual preference for other women. During various periods, "lesbian" was a term applied to any woman outside the mainstream. For instance, a woman whom we might now call by another Greek name—hetaera—was referred to as lesbian. Jeannette H. Foster pointed out that this distinction between prostitute and homosexual was blurred as far back as Roman times, because the only women known to engage in homosexual sex were, in fact, prostitutes. In this context, the essential meaning for the disciples of Sappho might be an accomplished woman with an obvious sexuality.

Due to the influence of the Christian disciple Paul and other ascetics, sexuality in our culture is suppressed. A prominent Sappho in a person's chart suggests someone who is interested in exploring sexuality, male or female, heterosexual, homosexual or bisexual. It is less an indicator of sexual preference than of sexual activity or interest.

Given the renown of Sappho's poetry, it is not surprising to find the asteroid Sappho prominent in the charts of persons involved in the fine arts. Henry Miller has Sappho in Scorpio conjunct the South Node; John Lennon had Sappho in Scorpio in the eighth house conjunct Mercury, the ruler of his ninth house; Aldous Huxley had Sappho in Leo sextile Jupiter; Allen Ginsberg's Sappho is in Virgo opposing Mars and sextile Saturn; Paul McCartney has Sappho in Gemini conjunct his Sun; Ian Fleming had Sappho in Leo square Moon in Taurus; Mishima had Sappho in Taurus quincunx Venus-Mercury; and Arthur Miller had Sappho in Libra sextile Mars in Leo (all orbs of 1 degree or less).

The key meaning on the sexual side is sexual availability, although this does not necessarily mean availability to all people equally. Nevertheless, there is an obvious sexual energy associated with Sappho. If Sappho is prominent in the chart, this sexual energy could be called sexual charisma. For example, Jim Jones had Sappho in Aries conjunct North Node and the Moon. While we know little of what happened sexually in his People's Temple, there is little doubt that Jones' sexual charisma helped to gather the flock together. John F. Kennedy, certainly the most sexually charismatic president in America's history, had heliocentric Sappho in a T-square, square to Venus in Gemini, opposite Moon in Virgo.

Biology/Symbology

When Al Morrison and I first studied the effects of Sappho, we noticed two hot spots: sex and work. While the sexual aspect was expected because of the life of the poet Sappho, it became clear that the work-related aspects could not be dismissed as coincidental, since they occurred too regularly. For example, I have seen several examples of people either starting or quitting jobs during Sappho transits to Mars.

This dual pattern of sex and work resembles Freud's major theme of sexual neurosis. Since the time of Freud, therapists have found many other etiologies for dysfunction; still, the channeling of sexual energy is a clear factor in many cases.

Freud did not distinguish between sexual and erotic; in fact, he used the terms interchangeably. Since his time, clear differentiation has become possible, especially since the erotic has proved so useful to those who wish to manipulate public opinion.

To the degree that we have any instincts in the matter, these instincts would be considered sexual. Sexual in this case means directly pertaining to the genitals or to specific erogenous zones which, when stimulated, are part of what Masters and Johnson would classify as part of the orgasmic cycle: excitation, plateau, orgasm and resolution. In astrological interpretation, the placement of Sappho in the natal chart may show something of the ability to integrate sex into the personality, and in some cases, whether there are sexual dysfunctions.

Freud was actually referring to the sexual energy represented by Sappho, while often confusing the erotic with that sexual energy. According to his model, the sexual was indistinguishable from the erotic; the erotic would be considered the natural outgrowth of the sexual biological function.

What is not obvious from Freud's work is that the charging of objects with sexual meaning, as exemplified by Eros, requires a symbolic capacity far outstripping the capabilities of early hominoids. Sex and the asteroid Sappho exist on a biological level (i.e., they function within the body), while Eros is a function of the intellect.

To further understand the biological meaning of Sappho, we might take a cue from ecology and consider what has become known as "r" and "k" selection. Ecologists have noted that plants and animals have adopted two main strategies for species success: the development of many, sexually precocious offspring as soon in the life cycle as possible, or the development of a strong individual who, though producing fewer offspring, is able to devote more energy to the survival of each offspring produced. We might distinguish these two strategies by calling them the quantity ("r") or the quality ("k") strategies.

The "r" strategy could be considered a pure Sappho strategy. All available energy goes directly into reproduction. It is essentially a temporary strategy, one adapted to new environments or new opportunities. The "k" strategy, on the other hand, means the reduction in the emphasis on reproduction. The available energy goes into producing a larger body rather than more reproductive tissue. This strategy is more individual-oriented.

Sappho, as a representation of the sexual energy, is impersonal. Sex requires a certain level of physical maturity and appropriate plumbing; little else. The lack of emotional involvement which may be present at the time of Sappho transits may appear to indicate a profound clarity on the part of the individual. In reality, Sappho represents a very simple state, one uncluttered by concerns for anything other than the next fifteen minutes. This is the simplicity of an early stage of development. Overemphasis of the Sappho sexual impulse is not a recipe for long-term success. The fire dies down, and there had better be something left.

When I studied transits of Sappho I observed that people experienced heightened periods of sexual activity during Sappho transits to the Sun, Moon, Venus, Mars or Ascendant, or acquired venereal disease when Sappho was transiting Saturn. (Since the time of acquisition of AIDS is almost impossible to determine, I have been unable to study this phenomenon with any degree of completeness. In one case, however, an AIDS patient died when transiting Sappho was exactly square a close friend's natal Amor.) One woman reported losing her virginity when Sappho transited her natal Eros; she experienced her first orgasm when solar arc Sappho conjoined her Ascendant. Another woman reported having sex with a partner who was heavily into sado-masochism (although this particular encounter was not) when transiting Sappho conjuncted her natal Pluto.

Metaphysics

While the sexual effects connected with the asteroid Sappho are spectacular, sexuality is not the only issue involved. After all, Mother Teresa has Sappho conjunct her North Node natally.[2] Few psychologists accept Freud's repression theory as the be-all and end-all of human psychology. Fortunately, the Eastern kundalini teachings begin where Freud left off.

Westerners are becoming familiar with the idea of the *chakras.* Briefly, the chakras represent seven energy levels in the body. They run from the Muladhara or first chakra, associated with the peritoneum, to the Sahasrara or seventh chakra, associated with the crown of the head. The chakras are "connected" by three nerves, of which the Sushumna is the principal. The *kundalini* is the spiritual energy which all entities have; the Sushumna is the road that the kundalini may follow as it travels "up" the body to a higher level of spiritual consciousness. Each chakra is associated with particular vibrations, from the gross survival level represented by the first chakra to the emotional/sexual energy of the second (localized at the genitals), through to the most sublime merging into the Nameless associated with the seventh.

In most people this current does not circulate; it is "blocked," so that it is seldom active above a particular chakra. This block represents the entity's level of development. In this hierarchy, simple survival comes first, then emotional/sexual considerations, then mental/intellectual processes, then faith/service, then psychic/spiritual realms, then the identification with the Self, followed finally by the complete dissolution of the separate self.

While most authors have expressed spiritual development as a hierarchy of acquired characters, it is also possible to look at the same evidence of those who have followed the road and to view the process as analogous to a series of concentric spheres. The advantage of this latter model is that each large sphere completely encloses or includes the smaller ones. The aspirant is not merely "conquering" an energy type and then rejecting it, but including it into a wider sphere of activity.

There are far fewer sources to draw on in the West than in the East to convey this process. One of them is alchemy. Carl Jung's work on alchemy, notably *Alchemical Studies* and *Mysterium Coniunctionis* have greatly clarified the true purpose of alchemy. While it is still taught that alchemy was a superstitious precursor to modern chemistry, Jung suggests that alchemy was actually a method for soul purification, expressed allegorically as the turning of a base substance into gold. That many people took the allegory literally is well documented. Nonetheless, the practices resemble to a considerable degree the practices of tantra.

Tantra, or sexual yoga, as a spiritual practice always represented a minority approach in the East. This was probably because of the great potential for abuse of sexuality among unenlightened spiritual practitioners. In this light, it is interesting to note that in the natal chart of Da Free John (Franklin Jones), a Western spiritual community leader discussed in chapter 4, who has recently been attacked by former followers for his sexual practices, there is an exact Sappho-Neptune conjunction in Virgo. Whether the particular charges are true, there is no question that Free John's own works advocate the use of sexual practices in spiritual practice, and that one phase of his teaching work, the so-called Garbage and the Goddess period, involved extensive sexual experimentation.

This tendency to mix the spiritual with the sexual is hardly a new trend. The celibacy advocated by most Hindu Vedanta groups is as much of an acknowledgment of the power of sexuality as the most sexually involved tantric practices. The very core of our psychology is reflected in the generative connection to sexuality: the drive to immortality. This mental/emotional component, which will be more fully elucidated in the Eros chapter following, is essential for the release of the personal self into the impersonal whole/void, the ultimate step in either spiritual practice or personal death. As Alain Danielou reminds us,

> "... pleasure is the image of the divine state .... The sexual organ has ... a double role: the lesser role of procreation, and the higher role by which it becomes a means of contact with the divine state, the ecstasy of sensual pleasure (ananda). Sensual enjoyment is a 'sensation of the divine.' "[3]

Another Western system that suggests the spiritual process or "quest" is Kabbalah. I doubt that it is coincidence that the apex in the Kabbalistic Tree of Life is known as Kether, or crown. This is presumably a veiled reference to the seventh chakra.

Esoteric spiritual life involves more than chanting a few mantras, singing a few hymns, or spending a few minutes on a meditation cushion. Further, one could make a case that all of human existence is a spiritual "quest," whether conscious or not.

This "quest" is manifested by how well or how poorly energy circulates within the body. Without going into any esoteric detail, suffice it to say that by *body* I mean something more inclusive than the biochemical engine we normally think of. The greater the circulation of energy, the more components the individual is consciously reaching. I suggest that the asteroid Sappho shows something about what we could call "energy cycling" or "energy internalization." In this respect we can contrast Sappho with Mars. Mars shows energy directed outward. Sappho is not energy directed inward, but energy utilized internally.

This distinction can be illustrated by contrasting the differences between transiting Sappho to natal Mars and transiting Mars to natal Sappho. Transiting Mars to natal Sappho may represent a time of heightened sexual tension. (The aspect is no guarantee of sexual satisfaction, only an indicator of enhanced interest!) Transiting Sappho to natal Mars may represent greater energy available for the person's usual Mars functions. Transiting Mars brings the outwardly-directed energy of Mars to the sexual desire of the natal Sappho; Sappho functions receive a transfusion at this time. With transiting Sappho, the inner power of the sexual system receives an outlet through natal Mars. If the individual uses Mars primarily in a sexual fashion, then this will be a time of heightened sexual desire. But if the person uses Mars primarily for work, the Sappho transit can represent a time when more work is done, or more physical activity is needed.

Just as the relationship to Mars shows how Sappho energy interacts with the personality, the connection of Sappho to the outer planets shows the transpersonal manifestations of this energy.

Sappho-Uranus connections produce more than the sudden intensity of sexual interest. Uranus rules electricity, and the experience of chakra energy in the body is electric. Several of the words evoked in the study of this energy are distinctly Uranian: polarity, conductivity, potential, transmission. Within the context of Western science, most descriptions of the movement of this energy read like textbooks in electricity. Often a person becomes aware of these sensations suddenly, as if affected by an electrical shock.

Sappho-Neptune contacts show the ways that the separate sexual identity may be dissolved, often through the orgasmic process. Often lovers express a desire for unity which goes beyond sexual coupling. This desire is another expression of the Neptunian universal solvent. The other side of Neptune is overidealism, and Sappho-Neptune aspects can indicate those regions of sexuality where the person has no objectivity. From the standpoint of the chakras, Sappho-Neptune represents the extension of chakra energy beyond the seven chakras, the delocalization of this energy.

Sappho-Pluto is the basis of what every Taurus and Scorpio wants. While Neptune may crave unity, Pluto craves annihilation. Even the words in our vocabulary for sex and orgasm have a Plutonic cast, for example, "little death." Sappho-Pluto represents total transformation of the chakra energy. With Pluto, there is no circuit left for the energy to utilize. Pluto has pierced the seventh chakra and extended into the universal.

Sappho and Work

Earlier I mentioned the dual connections of Sappho and sex, and Sappho and work. According to most Eastern writings, the vast majority of what passes for human "civilization" is the result of activities related to the two lowest chakras: gross survival and emotional/sexual. Given the natures of these two chakras, one could argue that historically, there was little activity beyond these two levels prior to the era of Classical Greece, because the ego simply had not developed sufficiently to allow the complete differentiation of the lower mental facilities.

This does not imply that there was no civilization before this time. What appears to be missing before this period is egocentric thinking. Written records to that point showed mainly the business accounting of the day, and when records were of a more liberal arts variety, the principal actors were the gods or goddesses rather than individual human beings. It appears that humans were not concerned with their "place" in the world. There was little evidence for the gross individualism which seems to have developed first during Classical Greece.

The idea of an evolution in consciousness at the time of Classical Greece can be traced in literature. The clue is that until this point, there did not seem to be the sort of ego development associated with self reflection. The actions of humans usually were attributed to the gods or to Fate if any causative principle was invoked. Sometime during Classical Greece the concept of the will developed. After that time, there was less of a focus on controlling entities outside the person, and the basis for action became human rather than divine.

For someone who is fixated in these two lower chakras, Freud's description of how sexual energy is sublimated into work or other outlets is indeed apt. What Freud did not understand was that it is possible to transcend these two levels. Before considering that transcendence, since Sappho is being related to energy cycling, we should examine what energy cycling looks like at the first two stages.

If Sappho represents the reproductive process of the second chakra, then how is it also a symbol of energy cycling? The way we use our sexuality is completely intertwined with energy cycling in general. In fact, many people only recognize energy cycling when it is manifested as sexuality.

The first chakra is predominant only under extreme conditions, such as starvation or severe stress. There is no energy cycling: one's total attention is on survival. We tend to accumulate a lot of tension in this general area. The image of the kundalini as a coiled serpent—coiled at the base or first chakra—is indeed apt. In the Taoist tradition, this chakra is called the lowest gate. The point or location is the "Jen Mo," an acupuncture point of considerable importance. Many of the exercises involve placing one's heel in the area of this chakra: this becomes a way to stabilize the energy field of the body. It is hard to move the energy throughout the body (i.e., allow it to cycle freely) without first releasing or relaxing this gate. When we are in a situation that we interpret in a stressful manner, we tend to tense in this area. This is the region of the fight-or-flight syndrome.

When the progression to the emotional/sexual level is made, there is excess energy which can be used, cycled or discarded. From a developmental standpoint, there is a primitive ego present—one which understands the pleasure principle. The *modus operandi* of this level is "If it feels good, do it!" Delayed gratification can be used as a prod, but the bribe must be there. In this case, the stress is not as overt, or has been repressed or transmuted, so that it does not appear to be a matter of survival; stress begins to seem tantalizing or appealing.

Sex is obviously one of the more attractive goodies. Sex becomes the outlet for excess energy. And outlet it is, for energy is literally tossed off during the orgasmic process. (Note the typical indicator of a "good" night of sex: "We did it until we dropped," i.e., physical exhaustion.) This is the basic rationale behind many of the religious proscriptions against sexual activities: It has nothing to do with sin, but with the release of energy which otherwise could be put to other purposes.

As an indicator of sexual activity, Sappho transits may show the times when we have "energy to burn," and hence times when we are more available to the diversions of sex. Because Sappho connections with Mars indicate the double energy configuration—"available" energy and the tendency to externalize energy—these times may be most opportune for work movement or sexual encounters. In Taoist terminology, this chakra is called the cauldron or furnace. This is the form of energy which, when concentrated, produces the sensation of "burning."

There is also a question of individual temperament, which will be indicated in large part by the rest of the chart. How does the person handle "excess" energy? Does the person put it into emotional issues (water), ideas or communicating with others (air), projects (earth) or activities (fire)? The sign in which Sappho is placed, as well as those of the Sun, Moon and Mercury, will help to answer this question.

These possibilities do not exhaust the potential of Sappho. We can transcend the second chakra after all. The look may be different, but the idea is the same.

The fact that a particular person has progressed beyond the second chakra does not mean that second chakra-type issues cease to come up. One does not necessarily give up sex just because one is no longer functioning primarily out of the second chakra. Any function or activity relating to a "lower" chakra is freely available, although one may choose to deal with it differently. Thus, you may simply remember Lao Tze if somebody tells you that s/he has transcended sex. "Those who know don't say, those who say don't know."

The third chakra has to do with specific kinds of mental processes. This is not the same thing as intelligence. It is perfectly possible to show native intelligence at either the first or second chakra level. What is characteristic of the third chakra level is symbolic representation, the highest level of development discussed by Piaget in his description of childhood development.

This level represents the ability to conceptualize, to theorize, and especially to be able to manipulate symbols or ideas. Virtually every adult has this ability. The question is how much the facility is used.

At this stage of development, Sappho transits may show as times of enhanced creativity. It may be easier to solve problems (as in the mathematical sense of the term problem-solving). One may have all sorts of good ideas for new projects, or may seem to have plenty of time to finish old ones. This will be especially true if Mars and Mercury also are involved. Without Mars involvement, the projects might stay in the mind and never be acted upon.

Sappho transits show relative amounts of free attention which may then be applied to whatever the person typically does in life. In a sense, "free attention" may be equated to concentration. The effect of the movement of the Kundalini from chakra to chakra is one of channeling. During Sappho transits, it may be easier for the individual to remain focused on his/her agenda.

At the higher chakras, the expression of Sappho energy—and everything else—becomes more diverse. It is difficult to extend astrology to these levels because so much of astrology (as presently incarnated) is mental stuff of which the third stage is composed.

The effect of Sappho may then be summarized as the indicator of "free" energy, and that energy may be utilized in a number of ways, such as through sexuality, through work or through creativity. Sapphic energy is not necessarily directed outward; externalization is a facet of Mars.

Sappho shows the amount of this energy potentially available at any given time. The presence of more energy than usual may be uncomfortable for some people and lead to activities which allow them to slough off this "excess." Alternately, a person may use Sappho energy consciously for any number of internal or external processes. This extra energy will come out somehow. The only question is how one chooses to use it.

Sappho Delineations

Sun-Sappho: Sexuality is of great importance to this person. However, s/he often puts much time and energy into things other than relationships. This individual is highly self-involved, though this is not necessarily self-love. Rather, such people are often oblivious to the wishes and desires of their partners.

Moon-Sappho: This person defines his/her self-worth in terms of sexual relationships. Consequently, a relationship break-up may bring on depression. Before getting into a sexual relationship, this individual may insist on strong emotional ties. Hard aspects might indicate that sex is regarded only as a duty.

Mercury-Sappho: A high premium is placed on communication within the sexual relationship. This person may also be more interested in talking about sex than doing it.

Venus-Sappho: This person is sexually magnetic, and attracts others through his/her sexuality. Aesthetics are an important factor in sexual desire.

Mars-Sappho: This person may climb every mountain and ford every stream in order to get the sex partners that s/he wants. The trick is not to become more interested in the *process* of getting the other person than in the person him/herself. Once involved, the individual has lots of energy available for sex. In extreme cases, s/he may only be comfortable giving sexually—not receiving.

Jupiter-Sappho: This person may be interested in some of the more histrionic aspects of sexuality, such as sexual fantasy dressing or props. Or, s/he simply might make a production out of sexual encounters.

Saturn-Sappho: The sex partner may be of a markedly different age. Or, this individual might try to work out a parental relationship through his/her sexual relationships. Occasionally, instead of a parental relationship, some other relationship is worked out.

Uranus-Sappho: This is *not* a good aspect for monogamy! This individual is probably going to sample many different people, but might not be pleased if his/her partner does the same!

Neptune-Sappho: Sexual fantasies are very important. Unfortunately, this individual may fantasize that the other person really cares. Sometimes drugs are necessary for arousal. Persons with this combination who have spiritual interests may believe that sexuality represents a form of spiritual initiation, and may seek to play out tantric principles with or without appropriate training in them.

Pluto-Sappho: This person may feel the need to control every facet of a sexual encounter—the time, the position, etc.

Ascendant-Sappho: This individual is often viewed as being more interested in sex than s/he actually is. Though others frequently believe this person is always available sexually, this is not necessarily the case! Such people might use the sexual charisma implied by this combination to get what they want.

Pandora-Sappho: This person may get into sexual affairs or relationships unexpectedly. When an affair starts with one partner, others may try to jump on the bandwagon. These affairs might stir up all kinds of unexpected results.

Vesta-Sappho: This person may see sex as a duty.

Sappho in Relationship Analysis

Sappho appears to mediate two facets of relationships: whether the relationship is likely to be sexualized and the tenacity with which the relationship remains sexualized. For those who experience "love at first sight," the aspects transiting Sappho makes to the natal chart show the particular sexual needs the individual hopes the partner will fulfill.

If a chart of the first sexual encounter is available, transiting Sappho shows the circumstances of the sexual encounter.

In synastry, a tight conjunction (less than one degree) of each person's Sappho with a planet or asteroid in the other's chart indicates a very strong sexual attraction.

If Sappho indicates sexual interaction, then it should also show a *lack* of sexual attraction in specific cases. For example, two lesbians, both of whom tend to perceive their female friends as potential lovers, met while transiting Sun was conjunct transiting Venus. Yet Sappho was conjunct Pluto and square the Moon. (The Moon-Pluto "square" was out of orb.) There was no sexual attraction between the two, despite excellent synastry. In the relationship chart, Sappho was conjunct Sun, opposite Saturn, square Uranus. In the composite, Sappho only made two aspects: quincunx Jupiter, and a loose cross-sign conjunction with Mercury.

In another case, a woman entered a long-term relationship when progressed Juno reached her Sappho-Venus midpoint. The symbolism with Juno is certainly apt, and the Venus-Sappho midpoint suggests that many people do not play their Sappho energy out in a "pure" way. Many try to combine the sexual energy of Sappho with the more emotional or attractional factors of the chart. In this particular case, the woman had natal Moon square her partner's natal Sappho, while the woman's Sappho was trine her partner's natal Moon. Thus we may see that Sappho can be used not only for synastry, but also for timing.

Sappho-Pluto can be interesting. In one case, a woman had a one-night stand with a partner who was into sado-masochism when transiting Sappho conjoined her Pluto.

Sappho in Non-Personal Charts

The connections between Sappho and work, which Al and I noted early in our research, spurred me to put Sappho in charts other than those for individuals. For example, I served on a jury in Criminal Court in New York City in April 1981. The chart shown is for the time that we reached a verdict.

The trial was of two individuals accused of robbing a couple who were engaging in oral sex in a parked car in Riverside Park after smoking dope at a Village club. We deliberated for only a short time before returning a guilty verdict on several counts. In the chart, Sappho is rising within one minute! Clearly sexual issues were involved in this case. Amor is also culminating, and this could be because the accused robbers apparently were lovers. Amor culminating may also show the violence attendant with armed robbery (see chapter 15). Several other items in the chart are certainly worth noting: the tight Sun-Mars-Venus conjunction in the tenth house and the Toro-Jupiter-Saturn conjunction at the I.C. Jupiter's involvement with the law certainly spoke from an angular position. Neptune

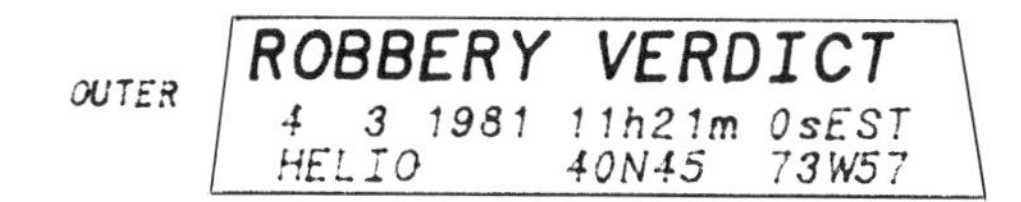

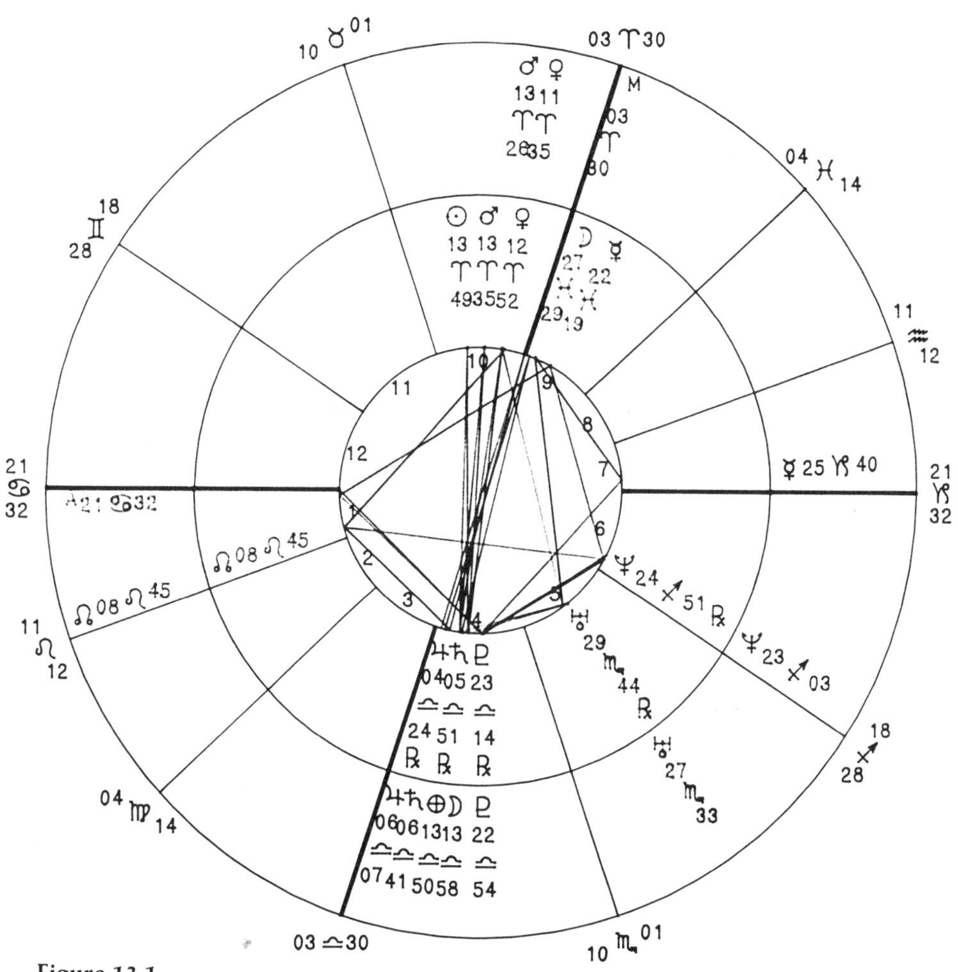

Figure 13.1

Asteroids

Amor	3	AR	52
Eros	0	PI	32
Sappho	22	CN	04
Toro	2	LI	25R

INNER CUSPS
ROBBERY VERDICT
4 3 1981 11h21m 0sEST
PLACIDUS 40N45 73W57

square Mercury could indicate how the victim couple's drug-induced state interfered with their judgment. (Anyone who knows New York City will appreciate that the 96th Street offramp of the West Side Highway is not exactly the world's safest location!)

Since this chart represents the verdict, the police motives may also be relevant. Although the amount stolen was less than $100, I was informed after the case had been settled that the district attorney had brought the case to trial because both defendants had previous convictions and the authorities were looking for a reason to put them away for a substantial sentence. The angular Sappho and Amor in this chart may represent examples of the expression "Up yours!" in much the same sense as did culminating Sappho in the chart of the first nuclear bombing of Japan.

Sappho in mundane charts may represent symbolic rape. In my files I have a chart of a mugging; it shows transiting Sappho conjunct the victim's Uranus.

The challenge of interpretation is that transiting Sappho is not always negative. Angular transiting Sappho may represent a power point, especially if it is hitting the natal chart. At the time of my first computer purchase, transiting Sappho was trine my natal Sappho, and two degrees off the Ascendant in the transiting first house. Since computers eventually became my profession, the chart shows a power point for work. Angular Sappho often indicates this kind of power point. The quality—positive or negative—varies according to the individual.

Notes

1. Hans Licht, *Sexual Life in Ancient Greece* (London: The Abbey Library, 1932), p. 318-319.
2. Zipporah Dobbyns, personal communication.
3. Alain Danielou, *Shiva and Dionysus* (New York: Inner Traditions International, 1979), p. 57.

14

Eros: Passion

"Since then Eros is acknowledged to be the oldest God, we owe to him the greatest blessings. For I cannot say what greater benefit can fall to the lot of a young man than a virtuous lover and to the virtuous lover than a beloved youth."

Plato

The asteroid Eros is the ruler of romance and passionate attachment. Eros' strength was made clear to me when I fell head-over-heels in love with two different persons, each of whom had the same Sun position. That Sun position didn't do anything else with my chart, unless I counted a quintile to the Sun, and the cross aspects between the two charts and mine left a lot to be desired. That Sun location, however, turned out to be the location of my Eros. While I was ready for romance, my love objects remarked that it was difficult for them to see me in a romantic light. Neither of these persons had Eros contacts with my chart!

Mythology

In Greek mythology, Eros was the god of love, the son of Aphrodite and Zeus. Unlike his fellow travelers, Amor and Cupid, Eros was fully grown. The major Eros myth is the story of Psyche. Eros fell in love with the mortal Psyche, and married her without telling her that he was a god. He imposed one condition on her: that she never see his face. She was happily married until her sisters tweaked her curiosity sufficiently so that she looked on his face while he was sleeping and realized that he was a god. However, a drop of wax from her candle hit Eros and awakened him; he left her because she had violated his condition. That started Psyche on a quest for her identity.

What can we learn about Eros from this? The most cogent modern interpretation of the Psyche-Eros myth appears in Erich Neumann's *Amor and Psyche,* which begins with a full translation of the Roman tale by Apuleius, and ends with a Jungian interpretation of the myth. In his introduction, Neumann clearly distinguished between Eros, Amor and Cupid:

> "In our interpretation we have not spoken of Cupid and Psyche as in Apuleius' text, which thus mixes Roman and Greek elements, but of Eros and Psyche, and we have uniformly translated the name back into its Greek form . . . . (I)n Our interpretation, which stresses mythical motifs, it seems more fitting to speak of the Eleusinian mysteries of Demeter, rather than of Ceres, and to call the Argive goddess Hera rather than Juno. Still more important, it is Aphrodite and not Venus, who in our eyes is associated with the Great Goddess, and in the myth Psyche's lover and husband is the mighty primordial god Eros, and not Amor or Cupid—the cunning little cherub known to us even from ancient works of art."[1]

To Neumann, Eros represented the original son-lover to the Great Goddess Aphrodite, the primordial pair in the matriarchal system. Eros is a fertility god. Part of the conflict in the Eros-Psyche tale occurs because Aphrodite is clearly threatened by this usurping outsider: Psyche is a threat to Aphrodite's relationship with Eros. Psyche represents Eros' assertion of independence. This period of development corresponds to the Great Goddess period and has its prime figure in threefold aspect: creator, sustainer and destroyer. As such, the love of Eros is a love which recognizes the inevitability of death. With Psyche, Eros has succeeded in personalizing love. The shift is from mother-son as the principal bond, to a love choice. The mother-son bond depends on nothing from the son's side: the mother would love him regardless. The love is impersonal in the sense that it does not depend on anything in his personality. By loving Psyche, Eros asserts the right to choose whom to love, and with it, he gives in to his passion.

Alain Danielou has noted the similarity of many of the Greek myths to the myths of the early Indian Shivaite religion. He connects Eros to the Indian god Kama:

"In Orphic cosmology, Eros, like Kama, has no mother, since he is antecedent to sexual union. He represents the original oneness to which all "separated" things aspire. The apparent world is the result of the separation of two contrary principles which exist only by and for each other, and aspire only to find each other. Eros is both the principle of existence and that of annihilation or death. Nothing exists without him; through him all things cease to exist. He represents the very nature of Shiva, the principle of life and death."[2]

If the connection to the Indian Kama is accepted, then again the significance of passion is foremost. Divine passion, or ecstasy, is the creative force in the world. The ecstatic dance of Shiva is the very act of creation. Passion most often can be experienced by mere mortals in a sexual context, but the capacity for ecstasy is just as significant in religion and spirituality.

Symbology/Metaphysics

Since Freud linked virtually everything to sex, the concept of Eros got muddled, partially because of Freud's inability to comprehend anything beyond the instinctual. In later years, Freud was greatly concerned with two "opposing" (that is polar) instincts, what he called Eros and Thanatos. Roughly, the two correspond to life and death. One way of thinking about this life-death polarity is to recognize Eros as the pleasure principle. In this model, death intrudes on the pleasure principle because of the inevitable awareness that pleasure is transient, and death is inevitable.

The typical response to the awareness of impending death is to grasp for everything at hand in the fervent hope that death will disappear. The tendency of humans when confronted with death is to deny it. Thus, we see Eros as grasping or clinging, the passion of awakened mortality. While part of this passion may be interpreted sexually, it also may be directed towards other human beings, animals, plants or inanimate objects. Eros sees the lover as an object, as an extension of himself.

To deny death means to accumulate life. Eros, as the god of love in this scheme, represents the accumulation/possession principal. The more I have, the more I am. Astrologically, we can reword this polarity from Eros-Thanatos to Eros-Pluto. Several facets of the Pluto myth are significant to the understanding of Eros. First, Pluto was the King of the Underworld; as the ruler of the dead, his connection with Thanatos is clear. Further, Pluto was possessive, as we know from the Persephone myth. Pluto took what he wanted with no concern for others. Persephone was treated as little more than a possession.

Eros is possessive; Pluto is possessive. Pluto extinguishes Eros (life); Eros extinguishes Pluto (death). Desire overpowers the other end of the polarity, or at least, so it appears. That which we have is a buffer between us and the knowledge that ultimately we have nothing. The Life-Death polarity is the strongest, the deepest-rooted, the most unconscious polarity we have to deal with. Few of us care to contemplate our own death or annihilation. Because this fear of death is so primary, the Eros instinct is one of the strongest we experience—as well as the least conscious. On the conscious level, we can scarcely perceive how we grasp at anything and anyone to preserve the illusion of immortality.

This grasping strategy is the origin of the desire to possess the loved one. This tendency to define the loved one as property gets us into trouble in our relationships. One is incapable of seeing the beloved's true needs, and this in turn leads to jealousy, to fear of loss, and a total inability to deal with the lover person to person. This attachment is the equivalent of the Eastern concept of karma, or bondage to our material nature through attachment to transient beings or objects. The *Bhagavad Gita* says:

> "The mode of passion is born of unlimited desires and longings . . . and because of this one is bound to material fruitive activities"[3]

In the Eastern traditions, passion is seen as a way of maintaining false attachment to this world. As Ramakrishna used to say, the two worst things to the spiritual aspirant were women and gold. (He told women aspirants men and gold.) He explained that it was not that either passion or money was bad in the sense of evil, but rather that each is so distracting that they interfered with the business of spiritual growth.

In the esoteric Christian tradition, Meister Eckhart preached:

> "All sorrow comes from love and from holding dear. Therefore, if I feel sorrow because of perishable things, my heart and I will still love and hold dear perishable things, and God still does not

> have the love of my whole heart, and I still do not love such things as God would have me love with him. Is it then any wonder that God decrees that I so justly suffer harm and sorrow?"[4]

Discussion

The asteroid Eros represents the pleasure principle, the tendency to grasp onto people or things. It is the polar opposite to death: passion. One might characterize Eros not as instinct to life, but as clinging to this physical existence to the exclusion of all else. Eros represents attachment and the passion which maintains attachment.

It is important to distinguish Eros and the way it works from those bodies, such as Venus, which have related functions. Venus is the Roman descendent of Aphrodite, and as such, has lost the primordial Great Goddess trappings. Venus enjoys pleasures and appreciates beauty. Most of all, Venus likes ease, or harmony. Venus is basically lazy, preferring to let Mars go out and conquer the world, which she will then be perfectly happy to reign over. Pathos is not an intrinsic part of Venus.

Eros is not so passive. Eros wants full blown passion, life lived to its fullest. Eros does not mind raw edges. Pleasure should be intense, the better to conceal the inevitable thread of pain: the poignancy of impending death.

Although Sappho represents sexuality *per se,* Eros and Sappho often become intertwined, but their respective spheres are clear. It is simply that most mere mortals are not able to separate passion and sex completely. Erotic implies the charging of objects or people with sexual overtones. Sexual refers to actual physical stimulus. Erotic involves the mental process of connecting things/people to the sexual cycle. Because of Eros, we don't simply couple with partners randomly in the street. From Sappho's standpoint, an itch is an itch, and the sooner it's scratched, the better. In contrast, Eros wants the correct ambiance for the scratch.

One of the most significant distinctions between Sappho and Eros is that Eros transcends Sappho in this sense: Eros represents the uniquely human ability to conceptualize attraction. Sex, as represented by Sappho, is a strictly biological affair of coupling by two or more people. While passion may be felt as an emotion, the passion is toward particular objects. Fetishes may be seen as Eros functions, when the *type* of object becomes more important than the *identity* of the object. John may think he loves

Janet, but really, he is attracted to blondes over 5′10″ tall who wear lots of purple. Often with Eros, the fetish or the fantasy is more important than the actual partner. Eros shows the ability to *objectify* one's attractions. Sappho does not especially care what you are wearing, so long as you take it off!

Eros cares about the ambiance, not for reasons of comfort as Venus does, but because the surroundings are part of the fantasy. Eros is erotic; Venus is sensual. The use of aphrodisiacs, slinky clothing, scents, lighting and music may be viewed as accoutrements to the erotic impulse. Venus may enjoy the texture of the clothing, or the smell of the perfume, but this enjoyment is purely pleasure for pleasure's sake. With Eros, the tendency is to try to get off on the fantasy regardless of the cost.

It is easy to see why the Eros principle plays such a significant part in advertising. Sexual fantasies are very powerful things, and if a way can be found to make that new car an extension of one's perceived sexual potency, then why not? These fantasies, of course, are bound up in the Eros fear of mortality. By surrounding ourselves with symbols of potency, be they fast cars, young sex objects or phallic symbols, we can repress our aging and eventual death.

Most of the negative emotional components of love can be seen as the failure of the attempt to completely paper over Pluto. One such component is the "You don't love me anymore" or "You love somebody else" syndrome. As many spiritual masters in the East have pointed out, our emotional lives are stunted by the obsessions that we inject into our relationships. Jealousy is one of those obsessions.

The fear of unlove is precisely the result of being unloved, which is the position of Eros. To "love" another as a possession is to worry that the "possession" would prefer to be possessed by someone else. From this perspective the "loved" one is just another goodie which protects one from eventual annihilation in Pluto/death.

Another side of Eros concerns the status of individuality, a Uranus issue. When we are passionately attracted to someone, we often say that s/he is "The One." We are practicing individuation, or pretending that I am here for you and you are here for me. In fact, you may just be the latest in my string of blondes, or Sun-Uranus squares. Yet I believe that I own you and you own me. This is one way of denying the fundamental fact that none of us is indispensable in the larger context. When we die the world will go on without us. Eros likes to play with uniqueness; Eros would prefer that the polarity be played out as Eros-Uranus rather than Eros-Pluto.

This identification with uniqueness is not the same as a genuine sense of connectedness with another. The challenge of connectedness is shown by the Eros-Neptune polarity. A love object, which is what Eros shows, is not the same as a soul mate or lover. A love object is there to remind you how wonderful you are. The Eros fascination with the other person as an abstraction or ideal is what distinguishes puppy love or infatuation from true love. Eros, in any case, shares similarities with Neptune because erotic infatuation is characterized by unrealistic expectations. Eros and Neptune are not interested in reality. Eros is essentially escapist.

The fascination with extramarital affairs, or promiscuity serves much the same purpose. Variety as the spice of life is simply an admission that boredom or habit dulls the senses, leaving the individual susceptible again to the "negative" side.

Power is an Eros issue, and so we have returned to Eros-Pluto. As Jung said, "An unconscious Eros always expresses itself as will to power."[5] Sex, fame, money, power, food—all these issues may come under the sway of Eros, because all may be a part of our strategy of denying death, of assuring our immortality.

Perhaps one of the most difficult things about dealing with Eros is the awareness that it is part of the polarity with death. Traditionally, we think that the opposite of love is hate, but this is most emphatically not the case. As astrologers, we are aware that the nature of a polarity is that the two energies must be balanced. One cannot deal exclusively with Cancer and not deal with Capricorn as its flip side. This is much the same as the psychological emphasis on the shadow or the unconscious. That which we repress is still there, merely altered in form.

Thus, the more repressed Pluto/Thanatos is, the more pronounced is the Eros game. The more we can come to terms with death, the less pronounced the more obsessive qualities of Eros will be.

Within the Eastern milieu, the acceptance of death is part of the spiritual process. Certain stages in the spiritual processes are called death: the death of the separate self. In *Up from Eden,* Ken Wilber traces the evolution of stages of consciousness. Eros is a part of what he calls the second, or typhonic stage. This is the stage in which the awareness of the separate self first emerges. Part and parcel of that self is the awareness, and then fear, of death of the self. In a later stage when the self merges into the

Self, or absolute, the death of the self occurs, but with changed significance. In the latter, ego death is experienced as liberation, not annihilation. In this state the ecstatic testimony of Jesus, "I and the Father are One" is not taken as applying exclusively to Jesus the Christ, the only Son of God, but to a state which is potential in all of us. This same perspective has also been enunciated in Vedanta, Sufism, Buddhism and elsewhere. In the words of the *Gita*:

> "The Blessed Lord said: He who does not hate illumination, attachment and delusion when they are present, nor longs for them when they disappear; who is seated like one unconcerned, being situated beyond these material reactions of the modes of nature, who remains firm, knowing that the modes alone are active; who regards alike pleasure and pain, and who looks on a clod, a stone and a piece of gold with an equal eye; who is wise and holds praise and blame to be the same; who is unchanged in honor and dishonor, who treats friend and foe alike, who has abandoned all fruitive undertakings—such a man is said to have transcended the modes of nature."[6]

Such a person has truly transcended Eros. For the rest of us, the Eros connection shows in our relationships to other people and things. Even without such absolute dispassion, we can at least become more aware of the process, and through awareness, adopt a more humorous attitude about our attachments.

Eros Delineations

Sun-Eros: This individual has a large ego involvement in perceived "love" relationships, but may not have the capacity to see the other person's perspective. Other people may be viewed primarily as objects to be manipulated or used.

Moon-Eros: This person tends to "fall in love" instantly, and expects the emotional involvement to happen immediately. It is difficult to distinguish between "love" and "in love."

Mercury-Eros: Interpersonal communication is an expression of attraction. Actual use of the words "I love you" may be more important than the validity of the sentiment. This person may write copiously about love.

Venus-Eros: The aspect of the true romantic! The trappings of romance are more important to this person than the substance.

Mars-Eros: This individual has trouble waiting for the love object to see his/her charm. While quite capable of romantic pursuit (i.e., the knight in shining armour), this person might not enjoy being pursued by another. This person may go to great lengths to demonstrate desire.

Jupiter-Eros: This person may actually believe in the reality of the perfect lover. However, that doesn't mean s/he has found this treasure. In fact, s/he may grossly overrate the value of passion.

Saturn-Eros: This person may alternate between rebelling against any limits to relationships (such as monogamy) and enforcing such constraints. Usually s/he believes that "restraint" is necessary and that romance should proceed slowly.

Uranus-Eros: This person is not inclined to long-lasting relationships unless there are other strong indicators in the chart! S/he wants intensity, good or bad, in love and most lovers are incapable of meeting these needs for long periods of time.

Neptune-Eros: Interestingly, this aspect seems to be a better predictor of "promiscuity" (whatever that may mean) than Uranus-Eros. Apparently this individual is capable of meeting the other person's fantasies and doesn't mind playing along. S/he feels swept along by the course of the romance and has no objectivity about the propriety of the course the relationship follows.

Pluto-Eros: This person is very fearful of not having the upper hand in a relationship. Therefore, s/he may put up barriers which define sexual/emotional encounters in ways that allow him/her to feel s/he is in control. Or s/he may take advantage of a partner's desire in order to get what s/he wants.

Ascendant-Eros: This person may appear to be emotionally open. Whether this is true is determined elsewhere in the chart.

Ceres-Eros: Nurturance is a part of romantic love for this person. When this aspect occurs between two people, the Ceres person is remarkably forgiving of the Eros person's exploits—so long as the Ceres person is not currently a lover! This can be a difficult aspect between lovers.

Hidalgo-Eros: Hidalgo wants to be in charge of any romantic entanglement. Desire becomes a means to be superior to others.

Sappho-Eros: S/he may automatically feel sexually attracted to anyone with whom an emotional bond develops. Those who become emotionally close may be sexually attracted to this person. Alternately, this individual may be unable to see that a sexual affair has no lasting meaning; the one-night stand might be mistaken for a real relationship.

Eros in Relationship Analysis

It is often hard in practice to separate the effects of Sappho and Eros. My theory is that the separation of these two concepts in our collective unconsciousness is still incomplete. Since the advent of The Pill and abortion on demand, the separation between sex and passion has become one of minutes in many cases.

One way to distinguish between the two asteroids is that Eros usually shows more possessiveness. Sappho is almost instinctual, and there is little possessiveness *per se* so long as the sexual desire is consummated.

Eros passions may last a long time, while sexual interest tends to ebb and flow. Eros relationships, therefore, may last longer than strictly Sappho ones.

Eros in Non-Personal Charts

Eros does not tend to show up in business or professional charts the way Sappho does. However, in mundane or electional work, Eros may appear angular or otherwise prominent if the situation is one indicating passionate attachment.

For example, at the time of my Ph.D. dissertation defense, there was a transiting Sappho-Eros conjunction square my natal Venus. The defense was certainly a culmination of a passionate attachment I had carried for a number of years!

I have noticed in my own chart that significant events often show transiting Eros in hard aspect to my natal Venus. In my natal chart, Venus is in the twelfth house, unaspected in the classical definition. Apparently, the Venus-Eros combination shows something about events to which we "attach" ourselves.

Notes

1. Erich Neumann, *Amor and Psyche* (Princeton University Press, 1956), p. 57.

2. Alain Danielou, *Shiva and Dionysus* (New York: Inner Traditions International, 1979), p. 162.

3. A. C. Bhaktivedanta Swami Prabhupada, trans. *Bhagavad Gita As It Is* (New York: The Bhaktivedanta Book Trust, 1972) p. 670.

4. E. Colledge and B. McGinn, trans. *Meister Eckhart: The Essential Sermons, Commentaries, Treatises and Defense* (New York: Paulist Press, 1981), p. 214.

5. Carl G. Jung, *Archetypes and the Collective Unconscious* (Princeton University Press, 1959), p. 88.

6. Prabhupada, p. 687.

15

Amor: Intimacy

The asteroid Amor was discovered in 1932. It has a sidereal period of 2.67 years and a diameter of two miles. Amor, like Eros, passes within the orbit of the Earth. Of the three asteroids presented here, the positions of Amor are the least accurate. The integration program developed by Mark Pottenger does not produce quite the same degree of accuracy for some of the inner asteroids, like Amor and Icarus.

Amor was the Roman god of love, a sanitized version of Eros. Synonymous with Cupid, he is the cute little cherub who goes around with his arrows putting people into the strangest circumstances. This adorable prankster has a case of arrested development. Can you imagine this cherub carrying on a passionate love relationship like Eros did with Psyche? By making him younger, the myth takes much of the sexual clout out of the god himself. While his arrows may get grownups into bed, he himself appears virtually sexless. (A number of critics have noticed how over the years Walt Disney cartoon characters have gotten younger, and hence more lovable. A comparison between the early adult Mickey Mouse—quite a nasty rogue—and the later, lovable younger Mickey shows the effectiveness of infantilization.)

The transformation of Eros into Amor changes the meaning of the symbol completely. Gone is the Eros-Thanatos pathos; what is left is love, pure and simple. There is an innocence to Amor, as if poor Cupid himself cannot understand what all the fuss is about. It is merely because we poor mortals have so much trouble with love pure and simple that Amor issues can get so convoluted.

Amor does not indicate anything about the sexual potential of a relationship. Instead, the person has the sensation that s/he has always known and loved the other person. There isn't the passion associated with Eros, or the need to make it physical associated with Sappho. With Amor, the connection can as easily be to a family member, a friend or a lover. Consequently, an Amor relationship is potentially more stable since the contact feels so deep, and there are no romantic trappings which can be destroyed.

An Amor connection between people inspires remarkable loyalty. Rather than the fury of spurned love, one often tries to understand the other person's perspective, giving her/him the benefit of the doubt no matter how ridiculous the situation looks to an outsider.

That does not mean one never gets angry at the Amor person. It simply means that not being connected to the person is almost unthinkable. Because of this extreme loyalty, the person with the activated Amor can be easily burned by the other person. The objective observer may marvel at how one walks over the other, but to the human doormat, this kind of imposition may be rationalized away. Multiple broken bones may be necessary before the Amor person sees the light. Even so, this is one of the most difficult of relationships to break off, because the individual intrinsically wants to forgive the other person so completely.

In contrast, Sappho connections may be broken by boredom, or when another Sappho contact comes along. Eros connections may be broken by a single circumstance which irrevocably destroys the romantic illusion (in other words, a circumstance in which Death wins out).

Amor represents a very "simple" emotion: love. The simplicity is shown mythologically through the child god of that name. In that same simplistic vein, the opposite of love is hate. There is little subtlety to Amor, hence the starkness of the contrast between the positive and negative sides. Unfortunately, in mundane work the negative side usually predominates.

On the negative side, Amor is the absence of love, almost a black hole with respect to love. If the person is not capable of dealing with the positive side of Amor, considerable antipathy may be shown. This antipathy is actually a projection of the individual's own negative feelings. Though it may be expressed as hatred, this energy also might find expression as a desire to destroy the potential love object so that the issue of love can be avoided. In a sense, this Amor expression is an extreme expression of the Eros-Thanatos split: If I cannot deal with love, then I must annihilate my love object.

In mundane work, Amor is often prominent at the time of violent happenings. The work of psychologist Carol Gilligan clarified the connection between love at one end of the spectrum and violence at the other. Gilligan has been especially concerned with the different ways that females and males mature. She has found that the developmental models proposed by the likes of Freud, Piaget and Erikson actually show the typical male developmental pattern and not the female one.

The primary difference is that boys tend to develop much more of a sense of independence, girls much more of a sense of interrelatedness. Hence, men find intimacy much harder as a rule than autonomy; women find autonomy much more difficult than intimacy.

An interesting experiment with regard to the Amor love-violence interface involved a study of college students. The students were shown a number of pictures, such as a man and a woman sitting together on a bench next to a river, or a man sitting alone at a desk, or a man and a woman on a trapeze. The students were then asked to write stories about the pictures. Fifty-one percent of the men wrote stories containing violence, while only 20 percent of the women included violence in their stories. Significantly, the violence in the men's stories tended to correspond to the pictures which showed clear interrelating; the violence in the women's stories tended to occur when only one person was shown. To quote Gilligan,

> "If aggression is conceived as a response to the perception of danger, the findings of the images of violence study suggest that men and women may perceive danger in different social situations and construe danger in different ways—men seeing danger more often in close personal affiliation . . . , women perceiving danger in impersonal achievement situations . . . . The danger men describe in their stories of intimacy is a danger of entrapment or betrayal, being caught in a smothering relationship or humiliated by rejection and deceit. In contrast, the danger women protray in their tales of achievement is a danger of isolation, a fear that in standing out or being set apart by success, they will be left alone.
>
> "As people are brought closer together in the pictures, the images of violence in the men's stories increase, while as people are set further apart, the violence in the women's stories increases . . . . Thus it appears that men and women may experience attachment and separation in different ways and that each sex perceives a danger which the other does not see—men in connection, women in separation."[1]

This sex-related difference in approach to the concept of intimacy—as a generalization—may explain some of the violent extremes of Amor. For women as a rule, one may posit that the opposite of love is loneliness, for men, the opposite is hate/violence.

Amor indicates the potential for intimacy, which is love. If the person finds this concept frightening, then s/he will exhibit the polar behavior, however s/he defines it. A weakly-aspected Amor in a woman's chart may be problematic; a strongly-aspected Amor in a man's chart may be problematic. This assumes that the individual in question adheres to the developmental pattern associated with her/his gender.

The interconnection between love and violence is shown in a number of maxims, such as "we only hurt the ones we love." The arrow of Cupid wounds the lover; the unrequited lover may die of a "broken" heart. Love hurts, and the absence of love hurts. Another way of interpreting the asteroid Amor would be to consider the Amor placement as the point of vulnerability to love. Vulnerability is a very scary concept. To allow oneself to be vulnerable is to open oneself to being hurt. To deny vulnerability is to wall up love and substitute hate.

The challenge of Amor is to get beyond the hornies of Sappho and the desires of Eros to the openness of Amor. To open up or expose oneself is one of the greatest risks humans take. Significantly, our saints of whatever religion are those who most often open themselves up to either other humans or to God. Because Amor reaches beyond the symbolism of sex and passion, he is most analogous to Mercury. There is even some physical resemblence between winged Mercury and winged Cupid, not to mention the role of both as tricksters.

When we open up, we risk being the fool. The process of being in love often appcars foolish to on-lookers, especially when we make allowances for the beloved. Amor contacts may indicate that the Amor person feels compelled to prove her/his love for the other person. The Amor person acts like the classic lover who will go on any quest to win the beloved. If the other reciprocates, both compassion and humility increase. If this behavior is carried far enough, one becomes a divine fool, or a person who cares so much for the divine beloved that little else matters. Thus, Amor can show agape, or spiritual love.

The emotions of Amor reach beyond the second chakra, which is involved with sex and reproduction, to the third and fourth chakras, the solar plexus and the heart. The opening of these centers may be associated with the Amor-Uranus polarity. The participation of Uranus brings the possibility for a new energy alignment. With this change comes the emotion of freedom or liberation of which the wise speak.

On the other side, if the love is unrequited, Amor ends up the victim in/to love. The Amor person will continue to love the other person regardless. The aspects to Amor may show what the individual expects in love.

For example, someone with an Amor-Pluto polarity may attempt to manipulate lovers. Control is always an issue. A Neptune-Amor polarity indicates that the person tends to idealize lovers. On the negative side, this polarity might indicate someone who attempts to drown in love or be a victim of love.

While Sappho indicates the basic biological urge to sex, and Eros shows the involvement of the mind in the integration of the procreative into the creative, Amor represents the potential for transcending the egoic attachment through love. Both the danger and the pleasure in Amor reside in the recognition of the other as an equal.

Amor Delineations

Sun-Amor: This person is tuned in to the importance of love. However, s/he may be quite unclear about how to actually find love. Thus, this individual might seek it in numerous sexual encounters before realizing the primacy of love over sex.

Moon-Amor: The Moon-Amor person knows better than the Sun-Amor person that love is the object. Often the highest expression of love is complete emotional support from another person.

Mercury-Amor: This person needs to be told that s/he is loved. The expression of love becomes more important than the actual presence of love. Such individuals may talk incessently about love with no evidence whatsoever that they have ever experienced it.

Venus-Amor: Peaceful surroundings are associated with love. This person often is described by others as being full of love for others. However, the desire for comfortable situations (and thus the setting for the expression of love) may mean that this person avoids conflict at all costs. In turn, this can inhibit others' self-expressiveness around this individual.

Mars-Amor: This person shifts between the desire for full loving expression and the need for self-assertion. Bursts of temper may endanger his/her relationships. S/he may feel that s/he must actively "do" the relationship. Love is expressed through activity.

Jupiter-Amor: The histrionic expression of love often can be more significant to this person than love itself. S/he may become obsessed with possessing the symbols of love rather than the love itself.

Saturn-Amor: This person alternates between giving and withholding love. The structure of the relationship may be more important than the relationship itself. Once a relationship gets beyond the initial stages, it has a good chance of lasting for a long time. Age differences in love relationships are not uncommon for someone with this aspect.

Uranus-Amor: Relationships become unstable if this individual gets bored. S/he finds it difficult to settle down into one long-lasting relationship unless s/he still has a considerable amount of freedom.

Neptune-Amor: This person may choose to love anyone, even when the circumstances are totally inappropriate. Love may be either addictive or a source of spiritual growth. Such persons have rich fantasy lives concerning lovers—real or imagined.

Pluto-Amor: The experience of love is an all or nothing one. This person may be completely transformed by lovers. Love relationships (or desired ones) may become obsessive. People with this aspect can be relationship junkies, or have occasional and very strong attachments. In some cases, they have no relationships at all, though this is rare. The person with this aspect expects to control the relationship absolutely. When the aspect is a difficult one, the partner may have a controlling personality. Love can be used as a method for controlling others.

Ascendant-Amor: This person often appears to be far more dependent on love relationships than s/he actually is.

Eros-Amor: Passionate attachments are the name of the game. With experience, these two forces can blend to nurture each other and produce success in long-term, highly satisfying attachments. Early in life, any short-term passion may be overdramatized.

Hidalgo-Amor: This individual is very possessive in love relationships.

Sappho-Amor: The expression of love through sexuality may become a driving force for this person, or s/he may believe that all loving relationships should be modeled *as if* they were sexual love relationships. A hard aspect between Amor and Sappho may indicate that the person has sexual difficulties with those s/he loves.

Examples of Amor in Personal Charts

Amor's energy can be very difficult to integrate. The love-hate polarity often winds up with hate ascendant. Goebbels had Amor opposite Pluto, which indicates his ruthless love of power. Director Sam Pekinpah, whose movies contain plenty of violence and hatred, has Amor trine Pluto from the ninth house. Interestingly, the Nazi hunter Simon Weisenthal also has Amor trine Pluto, as well as Amor square Mars.

To illustrate a more positive use of the Amor energy, I would like to present the natal charts for the Trappist monk Thomas Merton and the Hindu spiritual teacher Paramahansa Yogananda.

Merton had a geo-helio conjunction of Sappho trine natal Amor. His geocentric Amor was conjunct his Part of Fortune in the seventh house. His helio Amor was square helio Mars. Merton did exercise his Sappho in his youth, getting a girl pregnant "out-of-wedlock," as they said in those days. His geo Mars was quincunx Pluto and the Midheaven, suggesting that his assertiveness within relationships was an expression of power needs. His Sapphos did not make tight aspects to natal planets either heliocentrically or geocentrically, suggesting that his sexual desires were not in synch with his basic essence. Amor conjunct the Part of Fortune indicates that the lessons of love could be a means for synthesis of his deepest self. With helio Eros conjunct geo Saturn he certainly found normal passionate attachments difficult.

The expression of love through obedience, service and the discipline of a Trappist community gave him a means to express his love positively. He was able to have agape, spiritual love, as the realization of his Amor.

Paramahansa Yogananda came from a very different tradition to a very similar result. Yogananda's geo Amor was conjunct the descendant, opposite the Moon, sextile the Vertex and helio Jupiter. His helio Amor was in a wide conjunction to natal Sappho, quincunx Saturn, semi-sextile the Part of Fortune, and sextile geo Eros. Like Merton, Yogananda had a tenth house Pluto and Neptune, a seventh house helio Amor, and a fifth house Sun. Yogananda came from a spiritually-active family, and never got very involved in anything else.

The prominence of the angular geo Amor shows the centrality of this issue to his core being. The name Yogananda, given to him by his guru Sri Yukteswar, translates as the bliss of love. The helio Amor configurations show the importance of love as agape to his place in the world. The Saturn quincunx shows his long apprenticeship and his unwilling ascension to his teaching role in the West. On the other side, it shows his commitment to love as a means. In his writings, he indicates that love is the means to spiritual and human fulfilment.

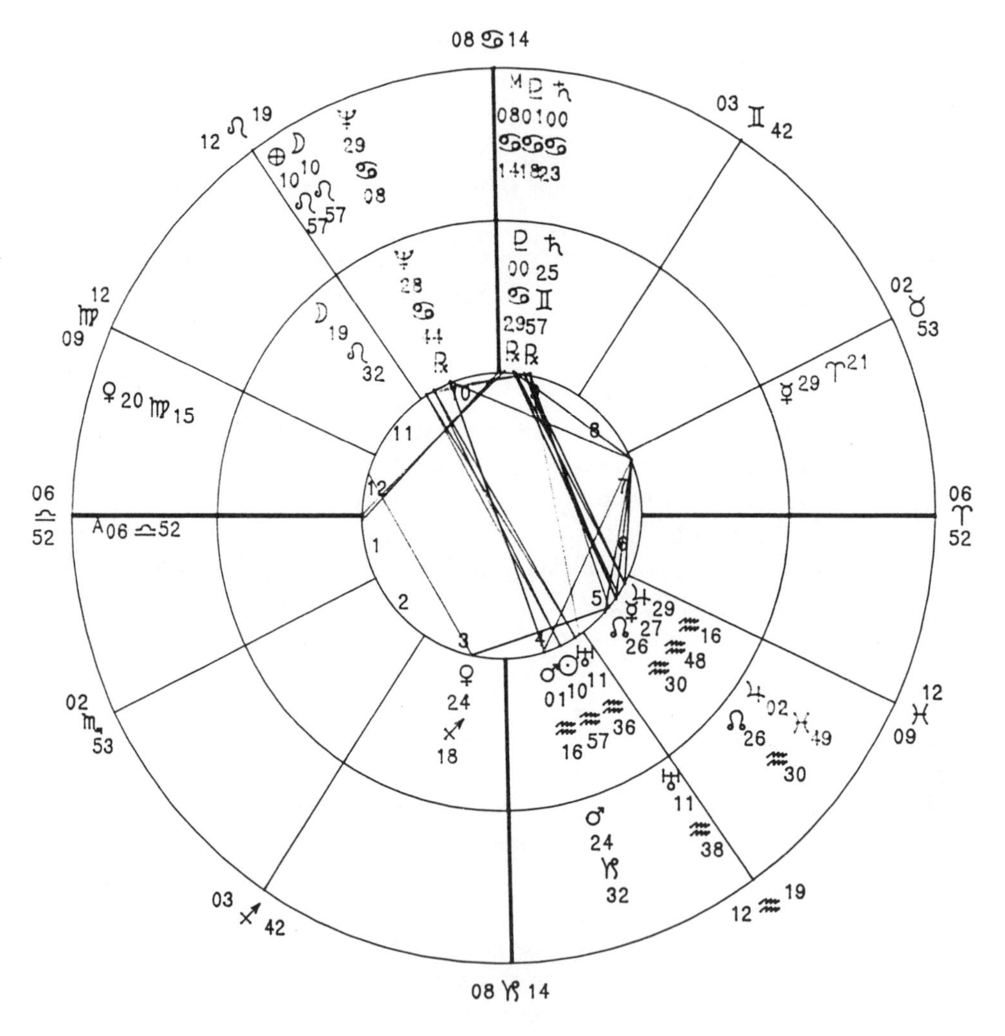

Figure 15.1

Geocentric Asteroids			
Amor	7	AR	47
Eros	0	TA	10
Sappho	5	LE	04R

Heliocentric Asteroids			
Amor	25	AR	29
Eros	26	GE	20
Sappho	7	LE	52

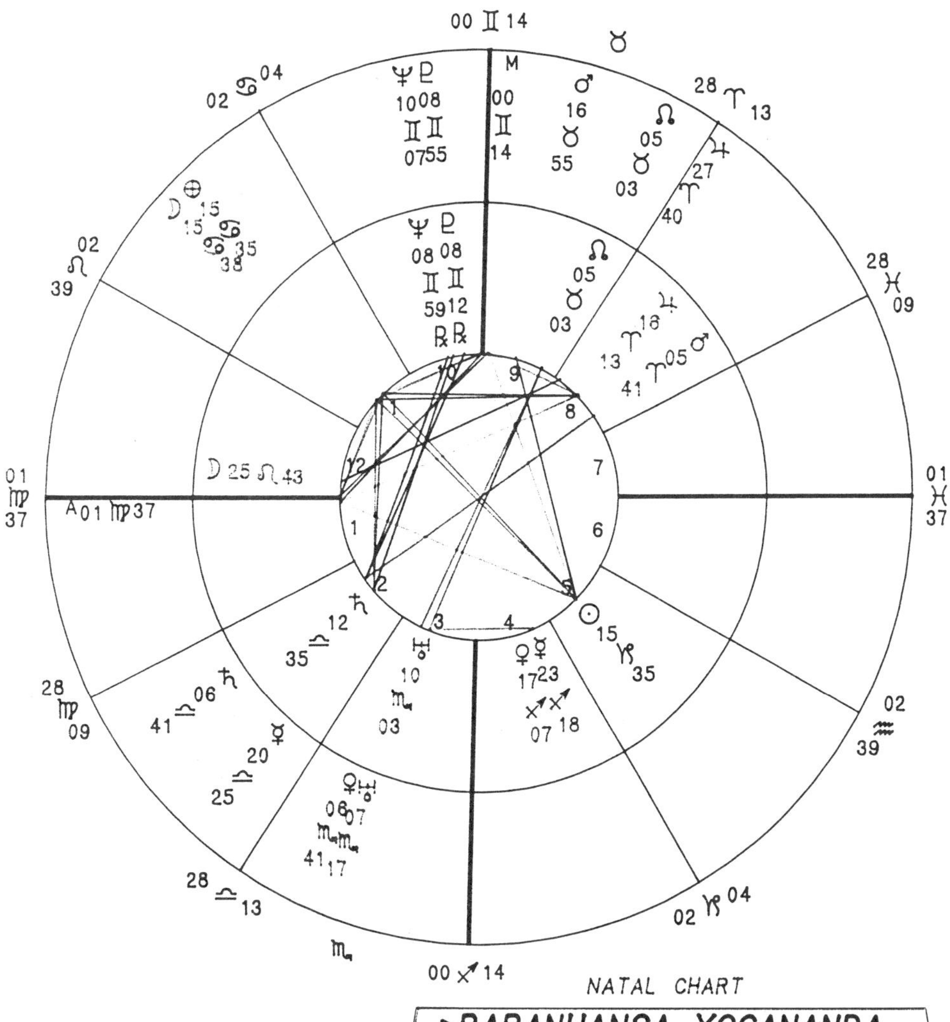

Figure 15.2

Geocentric Asteroids			
Amor	27	AQ	11
Eros	11	CP	20
Sappho	21	AQ	27

Heliocentric Asteroids			
Amor	12	PI	38
Eros	8	CP	55
Sappho	9	PI	47

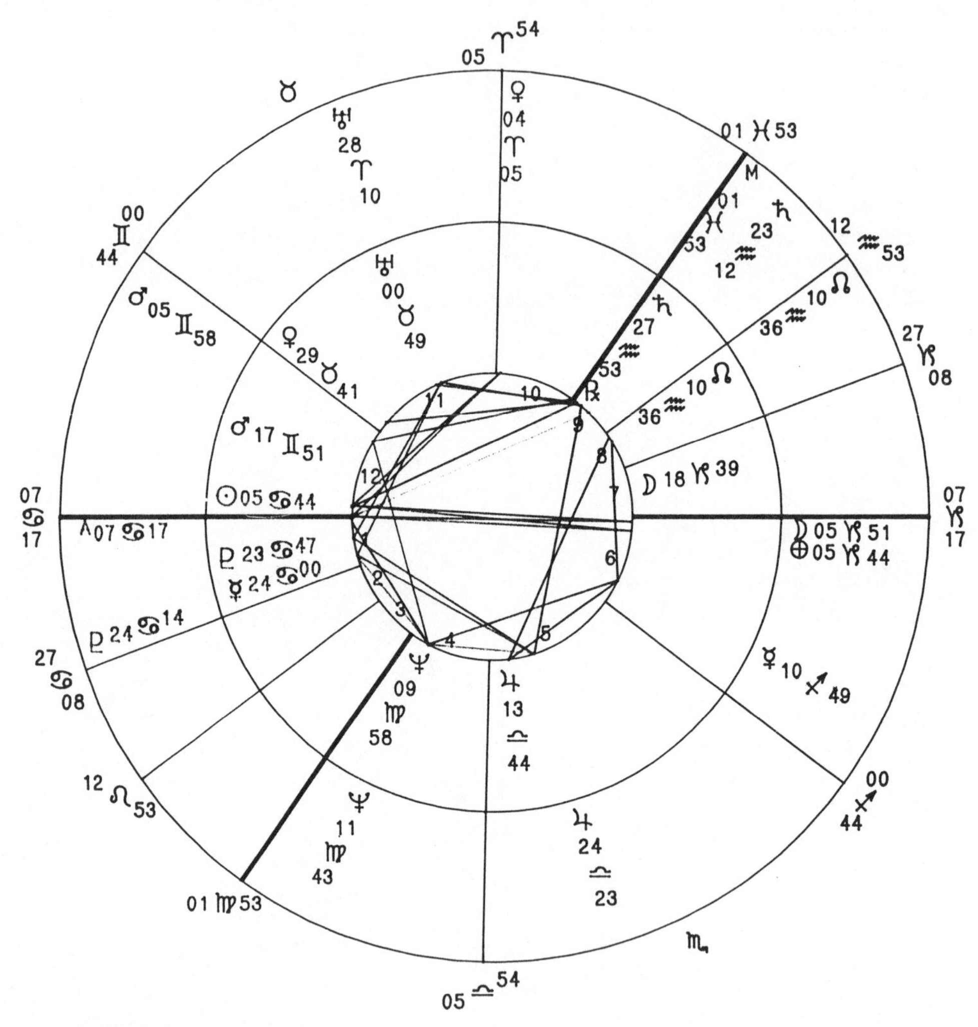

Figure 15.3

Geocentric Asteroids			
Amor	29	GE	43
Attila	25	SA	50R
Eros	6	CN	35
Europa	20	LE	18
Hidalgo	21	AR	55
Sappho	20	AQ	08

Heliocentric Asteroids			
Amor	26	GE	19
Attila	28	SA	17
Eros	7	CN	21
Europa	4	VI	43
Hidalgo	2	AR	23
Sappho	28	CP	57

Amor in Non-Personal Charts

It is unfortunate that the hate/violence side of Amor seems to predominate in non-personal charts. This was illustrated in the jury verdict chart in chapter 13.

The chart below is for an approximate time, alas. The event was the infamous Night of the Long Knives (also known as Roehm's Assassination). The particular time is for the arrest/assassination of Roehm, who was head of Hitler's Brown Shirts. Hitler needed to dispense with his Brown Shirts for two reasons: to clear the way for the domination of the Black Shirts (SS), and to get rid of "competition" for the Vermacht, the regular German Army, which Hitler wanted on his side. So on this particular night it was Nazi murdering Nazi.

Even without a precise time (and the various brutalities extended for most of the night past midnight anyway), the aspects are compelling. There are a number of conjuctions: geo and helio Eros, geo and helio Amor (loose), and Mercury-Pluto. The Amors were trine Saturn, indicating the long-term nature of these crimes. Himmler had been after Hitler for some time to eliminate Himmler's "rival" Roehm. The bloodiness was undoubtedly enhanced by Eros's conjunction to the Sun. The Amors are conjunct the fixed stars Polaris and Betelguesse. Polaris brings the possibility of tragedy and Betelguesse sudden loss. The quincunx to helio Sappho illustrates one of the excuses Hitler gave for the death of Roehm, namely Roehm's sexual excesses with boys. Geo Amor is approaching the square across the sign to Uranus, showing the suddenness of the events of the night.

Note

1. Carol Gilligan, *In a Different Voice* (Cambridge, MA: Harvard University Press, 1982), p. 42.

Sources for Data

Monica Furlong, *Merton: A Biography* (San Francisco: Harper & Row, Publishers, 1968).

Lois M. Rodden, *The American Book of Charts* (San Diego: Astro Computing Services, 1980).

16

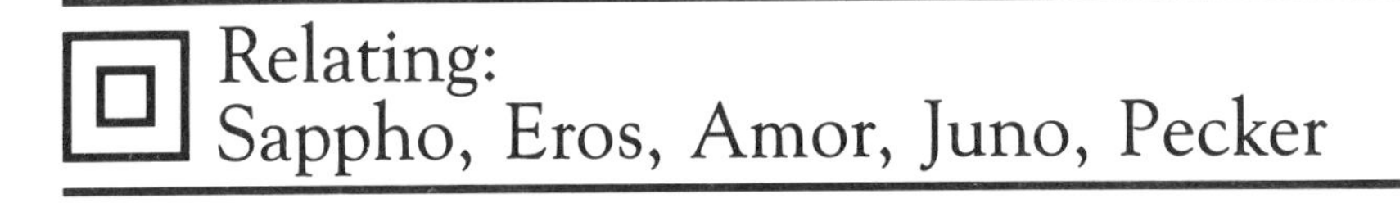

Relating: Sappho, Eros, Amor, Juno, Pecker

During the last two decades, there have been many changes in the way our society views sex and love. Astrologers have had to adjust, although many still cling to the old ways. From an astrological perspective, we must adapt our symbols as well. In growing numbers of astrological circles, a woman's chart is not automatically read primarily through the Moon and Venus, with the Sun and Mars showing the men in her life. This demonstrates some awareness that many women no longer choose to live out their lives through men. Similarly, many men no longer feel the need to project their Moons and Venuses onto the women in their lives.

There are at least a dozen different systems of inter-chart analysis. These include aspect comparisons, relationship charts, composite charts, midpoint synastry, 45 and 90 degree dials, date of meetings, marriage charts, etc. I have tried most of these, and unfortunately, they all seem to have some validity! Like many techniques in astrology, I eventually wind up asking the same question: not whether the technique works, but specifically what kind of information it shows.

There are at least three entities to consider when analyzing the interaction between two people: the two individuals and the relationship itself. I would hypothesize that the techniques directly concerned with the relationship itself—i.e., the composite and relationship charts—are probably more germaine if the people involved define themselves to a great degree in terms of the relationship. The traditional marriage was one such relationship, especially for the woman. The symbolism—Miss versus Mrs., rings, his and hers clothing—makes it very easy to define oneself in terms of the relationship. Persons living together in intimate relationships without the formality of a marriage license usually have less proprietary involvement in the relationship.

Since I belong to the generation that considers a relationship of five years long-term, I find that charting the relationship is most useful for people who are defining specific goals for the relationship. I don't consider companionship or sex to be goals in most cases. However, if there are specific goals, such as children or a business partnership, it is undoubtedly worth studying the relationship, which is going to have a life of its own in any case. Within this context, it is necessary to determine what each person brings to the relationship, in terms of agendas, attractions and past baggage. This may be studied through synastry, or chart comparison. Below are four examples which show how to use the asteroids in this manner. I have limited my discussion to direct chart-chart comparison. Other astrologers may wish to adapt this asteroid information to other methods, such as the 45 or 90 degree dial, or midpoints. In the third and fourth cases, I have thrown two additional asteroids into the pot: Juno and Pecker.

Often a client or friend in the throes of a new relationship calls me up and wants to know if this new love is The One. Some experimentation with synastry produced a very easy way to evaluate this situation. The technique is quite simple. I examine the cross aspects of Sappho, Eros and Amor between the two charts. (Of course, I also look at the planets. This asteroid examination is in addition to the usual checks for the Sun and Saturn, and all that.) The important thing is to see whose planets/asteroids are contacting whose chart.

In all comparisons, I use a maximum orb of three degrees. (I prefer one degree, but occasionally stretch it, especially in a preliminary reading.) For example, suppose I have Eros conjunct the other person's Sun. I will be romantically attracted to that person. If the other has no major aspects of her/his Eros to my chart, this romantic attraction will be one-sided. In other words, I will think this is the greatest thing since sliced bread; the other person will wonder what all the fuss is about.

Similarly, if your Sappho contacts something in my chart, say my Mars, you will be sexually attracted to me if I meet certain basic requirements. (Asteroids do not circumvent the usual filters we have, whether those filters are for a particular gender, age, race, ethnic background, or whatever.) If my Sappho contacts your Moon, or some other planet, then the attraction will be mutual. But if my Sappho doesn't hit anything in your chart, or it hits a planet that I don't like in myself, then I won't be attracted to you.

If my Amor hits your chart, I will feel like I've always loved you. This is not even likely to be a big deal unless one of the other asteroids is also involved, because it just doesn't fit the stereotypes of what "love" is about.

If there are cross aspects between each person's three asteroids and the other person's chart, then the relationship looks very hot—provided that the aspects are appropriate. If they aren't, it will show something of the problems that are likely to occur. For example, if my Sappho is on your Saturn, it can mean several things: that my sexual interest in you is as a parent substitute, or I would get a venereal disease from you, or that sex between us may not be memorable but it may be around for a long time. As usual in chart synastry, my aspects to your chart show a lot more about me than about you.

By examining both sides of the picture, the astrologer can get a pretty good idea of whether the attractions are mutual or one-sided. While it isn't likely that we can convince our Eros- or Sappho-struck clients/friends of the folly of their ways, perhaps the knowledge can help to put things in perspective after the break.

It is worth considering the synastry of these asteroids in relationships other than sexual/intimate ones. Unfortunately, I haven't found an alternate way to use cross Eros contacts. It seems that Eros love for passion and escape works best when there is sex involved. Otherwise, it tends to smolder in the background and do all sorts of things to the relationship. If the Eros contact is one-sided, the non-Eros person may begin to feel very manipulated by the Eros person. The only compensation is that the Eros person will play out the loyalty side of romance to the hilt—at least until something kills the passion.

On the other hand, Sappho contacts do very well under circumstances where sex is not appropriate. Perhaps this is because we are far better at sublimating our sexual energies because we've all had lots of practice at it. Under conditions of sublimation, Sappho simply adds zing to the relationship. And it may not be easy to sublimate forever unless circumstance and/or morals preclude it.

Amor, on the other hand, is as easy in a non-sexual relationship as a sexual one, because most people don't interpret the energy of Amor as sexual. Amor's energy can stabilize either kind of relationship. While there may be many ups and downs in an Amor relationship, they don't usually serve to weaken the relationship. With Eros, a particularly "bad" situation might destroy the romance and destabilize the relationship. With Amor, the "down" is more often transitory because there is less illusion in the relationship.

This is not to say that there is no illusion with Amor. One is just as capable of loving "blindly" with Amor as with Eros. The difference is that the Eros illusion tends to have more fantasy elements, the trappings of romance. The Eros illusion is only effective when the projection of these wonderful characteristics on the other person can hold up. If the partner violates the fantasy, the whole thing tumbles. With Amor, the illusion is that one is in love with love. Consequently, there is less substance to potentially destroy. The individual is more likely to be able to verbalize the exact nature and problems associated with the relationship. Regardless of how bad the relationship objectively is, however, the person still finds it difficult—if not impossible—to break off the contact.

One case worth mentioning is cross contacts to the same asteroid: Sappho to Sappho, Eros to Eros or Amor to Amor. In any of these, the effect is especially dramatic, because it says (in the case of the conjunction) that the two people deal with that energy through the same rose-colored glasses, so to speak. In the case of the opposition, a polarity is established. The other aspects may be interpreted similarly.

To summarize, the use of the asteroids in synastry can add a new dimension to the interpretation of how each person views the relationship. They can show when both people have the same kinds of attractions, and when they don't. By using the three asteroids Sappho, Eros and Amor, one can easily determine the nature of the attractions.

Case One: Mary and Bob, a College Affair

Mary met Bob during Bob's junior year of college. Mary had already graduated. After an initial period during which they were dating others, they embarked on an affair which lasted somewhat less than a year. Shortly thereafter, Bob became engaged to another woman, and Mary moved into a long-term relationship.

For the examination of this relationship, I have included geocentric and heliocentric versions of their charts. In the geocentric chart, Bob's Sun and Ascendant in Virgo don't interact with Mary's personal points, nor does Mary's Sun-Moon opposition interact with Bob's personal points. The only cross aspect between the personal points is Bob's Libra Moon about two degrees off Mary's Descendant. Adding Mars and Venus, the only additional aspect between them to the inner planets is Bob's Mars sextile Mary's Venus. While the Mars-Venus contact may be nice, the absence of inner planet activity—combined with Bob's Mercury square Mary's Mars—explains why the relationship didn't last any longer than it did.

	NATAL CHART		NATAL CHART
OUTER	MARY 6 27 1953 0h50m 0sEST HELIO 35N14 80W51	INNER CUSPS	MARY 6 27 1953 0h50m 0sEST PLACIDUS 35N14 80W51

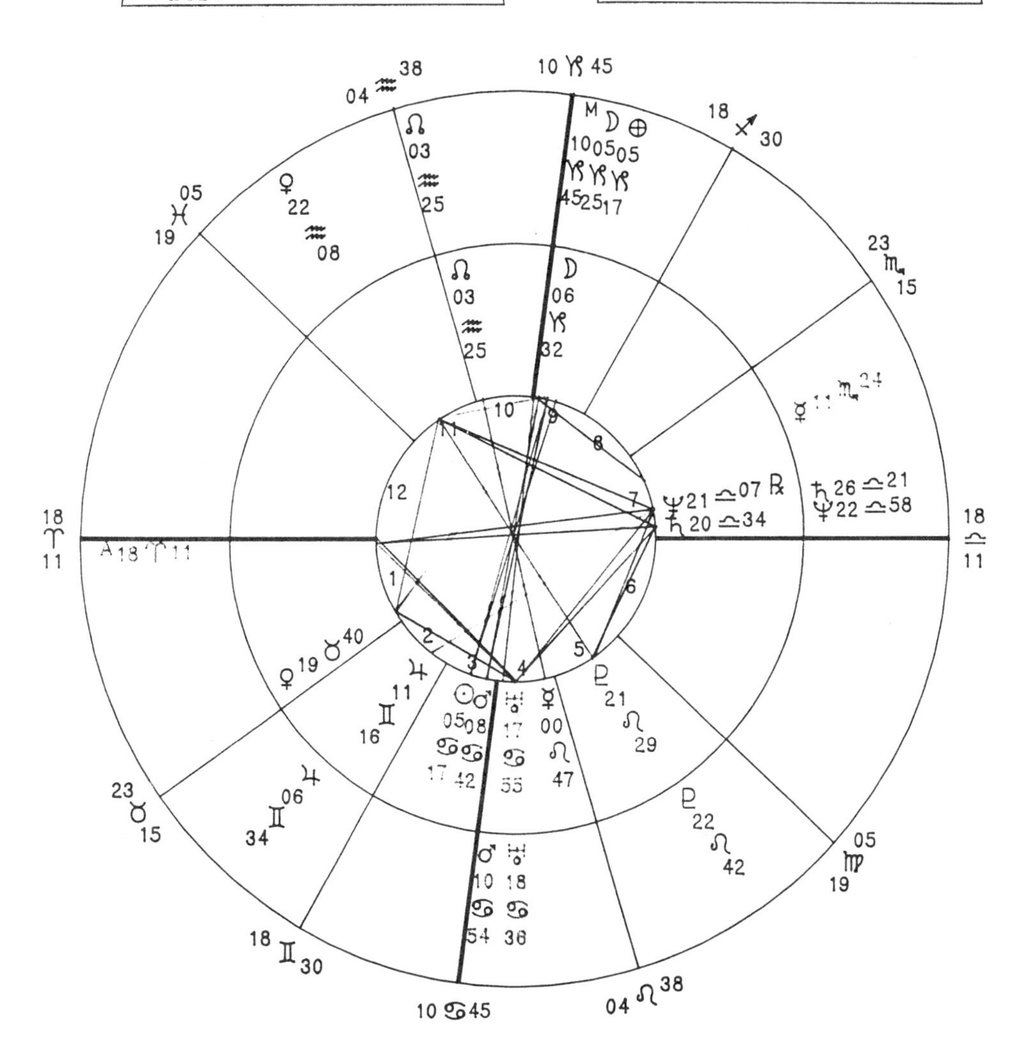

Figure 16.1

Geocentric Asteroids

Amor	9	LE	39
Eros	4	TA	27
Sappho	7	LE	12

Heliocentric Asteroids

Amor	9	VI	19
Eros	5	AR	40
Sappho	19	LE	13

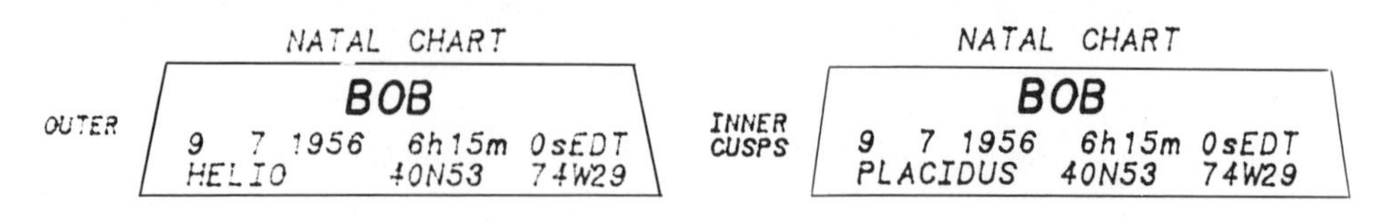

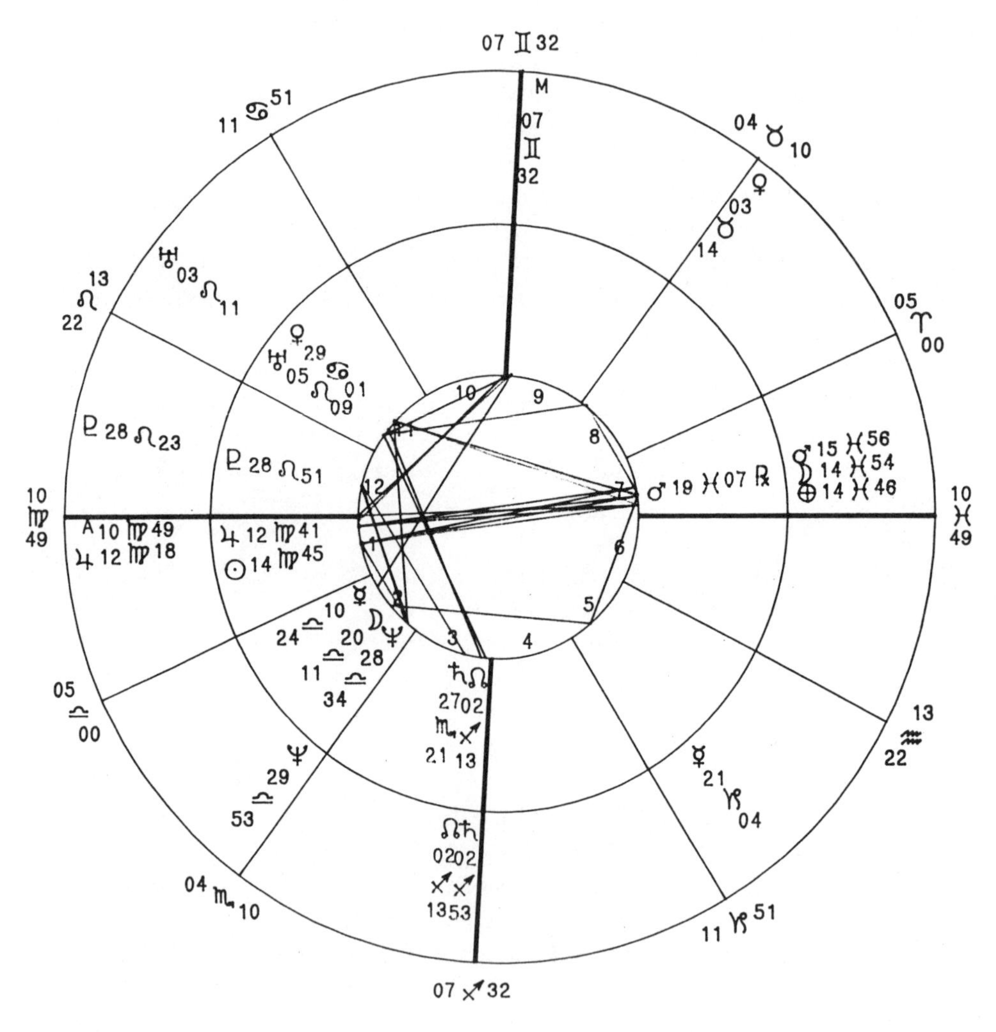

Figure 16.2

Geocentric Asteroids			
Amor	4	CP	46
Eros	18	CP	13S
Sappho	9	LE	32

Heliocentric Asteroids			
Amor	8	AQ	25
Eros	16	AQ	36
Sappho	25	CN	36

The close conjunction of Bob's Sappho and Mary's Amor may explain why they got together at all. There was a considerable sexual attraction between them, enhanced by Mary's Sappho in wide conjunction to her own Amor and hence to Bob's Sappho. Mary's Eros was square Bob's Sappho, and she did have a very romantic view of their lovemaking. Bob's Eros was opposite Mary's Uranus and conjunct her Chiron. Bob represented a "different" sort of passion: Mary had been in mainly lesbian relationships for several years prior to getting together with Bob! Bob's Amor was opposite Mary's Sun and in wide conjunction to her Moon.

Thus, all three asteroids in each person's chart were activated, though little activation was present to the inner planets. The asteroids were enough to bring them together, but not to keep them together. Both seemed to enjoy the relationship very much while it lasted and to care a great deal about each other. In the end, they simply drifted apart.

The heliocentric charts also had a dearth of cross contacts between the inner planets. In this case, Mary's Amor was quincnux Bob's Amor, adding to the eventual instability of the relationship. The asteroid which had cross aspects between the two charts was Vesta, with Mary's Vesta conjunct Bob's Amor and Bob's Vesta trine Mary's Sappho. They met through work, something about which both were very conscientious. In general, the heliocentric charts emphasize the ephemeral nature of the relationship.

Case Two: Bill and Nancy, a Marriage

Bill and Nancy met in college and lived together for several years before getting married. They have been happily married for over a decade.

The first thing I noticed about both charts is the emphasis on Libra. Is it mysterious that they both are very committed to their relationship? Nancy's Sun trines Bill's Mars-Venus, Bill's Sun is sextile Nancy's Mercury, Nancy's Moon is on Bill's Descendant, and Bill's Moon is trine Nancy's Saturn. In this relationship there are more inner planet aspects, which certainly gives a better prognosis to the relationship.

Bill's sexual nature is shown by the combination of the Mars-Venus conjunction in Virgo with Sappho in Sagittarius. He once embarked on an international adventure which included many sexual contacts. While the Sagittarius part certainly was having fun on a global basis, the Virgo part made sure that he carried plenty of condoms! This is probably just

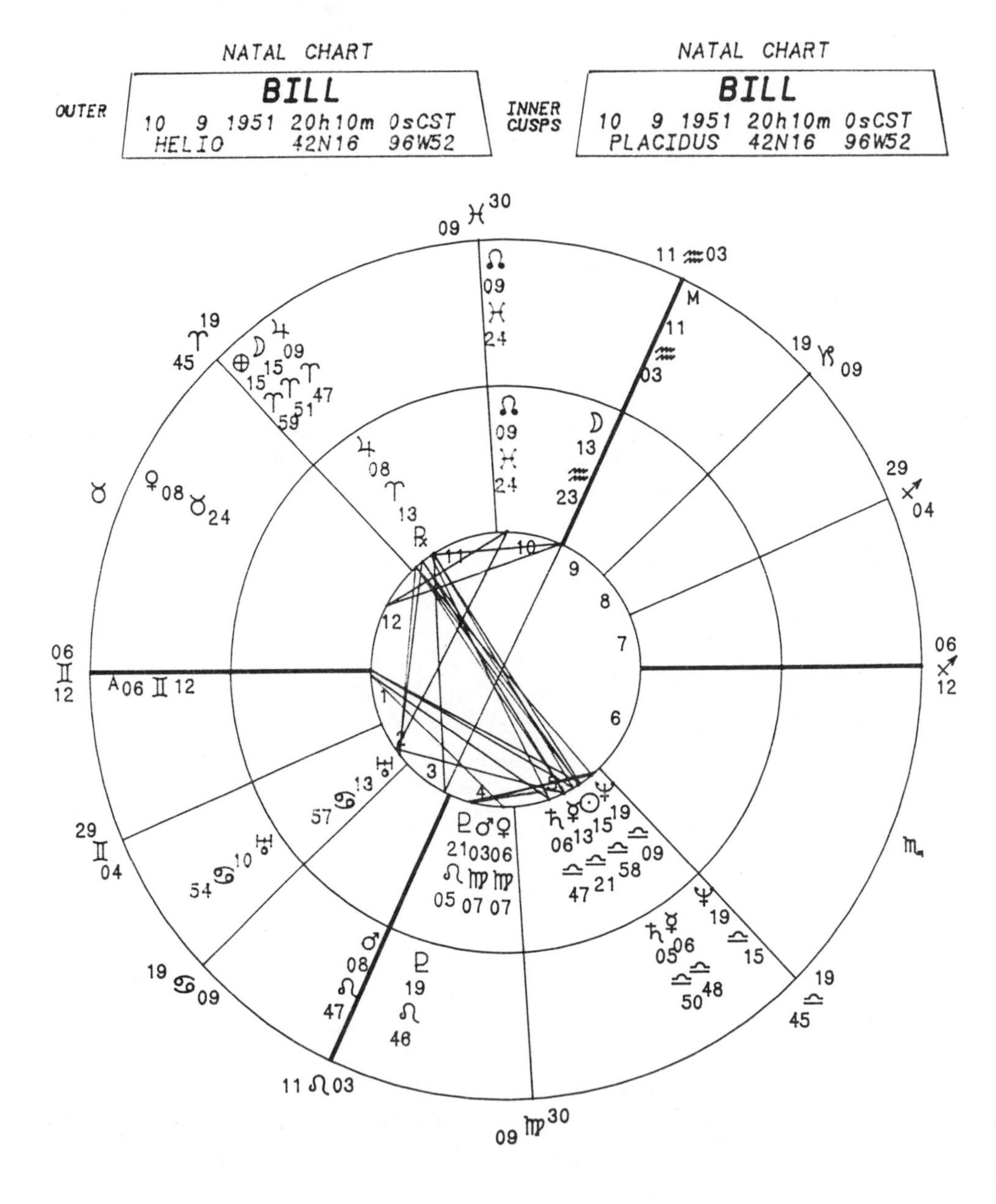

Figure 16.3

Geocentric Asteroids			
Amor	29	AQ	36R
Eros	11	AR	02R
Sappho	19	SA	02

Heliocentric Asteroids			
Amor	16	PI	25
Eros	14	AR	18
Sappho	14	CP	40

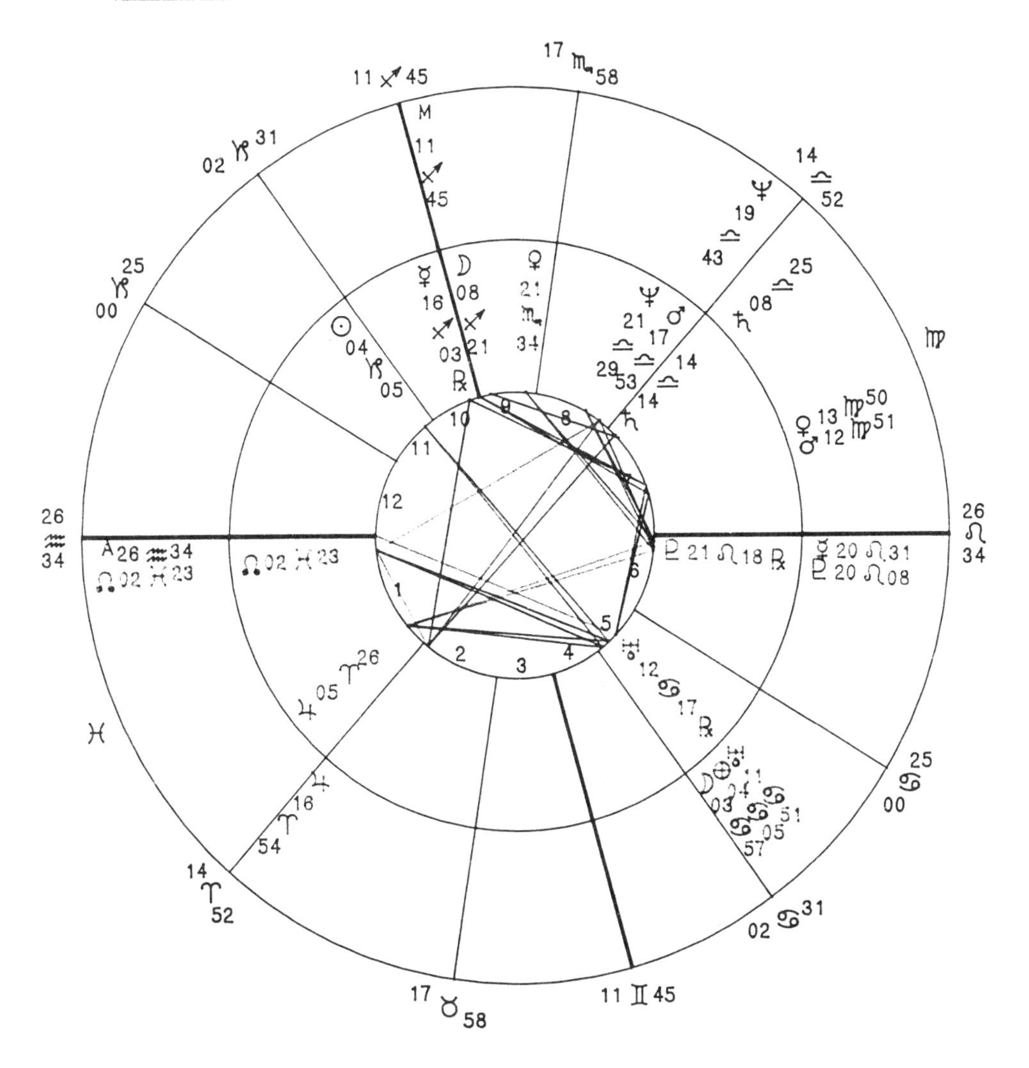

Figure 16.4

Geocentric Asteroids

Amor	9	PI	13
Eros	11	AR	24
Sappho	29	CP	58

Heliocentric Asteroids

Amor	0	AR	19
Eros	4	GE	12
Sappho	13	AQ	30

as well, because with his Sappho-Neptune sextile, he might have come down with something difficult to diagnose or treat if his Virgo conjunction had not exercised caution.

Other than the Eros-Eros conjunction, the asteroids are not as strong in this chart comparison as they were in Bob and Mary's. One may conclude that the sexual factor is less critical in this relationship: Bill and Nancy's focus is on in their mutual interests and activities.

The heliocentric charts show a number of cross aspects to the asteroids, not all of them easy. Their Sapphos are semi-sextile. In addition, Bill's Moon-Earth-Eros is square his own Sappho, suggesting that he may be a bit more passionate in situations that provide an element of excitement. Nancy's Eros is on Bill's Ascendant and her Sappho is on his Midheaven. They both help each other in their work.

These two examples show several things about the use of asteroids in relationship analysis. The major points are:

1. In order for two individuals to get close to each other in a relationship of short duration, cross asteroid contacts may be necessary.
2. The asteroid aspects are less crucial in a relationship that has plenty going for it. In a long-term relationship the asteroids may show how much emphasis is placed on sex and passion, not whether the relationship is stable.
3. The asteroids may be used in an individual's chart to show his/her own approach to an issue like sex, apart from any specific interaction to a partner's chart.
4. Because different people tend to emphasize either their geo placements or their helio placements, it may be necessary to find out which chart better describes the person before using the asteroids predictively.

Case Three: The Asteroid Juno

One of the first four asteroids discovered, Juno has been reasonably well researched by such asteroid workers as Bach, Donath, Dobyns and George. Because Juno was married to Jupiter, the asteroid that bears her name often is considered to be a significator of marriage. This may be true, but as these authors have pointed out, Juno is more likely the indicator for a "traditional" marriage than a nontraditional one.

Dobyns has advocated the use of all the asteroids in mundane work as well as in natal. Here I would like to emphasize that Juno has significance in relationships—but not necessarily marriage. The example I present here is for the airplane crash of Pan Am Flight 806 at Pago Pago on January 30, 1974. In that crash, all ten crew members and eighty-six of the ninety-one passengers died.

The National Transportation Safety Board issued a report on this crash which contained the time and location of the crash, the manufacture date of the Boeing 707-321B plane that crashed (number N454PA), and the birthdates for six of the flight attendants killed in that crash. In addition, I added the data for the first successful powered flight by the Wright brothers, and the time of the first fatal airplane crash. This data is given, along with selected asteroid positions, in Table 16.3.

Table 16.1 shows the aspects of natal geo and helio Juno of the flight attendants to the time of the crash and the manufacture date of the aircraft used. In all cases but one, both natal geo and helio positions of the flight attendants aspected the crash chart, the one exception being Yvonne Cotte's helio Juno, which was in the critical degree of 0 Libra. The hits were strong to the manufacture date, but not nearly as strong as to the crash time.

Table 16.2 shows the aspects at the time of the crash to the geo and helio Juno positions of the first powered flight, the first fatal crash, the first test flight of a Boeing 707, and the manufacture date of Boeing 707-321B, N454PA. Again, there are both helio and geo aspects to the crash in all cases, except with the first flight, in which the helio Juno is 0 Capricorn, critical again.

In this example, Juno is not indicating anything about a marriage, unless one wants to stretch the definition and say that these flight attendants and events were "married" to each other. The point is that Juno shows something of relationship, and especially the kind of relationship in which the individual does not hold the key to power. These flight attendants were not flying the plane. In addition, there was hazardous material (flammable chemicals) on board the aircraft that the crew didn't know about. The fire trucks had trouble reaching to the crash site. The NTSB rated this a survivable crash because the impact was not enough to cause fatalities. These people died because of factors beyond their own control.

Table 16.1: Geo and Helio Juno aspects of the flight attendants of Pan Am Flight 806 to the time of the crash and the manufacture of the aircraft involved in the crash.

Attendants to Crash Time	Geo Aspects	Helio Aspects
Elizabeth Givens	trine geo Mars quincunx helio Venus	trine geo Pluto
Gorda Rupp	conjunct helio Jupiter	trine helio Saturn
Gloria Olson	quincunx geo Uranus square true Node semi-sextile geo Mercury trine M.C.	trine helio Saturn
Patricia Reilly	sextile Sun	semi-sextile Chiron quincunx helio Venus sextile geo Mars
Kinuko Seko	square helio Pluto	square East Point
Yvonne Cotte	quincunx geo Jupiter	0 Libra
Elizabeth Givens	sextile helio Mars	quincunx geo Jupiter
Gorda Rupp	semi-sextile Ascendant	—
Gloria Olson	square M.C. conjunct East Point conjunct Anti-Vertex quincunx helio Saturn opposite helio Uranus	—
Patricia Reilly	—	conjunct helio Mars trine geo Venus

Attendants to Manufacture Date	Geo Aspects	Helio Aspects
Kinuko Seko	—	—
Yvonne Cotte	conjunct helio Pluto	0 Libra

Table 16.2: Geo and Helio Juno aspects of various dates to the time of the Pan Am crash.

	Geo Aspects	Helio Aspects
First Fatal Flight	sextile Ascendant square helio Pluto	semi-sextile geo Mercury semi-sextile mean Node
Manufacture Date	conjunct East Point	opposite geo Chiron quincunx helio Mars sextile helio Venus
1st 707 Test Flight	square geo Neptune	square helio Mars semi-sextile helio Uranus
1st Powered Flight	semi-sextile geo Venus opposite geo Saturn	0 Capricorn

Table 16.3: Asteroid Positions of people and dates, full data on dates used.

PA 707-321B (Crash Time)
1/31/1974 10 40 42 UT
Pago Pago, Am. Samoa 14 S 20.9 170 W 43.9

	Geocentric	Heliocentric
1862 APOLLO	29 GM 19R	19 CN 23
1566 ICARUS	17 CP 46	28 SA 30
3 JUNO	28 CP 2	23 CP 32
2228 SOYUZ-APOLLO	20 CN 23R	27 CN 44
1864 DAEDALUS	28 LE 12R	20 LE 34

First Powered Flight
12/17/1903 15 35 0 UT
Kitty Hawk, NC 36 N 02 75 W 42

	Geocentric	Heliocentric
1862 APOLLO	10 AQ 30	16 AR 58
1566 ICARUS	14 CP 18	24 CP 14
3 JUNO	29 SA 24	0 CP 57
2228 SOYUZ-APOLLO	15 SA 51	13 SA 33
1864 DAEDALUS	16 LI 38	21 VI 31

1ST "707" 367-80 Test Flight
7/15/1954 20 0 0 UT
Renton, WA 47 N 29 122 W 12

	Geocentric	Heliocentric
1862 APOLLO	24 CN 5	24 CN 44
1566 ICARUS	25 AQ 35R	7 AQ 15
3 JUNO	9 VI 18	24 VI 25
2228 SOYUZ-APOLLO	29 SA 36R	6 CP 1
1864 DAEDALUS	5 LE 8	11 LE 23

Aircraft #N454PA Manufacture Date
12/20/1967 20 0 0 UT
Everett, WA 47 N 58.8 122 W 12

	Geocentric	Heliocentric
1862 APOLLO	14 SC 13	16 LI 40
1566 ICARUS	23 CP 43	7 AQ 47
3 JUNO	1 SC 42	16 LI 15
2228 SOYUZ-APOLLO	1 GM 0R	10 GM 54
1864 DAEDALUS	8 SA 43	19 SC 13

F/A Elizabeth Givens
9/28/1943 14 0 0 UT

	Geocentric	Heliocentric
1862 APOLLO	12 LE 46	18 CN 28
1566 ICARUS	10 LI 57	1 SC 8
3 JUNO	15 CP	6 AQ 22
2228 SOYUZ-APOLLO	9 CP 1	26 CP 8
1864 DAEDALUS	18 LE 37	21 CN 30

F/A Gorda Rupp
9/12/1939 14 0 0 UT

	Geocentric	Heliocentric
1862 APOLLO	13 LI 18R	3 PI 33
1566 ICARUS	18 SA 59	21 CP 2
3 JUNO	22 AQ 35R	3 PI 14
2228 SOYUZ-APOLLO	11 GM 57	19 TA 25
1864 DAEDALUS	17 LI 3	13 SC 53

F/A Gloria Olson
6/4/1948 14 0 0 UT

	Geocentric	Heliocentric
1862 APOLLO	7 VI 18	1 SC 39
1566 ICARUS	12 AQ 15R	6 CP 27
3 JUNO	26 PI 57	2 PI 53
2228 SOYUZ-APOLLO	10 SA 36R	11 SA 29
1864 DAEDALUS	8 VI 17	22 LI 51

F/A Patricia Reilly
7/22/1948 14 0 0 UT

	Geocentric	Heliocentric
1862 APOLLO	20 VI 45	20 SA 0
1566 ICARUS	29 SA 7R	15 CP 54
3 JUNO	10 AR 41	15 PI 58
2228 SOYUZ-APOLLO	4 SA 28R	17 SA 37
1864 DAEDALUS	22 VI 34	23 SC 37

F/A Kinuko Seko
3/19/1945 14 0 0 UT

	Geocentric	Heliocentric
1862 APOLLO	27 AR 36	24 TA 54
1566 ICARUS	10 AQ 47	16 CP 42
3 JUNO	4 CN 29	1 LE 5
2228 SOYUZ-APOLLO	1 TA 1	12 TA 46
1864 DAEDALUS	8 AR 38	23 AR 51

F/A Yvonne Cotte
4/10/1950 14 0 0 UT

	Geocentric	Heliocentric
1862 APOLLO	28 CP 45	17 SC 25
1566 ICARUS	21 PI 45	26 AQ 43
3 JUNO	20 VI 53R	0 LI 34
2228 SOYUZ-APOLLO	2 AR 27	26 PI 23
1864 DAEDALUS	8 CP 40	8 SC 42

Take-Off PA#806 (Beginning time of fatal flight)
1/31/1974 7 14 0 UT
Auckland, New Zealand 36 S 52 174 E 46

	Geocentric	Heliocentric
1862 APOLLO	29 GM 20R	19 CN 21
1566 ICARUS	17 CP 41	28 SA 27
3 JUNO	27 CP 58	23 CP 31
2228 SOYUZ-APOLLO	20 CN 24R	27 CN 42
1864 DAEDALUS	28 LE 16R	20 LE 32

First Fatal Crash
9/17/1908 22 0 0 UT
Fort Myer, VA 38 N 53 77 W 4

	Geocentric	Heliocentric
1862 APOLLO	19 VI 16	16 VI 39
1566 ICARUS	2 CP 57R	10 AQ 19
3 JUNO	5 CP 41	26 CP 15
2228 SOYUZ-APOLLO	28 LI 5	6 SC 53
1864 DAEDALUS	29 LE 7	17 LE 57

Pecker

Pecker is one of those asteroids with a common name quality that is difficult to miss! In order to examine the effect of the asteroid Pecker, I used a small data set that was published by T. Patrick Davis: rape-murder victims, and perpetrators of sex murders. The results for these asteroids are shown in Table 16.4.

Table 16.4: Murder/sexual assault victims & perpetrators

A. Victims

Nancy Titterton 4/16/1902 20 13 0 UT
Georgetown, OH 38 N 52 83 W 54

	Geocentric		Heliocentric	
1221 AMOR	23 TA 44		6 GM 38	sq H. Mercury tr Vertex
1629 PECKER	22 VI 20R	tr H. Saturn qx Mars	8 LI 46	ssx Ascendant
80 SAPPHO	2 SC 55R	ssx H. Venus sq H. Jupiter	0 SC 15	
433 EROS	29 PI 8	op H. Juno sq Neptune	13 PI 12	ssx Jupiter ssx Mercury

Mary Case
10/3/1910 13 5 0 UT
Lancaster, PA 40 N 02 76 W 18

	Geocentric		Heliocentric	
1221 AMOR	13 VI 46	sx East Point	23 LE 8	qx H. Uranus sx H. Jupiter
1629 PECKER	24 TA 2R	tr H. Uranus qx H. Jupiter	7 TA 51	qx Mars
80 SAPPHO	25 TA 24R	tr Venus	3 TA 8	cj H. Saturn
433 EROS	23 SA 48	ssx H. Uranus sx H. Jupiter	26 CP 33	qx H. Pluto tr Venus-Mercury

Blondie
4/19/1915 21 15 0 UT
Detroit, MI 42 N 20 83 W 03

	Geocentric		Heliocentric	
1221 AMOR	6 TA 3		8 TA 55	sx H. Jupiter
1629 PECKER	25 CN 6	sx Ascendant	22 LE 2	ssx East Point
		op H. Venus		qx H. Venus
80 SAPPHO	2 LE 30	tr Mars	24 LE 54	ssx Ascendant
		ssx H. Pluto		ssx H. Juno
		ssx H. Saturn		qx H. Venus
433 EROS	9 CN 35	tr H. Jupiter	4 VI 19	
		tr H. Mercury		
		sx Juno		

Marion Parker
10/11/1915 10 30 0 UT
Los Angeles, CA 34 N 03 118 15

	Geocentric		Heliocentric	
1221 AMOR	23 CN 31	ssx H. Mars	21 GM 20	qx Moon
1629 PECKER	20 LI 8	qx Jupiter	21 LI 45	cj Juno
		sq Vertex		ssx Moon
80 SAPPHO	4 LI 5	ssx Ascendant	29 VI 25	tr M.C.
		qx H. Mercury		
433 EROS	27 SC 9		20 SA 8	sq Jupiter

Shirley Marie
1/14/1936 2 25 0 UT
San Bernadino, CA 34 N 07 117 W 09

	Geocentric		Heliocentric	
1221 AMOR	15 PI 53	op G/H Neptune	2 AR 50	ssx M.C.
1629 PECKER	7 SA 55	cj H. Jupiter	16 SC 9	sx Neptune
		sq Saturn		
80 SAPPHO	26 LE 58R	ssx G/H Pluto	14 LE 28	qx H. Neptune
433 EROS	26 PI 24	tr G/H Pluto	7 TA 54	qx H. Jupiter
				sx Saturn

Corene
7/1/1951 20 2 0 UT
Taylorville, IL 39 N 33 89 W 17

	Geocentric		Heliocentric	
1221 AMOR	21 PI 10	sq H. Mars	23 AQ 37	op Venus
1629 PECKER	28 GM 5	cj Mars	23 GM 39	sx Venus
80 SAPPHO	26 SC 28R	sx Saturn	13 SA 56	sq Mean Node
433 EROS	5 AR 52		29 AQ 41	sq H. Venus

Dawn
9/26/1954 17 33 0 UT
Riverton, WY 43 N 02 108 W 23

	Geocentric		Heliocentric	
1221 AMOR	6 AR 54R	qx Saturn op G/H Juno	5 AR 29	
1629 PECKER	7 CN 53	tr Saturn sq H Juno	14 GM 43	qx Mars
80 SAPPHO	8 SC 24	cj Saturn	21 SC 52	
433 EROS	18 SA 43		21 CP 55	

Rose
10/25/1965 2 2 0 UT
Woodbury, NJ 39 N 50 75 W 09

	Geocentric		Heliocentric	
1221 AMOR	29 AR 59R		0 TA 32	op Sun sx Jupiter
1629 PECKER	7 LI 2	ssx Moon	25 VI 21	sx Vertex
80 SAPPHO	4 CP 36		1 AQ 32	qx Jupiter sq Sun
433 EROS	1 PI 21	tr Jupiter tr Sun	6 AR 3	ssx H. Venus

Mark
11/25/1966 22 52 0 UT
Pekin, IL 40 N 35 89 W 40

	Geocentric		Heliocentric	
1221 AMOR	6 SC 35	ssx Venus	12 LI 27	sx H. Venus
1629 PECKER	1 AQ 56		24 AQ 55	qx H. Jupiter
				qx Mars
80 SAPPHO	24 CN 1R	cj H. Jupiter	4 CN 5	ssx Jupiter
				qx Sun
				tr Vertex
433 EROS	20 SA 6	sq Pluto	29 SA 50	
		sq H. Uranus		

Brenda
3/12/1970 23 28 0 UT
Woodbury, NJ 39 N 50 75 W 09

	Geocentric		Heliocentric	
1221 AMOR	5 PI 46	qx H. Uranus	28 AQ 25	tr H. Jupiter
		tr Jupiter		
		sx Saturn		
1629 PECKER	29 AQ 10	sq H. Neptune	18 AQ 44	cj H. Mercury
		sq Moon		sx H. Venus
80 SAPPHO	19 TA 42	ssx H. Venus	13 GM 22	
433 EROS	3 CP 41	sq Venus	22 SC 19	tr Sun
		tr Mars		tr Vertex

B. Perpetrators

All American Boy
10/29/1948 8 50 0 UT
East Orange, NJ 40 N 44 74 W 13

	Geocentric		Heliocentric	
1221 AMOR	24 CP 58	op H. Venus	22 AQ 57	tr H. Mercury
1629 PECKER	16 VI 39	ssx Pluto	26 LE 36	ssx Moon-Venus
				trine Jupiter
80 SAPPHO	5 PI 11	qx East Point	3 AR 23	
		tr Sun		
		sx H. Jupiter		
		sx Mean Node		
		qx Ascendant		
433 EROS	14 LI 8	cj Neptune	26 VI 33	cj Moon-Venus
		sx H. Pluto		sq Jupiter

Seducer/Quirk
5/31/1923 19 31 0 UT
Petersburg, IN 38 N 30 87 W 17

	Geocentric		Heliocentric	
1221 AMOR	23 TA 16		16 TA 42	op H. Jupiter tr Mean Node sx Uranus sq G/H Neptune
1629 PECKER	25 PI 55		29 AQ 39	
80 SAPPHO	8 SC 21R	qx Vertex qx Sun	20 SC 25	
433 EROS	12 AR 43	op Saturn ssx Venus	11 PI 58	sx Venus

Murdering Rapist
1/10/1937 4 30 0 UT
Bloomington, IN 40 N 29 89 W 00

	Geocentric		Heliocentric	
1221 AMOR	15 TA 13R		10 GM 58	
1629 PECKER	4 PI 48	cj Venus sx Uranus	22 PI 17	cj H. Saturn sq Mean Node
80 SAPPHO	16 SC 0		26 LI 43	sq G/H Pluto ssx H. Mars
433 EROS	12 CP 11		8 CP 2	cj Jupiter tr Uranus

Key to abbreviations in Table 16.4:

H. = Helio
G/H = both Geo & Helio Placements
cj = conjunct
op = opposite
sx = sextile
tr = trine
sq = square
qx = quincunx
ssx = semi-sextile

In addition to the aspects to planets, all but one of these people had cross-aspects between the asteroids listed. The cross-aspects bring into focus the nexus created by these sexual/relationship energies. When I substituted the asteroid Astarte for Pecker, I only noted six (instead of nine) such cross aspects.

The number of aspects of Pecker to the victim's planets is both higher and more evenly distributed than the number of aspects from the other asteroids. It is interesting that geo Amor aspects are the least common, given the intimacy issue of Amor. These individuals died at the hands of people for whom intimacy was the antithesis of their interests.

Among the perpetrators, the one called Seducer/Quirk has a dirth of aspects to these asteroids. He was not a rapist, but a seducer of women, who wanted to choke his lover during the sexual act. But he did not attempt to get women into bed against their will—simply without their knowledge of what he wanted to do once they got there.

The aspects to Pecker were somewhat stronger in the victims than the criminals. It is often difficult to tell the charts of murderers from those of their victims, so a larger sample might clarify the issue.

In the last two examples I have shown an arguably extreme form of relationship: relationship to death. In reality, our lives are lived as a series of relationships: to a person, place, thing, or even to a moment.

Sources for Data

T. Patrick Davis, *Sexual Assaults: Pre-Identifying those Vulnerable* (Windemere, FL: Davis Research Reports, 1978).

Charles Harvard Gibbs-Smith, *The Wright Brothers* (London: Her Majesty's Stationery Office, 1987).

National Transportation Safety Board, *Aircraft Accident Report: Pan American World Airways, Inc. Boeing 707-321B, N454PA Pago Pago, American Samoa January 30, 1974* (Washington, DC: NTSB Report Number NTSB-AAR-74-15, 1974).

Part Four

Mundane, Theme and Variations

17

Business

In many respects, the business side of the asteroids hardly needs to be separated from any other aspect of asteroid astrology. Business astrology in general is not that different from other branches except for one detail: unlike most natal astrology, business astrology seeks more overt prediction capability, much like horary or event astrology. Further, as is true with mundane astrology in general, interpretation can be much more literal, since there is less latitude for the choices.

Business astrology achieves its success in prediction (at least as far as markets are concerned) because its more successful practitioners are skilled at both technical analysis and astrological models. Since there is hardly space in one chapter even to begin to touch on these principles, I shall instead cover a related topic: the process of producing a model, which may then be of some predictive value. I will illustrate this process using business examples.

Before proceeding, I would like to cite the work of Bill Meridian as the most important work to date on business asteroids. Meridian works as a technical analyst for a major brokerage house. He has found a correlation between the four-year cycle in the Standard & Poors 500 index and the synodic cycle of Mars and Vesta. The synodic cycle of Mars and Vesta is 3.90 years, while the four-year technical cycle is 3.84 years. He tested this Mars-Vesta cycle by comparing a strategy of buying at the 240 degree points and selling at the 90 degree points to simply buying the stock at the beginning of the test and selling at the end. In comparing these two strategies over twenty-one and one-half cycles, the Mars-Vesta method had a yield almost double the buy and hold strategy.

Now back to our problem. Suppose you want to build a predictive model, or perhaps want to do an astrological comparison of some similar entities. The first thing to do is to design an experiment, but if you are really starting at ground zero, the first step is probably to gather some data and analyze it in order to develop a model which you can then test further.

If you are studying a business problem, one of the essential references in the United States is the two volumes by Carol S. Mull. One of these two volumes gives the corporate data for The S&P 500, the other covers 750 over-the-counter stocks. The data is well researched, but there are two cautions. The companies comprising the S&P 500 index change, so while Mull's book was accurate when it was published, the data has changed. The second caveat concerns which chart to use for a corporation.

Corporations, unlike individuals, experience multiple reincarnations without having to bother with a death in between each one. Citibank, one portion of Citicorp, is a good example. To quote Cleveland and Huertas:

> "The original name at the time of the 1812 incorporation was the City Bank of New York. It became the National City Bank of New York when the bank joined the national banking system in 1865. *First* was added to the name in 1955, as a result of the merger agreement whereby the bank took over The First National Bank of the City of New York. In 1962 the name was changed to First National City Bank, the words *The* and *New York* being dropped. In 1968 the bank became a subsidiary of the newly formed one-bank holding company First National City Corporation, and in 1974 the holding company's name was changed to Citicorp. Two years later the bank's name was changed to Citibank, N.A."[1]

Each of these incarnations has meaning, and yet a different flavor. The original bank was chartered by the State of New York during a period following the collapse of the Bank of the United States. It was a mercantile bank, since that was the dominant economic structure of the day. In the early days, the bank was principally a merchants' bank, until James Stillman added investment banking in the 1890s. The bank has gone through multiple federal banking systems and laws, panics, good times, bad times, and different corporate structures. The 1812 chart still works—and yet so do the current chart and certain important intermediate ones. For convenience, the astrologer may choose to use either the first or the most recent chart, but this does require some serious study.

Another difficulty is the question of birth date and time. With a corporation, the incorporation date is generally used. But with a retail establishment or restaurant, the astrologer may use the time when the doors are first opened for business. Another possibility is the moment of registering the company with the government.

The registration issue—whether in the form of incorporation or other type—may be sticky because not all states allow personal submission and immediate registration of papers. This means that there can be a considerable time gap between when the papers are submitted and when they are stamped, i.e., acknowledged legally.

Usually in event astrology, the time is taken to be the first irrevocable act which sets the process in motion, such as mailing a letter. Although nothing about that letter is going to have an effect until it is received, the act of mailing puts it beyond the source's power—presumably because s/he cannot pry it back out of the mailbox. However, with a corporation, the mailing date generally is not used because the corporation is a legal entity, and hence it is not "born" until the State says so. Because frequently the time of the incorporation is not known, the convention is to use a noon chart. Because the true chart is not known, in the examples below I have not used aspects to the angles and the Moon, even though many business astrologers do use these noon charts precisely like timed birth charts.

Back to our model question. Suppose I want to look at breweries. Is there a common thread among them? In Table 17.1 I have presented an asteroid study of the four breweries in the S&P 500 from Mull's first book. I looked at the position of four asteroids in these charts: Beer and Vinifera because of the clear name connection to the brewing process, Industria as a name generally applicable to the manufacturing process, and Vesta, because its effects in business cycles have been clearly demonstrated.

In the chart of Anheuser-Busch, geo Beer is quite active, opposing Pluto, squaring Venus, quincunx helio Uranus and the Mean Node, and trine Helio Neptune. But geo Beer makes no aspects in the chart for Adolph Coors! (I am tempted to apply my own taste preferences and say something here, but I will refrain!) In fact, with the exception of this one geo Beer cluster, the results are completely boring!

Table 17.1: Breweries

A. Data

1. Anheuser-Busch Companies, Inc.
   2 21 1979 17 0 0 UT
   Delaware 39 N 09 75 W 32

2. Adolph Coors Company
   6 12 1913 19 0 0 UT
   Colorado 39 N 45 104 W 59

3. G.Heileman Brewing Co., Inc.
   4 2 1918 18 0 0 UT
   Wisconsin 43 N 05 89 W 24

4. Pabst Brewing Company
   7 10 1924 17 0 0 UT
   Delaware 39 N 09 75 W 32

B. Asteroid Positions

		Beer	Industria	Vesta	Vinifera
Anheuser Busch	G	18 AR 40 op Pluto qx H. Uranus tr H. Neptune sq Venus qx Mean Node	05 SA 21 tr H. Jupiter ssx H. Venus	04 PI 05	05 LE 46 R sq H. Venus cj H. Jupiter
	H	10 TA 27 tr G/H Saturn	11 SC 33 sx Saturn	04 PI 47	13 LE 51 tr H. Mercury
Adolph Coors	G	14 TA 24	22 AR 26 sq H. Venus	17 AQ 19	29 AQ 14 ssx Mean Node tr G/H Pluto
	H	25 AR 31 sq H. Neptune	04 AR 03 sx H. Uranus sq Mercury	24 CP 45 op G/H Neptune	02 AQ 14

		Beer	Industria	Vesta	Vinifera
G. Heileman	G	07 LE 54 cj Saturn sx Jupiter	09 TA 39	09 TA 00	11 AR 53 cj Sun
	H	26 LE 56 op Venus-Uranus tr Mean Node	19 TA 28	19 TA 14	11 AR 46 cj Sun
Pabst	G	06 GE 49	02 SC 32 cj H. Saturn sq H. Mars	04 PI 47 R cj Mars tr Venus	25 LE 48 sx Saturn cj Mean Node

C. Summary

	Beer	Industria	Vesta	Vinifera	Totals
All Geo	10	5	3	8	26
All Helio	6	4	3	2	15

Key to abbreviations:

H. = Helio
G/H = both Geo & Helio Placements
cj = conjunct
op = opposite
sx = sextile
tr = trine
sq = square
qx = quincunx
ssx = semi-sextile

In fact, most such studies of natal positions are doomed to failure, probably because the group is actually far more heterogeneous than we might think. A business entity can form for a multitude of reasons, and represent a multitude of meanings for its customers, employees, managers and owner/ stockholders. If a common thread is to be found, it most likely would be with a large data sample (at least several hundred and perhaps several thousand), and the effect would probably be a statistical one: a small, but measurable, one which is of practically no use predictively, but of great use for verifying that you are on the right track.

The absence of a clearly demonstrable "Beer" effect for breweries in no way negates the ability to read the asteroid Beer in the chart of a brewery, or in the chart of a person for that matter. It simply indicates that "Beer" means different things to different breweries.

Perhaps a more interesting way to compare multiple charts is to postulate a relationship between that group and some other entity, and then the comparison can be in the nature of either transit analysis or synastry. This idea is illustrated with the S&P airlines of 1984, shown in Table 17.2.

Here one may ask, are the birthtimes of airlines related to the time of the first powered flight, since the first flight made available the technology necessary to create an airline in the first place. In Section C, the aspect comparisons are summarized, first for the asteroids within the corporate charts, and next for the comparison between the airlines' asteroids and the First Flight chart.

The natal results are comparable to the brewery results. But look at how much higher the helio aspects are for the synastry/transit example! Helio Icarus—one of the first flyers—and helio Vesta are much more strongly represented than in geo, perhaps illustrating T. Pat Davis' point about the public nature of helio bodies. Airlines are certainly public manifestations of flight!

Table 17.2: Airlines

A. Data

1. American Airlines, Inc.
   4 11 1934 17 0 0 UT
   Delaware 39 N 09 75 W 32

2. Delta Air Lines, Inc.
   6 28 1967 16 45 0 UT
   Delaware 39 N 09 75 W 32

3. Northwest Airlines, Inc.
   4 16 1934 18 0 0 UT
   Minnesota 44 N 57 93 W 06

4. Pan American World Airways, Inc.
   3 14 1927 17 0 0 UT
   New York 42 N 39 73 W 47

5. United Air Lines, Inc.
   7 20 1934 18 0 0 UT
   Delaware 39 N 09 75 W 32

6. First Powered Flight
   12 17 1903 15 35 0 UT
   Kitty Hawk, NC 36 N 02 75 W 42

B. Positions

		Daedalus	Icarus	Industria	Vesta
American	G	16 LI 40 R	25 AQ 51	24 PI 59	22 VI 48 R
	H	19 LI 37	00 AQ 08	15 PI 43	05 LI 00
Delta	G	28 LE 09	17 CP 03 R	24 AQ 42 R	17 SC 45
	H	19 VI 29	10 CP 15	08 AQ 13	08 SA 31
Northwest	G	12 LI 03 R	27 AQ 47	27 PI 00	22 VI 00 R
	H	21 LI 37	00 AQ 53	16 PI 47	06 LI 30
Pan Am	G	20 SC 02	06 AQ 21	09 GE 40	16 SC 44
	H	17 LI 34	09 CP 06	01 CN 26	25 LI 20
United	G	04 CN 25 R	02 AR 34	28 AR 15	09 LI 20
	H	20 AQ 56	19 AQ 22	06 AR 34	05 SC 44
1st Flight	G	16 LI 38	14 CP 18	06 CP 28	28 TA 01 R
	H	21 VI 31	24 CP 14	10 CP 56	07 GE 52

C. Summary

	Daedalus	Icarus	Industria	Vesta	Totals
Natal					
All Geo	4	5	7	6	22
All Helio	3	1	3	3	10
Airline ➡ 1st Flight					
All Geo	7	3	8	2	20
All Helio	7	7	5	8	27

The final illustration of this method is given with the New York bank S&P stocks. The New York banks were very much involved in the establishment— behind the scenes, of course—of the Federal Reserve Bank. The Federal Reserve is the regulatory body for the currency supply, and there is a strong interconnection between the banks and the Federal Reserve.

The data for these banks and the Federal Reserve is given in Table 17.3. Like the other two samples, the aspects in the natal chart—this time for asteroids Bourgeois, Midas, Vesta and Abundantia—are less than striking. Bourgeois was selected in this lot because banks are often cited as one of the major bourgeois institutions. Midas was selected because of the mythological Midas's famous avarice and connection to gold. Of these banks, only Citicorp stores precious metals for the commodity exchanges. The other two New York banks—Irving Trust and Republic National—are not on the S&P. Citicorp's geo Midas makes no aspects to the corporate chart, and helio Midas is conjunct the Descendant, which may be significant, especially since the time given is not a noon time. (But it is for 10:00 AM, which is almost certainly rounded off.) The placement of helio Midas on the Descendant could be especially si gnificant, because the vault is the depository of *other people's* gold and precious metals, certainly a classic Descendant description.

Table 17.3: New York Banks

A. Data

1. Bankers Trust New York Corporation
   5 12 1965 17 0 0 UT
   New York 42 N 39 73 W 47

2. The Chase Manhattan Corporation
   5 29 1969 15 00 0 UT
   Delaware 39 N 09 75 W 32

3. Chemical New York Corporation
   11 26 1968 17 0 0 UT
   Delaware 39 N 09 75 W 32

4. Citicorp
   12 04 1967 15 0 0 UT
   Delaware 39 N 09 75 W 32

5. Manufacturers Hanover Corporation
   12 05 1968 17 0 0 UT
   Delaware 39 N 09 75 W 32

6. J.P. Morgan & Company
   12 20 1968 17 0 0 UT
   New York 42 N 39 73 W 47

7. Federal Reserve Bank
   12 23 1913 23 2 0 UT
   Washington, DC 38 N 54 77 W 02

B. Positions

		Bourgeois	Midas	Vesta	Abundantia
Bankers Tr	G	12 LE 49	24 GE 12	16 TA 02	25 LI 26 R
	H	00 VI 02	08 CN 33	13 TA 46	05 SC 39
Chase	G	05 LE 28	02 GE 17	15 GE 46	02 LI 51
	H	19 LE 48	29 TA 34	18 GE 49	24 LI 23
Chemical	G	12 LE 29	02 PI 39	20 AR 22 R	28 VI 54
	H	26 CN 59	16 AR 53	06 TA 18	07 VI 42
Citicorp	G	22 TA 19 R	16 LE 00	14 CP 59	08 GE 08 R
	H	29 TA 40	20 CN 38	29 CP 10	09 GE 31
Manufact.	G	12 LE 19 R	08 PI 35	20 AR 01 R	02 LI 05
	H	28 CN 07	20 AR 11	08 TA 27	09 VI 59
JP Morgan	G	11 LE 04 R	17 PI 23	20 AR 45	06 LI 55
	H	00 LE 00	25 AR 07	12 TA 00	13 VI 49
Fed Reserv	G	04 VI 43 R	16 LI 00	08 PI 08	08 CN 00
	H	19 LE 56	02 VI 46	20 PI 57	05 CN 31

C. Summary

	Bourgeois	Midas	Vesta	Abundantia	Totals
Natal					
All Geo	5	3	3	6	17
All Helio	5	6	6	2	19
Bank ➡ Fed Reserve					
All Geo	4	4	8	5	21
All Helio	16	6	10	5	37
Fed Reserve ➡ Bank					
All Geo	7	4	5	4	20
All Helio	2	5	2	8	17

The aspects of the banks' asteroid helio aspects to the Federal Reserve chart is again high, like the situation of the relationship between the airlines and the first flight. All of these bank incorporations were actually *reincorporations* under a newer banking environment. The public intertwining of the banks with the Federal Reserve is illustrated well by the higher helio placements. Interestingly, the converse is not true, as shown by taking the asteroids in the Federal Reserve chart and comparing them to the bank charts.

In each of these three cases, I selected asteroids that either related directly to the nature of the business, or asteroids such as Vesta or Industria which could have relevence to a number of businesses. The common name aspect of asteroid astrology, thus, is useful in business investigation. However, as I have shown, this kind of study is likely to involve more than natal aspects in order to get interesting results.

Sources for Data Used

Charles Harvard Gibbs-Smith, *The Wright Brothers* (London: Her Majesty's Stationery Office, 1987).

Carol S. Mull, *Standard and Poor's 500* (Tempe, AZ: American Federation of Astrologers, 1984).

18

The Asteroid Nodes

Most astrologers have no experience with using the nodes of any body except the Moon. Therefore, some definitions are in order since the general ideas presented here apply equally to the planetary nodes as well as the asteroid nodes. The nodes of any orbiting body are the two points (locations) where the plane of that body's orbit intersects the plane of the Earth's orbit (also known as the ecliptic). You may recall from geometry that the intersection of two planes is a line. Then the two nodes, ascending (north) and descending (south), represent the axis of the intersection. Nodes may be expressed either geocentrically or heliocentrically. Heliocentric north and south nodes are always exactly opposite; geocentric ones are not. Further, the difference between the instantaneous true node and the mean true node over time may differ by slightly more than one degree.

Heliocentric planetary nodes advance through the zodiac by precession about one degree per century. The asteroid nodes presented here—the instantaneous true nodes—advance about three degrees per century. However, all of the nodes—planetary or asteroid—advance at differing rates. This is especially true for the asteroids, where significant perturbations occur in the orbits. Therefore, it is not advisable to project the current nodal positions very far into the past or future.

Since the Moon's nodes have been so extensively used over the centuries, it seemed appropriate to use the asteroid nodes, since that would be a fairly straightforward way to approach the two thousand-plus bodies which are the asteroids. Since we have already established that the names of the asteroids appear meaningful, the nodal positions are also worth exploring. Accordingly, I have included two node listings as Appendices A and B: a zodiacal sort, and an alphabetical listing. The nodes shown are heliocentric instantaneous positions of the ascending true nodes for 1 December 1988 00:00 ET published by the Leningrad Institute of Theoretical Astronomy.

What do these nodes mean? Dane Rudhyar's work suggests that the north node of any body represents intake, while the south node represents release. This may be connected to what in many spiritual teachings is referred to as the cycle of reception-release. The gist of this argument is that what goes in must come out, and what goes up must come down. In other words, reception is part of the same cycle as release.

Unfortunately, most of us are far better at reception than release. This is probably why the south lunar node has a considerably worse reputation than the north one: Since we don't tend to release voluntarily, if release is necessary, it is often under what appear to be unpleasant circumstances. Eastern religions have a reputation for attempting to turn inward while ignoring the outside world. The important lesson of the reception-release cycle is that we are continually "exposed" to all sorts of phenomena. The "method" is to let the flow continue, not to obstruct it. As Heraclitus noted, everything changes. The nodes represent "points" where the individual is challenged by the energy represented by that planet or asteroid. Ascending node positions represent the intake side; descending nodes the output side. The node itself does not indicate how well a person will be able to cope with the challenge. The aspects to the rest of the chart must be examined to determine this.

To get a general picture of how the nodes might work, I looked at some nodal positions in the charts of some famous people. All charts, unless otherwise indicated, come from Lois Rodden's *The American Book of Charts.*

The position of the ascending nodes of asteroids Kaiser and Bellerophon are at 13 Aries 07. In the chart of Otto von Bismarck, that position falls in the ninth house, near his natal Sun at 10 Aries 55. (Note that because of the precession and perturbation effects noted above, the asteroid nodes probably will fall a little before the Sun position, since Bismarck was born in 1815.) Ascending Kaiser on the Sun does seem like an appropriate position. Assuming a 4 to 6 degree difference between 1815 and 1985 positions, some other aspects in his chart include ascending Wodan (the head Teutonic god) conjunct natal Venus, descending Croatia conjunct natal Pluto (note Bismarck's involvement in the Serbo-Croatian War), descending Slavia opposite natal Sun (one of Bismarck's principal diplomatic achievements was coming up against Russia in Eastern Europe). Ascending Hannibal was conjunct natal Uranus, descending Tautonia opposite natal Moon, ascending Industria and Bohemia conjunct natal Moon (Bismarck was instrumental in the industrialization of Germany), ascending Brunhild conjunct natal Mars, and ascending Achilles conjunct natal Saturn. His natal ascendant was conjunct ascending Sevastopol. It is worth remembering that the asteroid research by the New York N.C.G.R. Asteroid Subcommitte indicated that the spelling of the asteroid name did not have to be exact.

The ascending node of asteroid Karl Marx is at 28 Aries 19. Nicolai Lenin had Jupiter at 26 Taurus and the M.C. at 25 Virgo. Since Lenin was born in 1870, the appropriate estimated movement of the nodes is three degrees. His natal Venus at 16 Pisces 53 was square ascending Leningrad.

Natalie Clifford Barney, dubbed the Amazon of Letters, was one of two key American expatriot women in Paris in the first half of the twentieth century who hosted salons at which the writers and artists of Paris were in attendance. (The other was Gertrude Stein.) Barney's exact birthtime is not known (it was in the morning sometime around 8:00 to 9:00 A.M. local time, according to her biography), but I am presenting here a chart with Sagittarius rising, since Barney was especially known for her "theatricals," one of them being Mata Hari (before the spying career) nude riding a horse as Lady Godiva.

The ascending node of the asteroid Natalie was approximately conjunct Barney's natal Neptune. (Again, three degrees is the estimated conversion.) The descending node of Amazone was near the degree of her very tight Moon-Mercury opposition. In the same degree location is the node of the asteroid Sidonia. One of Barney's best friends through the years was the writer Colette, née Sidonie-Gabrielle. We have a good chart for Colette because Colette was interested in astrology; one of her biographies reproduced a copy of her chart that was in Colette's possession. Colette's natal Pluto was conjunct ascending Amazone; ascending Natalie was close to square Colette's Uranus. Barney, a lifelong lesbian, had Jupiter in the degree of ascending Sappho. It is also worth mentioning in this case that Barney's second grande passion was the English poet Renee Vivien. The descending node of Salome was conjunct the ascendant I am proposing for Barney; this I believe is highly appropriate because Barney was such a flirt. The descending node of Dike was conjunct Barney's Sun. Again, the spelling may not be perfect, but it is suggestive.

When studying the nodes, sometimes it is interesting to note which nodes are at the same degree. For example, Perseverantia at 8 Taurus 32 is conjunct Abundantia at 8 Taurus 33. Anubis (a jackal-headed Egyptian god of the dead) at 16 Gemini 2 is conjunct Nemesis at 16 Gemini 17. Constantia at 11 Virgo 15 is conjunct Prudentia at 11 Virgo 23. Arachne at 24 Capricorn 17 is conjunct Charlotte at 24 Capricorn 54. Lucifer at 18 Aquarius 20 is conjunct Lumen at 18 Aquarius 27. Harvard at 15 Leo 30 is square Princetonia at 14 Taurus 17. Germania at 0 Capricorn 37 is opposite Fanatica at 0 Cancer 03.

A cursory reading of the asteroid names shows that, in addition to mythological characters, operatic characters, Knights of the Round Table, place names, common names and flowers, a number of the asteroids were named for specific people. While we may expect that these last names would apply to others with that last name, they should be especially appropriate for either the person so honored or the person who gave the asteroid its name. (It would be absolutely fascinating to obtain chart data on the astronomers who named the asteroids so we could study the bodies they discovered in their charts!)

Accordingly, we may note that the ascending asteroid node Darwin at 28 Aquarius 07 is within hailing range, minus 5 to 6 degrees, with Charles Darwin's natal Sun at 22 Aquarius 01. (Darwin was born in 1808.) Arthur Conan Doyle had natal Sun conjunct ascending Watsonia. (He was born in 1859, so we should adjust node positions by 3 to 4 degrees.)

When I was calculating the ephemerides of the asteroids, it became clear that many of the asteroids have rather exotic orbits. Bacchus, for instance, showed a distinctly asymmetric distribution in the signs. Accordingly, I wondered if the positions of the nodes were equally distributed through the signs. Table 18.1 shows the distribution for the 2,958 asteroids listed in the 1985 ephemeris. (The total adds up to less than 2,958 because some of the asteroids have been designated lost.)

In summary, it appears that the asteroid nodal positions have essentially the same properties as the asteroids themselves: The names or the meanings of the names are astrologically significant. Rudhyar's discussion of the nodes is essentially valid: All the nodes are involved with reception-release. This can be called karmic, of course, as many astrologers have said.

Table 18.1: Distribution of asteroid nodes through the signs

Sign	Number
Aries	272
Taurus	267
Gemini	234
Cancer	246
Leo	283
Virgo	278
Libra	271
Scorpio	187
Sagittarius	183
Capricorn	190
Aquarius	211
Pisces	280

19

Conclusion

This book represents, in part, a major theme that I have lived with for the last seven years. My friend Martha Wheelock tried unsuccessfully to get me into 'roids for two years before Al Morrison finally broke the barriers. The reason Martha failed was simple: I found the first four asteroids to be somewhat boring. I felt then—as I do now—that some astrologers got into 'roids mainly because it allowed them to introduce more unambiguously female influences. While this might be laudable, it isn't sufficient for my book. Since women and men differ by only one chromosome out of forty-six, I think the emphasis on polarity between the sexes is much overplayed. I have certainly seen plenty of women who can manage their Mars energy as well as any man.

I only became a convert to asteroids when I realized their usefulness in specific situations. Though I occasionally give readings using the asteroids exclusively, this has been mainly to demonstrate the potential of the 'roids. My preferred mode is to run the 'roids using Mark Pottenger's program in zodiacal order, and then pick out the degree areas of the chart which are critical from the perspective of the reading. In any two different readings, I will probably use entirely different asteroids because each person raises different issues.

Eleanor Bach's original work on the asteroids represented the dawn of asteroid astrology. Emma Belle Donath and Zip Dobyns echoed Eleanor's findings, extending the interpretations to encompass signs and aspects. Al Morrison and I came in to make over the whole field by going beyond the first four. Not to skip the more hilarious side, Batya Stark has given us such gems as Dudu and Dembowska. On a more serious tone, the New York National Council for Geocosmic Research Asteroids Subcommittee recognized that the name of the asteroid may be significant, and that led to work on common name asteroids. This work illuminated the psychic connection between name and meaning.

For those who think that the asteroids introduce too many new symbols, perhaps a brief demonstration of just how many different symbols we use may be helpful. The last paragraph is shown as Table 19.1, with all the asteroid names indicated!

In this book I have explored the astrological meanings of only a small number of the known asteroids. In fact, I could have used an entirely different set of 'roids. This is the strength—and the weakness—of the 'roids. We are each free to explore any of them, because any asteroid may be significant under the appropriate circumstances. The key is to let your imagination run free.

Symbols run through people's lives in meandering ways. For example, a friend of mine recently noticed that her Eros is conjunct both her daughter's and my Mercury. She is not going to play this out through an erotic relationship, but instead, she talks about her sex life with the two of us more than with any of her other friends!

The same friend has Fantasia and Kenya conjunct the Ascendant. She asked me if this means she should be fantasizing about Kenya! This is a reminder of something I have mentioned sporadically throughout the book: Not all asteroids are going to work in all people.

Asteroids are *not* exactly like planets because their meanings are not as universal as planets'. This is the reason I mentioned that I don't use asteroids with rulerships, nor do I use them with planetary patterns like T-squares and grand trines. I think it is a fair question to ask whether you are prepared to contemplate rulerships for at least three thousand bodies before you assign rulership to any one asteroid. I am perfectly prepared to admit that there are at least three thousand different symbols or ideas around, but not three thousand rulerships when there are only twelve signs!

Table 19.1: Asteroid names in paragraph. Asteroid names are in italics, with the asteroid numbers given after.

Eleanora (354) Bach's (1814) original work on the asteroids represented the *Dawn* (1618) of asteroid astrology. *Emma* (283) *Bella* (695) Donath and Zip Dobyns *Echo* ed (60) *Eleanora*'s findings,extending the interpretations to encompass *Signes* (459) and aspects. Al *Morrison* (2410) and I came in to *Makover* (1771) the whole field by going beyond the first four. Not to *Skip* (1884) the more *Hilaritas* (996) side, Batya Stark has given us such gems as *Dudu* (564) and *Dembowska* (349). On a more serious tone, the New York National Council for Geocosmic Research Asteroids Subcommittee recognized that the name of the asteroid *May* (348) be significant, and that led to work on common name asteroids. This work il*Lumen*ated (141) the *Psyche* (16) connection between name and meaning.

Years ago someone commented that part of the reason I liked asteroids was that something small and numerous appealed to my Virgo planets. This is undoubtedly true! The challenge with the 'roids is discriminate (Virgo?!!) use: capitalizing on their advantages while not getting bogged down in their weaknesses. If a particular 'roid doesn't work in a particular situation, pass it by. That image may not be a vital one to the event. I am perfectly confident that you can find enough good instances to develop a feel for the asteroids.

The appendices give the ascending nodal positions of a number of the 'roids, as well as ephemerides of several. It would not be practical to include them all!

If you wish to explore more asteroids than are given here, there are several sources available to you. Ephemerides of Ceres, Pallas, Juno and Vesta are available from several sources: My favorite is the TIA publication, given in the Resources at the end of this chapter. All of the asteroids with which I was involved are available from C.A.O. Times, which is also the source of the Chiron ephemeris. Batya Stark's book has ephemerides of Dudu, Dembowska, Frigga and Pittsburghia. Demetra George's book has ten-day positions of Ceres, Pallas, Juno, Vesta, Psyche, Eros, Lilith, Toro, Sappho, Amor, Pandora, Icarus, Diana, Hidalgo, Urania and Chiron. Most of the computing services will include at least the first four asteroids in a chart order. The first four asteroids and Chiron are available in the CCRS and Nova astrology programs, and even on the Astron Model 3 hand-held computer. Astro Computing Service will calculate the ephemeris of any asteroid that you absolutely *have* to have.

The Cadillac of programs—requiring a PC compatible and preferably a hard disk—is Mark Pottenger's asteroid program. This can give you the heliocentric and geocentric positions for about four hundred 'roids at any time during the twentieth century. It also interfaces with Mark's CCRS astrology program, so you don't have to type your data twice!

If you don't have a computer, but still want the 'roid positions for more than ten or so, please write me and I will send you the set from Mark's programs at a nominal charge.

Happy 'roiding!!

'Roid Resources

Astro Computing Services, for computed charts with asteroid option and custom asteroid ephemerides, also the publisher of Demetra George's book: P.O. Box 16430, San Diego, CA 92116-0430, (800) 525-1786 (California), (800) 826-1085 (U.S.), (619) 297-5648.

Astrolabe, for CCRS and Nova astrology programs: P.O. Box 28, Orleans, MA 02653, (617)255-0510.

Astron Astrology Pocket Computer, available from Hugh Martin, 154 South Nardo Avenue, Solana Beach, CA 92075, (619) 755-3928.

C.A.O. Times, publisher of ephemerides of Sappho, Eros, Amor, Lilith, Toro, Icarus, Pandora, Psyche, Urania, Hidalgo, Bacchus, Apollo, America, and Chiron: P.O. Box 75, Old Chelsea Station, New York, NY 10113.

Lehman, J. Lee, source for asteroid print-outs from Mark Pottenger's program. Specify alphabetical or zodaical sort, 760 Market Street, Suite 315, San Francisco, CA 94102.

Pottenger, Mark, author of the asteroid ephemeris programs for microcomputers (and the routines adapted by Rique Pottenger at ACS). MSDOS, with hard disk, $100 for all asteroids, partial rates also available by group, 838 Fifth Avenue, Los Angeles, CA 90005.

TIA Publications, publishing the Dobyns, Pottenger and Michelson ephemeris of Ceres, Pallas, Juno and Vesta: P.O. Box 45558, Los Angeles, CA 90045.

Appendix

ASTEROID NODE POSITIONS: 1988 Osculating Elements

Name	Position	Name	Position
2262 Mitidika	00 AR 10.2	2876 Aeschylus	02 AR 30.8
671 Carnegia	00 AR 18.1	705 Erminia	02 AR 31.3
645 Agrippina	00 AR 19.7	94 Aurora	02 AR 32.4
209 Dido	00 AR 21.4	571 Dulcinea	02 AR 44.1
2154 Underhill	00 AR 26.2	2452 Lyot	02 AR 48.8
2006 Polonskaya	00 AR 34.0	2023 Asaph	02 AR 49.7
1170 Siva	00 AR 34.3	109 Felicitas	02 AR 51.6
1996 Adams	00 AR 36.8	47 Aglaja	02 AR 54.7
1126 Otero	00 AR 40.6	734 Benda	02 AR 56.6
2070 Humason	00 AR 52.3	1160 Illyria	03 AR 18.2
1174 Marmara	00 AR 53.0	373 Melusina	03 AR 30.0
2114 Wallenquist	00 AR 56.6	843 Nicolaia	03 AR 47.7
81 Terpsichore	00 AR 57.0	93 Minerva	03 AR 56.6
1192 Prisma	00 AR 57.4	1622 Chacornac	03 AR 59.6
77 Frigga	00 AR 58.2	2640 Hallstrom	04 AR 02.6
1541 Estonia	01 AR 11.1	2004 Lexell	04 AR 10.3
746 Marlu	01 AR 29.8	208 Lacrimosa	04 AR 15.6
139 Juewa	01 AR 33.5	2730 Barks	04 AR 33.1
1080 Orchis	01 AR 35.2	3508 Pasternak	04 AR 34.7
1790 Volkov	01 AR 36.7	1188 Gothlandia	05 AR 02.3
1035 Amata	01 AR 38.2	2386 Nikonov	05 AR 06.1
828 Lindemannia	01 AR 44.2	2835 Ryoma	05 AR 10.9
994 Otthild	02 AR 05.0	2675 Tolkien	05 AR 30.2
2022 West	02 AR 09.8	842 Kerstin	05 AR 37.6
1227 Geranium	02 AR 15.9	1572 Posnania	05 AR 43.8

Name	Position	Name	Position
1707 Chantal	05 AR 44.8	160 Una	08 AR 26.2
2234 Schmadel	05 AR 49.3	1016 Anitra	08 AR 27.8
1134 Kepler	05 AR 49.7	2428 Kamenyar	08 AR 31.7
359 Georgia	05 AR 53.5	2362 Mark Twain	08 AR 39.7
1864 Daedalus	06 AR 09.4	1729 Beryl	08 AR 41.7
551 Ortrud	06 AR 11.8	422 Berolina	08 AR 43.7
2323 Zverev	06 AR 12.0	3092 Herodotus	08 AR 45.1
1686 De Sitter	06 AR 16.1	2786 Grinevia	08 AR 52.5
497 Iva	06 AR 19.0	915 Cosette	08 AR 54.8
312 Pierretta	06 AR 24.4	2010 Chebyshev	08 AR 55.4
3053 Dresden	06 AR 31.8	2037 Tripaxeptalis	09 AR 06.4
1453 Fennia	06 AR 34.0	1735 ITA	09 AR 13.3
2520 Novorossijsk	06 AR 34.8	3293 Rontaylor	09 AR 16.5
73 Klytia	06 AR 45.5	2905 Plaskett	09 AR 36.6
3294 Carlvesely	06 AR 45.9	1351 Uzbekistania	09 AR 37.7
195 Eurykleia	06 AR 50.2	2504 Gaviola	09 AR 45.1
642 Clara	06 AR 50.3	2757 Crisser	09 AR 50.3
3013 Dobrovoleva	06 AR 51.1	1512 Oulu	09 AR 53.5
1317 Silvretta	06 AR 53.0	280 Philia	09 AR 55.1
812 Adele	06 AR 57.8	2941 Alden	09 AR 57.6
37 Fides	07 AR 12.4	1552 Bessel	09 AR 59.3
1204 Renzia	07 AR 21.1	290 Bruna	10 AR 03.6
302 Clarissa	07 AR 23.1	55 Pandora	10 AR 10.1
802 Epyaxa	07 AR 24.2	2370 Van Altena	10 AR 12.2
66 Maja	07 AR 26.6	477 Italia	10 AR 13.4
2432 Soomana	07 AR 32.8	91 Aegina	10 AR 23.7
2384 Schulhof	07 AR 39.6	1046 Edwin	10 AR 29.0
298 Baptistina	07 AR 48.8	2551 Decabrina	10 AR 40.4
766 Moguntia	07 AR 51.1	2045 Peking	10 AR 49.0
3439 Lebofsky	07 AR 51.3	2501 Lohja	10 AR 59.1
2236 Austrasia	07 AR 54.8	2178 Kazakhstania	11 AR 02.5
2705 Wu	07 AR 55.3	3056 INAG	11 AR 09.7
927 Ratisbona	07 AR 57.4	453 Tea	11 AR 19.3
2464 Nordenskiold	08 AR 06.8	2134 Dennispalm	11 AR 21.1
3006 Livadia	08 AR 07.9	3193 Elliot	11 AR 23.8
3009 Coventry	08 AR 08.1	2503 Liaoning	11 AR 26.9
33 Polyhymnia	08 AR 14.8	604 Tekmessa	11 AR 47.4

Name	Position	Name	Position
1225 Ariane	11 AR 59.6	2007 McCuskey	16 AR 44.5
1768 Appenzella	12 AR 08.5	2156 Kate	16 AR 44.7
2406 Orelskaya	12 AR 10.3	855 Newcombia	16 AR 52.7
1521 Seinajoki	12 AR 15.0	2181 Fogelin	16 AR 54.1
3432 Kobuchizawa	12 AR 16.1	2143 Jimarnold	16 AR 59.5
1235 Schorria	12 AR 26.3	1226 Golia	17 AR 04.3
1495 Helsinki	12 AR 45.9	795 Fini	17 AR 04.7
1808 Bellerophon	13 AR 05.7	2042 Sitarski	17 AR 05.2
1694 Kaiser	13 AR 06.0	1446 Sillanpaa	17 AR 07.6
615 Roswitha	13 AR 25.7	428 Monarchia	17 AR 13.0
255 Oppavia	13 AR 28.7	2016 Heinemann	17 AR 14.2
1332 Marconia	13 AR 44.5	1053 Vigdis	17 AR 27.1
1325 Inanda	13 AR 55.9	2900 Lubos Perek	17 AR 31.7
2383 Bradley	14 AR 02.6	1479 Inkeri	17 AR 57.7
2701 Cherson	14 AR 02.8	1368 Numidia	17 AR 59.0
1908 Pobeda	14 AR 12.6	2819 Ensor	18 AR 07.6
1508 Kemi	14 AR 12.8	161 Athor	18 AR 13.1
450 Brigitta	14 AR 20.8	952 Caia	18 AR 19.0
186 Celuta	14 AR 21.5	2632 Guizhou	18 AR 21.2
2502 Nummela	14 AR 24.6	1669 Dagmar	18 AR 39.2
1183 Jutta	14 AR 47.1	2034 Bernoulli	18 AR 40.0
932 Hooveria	14 AR 48.0	361 Bononia	18 AR 40.4
2179 Platzeck	14 AR 56.1	1846 Bengt	18 AR 41.5
1824 Haworth	14 AR 57.4	2165 Young	19 AR 02.8
2057 Rosemary	15 AR 00.6	1500 Jyvaskyla	19 AR 30.2
784 Pickeringia	15 AR 04.8	2335 James	19 AR 30.4
2008 Konstitutsiya	15 AR 08.6	1814 Bach	19 AR 53.9
697 Galilea	15 AR 17.2	610 Valeska	20 AR 04.1
1962 Dunant	15 AR 24.9	1207 Ostenia	20 AR 05.9
1519 Kajaani	15 AR 39.8	1008 La Paz	20 AR 24.1
1527 Malmquista	15 AR 50.3	1696 Nurmela	20 AR 39.6
2471 Ultrajectum	15 AR 55.6	2803 Vilho	20 AR 40.0
1786 Raahe	16 AR 10.6	2239 Paracelsus	20 AR 42.7
2792 Ponomarev	16 AR 18.6	175 Andromache	20 AR 57.9
1631 Kopff	16 AR 30.1	1452 Hunnia	20 AR 58.5
1501 Baade	16 AR 32.5	2492 Kutuzov	21 AR 02.3
2526 Alisary	16 AR 38.3	944 Hidalgo	21 AR 02.5

Name	Position	Name	Position
3068 Khanina	21 AR 08.5	1654 Bojeva	25 AR 13.3
1637 Swings	21 AR 13.7	2931 Mayakovsky	25 AR 13.7
1358 Gaika	21 AR 20.3	82 Alkmene	25 AR 20.9
1770 Schlesinger	21 AR 25.1	2072 Kosmodemyanskaya	
546 Herodias	21 AR 33.5		25 AR 53.8
187 Lamberta	21 AR 36.6	1190 Pelagia	26 AR 02.7
468 Lina	21 AR 47.6	961 Gunnie	26 AR 26.1
2446 Lunacharsky	21 AR 52.3	1927 Suvanto	26 AR 42.0
331 Etheridgea	21 AR 58.9	340 Eduarda	26 AR 51.8
2574 Ladoga	22 AR 00.6	362 Havnia	26 AR 52.4
757 Portlandia	22 AR 04.3	852 Wladilena	27 AR 03.4
1960 Guisan	22 AR 06.7	1744 Harriet	27 AR 04.0
2243 Lonnrot	22 AR 10.4	1698 Christophe	27 AR 12.7
1124 Stroohantia	22 AR 11.8	1013 Tombecka	27 AR 14.9
1205 Ebella	22 AR 42.4	83 Beatrix	27 AR 18.0
435 Ella	22 AR 46.8	2434 Bateson	27 AR 19.0
2113 Ehrdni	22 AR 55.4	2641 Lipschutz	27 AR 19.3
126 Velleda	23 AR 00.5	1518 Rovaniemi	27 AR 22.7
1099 Figneria	23 AR 01.7	2151 Hadwiger	27 AR 41.0
1948 Kampala	23 AR 02.0	229 Adelinda	27 AR 55.7
1983 Bok	23 AR 22.2	2871 Schober	27 AR 56.9
2080 Jihlava	23 AR 26.4	2212 Hephaistos	28 AR 00.9
2293 Guernica	23 AR 28.8	254 Augusta	28 AR 05.6
1861 Komensky	23 AR 32.5	2807 Karl Marx	28 AR 18.9
761 Brendelia	23 AR 36.0	1354 Botha	28 AR 20.3
3036 Krat	23 AR 57.7	341 California	28 AR 42.5
250 Bettina	24 AR 00.5	207 Hedda	28 AR 49.5
1366 Piccolo	24 AR 01.4	1390 Abastumani	28 AR 51.8
595 Polyxena	24 AR 19.9	1027 Aesculapia	28 AR 59.9
1267 Geertruida	24 AR 27.6	459 Signe	29 AR 11.6
2631 Zhejiang	24 AR 33.9	3309 Brorfelde	29 AR 16.5
3414 Champollion	24 AR 36.7	1797 Schaumasse	29 AR 18.2
1462 Zamenhof	24 AR 39.8	578 Happelia	29 AR 19.2
1851 Lacroute	24 AR 41.3	1961 Dufour	29 AR 19.7
215 Oenone	24 AR 48.3	1675 Simonida	29 AR 45.7
1989 Tatry	24 AR 54.0	1029 La Plata	29 AR 46.0
2305 King	25 AR 02.0	2032 Ethel	29 AR 48.1

Name	Position	Name	Position
1132 Hollandia	29 AR 49.1	1447 Utra	05 TA 22.8
2424 Tautenburg	29 AR 52.3	1439 Vogtia	05 TA 28.4
2913 Horta	00 TA 00.8	720 Bohlinia	05 TA 32.2
2315 Czechoslovakia		859 Bouzareah	05 TA 39.4
	00 TA 03.8	24 Themis	05 TA 40.1
1087 Arabis	00 TA 07.8	793 Arizona	05 TA 49.1
2787 Tovarishch	00 TA 13.9	3045 Alois	05 TA 59.8
1081 Reseda	00 TA 27.8	401 Ottilia	06 TA 00.3
964 Subamara	00 TA 30.7	162 Laurentia	06 TA 01.9
31 Euphrosyne	00 TA 40.1	2047 Smetana	06 TA 04.9
2761 Eddington	00 TA 54.7	463 Lola	06 TA 10.0
127 Johanna	00 TA 56.5	597 Bandusia	06 TA 13.1
281 Lucretia	00 TA 58.4	953 Painleva	06 TA 14.0
332 Siri	01 TA 21.2	2777 Shukshin	06 TA 32.7
2300 Stebbins	01 TA 29.8	3292 Sather	06 TA 35.6
326 Tamara	01 TA 49.3	1072 Malva	06 TA 36.7
454 Mathesis	01 TA 58.9	154 Bertha	06 TA 41.9
2955 Newburn	02 TA 03.1	2193 Jackson	06 TA 42.8
349 Dembrowska	02 TA 09.2	3133 Sendai	06 TA 44.0
210 Isabella	02 TA 22.6	905 Universitas	06 TA 45.7
2063 Bacchus	02 TA 41.2	2734 Hasek	06 TA 55.2
3078 Horrocks	02 TA 45.1	272 Antonia	07 TA 10.5
796 Sarita	02 TA 48.8	448 Natalie	07 TA 28.4
2826 Ahti	03 TA 22.9	2603 Taylor	07 TA 38.8
2248 Kanda	03 TA 31.9	1074 Beljawskya	07 TA 47.3
2314 Field	03 TA 36.8	651 Antikleia	07 TA 52.3
2238 Steshenko	03 TA 44.3	3015 Candy	07 TA 55.7
520 Franziska	04 TA 09.5	2360 VolgoDon	08 TA 04.6
2985 Shakespeare	04 TA 12.7	718 Erida	08 TA 07.3
2308 Schilt	04 TA 15.4	494 Virtus	08 TA 08.9
3437 Kapitsa	04 TA 16.3	262 Valda	08 TA 13.9
912 Maritima	04 TA 16.4	343 Ostara	08 TA 20.8
257 Silesia	04 TA 30.2	2770 Tsvet	08 TA 21.1
636 Erika	04 TA 39.5	769 Tatjana	08 TA 25.0
475 Ocllo	04 TA 43.6	975 Perseverantia	08 TA 26.9
1813 Imhotep	04 TA 49.1	2155 Wodan	08 TA 28.3
1862 Apollo	05 TA 22.0	1715 Salli	08 TA 33.1

Name	Position	Name	Position
151 Abundantia	08 TA 33.2	1421 Esperanto	12 TA 45.8
1155 Aenna	08 TA 42.5	1424 Sundmania	12 TA 47.4
721 Tabora	08 TA 45.4	2982 Muriel	12 TA 47.7
768 Struveana	08 TA 56.5	1186 Turnera	12 TA 48.4
1757 Porvoo	09 TA 05.9	907 Rhoda	12 TA 51.4
1073 Gellivara	09 TA 06.5	735 Marghanna	12 TA 51.8
2880 Nihondaira	09 TA 18.4	300 Geraldina	12 TA 52.4
2040 Chalonge	09 TA 32.3	845 Naema	12 TA 54.5
1162 Larissa	09 TA 34.7	292 Ludovica	13 TA 08.8
1253 Frisia	09 TA 36.2	3343 Nedzel	13 TA 09.1
680 Genoveva	09 TA 51.0	1378 Leonce	13 TA 12.3
2333 Porthan	09 TA 51.5	1599 Giomus	13 TA 14.5
152 Atala	09 TA 51.7	3254 Bus	13 TA 37.1
906 Repsolda	09 TA 57.5	1869 Philoctetes	13 TA 38.6
321 Florentina	10 TA 03.5	2849 Shklovskij	13 TA 43.7
1939 Loretta	10 TA 07.0	617 Patroclus	13 TA 44.9
1840 Hus	10 TA 12.8	1738 Oosterhoff	13 TA 46.3
2543 Machado	10 TA 30.5	1664 Felix	13 TA 49.1
698 Ernestina	10 TA 33.8	1798 Watts	13 TA 51.3
155 Scylla	10 TA 37.6	2910 YoshkarOla	13 TA 52.5
1223 Neckar	10 TA 38.8	68 Leto	13 TA 55.9
641 Agnes	10 TA 39.8	2839 Annette	14 TA 08.5
1343 Nicole	10 TA 49.7	1781 Van Biesbroeck	14 TA 14.6
2255 Qinghai	10 TA 49.9	508 Princetonia	14 TA 15.9
965 Angelica	10 TA 59.2	770 Bali	14 TA 17.4
1459 Magnya	11 TA 04.4	599 Luisa	14 TA 18.3
2884 Reddish	11 TA 04.7	519 Sylvania	14 TA 31.9
99 Dike	11 TA 14.2	806 Gyldenia	14 TA 31.9
104 Klymene	11 TA 44.8	3131 MasonDixon	14 TA 35.3
446 Aeternitas	11 TA 49.9	1448 Lindbladia	14 TA 38.7
988 Appella	11 TA 52.2	1472 Muonio	14 TA 47.8
1219 Britta	12 TA 09.1	3275 Oberndorfer	14 TA 57.0
1982 Cline	12 TA 09.5	1838 Ursa	14 TA 59.7
2364 Seillier	12 TA 27.2	1841 Masaryk	15 TA 07.2
2361 Gogol	12 TA 40.0	1895 Larink	15 TA 15.7
2226 Cunitza	12 TA 42.0	722 Frieda	15 TA 19.6
13 Egeria	12 TA 45.4	2750 Loviisa	15 TA 21.4

Name	Position	Name	Position
2430 Bruce Helin	15 TA 27.8	2703 Rodari	19 TA 01.3
26 Proserpina	15 TA 34.4	1193 Africa	19 TA 20.2
2321 Luznice	15 TA 38.6	264 Libussa	19 TA 23.2
1323 Tugela	15 TA 43.4	2584 Turkmenia	19 TA 40.3
2558 Viv	15 TA 45.5	910 Anneliese	19 TA 45.0
715 Transvaalia	15 TA 49.6	2538 Vanderlinden	19 TA 55.9
981 Martina	16 TA 02.9	528 Rezia	19 TA 57.8
731 Sorga	16 TA 10.0	2657 Bashkiria	20 TA 02.7
3273 Drukar	16 TA 17.6	1202 Marina	20 TA 05.6
1972 Yi Xing	16 TA 20.8	1849 Kresak	20 TA 21.2
492 Gismonda	16 TA 24.0	1957 Angara	20 TA 34.1
867 Kovacia	16 TA 28.5	2401 Aehlita	20 TA 42.1
1440 Rostia	16 TA 32.9	1839 Ragazza	20 TA 43.5
1995 Hajek	16 TA 44.0	178 Belisana	20 TA 51.3
2194 Arpola	16 TA 46.1	2224 Tucson	21 TA 00.7
2124 Nissen	16 TA 51.1	2566 Kirghizia	21 TA 11.0
2069 Hubble	17 TA 02.0	1567 Alikoski	21 TA 13.6
118 Peitho	17 TA 11.2	2182 Semirot	21 TA 19.3
1429 Pemba	17 TA 26.6	1238 Predappia	21 TA 30.6
2153 Akiyama	17 TA 31.1	813 Baumeia	21 TA 31.4
384 Burdigala	17 TA 34.9	1899 Crommelin	21 TA 43.9
70 Panopaea	17 TA 37.2	2986 Mrinalini	21 TA 45.7
2369 Chekhov	17 TA 39.5	2438 Oleshko	21 TA 53.1
3080 Moisseiev	17 TA 53.6	1545 Thernoe	22 TA 02.5
2897 Ole Romer	17 TA 54.8	2177 Oliver	22 TA 04.0
223 Rosa	17 TA 55.2	1042 Amazone	22 TA 09.1
1208 Troilus	17 TA 56.5	941 Murray	22 TA 11.8
1616 Filipoff	17 TA 59.9	1071 Brita	22 TA 16.7
947 Monterosa	18 TA 04.0	1375 Alfreda	22 TA 17.9
2883 Barabashov	18 TA 08.1	1540 Kevola	22 TA 19.3
1236 Thais	18 TA 11.9	1048 Feodosia	22 TA 35.1
344 Desiderata	18 TA 16.4	2768 Gorky	22 TA 52.0
2830 Greenwich	18 TA 32.3	3176 Paolicchi	22 TA 58.9
438 Zeuxo	18 TA 47.3	1563 Noel	23 TA 13.8
2805 Kalle	18 TA 54.1	2044 Wirt	23 TA 24.4
1515 Perrotin	18 TA 59.8	2678 Aavasaksa	23 TA 38.0
926 Imhilde	18 TA 59.9	2355 Uei Monggol	23 TA 44.5

Name	Position
2573 Hannu Olavi	23 TA 49.0
1978 Patrice	23 TA 57.8
1088 Mitaka	24 TA 05.5
1388 Aphrodite	24 TA 10.1
1676 Kariba	24 TA 10.9
2263 Shaanxi	24 TA 12.5
1630 Milet	24 TA 18.1
138 Tolosa	24 TA 39.4
2027 Shen Guo	25 TA 01.0
2867 Steins	25 TA 04.3
1093 Freda	25 TA 14.3
1741 Giclas	25 TA 14.7
2578 SaintExupery	25 TA 29.5
1889 Pakhmutova	25 TA 31.6
2515 Gansu	25 TA 38.0
2582 HarimayaBashi	26 TA 03.6
2489 Suvorov	26 TA 08.0
1393 Sofala	26 TA 25.4
110 Lydia	26 TA 39.6
2197 Shanghai	26 TA 42.9
815 Coppelia	26 TA 49.7
1119 Euboea	26 TA 59.0
1967 Menzel	27 TA 22.9
1237 Genevieve	27 TA 34.0
1327 Namaqua	27 TA 41.0
1038 Tuckia	27 TA 43.1
416 Vaticana	27 TA 49.9
1133 Lugduna	27 TA 51.2
567 Eleutheria	28 TA 02.5
2129 Cosicosi	28 TA 04.9
674 Rachele	28 TA 10.1
1753 Mieke	28 TA 17.6
1423 Jose	28 TA 19.2
3452 Hawke	28 TA 23.8
1024 Hale	28 TA 35.9
2840 Kallavesi	28 TA 41.4
2505 Hebei	28 TA 53.0
1670 Minnaert	28 TA 58.7
1128 Astrid	29 TA 03.4
1902 Shaposhnikov	29 TA 04.4
2286 Fesenkov	29 TA 07.4
536 Merapi	29 TA 07.7
959 Arne	29 TA 09.9
2117 Danmark	29 TA 20.0
2693 Yan'An	29 TA 24.2
1544 Vinterhansenia	29 TA 32.8
1044 Teutonia	29 TA 36.7
1344 Caubeta	29 TA 39.8
1041 Asta	29 TA 59.4
1600 Vyssotsky	00 GE 04.9
1522 Kokkola	00 GE 10.5
1941 Wild	00 GE 16.9
2185 Guangdong	00 GE 25.4
2240 Tsai	00 GE 26.8
886 Washingtonia	00 GE 31.9
1713 Bancilhon	00 GE 46.4
425 Cornelia	00 GE 59.6
753 Tiflis	00 GE 59.9
245 Vera	01 GE 06.8
1619 Ueta	01 GE 06.8
2332 Kalm	01 GE 07.3
2341 Aoluta	01 GE 10.0
1701 Okavango	01 GE 12.2
293 Brasilia	01 GE 12.5
2725 David Bender	01 GE 35.0
2837 Griboedov	01 GE 35.8
157 Dejanira	01 GE 36.9
278 Paulina	01 GE 39.7
1580 Betulia	01 GE 42.2
2376 Martynov	01 GE 48.7
936 Kunigunde	01 GE 50.4
2312 Duboshin	01 GE 51.1
106 Dione	01 GE 59.0
2488 Bryan	02 GE 00.3

Name	Position	Name	Position
963 Iduberga	02 GE 02.6	2273 Yarilo	06 GE 01.6
2633 Bishop	02 GE 02.8	2081 Sazava	06 GE 05.5
1534 Nasi	02 GE 07.6	2269 Efremiana	06 GE 26.6
2149 Schwambraniya	02 GE 11.5	481 Emita	06 GE 31.0
1305 Pongola	02 GE 40.7	2571 Geisei	06 GE 33.4
2695 Christabel	02 GE 44.9	23 Thalia	06 GE 41.5
1122 Neith	02 GE 58.1	2681 Ostrovskij	06 GE 46.3
692 Hippodamia	03 GE 02.3	3215 Lapko	06 GE 49.0
1897 Hind	03 GE 03.5	858 El Djezair	06 GE 54.7
1866 Sisyphus	03 GE 04.7	1261 Legia	06 GE 56.9
1480 Aunus	03 GE 15.1	621 Werdandi	06 GE 59.6
1218 Aster	03 GE 25.9	394 Arduina	06 GE 59.6
2206 Gabrova	03 GE 31.3	2812 Scaltriti	07 GE 44.9
1920 Sarmiento	03 GE 31.5	3007 Reaves	08 GE 27.5
116 Sirona	03 GE 31.6	9 Metis	08 GE 28.9
772 Tanete	03 GE 32.6	1850 Kohoutek	08 GE 31.3
1517 Beograd	03 GE 39.1	725 Amanda	08 GE 33.0
991 McDonalda	03 GE 46.3	503 Evelyn	08 GE 40.0
1364 Safara	03 GE 46.6	2024 McLaughlin	08 GE 52.3
2593 Buryatia	03 GE 49.4	2077 Kiangsu	09 GE 01.7
2699 Kalinin	03 GE 51.9	1594 Danjon	09 GE 03.5
1028 Lydina	03 GE 53.3	2512 Tavastia	09 GE 05.8
1359 Prieska	03 GE 58.3	423 Diotima	09 GE 10.0
742 Edisona	04 GE 10.4	940 Kordula	09 GE 13.3
1893 Jakoba	04 GE 17.1	3327 Campins	09 GE 17.3
3194 Dorsey	04 GE 17.8	946 Poesia	09 GE 18.5
2652 Yabuuti	04 GE 18.0	2797 Teucer	09 GE 24.7
3164 Prast	04 GE 24.4	750 Oskar	09 GE 28.9
363 Padua	04 GE 30.1	1950 Wempe	09 GE 31.6
2003 Harding	04 GE 32.6	1356 Nyanza	09 GE 35.4
1550 Tito	04 GE 43.1	1693 Hertzsprung	09 GE 46.3
529 Preziosa	05 GE 01.0	1890 Konoshenkova	09 GE 58.4
1929 Kollaa	05 GE 04.6	596 Scheila	10 GE 31.3
131 Vala	05 GE 14.8	562 Salome	10 GE 32.2
1412 Lagrula	05 GE 41.5	90 Antiope	10 GE 33.1
22 Kalliope	05 GE 48.4	818 Kapteynia	10 GE 33.9
235 Carolina	05 GE 49.9	564 Dudu	10 GE 35.3

Name	Position	Name	Position
2215 Sichuan	10 GE 48.4	2828 IkuTurso	14 GE 01.0
1765 Wrubel	10 GE 49.6	1953 Rupertwildt	14 GE 05.2
1482 Sebastiana	10 GE 51.6	121 Hermione	14 GE 05.3
942 Romilda	11 GE 00.7	2046 Leningrad	14 GE 14.8
1089 Tama	11 GE 03.6	1601 Patry	14 GE 16.0
1968 Mehltretter	11 GE 15.0	1773 Rumpelstilz	14 GE 19.2
2169 Taiwan	11 GE 29.7	1883 Rimito	14 GE 24.5
1185 Nikko	11 GE 29.9	1450 Raimonda	14 GE 28.5
1320 Impala	11 GE 37.3	2999 Dante	14 GE 32.2
447 Valentine	11 GE 37.4	1259 Ogyalla	14 GE 41.1
1303 Luthera	11 GE 47.4	1602 Indiana	14 GE 43.4
1483 Hakoila	11 GE 49.3	3057 Malaren	14 GE 54.8
1140 Crimea	11 GE 51.6	977 Philippa	15 GE 33.5
785 Zwetana	11 GE 54.6	593 Titania	15 GE 45.7
3174 Alcock	11 GE 55.3	2288 Karolinum	15 GE 49.6
553 Kundry	11 GE 58.2	1032 Pafuri	15 GE 55.3
2050 Francis	12 GE 02.5	2270 Yazhi	15 GE 58.1
2611 Boyce	12 GE 03.0	1912 Anubis	16 GE 01.3
2092 Sumiana	12 GE 05.3	128 Nemesis	16 GE 03.6
196 Philomela	12 GE 06.2	2569 Madeline	16 GE 06.7
1115 Sabauda	12 GE 07.9	144 Vibilia	16 GE 08.3
966 Muschi	12 GE 09.6	455 Bruchsalia	16 GE 08.3
1831 Nicholson	12 GE 11.7	2201 Oljato	16 GE 23.6
2219 Mannucci	12 GE 18.7	2035 Stearns	16 GE 30.2
1792 Reni	12 GE 31.8	1761 Edmondson	16 GE 36.8
1161 Thessalia	12 GE 37.0	164 Eva	16 GE 37.0
1484 Postrema	12 GE 39.7	3234 Hergiani	16 GE 39.5
1012 Sarema	12 GE 45.7	145 Adeona	17 GE 04.4
1239 Queteleta	12 GE 46.8	1397 Umtata	17 GE 09.8
3262 Miune	12 GE 55.6	2421 Nininger	17 GE 29.1
2038 Bistro	13 GE 05.5	1801 Titicaca	17 GE 30.0
87 Sylvia	13 GE 05.9	3344 Modena	17 GE 33.8
279 Thule	13 GE 09.7	1047 Geisha	17 GE 52.4
267 Tirza	13 GE 32.0	1137 Raissa	18 GE 02.7
2617 Jiangxi	13 GE 38.5	2133 Franceswright	18 GE 11.2
1578 Kirkwood	13 GE 39.9	1952 Hesburgh	18 GE 11.6
1460 Haltia	13 GE 57.8	1427 Ruvuma	18 GE 15.7

Name	Position	Name	Position
751 Faina	18 GE 21.4	3041 Webb	22 GE 08.0
1745 Ferguson	18 GE 25.9	2132 Zhukov	22 GE 11.5
1805 Dirikis	18 GE 37.4	728 Leonisis	22 GE 14.5
1937 Locarno	18 GE 38.9	857 Glasenappia	22 GE 30.6
2874 Jim Young	18 GE 45.5	579 Sidonia	22 GE 31.4
2639 Planman	18 GE 51.7	1154 Astronomia	22 GE 37.3
3173 McNaught	18 GE 54.1	1300 Marcelle	22 GE 39.0
1248 Jugurtha	19 GE 02.3	1663 Van Den Bos	22 GE 51.5
2208 Pushkin	19 GE 06.2	3172 Hirst	22 GE 57.2
2470 Agematsu	19 GE 12.2	367 Amicitia	23 GE 02.3
2904 Millman	19 GE 18.3	1680 Per Brahe	23 GE 06.2
2301 Whitford	19 GE 26.6	3198 Wallonia	23 GE 09.0
2707 Ueferji	19 GE 29.6	971 Alsatia	23 GE 16.9
1123 Shapleya	19 GE 29.9	1357 Khama	23 GE 22.1
776 Berbericia	19 GE 32.2	2519 Annagerman	23 GE 23.2
767 Bondia	19 GE 44.5	146 Lucina	23 GE 43.7
2779 Mary	19 GE 51.2	471 Papagena	23 GE 51.4
222 Lucia	19 GE 58.5	237 Coelestina	23 GE 56.6
3116 Goodricke	19 GE 58.9	2906 Caltech	24 GE 03.5
782 Montefiore	20 GE 01.6	42 Isis	24 GE 06.5
1 Ceres	20 GE 02.4	2172 Plavsk	24 GE 10.1
566 Stereoskopia	20 GE 09.4	535 Montague	24 GE 19.1
1091 Spiraea	20 GE 21.1	2417 McVittie	24 GE 19.1
1083 Salvia	20 GE 25.1	3277 Aaronson	24 GE 30.8
21 Lutetia	20 GE 26.8	488 Kreusa	24 GE 33.4
1667 Pels	20 GE 27.1	2073 Janacek	24 GE 41.3
2877 Likhachev	20 GE 37.1	752 Sulamitis	24 GE 43.1
311 Claudia	20 GE 44.1	2721 Vsekhsvyatskij	24 GE 44.9
1294 Antwerpia	20 GE 56.6	2245 Hekatostos	24 GE 49.3
1096 Reunerta	20 GE 57.4	1549 Mikko	24 GE 51.8
3449 Abell	21 GE 04.8	563 Suleika	24 GE 58.5
2984 Chaucer	21 GE 20.0	2511 Patterson	25 GE 07.8
1511 Dalera	21 GE 25.5	908 Buda	25 GE 09.5
197 Arete	21 GE 25.7	347 Pariana	25 GE 20.3
967 Helionape	21 GE 59.0	2909 HosheNoIe	25 GE 20.3
1886 Lowell	22 GE 01.1	449 Hamburga	25 GE 32.7
2476 Andersen	22 GE 04.4	2870 Haupt	25 GE 35.0

Name	Position	Name	Position
2445 Blazhko	25 GE 41.9	2268 Szmytowna	29 GE 20.5
1054 Forsytia	25 GE 47.0	1209 Pumma	29 GE 22.4
652 Jubilatrix	25 GE 47.5	786 Bredichina	29 GE 29.0
2894 Kakhovka	25 GE 50.5	521 Brixia	29 GE 29.8
2260 Neoptolemus	25 GE 55.4	2176 Donar	29 GE 45.5
1771 Makover	25 GE 58.5	348 May	29 GE 49.3
1934 Jeffers	26 GE 10.7	1302 Werra	29 GE 51.8
14 Irene	26 GE 12.6	350 Ornamenta	29 GE 55.2
974 Lioba	26 GE 13.9	2082 Galahad	29 GE 56.8
2525 O'Steen	26 GE 35.8	1278 Kenya	29 GE 59.2
1464 Armisticia	26 GE 36.5	1589 Fanatica	00 CN 01.7
86 Semele	26 GE 42.7	1591 Baize	00 CN 05.0
1304 Arosa	26 GE 58.2	1634 Ndola	00 CN 18.5
2336 Xinjiang	27 GE 01.5	1976 Kaverin	00 CN 28.7
2568 Maksutov	27 GE 07.4	505 Cava	00 CN 34.4
259 Aletheia	27 GE 10.3	1156 Kira	00 CN 43.3
629 Bernardina	27 GE 11.6	866 Fatme	00 CN 53.3
2587 Gardner	27 GE 19.6	1947 IsoHeikkila	01 CN 08.0
1348 Michel	27 GE 23.0	1665 Gaby	01 CN 09.4
1566 Icarus	27 GE 29.9	2841 Puijo	01 CN 10.3
2398 Jilin	27 GE 43.5	1061 Paeonia	01 CN 15.3
1965 Van De Kamp	27 GE 50.9	1556 Wingolfia	01 CN 32.0
2225 Serkowski	27 GE 52.3	598 Octavia	01 CN 41.1
2655 Guangxi	27 GE 55.4	346 Hermentaria	01 CN 47.6
691 Lehigh	27 GE 55.4	404 Arsinoe	02 CN 07.6
1180 Rita	27 GE 57.8	452 Hamiltonia	02 CN 13.1
1772 Gargarin	28 GE 04.1	274 Philagoria	02 CN 24.8
3032 Evans	28 GE 07.0	986 Amelia	02 CN 34.4
814 Tauris	28 GE 14.1	383 Janina	02 CN 46.9
3507 Vilas	28 GE 14.2	2287 Kalmykia	02 CN 52.3
3165 Mikawa	28 GE 16.8	2822 Sakajawea	03 CN 00.7
1817 Katanga	28 GE 19.4	3117 Niepce	03 CN 01.2
3417 Tamblyn	28 GE 22.3	1687 Glarona	03 CN 14.0
432 Pythia	28 GE 24.9	1926 Demiddelaer	03 CN 14.7
451 Patientia	29 GE 00.1	1078 Mentha	03 CN 26.1
199 Byblis	29 GE 02.4	2816 Pien	03 CN 33.3
1445 Konkolya	29 GE 09.5	1582 Martir	03 CN 34.0

Name	Position	Name	Position
2102 Tantalus	03 CN 43.1	498 Tokio	07 CN 06.9
40 Harmonia	03 CN 47.3	1487 Boda	07 CN 11.6
486 Cremona	03 CN 47.5	1270 Datura	07 CN 26.6
534 Nassovia	03 CN 51.8	3061 Cook	07 CN 37.7
369 Aeria	03 CN 57.8	2372 Proskurin	07 CN 52.6
1681 Steinmetz	04 CN 06.8	3459 Bodil	08 CN 11.2
27 Euterpe	04 CN 21.6	1588 Descamisada	08 CN 29.9
1504 Lappeenranta	04 CN 23.6	2636 Lassell	08 CN 30.8
1901 Moravia	04 CN 26.2	1010 Marlene	08 CN 34.4
913 Otila	04 CN 31.6	424 Gratia	08 CN 54.7
2220 Hicks	04 CN 36.5	351 Yrsa	08 CN 58.3
2222 Lermontov	04 CN 38.5	2395 Aho	08 CN 59.3
730 Athanasia	04 CN 41.7	580 Selene	08 CN 59.3
380 Fiducia	04 CN 50.5	2344 Xizang	09 CN 02.2
1784 Benguella	04 CN 51.5	1052 Belgica	09 CN 12.6
1063 Aquilegia	04 CN 56.6	1809 Prometheus	09 CN 15.1
2058 Roka	05 CN 05.7	1844 Susilva	09 CN 16.3
1658 Innes	05 CN 10.0	1569 Evita	09 CN 16.7
2087 Kochera	05 CN 12.4	2462 Nehalennia	09 CN 20.3
1852 Carpenter	05 CN 16.0	171 Ophelia	10 CN 08.0
2688 Halley	05 CN 16.3	1131 Poziar	10 CN 19.9
2600 Lumme	05 CN 19.1	741 Botolphia	10 CN 21.5
2656 Evenkia	05 CN 32.5	1529 Oterma	10 CN 23.2
2039 PayneGaposchkin		1196 Sheba	10 CN 32.0
	05 CN 32.9	2195 Tengstrom	10 CN 33.2
1674 Groeneveld	05 CN 33.1	2873 Binzel	10 CN 33.3
2068 Dangreen	05 CN 39.8	1075 Helina	10 CN 34.5
2621 Goto	05 CN 40.1	307 Nike	10 CN 48.2
1417 Walinskia	05 CN 58.1	825 Tanina	10 CN 59.0
2013 Tucapel	06 CN 10.5	3310 Patsy	11 CN 22.7
2218 Woltho	06 CN 13.9	1326 Losaka	11 CN 28.2
261 Prymno	06 CN 14.3	92 Undina	11 CN 30.8
700 Auravictrix	06 CN 25.6	2171 Kiev	11 CN 38.3
410 Chloris	06 CN 46.0	1789 Dobrovolsky	11 CN 39.5
1125 China	06 CN 48.1	2769 Mendeleev	11 CN 56.6
323 Brucia	06 CN 56.8	3128 Obruchev	11 CN 58.0
1336 Zeelandia	07 CN 05.5	464 Megaira	12 CN 10.7

Name	Position	Name	Position
2089 Cetacea	12 CN 15.5	1166 Sakuntala	16 CN 17.9
581 Tautonia	12 CN 20.0	2244 Tesla	16 CN 18.1
353 RupertoCarola	12 CN 22.6	2127 Tanya	16 CN 26.5
2510 Shandong	12 CN 30.2	1963 Bezovec	16 CN 28.0
1618 Dawn	12 CN 36.1	1210 Morosovia	16 CN 34.7
2482 Perkin	12 CN 46.6	1778 Alfven	16 CN 40.0
638 Moira	13 CN 06.1	512 Taurinensis	16 CN 41.9
1391 Carelia	13 CN 17.2	1847 Stobbe	16 CN 42.4
4 Vesta	13 CN 22.2	182 Elsa	16 CN 48.6
1702 Kalahari	13 CN 38.6	140 Siwa	16 CN 49.3
1056 Azalea	13 CN 44.5	3352 McAuliffe	16 CN 54.2
413 Edburga	13 CN 57.4	1551 Argelander	16 CN 55.1
2531 Cambridge	14 CN 01.3	3286 Anatoliya	17 CN 06.5
504 Cora	14 CN 20.5	2009 Voloshina	17 CN 08.8
1581 Abanderada	14 CN 26.3	411 Xanthe	17 CN 14.7
1461 JeanJacques	14 CN 35.6	1341 Edmee	17 CN 15.7
3467 Bernheim	14 CN 57.6	511 Davida	17 CN 19.3
2277 Moreau	14 CN 58.1	532 Herculina	17 CN 23.5
462 Eriphyla	14 CN 58.2	2784 Domeyko	17 CN 39.8
1657 Roemera	15 CN 00.6	3216 Harrington	17 CN 41.0
2122 Pyatiletka	15 CN 01.2	3356 Resnik	17 CN 53.3
560 Delila	15 CN 03.2	548 Kressida	18 CN 01.7
1609 Brenda	15 CN 03.9	2062 Aten	18 CN 02.2
364 Isara	15 CN 07.3	2938 Hopi	18 CN 41.0
1141 Bohmia	15 CN 09.5	3318 Blixen	18 CN 49.5
2614 Torrence	15 CN 11.2	3201 Sijthoff	19 CN 02.5
630 Euphemia	15 CN 12.8	2804 Yrjo	19 CN 07.8
2138 Swissair	15 CN 14.0	2627 Churyumov	19 CN 11.3
1684 Iguassu	15 CN 17.8	749 Malzovia	19 CN 25.0
891 Gunhild	15 CN 31.9	644 Cosima	19 CN 28.4
758 Mancunia	15 CN 47.5	3062 Wren	20 CN 09.2
590 Tomyris	15 CN 53.6	887 Alinda	20 CN 13.4
1988 Delores	15 CN 55.7	1449 Virtanen	20 CN 21.4
1592 Mathieu	15 CN 58.1	3119 Dobronravin	20 CN 26.6
2280 Kunikov	16 CN 04.4	1107 Lictoria	20 CN 27.8
1904 Massevitch	16 CN 05.9	2924 MitakeMura	20 CN 28.7
412 Elisabetha	16 CN 06.8	8 Flora	20 CN 29.0

Name	Position
2404 Antarctica	20 CN 30.0
1558 Jarnefelt	20 CN 35.3
1553 Bauersfelda	20 CN 36.2
618 Elfriede	20 CN 42.3
414 Liriope	20 CN 48.8
1815 Beethoven	20 CN 49.5
1655 Comas Sola	20 CN 53.0
3197 Weissman	21 CN 20.3
1022 Olympiada	21 CN 35.3
2708 Burns	21 CN 38.9
628 Christine	21 CN 40.6
559 Nanon	21 CN 40.6
2075 Martinez	21 CN 44.3
1595 Tanga	21 CN 49.7
1703 Barry	21 CN 50.4
2956 Yeomans	21 CN 57.2
679 Pax	22 CN 10.5
2338 Bokhan	22 CN 18.2
931 Whittemora	22 CN 20.0
2163 Korczak	22 CN 53.4
2439 Ulugbek	22 CN 53.6
2720 Pyotr Pervyi	23 CN 03.3
1220 Crocus	23 CN 07.4
3132 Landgraf	23 CN 08.5
1925 FranklinAdams	23 CN 14.8
1758 Naantali	23 CN 22.4
943 Begonia	23 CN 43.0
2563 Boyarchuk	24 CN 15.6
1276 Ucclia	24 CN 16.3
2091 Sampo	24 CN 20.6
487 Venetia	24 CN 34.8
861 Aida	24 CN 39.1
1333 Cevenola	24 CN 41.8
1385 Gelria	24 CN 43.7
240 Vanadis	24 CN 44.6
1633 Chimay	24 CN 46.5
2891 McGetchin	24 CN 53.3
1623 Vivian	25 CN 18.7
1021 Flammario	25 CN 22.4
1428 Mombasa	25 CN 29.8
2164 Lyalya	25 CN 33.8
868 Lova	25 CN 34.8
740 Cantabia	25 CN 44.3
877 Walkure	25 CN 55.1
863 Benkoela	26 CN 34.4
1880 McCrosky	26 CN 34.9
1548 Palomaa	26 CN 40.5
2066 Palala	26 CN 44.6
1105 Fragaria	26 CN 50.9
2450 Ioannisiani	26 CN 54.2
522 Helga	26 CN 54.4
431 Nephele	27 CN 01.6
3082 Dzhalil	27 CN 16.6
3098 Van Sprang	27 CN 22.4
1431 Luanda	27 CN 24.4
2085 Henan	27 CN 49.3
2237 Melnikov	28 CN 02.2
820 Adriana	28 CN 21.1
2443 Tomeileen	28 CN 25.6
1017 Jacqueline	28 CN 36.1
2887 Krinov	28 CN 37.9
938 Chlosinde	28 CN 45.6
1725 CrAO	28 CN 54.4
2815 Soma	29 CN 25.9
1593 Fagnes	29 CN 37.9
1646 Rosseland	29 CN 38.9
2597 Arthur	29 CN 54.1
2553 Viljev	29 CN 55.3
288 Glauke	00 LE 05.5
537 Pauly	00 LE 09.1
1914 Hartbeespoortdam	00 LE 12.5
527 Euryanthe	00 LE 15.9
2863 Ben Mayer	00 LE 19.9

Name	Position	Name	Position
870 Manto	00 LE 25.1	11 Parthenope	05 LE 04.1
268 Adorea	00 LE 30.5	817 Annika	05 LE 11.0
2334 Cuffey	00 LE 33.4	1966 Tristan	05 LE 13.4
3129 Bonestell	00 LE 36.2	62 Erato	05 LE 15.1
1015 Christa	00 LE 41.2	1586 Thiele	05 LE 22.3
1362 Griqua	00 LE 49.0	1748 Mauderli	05 LE 35.8
2959 Scholi	00 LE 56.9	745 Mauritia	05 LE 39.5
850 Altona	01 LE 00.0	3338 Richter	05 LE 42.1
2933 Amber	01 LE 00.5	1516 Henry	05 LE 48.4
2733 Hamina	01 LE 03.2	2556 Louise	06 LE 02.3
296 Phaetusa	01 LE 08.6	3478 Fanale	06 LE 02.3
1216 Askania	01 LE 09.5	2410 Morrison	06 LE 05.2
515 Athalia	01 LE 19.0	2161 Grissom	06 LE 14.5
1331 Solvejg	01 LE 19.1	2213 Meeus	06 LE 30.2
3106 Morabito	01 LE 38.2	2651 Karen	06 LE 32.6
213 Lilaea	01 LE 40.1	3010 Ushakov	06 LE 46.5
1819 Laputa	02 LE 19.8	2682 Soromundi	06 LE 50.3
1607 Mavis	02 LE 28.6	472 Roma	06 LE 53.2
2742 Gibson	02 LE 29.9	415 Palatia	06 LE 53.6
1171 Rusthawelia	02 LE 33.1	484 Pittsburghia	07 LE 00.1
3123 Dunham	02 LE 39.5	2394 Nadeev	07 LE 07.7
1432 Ethiopia	02 LE 55.6	1271 Isergina	07 LE 13.3
2516 Roman	02 LE 56.3	100 Hekate	07 LE 14.3
113 Amalthea	03 LE 01.6	625 Xenia	07 LE 15.4
2210 Lois	03 LE 04.8	1720 Niels	07 LE 20.0
1689 FlorisJan	03 LE 06.7	2844 Hess	07 LE 20.9
1577 Reiss	03 LE 10.0	816 Juliana	07 LE 30.0
1215 Boyer	03 LE 17.1	1163 Saga	07 LE 42.1
316 Goberta	03 LE 38.4	2217 Eltigen	07 LE 48.0
1800 Aguilar	03 LE 46.9	3209 Buchwald	07 LE 48.1
888 Parysatis	03 LE 48.0	387 Aquitania	07 LE 54.3
2310 Olshaniya	03 LE 58.1	1455 Mitchella	07 LE 57.4
1262 Sniadeckia	04 LE 07.5	2405 Welch	08 LE 09.6
729 Watsonia	04 LE 12.1	2162 Anhui	08 LE 10.5
381 Myrrha	04 LE 57.4	3210 Lupiskho	08 LE 12.8
856 Backlunda	05 LE 01.4	2625 Jack London	08 LE 18.2
17 Thetis	05 LE 02.4	2189 Zaragoza	08 LE 23.3

Name	Position	Name	Position
1127 Mimi	08 LE 24.3	502 Sigune	12 LE 29.2
2672 Pisek	08 LE 35.7	1727 Mette	12 LE 29.7
530 Turandot	08 LE 37.1	727 Nipponia	12 LE 37.2
52 Europa	08 LE 37.5	1627 Ivar	12 LE 39.3
1104 Syringa	08 LE 46.8	653 Berenike	12 LE 45.2
166 Rhodope	08 LE 48.7	2349 Kurchenko	12 LE 49.2
1312 Vassar	08 LE 58.7	2265 Verbaandert	12 LE 55.1
2802 Weisell	09 LE 02.0	1334 Lundmarka	13 LE 03.2
1562 Gondolatsch	09 LE 03.2	634 Ute	13 LE 04.4
402 Chloe	09 LE 08.5	2722 Abalakin	13 LE 08.3
928 Hildrun	09 LE 22.6	2895 Memnon	13 LE 21.4
791 Ani	09 LE 30.0	1097 Vicia	13 LE 27.1
1211 Bressole	09 LE 35.4	662 Newtonia	13 LE 27.3
747 Winchester	09 LE 35.4	1624 Rabe	13 LE 27.6
655 Briseis	09 LE 44.3	2041 Lancelot	13 LE 40.7
1603 Neva	09 LE 45.8	2714 Matti	13 LE 45.1
2507 Bobone	09 LE 46.4	275 Sapientia	13 LE 45.7
1648 Shajna	10 LE 02.5	159 Aemilia	13 LE 47.5
1951 Lick	10 LE 08.4	2634 James Bradley	13 LE 50.4
334 Chicago	10 LE 08.6	3125 Hay	13 LE 57.4
2772 Dugan	10 LE 13.0	2160 Spitzer	14 LE 02.1
2960 Ohtaki	10 LE 21.2	3071 Nesterov	14 LE 03.8
555 Norma	10 LE 23.8	1269 Rollandia	14 LE 19.5
2411 Zellner	10 LE 33.9	2622 Bolzano	14 LE 24.8
811 Nauheima	10 LE 36.2	458 Hercynia	14 LE 25.8
2146 Stentor	10 LE 37.2	1191 Alfaterna	14 LE 26.7
44 Nysa	11 LE 03.4	442 Eichsfeldia	14 LE 27.2
1329 Eliane	11 LE 42.8	2086 Newell	14 LE 42.2
738 Alagasta	11 LE 53.5	1711 Sandrine	14 LE 43.0
807 Ceraskia	11 LE 58.6	226 Weringia	14 LE 44.5
1011 Laodamia	12 LE 08.1	2864 Soderblom	14 LE 51.8
360 Carlova	12 LE 08.6	1903 Adzhimushkaj	14 LE 59.6
1111 Reinmuthia	12 LE 15.8	2918 Salazar	15 LE 17.6
1629 Pecker	12 LE 15.9	3317 Paris	15 LE 18.8
889 Erynia	12 LE 22.8	736 Harvard	15 LE 27.8
2612 Kathryn	12 LE 26.5	1513 Matra	15 LE 44.5
1860 Barbarossa	12 LE 27.3	294 Felicia	15 LE 45.4

Name	Position	Name	Position
103 Hera	15 LE 50.0	3355 Onizuka	20 LE 25.7
2690 Ristiina	15 LE 56.6	1252 Celestia	20 LE 32.4
479 Caprera	15 LE 56.8	2350 Von Lude	20 LE 37.9
129 Antigone	15 LE 57.4	851 Zeissia	20 LE 43.1
2271 Kiso	16 LE 09.9	2392 Jonathan Murray	
2585 Irpedina	16 LE 09.9		20 LE 48.9
739 Mandeville	16 LE 22.9	933 Susi	20 LE 58.7
3208 Lunn	16 LE 35.2	538 Friederike	20 LE 59.3
330 Adalberta	16 LE 38.2	5 Astraea	21 LE 08.0
202 Chryseis	16 LE 39.2	306 Unitas	21 LE 27.8
2187 La Silla	17 LE 00.8	824 Anastasia	21 LE 28.3
1345 Potomac	17 LE 02.0	221 Eos	21 LE 37.7
1492 Oppolzer	17 LE 19.4	3488 Brahic	21 LE 38.3
526 Jena	17 LE 23.6	183 Istria	21 LE 39.7
357 Ninina	17 LE 52.1	783 Nora	21 LE 43.0
781 Kartvelia	18 LE 04.3	622 Esther	21 LE 51.9
2144 Marietta	18 LE 14.2	287 Nephthys	21 LE 53.4
6 Hebe	18 LE 21.7	1069 Planckia	22 LE 06.5
2596 Vainu Bappu	18 LE 46.3	301 Bavaria	22 LE 07.4
1528 Conrada	18 LE 52.4	744 Aguntina	22 LE 13.1
1003 Lilofee	18 LE 58.4	3341 Hartmann	22 LE 14.5
600 Musa	19 LE 00.8	627 Charis	22 LE 14.7
1142 Aetolia	19 LE 02.2	1802 Zhang Heng	22 LE 18.6
1350 Rosselia	19 LE 15.0	1845 Helewalda	22 LE 21.1
2279 Barto	19 LE 34.7	2567 Elba	22 LE 21.1
2441 Hibbs	19 LE 39.3	2929 Harris	22 LE 22.0
1085 Amaryllis	19 LE 40.9	1945 Wesselink	22 LE 22.6
2325 Chernykh	19 LE 43.9	1539 Borrelly	22 LE 34.1
710 Gertrud	19 LE 44.4	3040 Kozal	22 LE 55.5
2110 MooreSitterly	19 LE 57.4	53 Kalypso	23 LE 24.7
2228 SoyuzApollo	19 LE 57.8	1414 Jerome	23 LE 25.9
354 Eleonora	20 LE 02.6	1076 Viola	23 LE 26.9
2188 Orlenok	20 LE 05.0	558 Carmen	23 LE 29.5
2297 Daghestan	20 LE 09.6	2253 Espinette	23 LE 31.4
2409 Chapman	20 LE 18.6	181 Eucharis	23 LE 32.8
1251 Hedera	20 LE 23.0	1019 Strackea	23 LE 53.3
2330 Ontake	20 LE 23.7	2710 Veverka	23 LE 59.6

Name	Position	Name	Position
28 Bellona	24 LE 01.1	45 Eugenia	27 LE 28.9
2845 Franklinken	24 LE 01.4	1090 Sumida	27 LE 34.5
234 Barbara	24 LE 04.9	2249 Yamamoto	27 LE 36.1
2846 Ylppo	24 LE 24.7	1082 Pirola	27 LE 44.9
282 Clorinde	24 LE 29.6	1999 Hirayama	27 LE 57.2
2147 Kharadze	24 LE 29.9	335 Roberta	27 LE 58.8
780 Armenia	24 LE 38.9	1217 Maximiliana	27 LE 59.9
893 Leopoldina	24 LE 41.0	173 Ino	28 LE 02.4
148 Gallia	24 LE 45.8	2595 Gudiachvili	28 LE 11.8
1820 Lohmann	24 LE 46.2	2416 Sharonov	28 LE 21.8
2732 Witt	24 LE 50.9	2316 JoAnn	28 LE 26.1
206 Hersilia	24 LE 52.7	2979 Murmansk	28 LE 30.6
1871 Astyanax	25 LE 04.7	2589 Daniel	28 LE 35.8
1148 Rarahu	25 LE 15.3	2020 Ukko	28 LE 36.6
2121 Sevastopol	25 LE 16.3	2136 Jugta	28 LE 38.9
962 Aslog	25 LE 16.4	1094 Siberia	28 LE 39.9
1649 Fabre	25 LE 17.0	2084 Okayama	28 LE 41.1
130 Elektra	25 LE 17.7	885 Ulrike	28 LE 43.9
2397 Lappajarvi	25 LE 23.2	286 Iclea	28 LE 44.0
1514 Ricouxa	25 LE 23.4	2879 Shimizu	29 LE 00.2
2284 San Juan	25 LE 29.8	1212 Francette	29 LE 07.7
3420 Standish	25 LE 47.5	2921 Sophocles	29 LE 10.8
1568 Aisleen	25 LE 47.7	1723 Klemola	29 LE 35.2
716 Berkeley	25 LE 49.5	16 Psyche	29 LE 52.3
2399 Terradas	25 LE 51.1	1585 Union	29 LE 58.0
3054 Strugaltskia	26 LE 03.1	873 Mechthild	29 LE 59.0
2542 Calpurnia	26 LE 06.1	2927 Alamosa	00 VI 02.0
2616 Lesya	26 LE 16.6	18 Melpomene	00 VI 03.2
2447 Kronstadt	26 LE 24.9	676 Melitta	00 VI 03.4
3270 Dudley	26 LE 29.5	2381 Landi	00 VI 03.7
3315 Chant	26 LE 30.4	924 Toni	00 VI 08.0
2199 Klet	26 LE 34.6	2067 Aksnes	00 VI 19.4
909 Ulla	26 LE 35.9	2379 Heiskanen	00 VI 35.3
1055 Tynka	26 LE 42.8	2209 Tianjin	00 VI 38.7
1117 Reginita	26 LE 47.0	2684 Douglas	00 VI 45.1
1830 Pogson	27 LE 01.3	3222 Liller	00 VI 51.5
633 Zelima	27 LE 03.4	876 Scott	00 VI 53.2

Name	Position	Name	Position
317 Roxane	00 VI 55.9	594 Mireille	04 VI 40.3
3291 Dunlap	00 VI 58.2	1617 Alschmitt	04 VI 43.4
2731 Cucula	01 VI 00.2	1489 Attila	04 VI 44.1
2741 Valdivia	01 VI 03.0	869 Mellena	04 VI 52.0
1245 Canvinia	01 VI 30.5	2142 Landau	04 VI 52.0
2106 Hugo	01 VI 35.6	2246 Bowell	05 VI 19.1
1907 Rudneva	01 VI 47.2	582 Olympia	05 VI 19.4
1764 Cogshall	01 VI 51.8	65 Cybele	05 VI 24.0
1742 Schaifers	01 VI 58.1	1732 Heike	05 VI 29.3
3362 Khufu	02 VI 01.3	2216 Kerch	05 VI 35.0
3072 Vilnius	02 VI 08.3	2326 Tololo	05 VI 37.0
2469 Tadjikistan	02 VI 10.1	251 Sophia	05 VI 53.0
232 Russia	02 VI 10.9	269 Justitia	06 VI 20.0
810 Atossa	02 VI 14.9	1799 Koussevitzky	06 VI 28.5
1434 Margot	02 VI 21.7	1164 Kobolda	06 VI 29.3
1731 Smuts	02 VI 34.8	1533 Saimaa	06 VI 35.1
2723 Gorshkov	02 VI 41.2	201 Penelope	06 VI 38.9
1384 Kniertje	02 VI 44.5	660 Crescentia	06 VI 40.0
667 Denise	02 VI 45.4	39 Laetitia	06 VI 45.1
542 Susanna	02 VI 46.4	1403 Idelsonia	06 VI 46.3
2385 Mustel	02 VI 48.2	1755 Lorbach	06 VI 49.8
1615 Bradwell	02 VI 48.4	2659 Millis	06 VI 53.3
2318 Lubarsky	02 VI 49.0	1782 Schneller	06 VI 57.8
1816 Liberia	02 VI 52.6	1144 Oda	06 VI 59.0
2564 Kayala	03 VI 03.0	2998 Berendeya	06 VI 59.2
1842 Hynek	03 VI 03.2	461 Saskia	07 VI 02.5
1004 Belopolskya	03 VI 20.4	2048 Dwornik	07 VI 06.5
185 Eunike	03 VI 27.5	2309 Mr. Spock	07 VI 11.1
2230 Yunnan	03 VI 33.6	2285 Ron Helin	07 VI 24.2
1956 Artek	03 VI 35.0	1283 Komsomolia	07 VI 24.6
2724 Orlov	03 VI 35.5	2711 Aleksandrov	07 VI 32.8
2814 Vieira	03 VI 45.0	2250 Stalingrad	07 VI 33.9
809 Lundia	04 VI 08.4	871 Amneris	07 VI 36.4
3101 Goldberger	04 VI 22.4	2602 Moore	07 VI 47.2
2093 Genichesk	04 VI 30.1	2311 El Leoncito	07 VI 52.2
1466 Mundleria	04 VI 32.4	1833 Shmakova	07 VI 54.8
2021 Poincare	04 VI 36.6	1993 Guacolda	07 VI 58.6

Name	Position	Name	Position
1263 Varsavia	08 VI 09.5	315 Constantia	11 VI 14.2
2550 Houssay	08 VI 11.3	474 Prudentia	11 VI 19.7
1120 Cannonia	08 VI 11.8	1788 Kiess	11 VI 23.7
2796 Kron	08 VI 14.2	3516 Rusheva	11 VI 28.3
2396 Kochi	08 VI 22.2	2624 Samitchell	11 VI 33.5
1597 Laugier	08 VI 26.2	3104 Durer	11 VI 49.6
1759 Kienle	08 VI 26.7	2109 Dhotel	11 VI 53.0
1330 Spiridonia	08 VI 27.5	246 Asporina	12 VI 03.4
2207 Antenor	08 VI 31.3	2919 Dali	12 VI 05.6
273 Atropos	08 VI 37.2	2435 Horemheb	12 VI 07.4
304 Olga	08 VI 43.3	1614 Goldschmidt	12 VI 08.6
1733 Silke	08 VI 55.1	2528 Mohler	12 VI 19.3
194 Prokne	09 VI 00.0	1247 Memoria	12 VI 23.2
1043 Beate	09 VI 03.2	1915 Quetzacoatl	12 VI 24.4
191 Kolga	09 VI 05.9	2832 Lada	12 VI 26.8
149 Medusa	09 VI 06.3	217 Eudora	12 VI 37.6
1736 Floirac	09 VI 20.7	1754 Cunningham	12 VI 40.8
2466 Golson	09 VI 20.9	864 Aase	12 VI 41.0
1874 Kacivelia	09 VI 25.1	2949 Kaverznev	12 VI 44.0
903 Nealley	09 VI 26.5	2496 Fernandus	12 VI 45.2
97 Klotho	09 VI 31.8	1025 Riema	12 VI 50.5
2422 Perovskaya	09 VI 32.4	1376 Micelle	13 VI 04.3
163 Erigone	09 VI 48.2	1898 Cowell	13 VI 09.1
561 Ingwelde	09 VI 59.5	723 Hammonia	13 VI 12.2
1337 Gerarda	10 VI 01.7	2166 Handahl	13 VI 31.6
794 Irenaea	10 VI 04.4	954 LI	13 VI 36.3
1762 Russell	10 VI 23.1	1724 Vladimir	13 VI 45.8
2204 Lyyli	10 VI 33.0	114 Kassandra	13 VI 51.1
3124 Kansas	10 VI 37.3	2954 Delsemme	13 VI 59.1
890 Waltraut	10 VI 38.8	2408 Astapovich	14 VI 20.3
1668 Hanna	10 VI 40.2	799 Gudula	14 VI 20.6
58 Concordia	10 VI 51.8	2717 Tellervo	14 VI 27.6
318 Magdalena	11 VI 08.4	1933 Tinchen	14 VI 31.0
291 Alice	11 VI 09.3	2506 Pirogov	14 VI 31.3
1301 Yvonne	11 VI 10.9	1361 Leuschneria	14 VI 36.3
1386 Storeria	11 VI 12.1	2859 Paganini	14 VI 53.1
1399 Teneriffa	11 VI 13.6	1070 Tunica	15 VI 07.4

Name	Position	Name	Position
609 Fulvia	15 VI 18.9	3186 Manuilova	19 VI 30.4
1299 Mertona	15 VI 19.4	59 Elpis	19 VI 43.7
260 Huberta	15 VI 36.0	2202 Pele	19 VI 44.8
2561 Margolin	15 VI 41.7	1178 Irmela	19 VI 45.9
1346 Uotha	15 VI 55.3	3 Juno	19 VI 51.7
805 Hormuthia	15 VI 56.0	314 Rosalia	20 VI 05.0
167 Urda	16 VI 03.2	1975 Pikelner	20 VI 05.9
2586 Matson	16 VI 11.0	2977 Chivilikhin	20 VI 16.1
3033 Holbaek	16 VI 28.0	2100 RaShalom	20 VI 16.9
386 Siegena	16 VI 30.4	218 Bianca	20 VI 19.0
489 Comacina	16 VI 46.3	669 Kypria	20 VI 26.9
1576 Fabiola	16 VI 56.3	2116 Mtskheta	20 VI 28.5
2111 Tselina	17 VI 04.6	3090 Tjossem	20 VI 29.5
2987 Sarabhai	17 VI 04.6	2759 Idomeneus	20 VI 33.2
1465 Autonoma	17 VI 13.4	688 Melanie	20 VI 38.6
2664 Everhart	17 VI 22.7	1221 Amor	20 VI 52.3
2419 Moldavia	17 VI 34.3	1410 Margret	20 VI 53.1
2534 Houzeau	17 VI 43.4	2017 Wesson	20 VI 56.3
689 Zita	17 VI 44.3	2700 Baikonur	21 VI 01.6
1747 Wright	17 VI 49.3	1730 Marceline	21 VI 16.4
1722 Goffin	17 VI 53.0	1938 Lausanna	21 VI 20.0
1636 Porter	17 VI 56.9	2605 Sahade	21 VI 42.5
592 Bathseba	18 VI 00.2	2400 Derevskaya	21 VI 44.8
1974 Caupolican	18 VI 00.5	379 Huenna	21 VI 45.2
3008 Nojiri	18 VI 19.6	358 Apollonia	22 VI 00.2
1811 Bruwer	18 VI 22.1	3094 Chukokkala	22 VI 04.9
3455 Kristensen	18 VI 22.6	2292 Seili	22 VI 07.7
125 Liberatrix	18 VI 47.0	1335 Demoulina	22 VI 08.4
2031 BAM	18 VI 50.5	2907 Nekrasov	22 VI 20.0
2329 Orthos	18 VI 53.6	2535 Hameenlinna	22 VI 20.0
2291 Kevo	19 VI 10.6	827 Wolfiana	22 VI 28.4
2608 Seneca	19 VI 11.5	2 Pallas	22 VI 37.8
2932 Kempchinsky	19 VI 17.3	732 Tjilaki	22 VI 49.8
1379 Lomonossova	19 VI 21.4	470 Kilia	22 VI 53.8
2030 Belyaev	19 VI 22.3	50 Virginia	23 VI 14.9
2995 Taratuta	19 VI 24.0	339 Dorothea	23 VI 23.8
601 Nerthus	19 VI 26.6	1605 Milankovitch	23 VI 28.0

Name	Position	Name	Position
107 Camilla	23 VI 29.9	329 Svea	27 VI 57.2
1691 Oort	23 VI 49.7	2902 Westerlund	27 VI 58.8
2299 Hanko	23 VI 53.1	490 Veritas	27 VI 59.7
3256 Daguerre	24 VI 05.2	1564 Srbija	28 VI 15.5
483 Seppina	24 VI 17.3	788 Hohensteina	28 VI 18.3
1389 Onnie	24 VI 23.0	1095 Tulipa	28 VI 19.6
670 Ottegebe	24 VI 25.3	2002 Euler	28 VI 21.3
2327 Gershberg	24 VI 29.0	122 Gerda	28 VI 25.1
2173 Maresjev	24 VI 30.6	1394 Algoa	28 VI 25.3
1443 Ruppina	24 VI 40.9	1812 Gilgamesh	28 VI 27.8
434 Hungaria	24 VI 48.4	2227 Otto Struve	28 VI 29.3
1451 Grano	24 VI 52.5	1413 Roucarie	28 VI 42.7
1774 Kulikov	24 VI 57.6	2357 Phereclos	28 VI 43.6
443 Photographica	25 VI 00.6	482 Petrina	29 VI 10.9
1656 Suomi	25 VI 02.0	2674 Pandarus	29 VI 11.9
491 Carina	25 VI 13.0	253 Mathilde	29 VI 26.1
664 Judith	25 VI 28.5	754 Malabar	29 VI 44.6
892 Seeligeria	25 VI 30.0	585 Bilkis	29 VI 47.5
51 Nemausa	25 VI 35.0	533 Sara	00 LI 17.5
2460 Mitlincoln	25 VI 40.1	1928 Summa	00 LI 20.6
1870 Glaukos	25 VI 42.1	1672 Gezelle	00 LI 20.8
2354 Lavrov	25 VI 43.5	1020 Arcadia	00 LI 23.2
190 Ismene	25 VI 52.4	239 Adrastea	00 LI 36.7
2148 Epeios	25 VI 53.6	46 Hestia	00 LI 38.5
313 Chaldaea	26 VI 19.4	1369 Ostanina	00 LI 45.7
3114 Ercilla	26 VI 32.0	1051 Merope	00 LI 52.4
1776 Kuiper	26 VI 40.0	808 Merxia	00 LI 52.5
865 Zubaida	26 VI 43.4	1628 Strobel	00 LI 54.2
2746 Hissao	26 VI 46.4	950 Ahrensa	01 LI 20.2
755 Quintilla	26 VI 52.9	1621 Druzhba	01 LI 27.1
1409 Isko	27 VI 20.2	1458 Mineura	01 LI 30.4
1671 Chaika	27 VI 20.4	308 Polyxo	01 LI 32.8
1679 Nevanlinna	27 VI 23.8	1896 Beer	01 LI 45.2
589 Croatia	27 VI 35.5	289 Nenetta	01 LI 46.1
831 Stateira	27 VI 38.0	2328 Robeson	01 LI 47.2
41 Daphne	27 VI 42.6	1734 Zhongolovich	01 LI 50.1
681 Gorgo	27 VI 49.8	2577 Litva	02 LI 04.7

Name	Position	Name	Position
900 Rosalinde	02 LI 16.4	495 Eulalia	06 LI 06.7
1992 Galvarino	02 LI 26.1	319 Leona	06 LI 15.0
853 Nansenia	02 LI 28.6	1084 Tamariwa	06 LI 40.0
256 Walpurga	02 LI 39.4	1796 Riga	06 LI 52.2
3079 Schiller	02 LI 46.4	2317 Galya	07 LI 01.3
1973 Colocolo	03 LI 06.5	619 Triberga	07 LI 01.3
635 Vundtia	03 LI 10.8	1651 Behrens	07 LI 02.1
834 Burnhamia	03 LI 12.5	421 Zahringia	07 LI 04.6
48 Doris	03 LI 22.6	2776 Baikal	07 LI 13.1
238 Hypatia	03 LI 43.9	3158 Anga	07 LI 15.3
656 Beagle	03 LI 48.9	1878 Hughes	07 LI 28.7
787 Moskva	03 LI 49.4	2922 Dikanka	07 LI 32.5
993 Moultona	04 LI 01.6	2981 Chagall	07 LI 37.4
34 Circe	04 LI 07.9	1030 Vitja	07 LI 41.9
1635 Bohrmann	04 LI 15.1	2557 Putnam	07 LI 47.6
1579 Herrick	04 LI 15.8	1917 Cuyo	07 LI 50.9
513 Centesima	04 LI 23.2	3149 Okudjeva	07 LI 51.7
737 Arequipa	04 LI 28.0	124 Alkeste	07 LI 52.1
1828 Kashirina	04 LI 29.5	105 Artemis	07 LI 55.8
2992 Vondel	04 LI 50.6	1275 Cimbria	08 LI 09.8
1295 Deflotte	04 LI 53.6	3480 Abante	08 LI 10.6
365 Corduba	04 LI 57.4	1705 Tapio	08 LI 14.2
1371 Resi	05 LI 13.6	2433 Sootiyo	08 LI 14.9
2473 Heyerdahl	05 LI 13.9	2861 Lambrecht	08 LI 18.8
69 Hesperia	05 LI 16.9	3225 Hoag	08 LI 18.8
1856 Ruzena	05 LI 28.3	3046 Moliere	08 LI 22.0
1425 Tuorla	05 LI 34.6	1766 Slipher	08 LI 25.4
2018 Schuster	05 LI 38.8	1932 Jansky	08 LI 30.9
1984 Fedynskij	05 LI 43.1	1872 Helenos	08 LI 34.5
801 Helwerthia	05 LI 44.0	2356 Hirons	08 LI 40.1
236 Honoria	05 LI 44.9	1469 Linzia	08 LI 47.2
2696 Magion	05 LI 53.3	1033 Simona	08 LI 55.0
3351 Smith	05 LI 54.4	1854 Skvortsov	08 LI 59.8
1352 Wawel	05 LI 56.0	2214 Carol	09 LI 00.3
136 Austria	06 LI 03.1	1743 Schmidt	09 LI 09.4
2527 Gregory	06 LI 04.0	1783 Albitskij	09 LI 11.7
2348 Michkovitch	06 LI 05.4	1435 Garlena	09 LI 15.9

Name	Position	Name	Position
611 Valeria	09 LI 25.1	572 Rebekka	14 LI 08.5
1570 Brunonia	09 LI 43.0	1023 Thomana	14 LI 09.6
3039 Yangel	10 LI 10.0	3027 Shavarsh	14 LI 19.3
2442 Corbett	10 LI 11.3	2322 Kitt Peak	14 LI 21.6
2689 Bruxelles	10 LI 11.4	872 Holda	14 LI 28.4
1606 Jekhovsky	10 LI 15.9	1494 Savo	14 LI 33.9
854 Frostia	10 LI 22.6	2303 Retsina	15 LI 03.5
2367 Praha	10 LI 23.9	1918 Aiguillon	15 LI 08.4
894 Erda	10 LI 24.5	1536 Pielinen	15 LI 11.8
874 Rotraut	10 LI 31.4	2549 Baker	15 LI 16.2
1855 Korolev	10 LI 35.8	1114 Lorraine	15 LI 27.4
682 Hagar	10 LI 46.4	444 Gyptis	15 LI 29.5
2135 Aristaeus	10 LI 47.2	2304 Slavia	15 LI 42.3
1546 Izsak	10 LI 49.8	1775 Zimmerwald	15 LI 49.9
3242 Bakhchisaraj	11 LI 09.6	875 Nymphe	15 LI 51.2
2508 Alupka	11 LI 11.2	3479 Malaparte	16 LI 19.0
60 Echo	11 LI 25.1	2283 Bunke	16 LI 23.6
1383 Limburgia	11 LI 47.2	3097 Tacitus	16 LI 37.5
3232 Brest	12 LI 03.3	225 Henrietta	16 LI 45.1
1767 Lampland	12 LI 11.3	74 Galatea	17 LI 00.9
2499 Brunk	12 LI 17.9	1868 Thersites	17 LI 06.2
956 Elisa	12 LI 18.9	2606 Odessa	17 LI 07.9
1708 Polit	12 LI 31.6	1873 Agenor	17 LI 14.1
920 Rogeria	12 LI 31.9	923 Herluga	17 LI 14.2
1795 Woltjer	12 LI 36.7	531 Zerlina	17 LI 20.0
2251 Tikhov	12 LI 44.0	2598 Merlin	17 LI 34.4
1289 Kutaissi	12 LI 50.3	2783 Chernyshevskij	17 LI 36.4
2810 Lev Tolstoj	12 LI 50.6	904 Rockefellia	17 LI 44.9
2539 Ningxia	12 LI 51.4	2298 Cindijon	18 LI 14.1
547 Praxedis	12 LI 53.9	2530 Shipka	18 LI 36.8
2053 Nuki	12 LI 55.7	1791 Patsayev	18 LI 41.8
1990 Pilcher	13 LI 09.4	57 Mnemosyne	18 LI 53.5
56 Melete	13 LI 10.5	2495 Noviomagum	19 LI 13.4
3454 Lieske	13 LI 14.3	417 Suevia	19 LI 15.4
2342 Lebedev	13 LI 17.3	836 Jole	19 LI 20.3
485 Genua	13 LI 17.4	1650 Heckmann	19 LI 23.6
2653 Principia	14 LI 07.7	1692 Subbotina	19 LI 25.4

Name	Position	Name	Position
1632 Siebohme	19 LI 33.2	2934 Aristophanes	22 LI 41.8
2049 Grietje	19 LI 33.7	1387 Kama	22 LI 44.6
837 Schwarzschilda	19 LI 37.4	510 Mabella	22 LI 50.4
1229 Tilia	19 LI 40.1	85 Io	22 LI 51.5
1935 Lucerna	19 LI 43.7	1201 Strenua	22 LI 52.1
1475 Yalta	19 LI 44.2	1739 Meyermann	22 LI 55.3
1059 Mussorgskia	20 LI 06.2	525 Adelaide	22 LI 59.0
3316 Herzberg	20 LI 11.3	189 Phthia	23 LI 12.7
1037 Davidweilla	20 LI 14.7	119 Althaea	23 LI 16.8
1638 Ruanda	20 LI 15.5	1165 Imprinetta	23 LI 20.9
1230 Riceia	20 LI 21.5	612 Veronika	23 LI 28.7
176 Iduna	20 LI 25.0	3073 Kursk	23 LI 41.6
219 Thusnelda	20 LI 26.3	2660 Wasserman	23 LI 53.3
1286 Banachiewicza	20 LI 27.2	518 Halawe	23 LI 54.7
2715 Mielikki	20 LI 27.4	2014 Vasilevskis	23 LI 55.3
2059 Baboquivari	20 LI 28.3	76 Freia	24 LI 01.8
2666 Gramme	20 LI 35.4	3403 Tammy	24 LI 02.8
3000 Leonardo	20 LI 36.9	2592 Hunan	24 LI 03.4
2642 Vesale	20 LI 40.0	2619 Skalnate Pleso	24 LI 08.5
2728 Yatskiv	20 LI 50.4	2484 Parenago	24 LI 11.3
1910 Mikhailov	20 LI 56.3	1502 Arenda	24 LI 11.5
1905 Ambartsumian	21 LI 00.0	2235 Vittore	24 LI 45.8
1882 Rauma	21 LI 10.0	2610 Tuva	24 LI 51.8
1422 Stromgrenia	21 LI 10.9	460 Scania	24 LI 56.5
2192 Pyatigoriya	21 LI 24.8	1200 Imperatrix	24 LI 58.6
1408 Trusanda	21 LI 28.3	922 Schlutia	24 LI 59.5
439 Ohio	21 LI 31.8	204 Kallisto	25 LI 02.2
1101 Clematis	21 LI 38.8	2518 Rutllant	25 LI 11.8
540 Rosamunde	21 LI 43.3	2420 Ciurlionis	25 LI 12.5
137 Meliboea	21 LI 52.1	921 Jovita	25 LI 18.1
252 Clementina	21 LI 54.4	2359 Debehogne	25 LI 38.8
1573 Vaisala	21 LI 55.9	2493 Elmer	25 LI 39.7
2727 Paton	21 LI 59.0	1309 Hyperborea	25 LI 45.5
1287 Lorcia	22 LI 13.6	20 Massalia	26 LI 03.1
1979 Sakharov	22 LI 16.3	150 Nuwa	26 LI 08.2
67 Asia	22 LI 21.5	3126 Davydov	26 LI 08.3
1718 Namibia	22 LI 39.5	1150 Achaia	26 LI 08.9

Name	Position	Name	Position
2698 Azerbajdzhan	26 LI 19.3	1400 Tirela	00 SC 20.3
79 Eurynome	26 LI 21.1	102 Miriam	00 SC 39.4
1575 Winifred	26 LI 23.5	276 Adelheid	00 SC 51.1
2654 Ristenpart	26 LI 28.4	19 Fortuna	00 SC 56.0
2628 Kopal	26 LI 34.2	2340 Hathor	01 SC 00.2
2074 Shoemaker	26 LI 43.6	2363 Cebriones	01 SC 11.9
242 Kriemhild	26 LI 49.5	205 Martha	01 SC 31.7
2529 Rockwell Kent	26 LI 58.6	1353 Maartje	01 SC 48.7
168 Sibylla	26 LI 59.8	1542 Schalen	01 SC 54.4
2704 Julian Loewe	27 LI 03.2	992 Swasey	02 SC 00.1
496 Gryphia	27 LI 10.9	3151 Talbot	02 SC 06.7
2976 Lautaro	27 LI 11.3	1944 Gunter	02 SC 08.4
2282 Andres Bello	27 LI 18.5	2490 Bussolini	02 SC 08.9
2061 Anza	27 LI 20.2	345 Tercidina	02 SC 14.3
258 Tyche	27 LI 23.5	2196 Ellicott	02 SC 19.0
3058 Delmary	27 LI 25.0	1865 Cerberus	02 SC 23.5
848 Inna	27 LI 31.0	1660 Wood	02 SC 29.0
756 Lilliana	27 LI 35.9	391 Ingeborg	02 SC 29.4
72 Feronia	27 LI 36.5	393 Lampetia	02 SC 42.1
3175 Netto	27 LI 44.7	1139 Atami	02 SC 50.7
2679 Kittisvaara	27 LI 47.3	2795 Lepage	02 SC 58.9
2477 Biryukov	27 LI 53.6	1419 Danzig	03 SC 07.9
2834 Christy Carol	28 LI 14.9	703 Noemi	03 SC 18.6
1673 Van Houten	28 LI 18.7	1257 Mora	03 SC 23.1
244 Sita	28 LI 31.5	2052 Tamriko	03 SC 34.3
2060 Chiron	28 LI 37.5	25 Phocaea	03 SC 42.2
2948 Amosov	28 LI 42.4	798 Ruth	04 SC 04.6
305 Gordonia	28 LI 43.1	668 Dora	04 SC 22.9
968 Petunia	28 LI 43.6	1146 Biarmia	04 SC 25.9
1136 Mercedes	29 LI 03.8	3312 Pedersen	04 SC 27.1
1969 Alain	29 LI 05.3	999 Zachia	04 SC 35.3
821 Fanny	29 LI 28.7	1363 Herberta	04 SC 38.2
822 Lalage	29 LI 32.8	2685 Masursky	04 SC 49.9
392 Wilhelmina	29 LI 33.2	650 Amalasuntha	04 SC 57.6
1281 Jeanne	29 LI 37.0	2051 Chang	04 SC 58.0
377 Campania	29 LI 44.3	2276 Warck	04 SC 58.0
2475 Semenov	29 LI 54.7	666 Desdemona	05 SC 05.8

Name	Position	Name	Position
216 Kleopatra	05 SC 09.6	2926 Caldeira	10 SC 33.1
2658 Gingerich	05 SC 21.8	1143 Odysseus	10 SC 35.0
1291 Phryne	05 SC 22.6	2961 Katsurahama	10 SC 37.9
1036 Ganymed	05 SC 37.6	1583 Antilochus	10 SC 39.5
3147 Samantha	05 SC 46.6	3181 Ahnert	10 SC 40.9
1130 Skuld	05 SC 49.0	1822 Waterman	10 SC 42.1
978 Aidamina	06 SC 08.2	1442 Corvina	10 SC 46.0
2459 Spellmann	06 SC 09.1	1060 Magnolia	10 SC 52.6
1473 Ounas	06 SC 12.8	2107 Ilmari	11 SC 00.2
263 Dresda	06 SC 16.9	1794 Finsen	11 SC 12.2
2820 Iisalmi	06 SC 18.4	3268 De Sanctis	11 SC 17.5
1102 Pepita	06 SC 19.2	1058 Grubba	11 SC 25.3
3387 Greenberg	06 SC 32.8	995 Sternberga	11 SC 26.9
2925 Beatty	06 SC 45.2	1039 Sonneberga	11 SC 28.1
1554 Yugoslavia	06 SC 45.4	2175 Andrea Doria	11 SC 43.9
614 Pia	06 SC 52.9	233 Asterope	11 SC 50.8
509 Iolanda	07 SC 27.1	2120 Tyumenia	12 SC 31.4
1881 Shao	07 SC 40.1	2198 Ceplecha	12 SC 44.9
771 Libera	07 SC 41.7	1377 Roberbauxa	12 SC 47.8
2343 Siding Spring	07 SC 45.4	3159 Prokofiev	12 SC 49.0
2232 Altaj	08 SC 05.9	2590 Mourao	13 SC 16.3
2716 Tuulikki	08 SC 06.7	1328 Devota	13 SC 19.0
3288 Seleucus	08 SC 09.0	631 Philippina	14 SC 26.2
1168 Brandia	08 SC 12.1	1203 Nanna	14 SC 32.8
1695 Walbeck	08 SC 16.2	1355 Magoeba	14 SC 42.4
80 Sappho	08 SC 21.0	1167 Dubiago	14 SC 57.5
2099 Opik	08 SC 22.6	1151 Ithaka	15 SC 04.0
2437 Amnestia	08 SC 32.2	2056 Nancy	15 SC 19.5
1031 Arctica	08 SC 42.2	570 Kythera	15 SC 23.4
374 Burgundia	08 SC 45.2	2862 Vavilov	15 SC 24.2
713 Luscinia	08 SC 55.6	1793 Zoya	15 SC 30.2
3469 Bulgakov	09 SC 07.2	565 Marbachia	15 SC 39.0
429 Lotis	09 SC 41.0	2686 Linda Susan	15 SC 45.2
320 Katharina	09 SC 49.3	1922 Zulu	15 SC 53.9
32 Pomona	09 SC 57.2	1590 Tsiolkovskaja	16 SC 09.4
2223 Sarpedon	10 SC 26.0	1296 Andree	16 SC 30.9
1827 Atkinson	10 SC 30.5	673 Edda	16 SC 40.0

Name	Position
2440 Educatio	16 SC 48.7
368 Haidea	17 SC 02.9
1909 Alekhin	17 SC 04.9
2427 Kobzar	17 SC 20.1
3224 Irkutsk	17 SC 37.6
2205 Glinka	17 SC 40.4
153 Hilda	17 SC 46.9
397 Vienna	17 SC 48.2
2671 Abkhazia	17 SC 53.8
3018 Godiva	18 SC 04.7
3091 Van Den Heuvel	18 SC 08.8
2478 Tokai	18 SC 19.9
849 Ara	18 SC 29.1
743 Eugenisis	18 SC 47.3
586 Thekla	18 SC 53.5
456 Abnoba	18 SC 55.8
1347 Patria	18 SC 59.6
1009 Sirene	19 SC 09.3
310 Margarita	19 SC 16.0
419 Aurelia	19 SC 19.1
2266 Tchaikovsky	19 SC 23.2
919 Ilsebill	19 SC 35.0
1537 Transylvania	19 SC 53.6
826 Henrika	20 SC 07.0
694 Ekard	20 SC 07.3
3023 Heard	20 SC 11.7
1690 Mayrhofer	20 SC 17.9
2920 Automedon	20 SC 26.3
712 Boliviana	20 SC 31.6
3333 Shaber	20 SC 47.8
1726 Hoffmeister	20 SC 53.5
929 Algunde	20 SC 56.8
277 Elvira	21 SC 15.2
979 Ilsewa	21 SC 23.1
2813 Zappala	21 SC 46.2
1561 Fricke	21 SC 54.7
2352 Kurchatov	22 SC 04.3
342 Endymion	22 SC 13.1
1760 Sandra	22 SC 13.7
378 Holmia	22 SC 18.9
2755 Avicenna	22 SC 20.1
789 Lena	22 SC 22.3
957 Camelia	22 SC 29.3
663 Gerlinde	22 SC 30.5
2908 Shimoyama	23 SC 08.0
2180 Marjaleena	23 SC 08.3
284 Amalia	23 SC 25.4
1307 Cimmeria	23 SC 26.0
478 Tergeste	23 SC 33.8
714 Ulula	23 SC 39.4
2436 Hatshepsut	23 SC 50.0
2257 Kaarina	23 SC 53.8
1108 Demeter	24 SC 03.1
1506 Xosa	24 SC 20.9
640 Brambilla	24 SC 36.9
336 Lacadiera	24 SC 37.7
1264 Letaba	24 SC 48.8
2371 Dimitrov	24 SC 51.8
685 Hermia	24 SC 55.9
12 Victoria	25 SC 07.7
266 Aline	25 SC 33.7
1857 Parchomenko	25 SC 37.0
1199 Geldonia	25 SC 41.0
2692 Chkalov	25 SC 53.2
1807 Slovakia	25 SC 57.8
1315 Bronislawa	26 SC 04.3
1293 Sonja	26 SC 12.8
1752 Van Herk	26 SC 32.9
480 Hansa	26 SC 53.9
1175 Margo	26 SC 56.4
2426 Simonov	26 SC 57.5
1611 Beyer	27 SC 15.8
2474 Ruby	27 SC 40.6

Name	Position	Name	Position
1712 Angola	27 SC 41.7	1888 Zu ChongZhi	04 SA 19.6
1255 Schilowa	27 SC 43.7	2346 Lilio	04 SA 25.9
1316 Kasan	27 SC 45.4	1994 Shane	04 SA 40.9
1256 Normannia	27 SC 53.9	1311 Knopfia	04 SA 51.3
797 Montana	27 SC 58.6	976 Benjamina	05 SA 01.6
1438 Wendeline	28 SC 04.1	2649 Oongaq	05 SA 02.3
3052 Herzen	28 SC 10.1	1395 Aribeda	05 SA 02.4
1964 Luyten	28 SC 15.2	3353 Jarvis	05 SA 03.4
1499 Pori	29 SC 22.3	2638 Gadolin	05 SA 23.7
230 Athamantis	29 SC 23.2	1688 Wilkens	05 SA 24.2
3050 Carrera	29 SC 36.4	1243 Pamela	05 SA 39.8
1647 Menelaus	29 SC 43.6	2382 Nonie	05 SA 41.7
838 Seraphina	29 SC 57.5	3285 Ruth Wolfe	05 SA 42.6
1751 Herget	00 SA 04.9	2221 Chilton	05 SA 44.8
1728 Goethe Link	00 SA 25.3	1943 Anteros	05 SA 45.8
2709 Sagan	00 SA 38.4	2751 Campbell	05 SA 50.9
299 Thora	00 SA 54.5	1222 Tina	05 SA 53.9
188 Menippe	00 SA 57.5	1980 Tezcatlipoca	06 SA 05.6
2872 Gentelec	01 SA 06.7	1574 Meyer	06 SA 14.0
2115 Irakli	01 SA 11.8	2264 Sabrina	06 SA 17.9
1110 Jaroslawa	01 SA 26.9	248 Lameia	06 SA 31.6
156 Xanthippe	01 SA 37.7	1172 Aneas	06 SA 46.5
898 Hildegard	01 SA 40.1	2233 Kuznetsov	06 SA 49.2
409 Aspasia	01 SA 58.7	352 Gisela	06 SA 50.5
726 Joella	02 SA 11.8	997 Priska	06 SA 57.6
699 Hela	02 SA 14.8	1277 Dolores	06 SA 59.1
2588 Flavia	02 SA 19.3	2454 Olaus Magnus	07 SA 15.3
2028 Janequeo	02 SA 24.0	2191 Uppsala	07 SA 17.3
95 Arethusa	03 SA 02.0	2983 Poltava	07 SA 36.6
989 Schwassmannia	03 SA 02.9	3182 Shimanto	08 SA 00.3
686 Gersuind	03 SA 15.7	1505 Koranna	08 SA 01.0
2112 Ulyanov	03 SA 15.7	2823 Van Der Laan	08 SA 11.9
937 Bethgea	03 SA 18.2	147 Protogeneia	08 SA 30.2
1716 Peter	03 SA 18.5	418 Alemannia	08 SA 32.2
701 Oriola	03 SA 46.7	960 Birgit	08 SA 46.7
420 Bertholda	03 SA 51.4	1596 Itzigsohn	08 SA 53.8
403 Cyane	04 SA 16.4	1818 Brahms	09 SA 04.3

Name	Position	Name	Position
430 Hybris	09 SA 10.3	823 Sisigambis	14 SA 35.8
457 Alleghenia	09 SA 19.2	1169 Alwine	14 SA 39.7
2694 Pino Torinese	09 SA 28.2	405 Thia	14 SA 43.7
1879 Broederstroom	09 SA 29.4	2736 Ops	15 SA 06.8
568 Cheruskia	09 SA 31.3	2331 Parvulesco	15 SA 10.9
774 Armor	09 SA 47.2	914 Palisana	15 SA 22.7
396 Aeolia	09 SA 50.9	1197 Rhodesia	15 SA 29.9
983 Gunila	10 SA 26.8	2170 Byelorussia	15 SA 31.3
2152 Hannibal	10 SA 45.3	499 Venusia	15 SA 55.6
803 Picka	10 SA 55.2	2139 Makharadze	16 SA 19.0
2953 Vysheslavia	10 SA 59.8	1319 Disa	16 SA 23.1
1652 Herge	11 SA 17.6	2184 Fujian	16 SA 23.2
3299 Hall	11 SA 18.1	3134 Kostinsky	16 SA 37.6
790 Pretoria	11 SA 33.8	882 Swetlana	17 SA 14.8
1014 Semphyra	11 SA 37.3	2513 Baetsle	17 SA 15.2
2483 Guinevere	11 SA 38.4	897 Lysistrata	17 SA 27.1
1177 Gonnessia	11 SA 54.3	583 Klotilde	17 SA 30.3
179 Klytaemnestra	11 SA 54.4	220 Stephania	17 SA 39.7
2104 Toronto	12 SA 05.6	1224 Fantasia	17 SA 45.6
643 Scheherezade	12 SA 05.9	1955 McMath	17 SA 53.8
899 Jokaste	12 SA 19.6	1894 Haffner	17 SA 54.2
322 Phaeo	12 SA 19.9	3367 Alex	17 SA 55.3
2167 Erin	12 SA 30.3	1365 Henyey	18 SA 07.2
951 Gaspra	12 SA 40.3	132 Aethra	18 SA 25.5
1322 Coppernicus	12 SA 44.2	3115 Baily	18 SA 27.8
690 Wratislavia	12 SA 45.1	1249 Rutherfordia	18 SA 35.0
1810 Epimetheus	13 SA 05.2	2083 Smither	18 SA 38.9
1520 Imatra	13 SA 06.0	1057 Wanda	18 SA 54.4
441 Bathilde	13 SA 19.6	764 Gedania	19 SA 02.4
896 Sphinx	13 SA 44.5	1704 Wachmann	19 SA 04.2
1441 Bolyai	13 SA 51.4	395 Delia	19 SA 06.0
1779 Parana	13 SA 56.3	1001 Gaussia	19 SA 08.6
1490 Limpopo	13 SA 57.9	7 Iris	19 SA 19.9
270 Anahita	13 SA 59.5	1198 Atlantis	19 SA 26.3
832 Karin	14 SA 17.4	683 Lanzia	19 SA 33.7
2817 Perec	14 SA 24.7	1436 Salonta	20 SA 18.3
647 Adelgunde	14 SA 25.0	1181 Lilith	20 SA 25.6

Name	Position	Name	Position
1420 Radcliffe	20 SA 47.2	541 Deborah	27 SA 33.6
523 Ada	20 SA 47.7	2119 Schwall	27 SA 47.5
1565 Lemaitre	20 SA 55.2	198 Ampella	28 SA 07.0
1661 Granule	21 SA 17.9	1109 Tata	28 SA 29.2
1149 Volga	21 SA 22.1	514 Armida	28 SA 38.2
846 Lipperta	21 SA 32.5	1407 Lindelof	29 SA 02.2
2190 Coubertin	21 SA 53.2	1129 Neujmina	29 SA 14.3
1232 Cortusa	22 SA 14.1	879 Ricarda	29 SA 48.2
2377 Shcheglov	22 SA 30.4	550 Senta	00 CP 15.5
1666 Van Gent	22 SA 46.3	1367 Nongoma	00 CP 23.9
675 Ludmilla	23 SA 00.1	1644 Rafita	00 CP 25.8
437 Rhodia	23 SA 06.4	241 Germania	00 CP 34.7
880 Herba	23 SA 09.7	1806 Derice	00 CP 37.5
211 Isolda	23 SA 25.3	847 Agnia	00 CP 52.0
1940 Whipple	23 SA 47.2	3107 Weaver	01 CP 13.9
895 Helio	24 SA 12.2	1292 Luce	01 CP 16.1
1314 Paula	24 SA 14.5	1213 Algeria	01 CP 30.7
2145 Blaauw	24 SA 15.3	2141 Simferopol	01 CP 37.0
43 Ariadne	24 SA 23.2	2229 Mezzarco	01 CP 47.9
2423 Ibarruri	24 SA 44.9	1858 Lobachevskij	01 CP 49.4
1498 Lahti	24 SA 46.7	1176 Lucidor	01 CP 57.7
1147 Stavropolis	24 SA 46.8	2697 Albina	02 CP 28.6
901 Brunsia	24 SA 48.0	1836 Komarov	02 CP 35.4
1949 Messina	24 SA 49.1	677 Aaltje	02 CP 46.4
3200 Paethon	24 SA 56.5	840 Zenobia	03 CP 02.9
1936 Lugano	25 SA 01.3	2105 Gudy	03 CP 07.1
792 Metcalfia	25 SA 02.3	1699 Honkasalo	03 CP 10.7
748 Simeisa	25 SA 19.6	1750 Eckert	03 CP 15.6
2937 Gibbs	25 SA 22.7	2940 Bacon	03 CP 37.9
1535 Paijanne	26 SA 03.1	1685 Toro	03 CP 45.6
1645 Waterfield	26 SA 06.9	1826 Miller	04 CP 00.1
1843 Jarmila	26 SA 35.3	1306 Scythia	04 CP 04.4
1402 Eri	26 SA 43.6	517 Edith	04 CP 23.0
1103 Sequoia	27 SA 08.2	539 Pamina	04 CP 27.1
552 Sigelinde	27 SA 11.5	2780 Monnig	04 CP 32.9
1045 Michela	27 SA 25.7	2140 Kemerovo	04 CP 35.3
3369 Freuchen	27 SA 26.3	2536 Kozyrev	04 CP 52.4

Name	Position	Name	Position
1189 Terentia	05 CP 10.0	584 Semiramis	11 CP 50.6
695 Bella	05 CP 27.8	389 Industria	11 CP 59.0
295 Theresia	05 CP 54.7	972 Cohnia	12 CP 17.0
88 Thisbe	06 CP 16.9	2365 Interkosmos	12 CP 26.2
2635 Huggins	06 CP 31.0	2799 Justus	12 CP 36.7
1244 Deira	06 CP 38.8	1785 Wurm	12 CP 51.2
2012 Guo ShouJing	06 CP 47.6	1867 Deiphobus	12 CP 57.7
881 Athene	06 CP 56.1	1509 Esclangona	12 CP 58.6
1401 Lavonne	07 CP 09.8	10 Hygiea	13 CP 03.2
2613 Plzen	07 CP 23.9	2494 Inge	13 CP 12.9
2029 Binomi	07 CP 39.4	1173 Anchises	13 CP 14.2
654 Zelinda	07 CP 58.8	371 Bohemia	13 CP 15.1
2524 Budovicium	08 CP 07.4	2764 Moeller	13 CP 21.2
158 Koronis	08 CP 08.0	779 Nina	13 CP 30.2
1954 Kukarkin	08 CP 11.4	2555 Thomas	13 CP 32.0
1531 Hartmut	08 CP 41.5	1138 Attica	13 CP 34.5
1626 Sadeya	09 CP 03.5	777 Gutemberga	14 CP 02.9
1525 Savonlinna	09 CP 04.2	1456 Saldanha	14 CP 15.2
1706 Dieckvoss	09 CP 15.8	1411 Brauna	14 CP 28.2
2648 Owa	09 CP 31.0	2670 Chuvashia	14 CP 35.5
3510 Veeder	09 CP 36.8	1959 Karbyshev	14 CP 38.6
398 Admete	09 CP 38.6	607 Jenny	14 CP 58.7
1040 Klumpkea	09 CP 42.8	883 Matterania	15 CP 09.1
639 Latona	09 CP 51.4	1911 Schubart	15 CP 12.8
2259 Sofievka	09 CP 53.8	2186 Keldysh	15 CP 13.5
1153 Wallenbergia	10 CP 05.2	1530 Rantaseppa	15 CP 29.0
1064 Aethusa	10 CP 09.1	49 Pales	15 CP 35.0
3237 Viktorplatt	10 CP 21.4	980 Anacostia	15 CP 37.7
704 Interamnia	10 CP 21.8	556 Phyllis	15 CP 37.8
2579 Spartacus	10 CP 30.2	1214 Richilde	15 CP 39.8
1837 Osita	10 CP 33.8	2444 Lederle	15 CP 49.9
1195 Orangia	10 CP 51.1	476 Hedwig	15 CP 58.6
678 Fredegundis	11 CP 17.2	2078 Nanking	16 CP 55.1
2618 Coonabarabran	11 CP 23.8	3063 Makhaon	17 CP 11.0
1900 Katyusha	11 CP 28.0	2368 Beltrovata	17 CP 11.1
707 Steina	11 CP 29.8	338 Budrosa	17 CP 13.4
2094 Magnitka	11 CP 32.4	1254 Erfordia	17 CP 46.7

Name	Position	Name	Position
1543 Bourgeois	17 CP 52.2	445 Edna	22 CP 41.9
969 Leocadia	17 CP 53.8	15 Eunomia	22 CP 53.8
2601 Bologna	18 CP 03.3	1829 Dawson	23 CP 00.5
1643 Brown	18 CP 07.6	2054 Gawain	23 CP 08.6
1825 Klare	18 CP 26.7	3025 Higson	23 CP 08.9
2676 Aarhus	19 CP 09.7	1280 Baillauda	23 CP 19.1
1560 Strattonia	19 CP 10.7	3127 Bagration	23 CP 31.0
763 Cupido	19 CP 28.6	507 Laodica	23 CP 51.8
1067 Lunaria	19 CP 30.4	1496 Turku	23 CP 55.7
3219 Komaki	19 CP 36.7	608 Adolfine	23 CP 57.3
702 Alauda	19 CP 37.9	407 Arachne	24 CP 15.7
500 Selinur	19 CP 48.7	543 Charlotte	24 CP 51.8
985 Rosina	20 CP 04.6	554 Peraga	25 CP 08.4
2403 Sumava	20 CP 05.0	38 Leda	25 CP 27.7
1246 Chaka	20 CP 14.4	3240 Laocoon	25 CP 32.5
3228 Pire	20 CP 17.7	1740 Paavo Nurmi	25 CP 33.2
370 Modestia	20 CP 33.5	1297 Quadea	25 CP 48.1
142 Polana	20 CP 40.4	2554 Skiff	25 CP 59.0
466 Tisiphone	20 CP 43.6	1457 Ankara	26 CP 03.8
2774 Tenojoki	20 CP 44.5	1273 Helma	26 CP 19.7
1339 Desagneauxa	20 CP 48.2	1835 Gajdariya	26 CP 39.8
1756 Giacobini	20 CP 52.9	1398 Donnera	26 CP 41.5
1233 Kobresia	20 CP 55.6	1373 Cincinnati	26 CP 58.0
2005 Hencke	20 CP 55.6	3077 Henderson	26 CP 59.2
1194 Aletta	21 CP 00.5	1821 Aconcagua	27 CP 00.8
1780 Kippes	21 CP 01.7	3001 Michelangelo	27 CP 16.6
1547 Nele	21 CP 02.8	427 Galene	27 CP 26.5
549 Jessonda	21 CP 14.1	657 Gunlod	27 CP 29.2
1250 Galanthus	21 CP 28.7	1485 Isa	27 CP 33.8
648 Pippa	21 CP 36.5	775 Lumiere	27 CP 35.0
2000 Herschel	21 CP 36.7	1598 Paloque	27 CP 36.1
440 Theodora	21 CP 37.9	1006 Lagrangea	27 CP 41.2
2669 Shostakovich	22 CP 21.2	2668 Tataria	27 CP 41.2
3030 Vehrenberg	22 CP 28.8	544 Jetta	27 CP 58.3
1507 Vaasa	22 CP 32.0	408 Fama	28 CP 06.2
1571 Cesco	22 CP 34.4	1313 Berna	28 CP 07.3
557 Violetta	22 CP 41.2	1853 McElroy	28 CP 32.1

Name	Position	Name	Position
665 Sabine	28 CP 40.1	1653 Yakhontovia	03 AQ 54.2
1288 Santa	28 CP 44.1	1324 Knysna	03 AQ 57.1
696 Leonora	29 CP 01.4	1876 Napolitania	03 AQ 58.0
925 Alphonsina	29 CP 14.6	283 Emma	04 AQ 01.1
982 Franklina	29 CP 19.2	1763 Williams	04 AQ 05.6
1258 Sicilia	29 CP 30.1	1100 Arnica	04 AQ 06.7
1298 Nocturna	29 CP 33.7	1034 Mozartia	04 AQ 10.6
576 Emanuela	29 CP 36.4	1234 Elyna	04 AQ 24.4
862 Franzia	29 CP 48.4	1260 Walhalla	04 AQ 27.4
1709 Ukraina	29 CP 49.8	2096 Vaino	04 AQ 28.1
1971 Hagihara	00 AQ 01.3	390 Alma	04 AQ 49.1
465 Alekto	00 AQ 09.7	1584 Fuji	04 AQ 56.3
1497 Tampere	00 AQ 13.9	2544 Gubarev	05 AQ 18.4
3086 Kalbaugh	00 AQ 14.6	1985 Hopmann	05 AQ 22.3
1714 Sy	00 AQ 24.8	111 Ate	05 AQ 25.9
3453 Dostoevsky	00 AQ 29.4	2131 Mayall	05 AQ 29.2
2455 Somville	00 AQ 31.3	762 Pulcova	05 AQ 35.9
170 Maria	00 AQ 53.5	1370 Hella	05 AQ 36.4
884 Priamus	01 AQ 00.7	2497 Kulikovskij	06 AQ 34.4
998 Bodea	01 AQ 22.0	3413 Andriana	06 AQ 35.8
569 Misa	01 AQ 34.6	1787 Chiny	06 AQ 50.9
376 Geometria	01 AQ 38.3	1007 Pawlowia	07 AQ 05.2
2064 Thomsen	01 AQ 42.3	1290 Albertine	07 AQ 05.6
2071 Nadezhda	01 AQ 49.6	2258 Viipuri	07 AQ 14.8
1374 Isora	02 AQ 13.0	1349 Bechuana	07 AQ 17.0
646 Kastalia	02 AQ 29.5	30 Urania	07 AQ 18.1
1284 Latvia	02 AQ 34.7	2302 Florya	07 AQ 18.4
165 Loreley	02 AQ 35.7	1092 Lilium	07 AQ 31.5
3065 Sarahill	02 AQ 47.6	123 Brunhild	07 AQ 31.7
1444 Pannonia	02 AQ 48.9	1228 Scabiosa	07 AQ 41.3
1112 Polonia	02 AQ 54.5	2366 Aaryn	07 AQ 50.5
1832 Mrkos	03 AQ 02.7	623 Chimaera	07 AQ 50.6
2412 Wil	03 AQ 16.4	2763 Jeans	08 AQ 19.9
2609 KirilMetodi	03 AQ 29.3	1468 Zomba	08 AQ 30.1
2762 Fowler	03 AQ 38.6	115 Thyra	08 AQ 36.4
2345 Fucik	03 AQ 43.7	2467 Kollontai	08 AQ 46.5
433 Eros	03 AQ 45.1	64 Angelina	09 AQ 02.2

Name	Position	Name	Position
1604 Tombaugh	09 AQ 05.1	588 Achilles	15 AQ 52.6
860 Ursina	09 AQ 10.3	1503 Kuopio	16 AQ 34.6
1823 Gliese	09 AQ 44.7	3280 Gretry	17 AQ 01.7
835 Olivia	09 AQ 46.6	1721 Wells	17 AQ 10.1
3160 Angerhofer	10 AQ 19.7	3047 Goethe	17 AQ 16.1
2916 Voronveliya	10 AQ 39.8	3031 Houston	17 AQ 23.8
285 Regina	10 AQ 53.7	1321 Majuba	17 AQ 40.0
2026 Cottrell	11 AQ 00.7	945 Barcelona	17 AQ 50.8
89 Julia	11 AQ 04.2	1285 Julietta	17 AQ 53.0
2576 Yesenin	11 AQ 05.3	759 Vinifera	18 AQ 02.2
426 Hippo	11 AQ 10.5	2878 Panacea	18 AQ 03.2
970 Primula	11 AQ 18.0	1478 Vihuri	18 AQ 05.8
1405 Sibelius	11 AQ 40.7	1555 Dejan	18 AQ 08.4
2123 Vitava	11 AQ 46.8	606 Brangane	18 AQ 15.7
180 Garumna	12 AQ 08.9	1930 Lucifer	18 AQ 18.5
3431 Nakano	12 AQ 14.1	141 Lumen	18 AQ 26.0
2824 Franke	12 AQ 22.7	133 Cyrene	18 AQ 38.9
1342 Brabantia	12 AQ 34.2	1118 Hanskya	18 AQ 44.3
506 Marion	12 AQ 42.5	1068 Nofretete	18 AQ 46.7
1086 Nata	12 AQ 54.6	1769 Carlostorres	18 AQ 53.8
228 Agathe	12 AQ 55.5	1612 Hirose	19 AQ 08.0
382 Dodona	13 AQ 05.2	2137 Priscilla	19 AQ 58.8
54 Alexandra	13 AQ 05.5	1477 Bonsdorffia	20 AQ 29.5
212 Medea	13 AQ 17.5	3142 Kilopi	20 AQ 36.1
2168 Swope	13 AQ 37.1	3354 McNair	20 AQ 47.3
1987 Kaplan	13 AQ 49.7	1613 Smiley	20 AQ 49.4
1491 Balduinus	13 AQ 55.8	2324 Janice	20 AQ 57.5
984 Gretia	14 AQ 05.6	949 Hel	21 AQ 00.7
1265 Schweikardia	14 AQ 08.7	1266 Tone	21 AQ 09.3
2479 Sodankyla	14 AQ 23.9	1272 Gefion	21 AQ 09.8
1437 Diomedes	15 AQ 08.0	2033 Basilea	21 AQ 19.8
1892 Lucienne	15 AQ 14.2	1433 Geramtina	21 AQ 19.8
2744 Birgitta	15 AQ 18.2	1891 Gondola	21 AQ 32.8
2882 Tedesco	15 AQ 28.5	2575 Bulgaria	21 AQ 34.5
2125 KarlOntjes	15 AQ 39.5	1471 Tornio	21 AQ 43.6
406 Erna	15 AQ 45.4	2043 Ortutay	21 AQ 45.1
71 Niobe	15 AQ 46.1	96 Aegle	21 AQ 45.4

Name	Position	Name	Position
1241 Dysona	21 AQ 54.1	2892 Filipenko	26 AQ 28.2
1746 Brouwer	22 AQ 00.6	524 Fidelio	26 AQ 28.9
773 Irmintraud	22 AQ 07.5	939 Isberga	26 AQ 42.1
467 Laura	22 AQ 14.8	1274 Delportia	26 AQ 47.3
2545 Verbiest	22 AQ 17.8	400 Ducrosa	26 AQ 48.1
2952 Lilliputia	22 AQ 57.2	372 Palma	26 AQ 50.8
987 Wallia	22 AQ 59.2	2456 Palamedes	26 AQ 52.0
778 Theobalda	23 AQ 10.6	1187 Afra	26 AQ 55.8
1719 Jens	23 AQ 14.8	1430 Somalia	27 AQ 06.4
112 Iphigenia	23 AQ 22.1	1737 Severny	27 AQ 08.7
1474 Beira	23 AQ 26.2	84 Klio	27 AQ 13.7
1000 Piazzia	23 AQ 27.7	1372 Haremari	27 AQ 13.9
1625 The NORC	23 AQ 32.8	2126 Gerasimovich	27 AQ 23.3
1240 Centenaria	23 AQ 37.0	1523 Pieksamaki	27 AQ 24.0
243 Ida	23 AQ 53.9	174 Phaedra	27 AQ 26.2
1282 Utopia	23 AQ 58.5	2108 Otto Schmidt	27 AQ 27.7
2388 Gase	23 AQ 58.6	227 Philosophia	27 AQ 30.8
1639 Bower	23 AQ 59.0	1559 Kustaanheimo	27 AQ 48.8
2570 Porphyro	24 AQ 04.4	324 Bamberga	27 AQ 51.2
587 Hypsipyle	24 AQ 06.5	1106 Cydonia	27 AQ 54.1
709 Fringilla	24 AQ 11.2	1991 Darwin	28 AQ 06.1
1113 Katja	24 AQ 11.5	2267 Agassiz	28 AQ 09.1
2380 Heilongjiang	24 AQ 12.5	2065 Spicer	28 AQ 09.9
1206 Numerowia	24 AQ 14.6	2076 Levin	28 AQ 32.9
200 Dynamene	24 AQ 15.5	2431 Skovoroda	28 AQ 38.4
3464 Owensby	24 AQ 17.9	516 Amherstia	28 AQ 44.5
800 Kressmannia	24 AQ 44.1	1098 Hakone	28 AQ 45.0
706 Hirundo	25 AQ 09.6	1415 Malautra	28 AQ 53.6
1338 Duponta	25 AQ 13.6	577 Rhea	29 AQ 09.4
934 Thuringia	25 AQ 14.9	1079 Mimosa	29 AQ 21.4
1804 Chebotarev	25 AQ 15.2	916 America	29 AQ 27.7
1682 Karel	25 AQ 20.9	1065 Amundsenia	29 AQ 58.1
2390 Nezarka	25 AQ 25.5	1476 Cox	00 PI 11.9
1885 Herero	26 AQ 00.2	2794 Kulik	00 PI 15.1
1467 Mashona	26 AQ 25.6	918 Itha	00 PI 16.0
1683 Castafiore	26 AQ 26.4	1493 Sigrid	00 PI 19.7
765 Mattiaca	26 AQ 27.2	1532 Inari	00 PI 29.5

Name	Position	Name	Position
1662 Hoffmann	00 PI 36.7	1426 Riviera	04 PI 55.3
1463 Nordenmarkia	00 PI 59.3	265 Anna	05 PI 00.1
1510 Charlois	01 PI 07.2	661 Cloelia	05 PI 03.2
2118 Flagstaff	01 PI 08.5	1279 Uganda	05 PI 20.9
1360 Tarka	01 PI 11.2	2758 Cordelia	05 PI 21.0
1697 Koskenniemi	01 PI 16.0	271 Penthesilea	05 PI 31.9
1641 Tana	01 PI 30.4	2791 Paradise	05 PI 35.9
760 Massinga	01 PI 35.8	1157 Arabia	05 PI 44.9
297 Caecilia	01 PI 37.6	1182 Ilona	05 PI 56.9
2451 Dollfus	01 PI 38.9	684 Hildburg	06 PI 06.8
172 Baucis	01 PI 39.7	375 Ursula	06 PI 12.8
1152 Pawona	01 PI 41.2	574 Reginhild	06 PI 28.1
2498 Tsesevich	01 PI 46.3	2973 Paola	06 PI 32.8
1848 Delvaux	01 PI 46.9	1526 Mikkeli	06 PI 38.8
602 Marianna	01 PI 54.7	1620 Geographos	06 PI 42.9
184 Dejopeja	02 PI 05.0	2833 Radishchev	06 PI 48.3
1977 Shura	02 PI 05.6	1803 Zwicky	06 PI 54.2
2095 Parsifal	02 PI 12.8	911 Agamemnon	07 PI 17.8
1404 Ajax	02 PI 16.3	2402 Satpaev	07 PI 19.8
2351 O'Higgins	02 PI 47.9	2098 Zyskin	07 PI 20.0
819 Barnardiana	02 PI 48.1	1677 Tycho Brahe	07 PI 34.4
143 Adria	02 PI 48.5	63 Ausonia	07 PI 38.1
1406 Komppa	02 PI 51.2	2254 Requiem	07 PI 47.6
1486 Marilyn	03 PI 03.5	839 Valborg	07 PI 57.6
78 Diana	03 PI 11.5	1659 Punkaharju	08 PI 10.7
61 Danae	03 PI 23.9	2011 Veteraniya	08 PI 16.6
2389 Dibaj	03 PI 43.3	3087 Beatrice Tinsley	08 PI 20.5
591 Irmgard	03 PI 55.1	1642 Hill	09 PI 00.0
545 Messalina	04 PI 02.9	2662 Kandinsky	09 PI 00.1
2996 Bowman	04 PI 06.0	2737 Kotka	09 PI 25.9
2860 Pasacentennium	04 PI 06.2	3199 Nefertiti	09 PI 26.6
469 Argentina	04 PI 13.2	2713 Luxembourg	09 PI 32.0
687 Tinette	04 PI 14.5	2090 Mizuho	09 PI 41.8
249 Ilse	04 PI 23.8	2875 Lagerkvist	10 PI 03.1
2785 Sedov	04 PI 25.6	1717 Arlon	10 PI 04.4
1777 Gehrels	04 PI 27.5	2055 Dvorak	10 PI 08.6
2537 Gilmore	04 PI 45.7	1916 Boreas	10 PI 15.8

Name	Position	Name	Position
1749 Telamon	10 PI 29.5	1958 Chandra	14 PI 37.5
930 Westphalia	10 PI 35.5	2532 Sutton	14 PI 42.5
120 Lachesis	11 PI 00.2	325 Heidelberga	14 PI 47.7
733 Mocia	11 PI 04.1	1066 Lobelia	14 PI 48.4
626 Notburga	11 PI 20.0	385 Ilmatar	14 PI 56.3
830 Petropolitana	11 PI 22.6	2200 Pasadena	15 PI 05.6
1062 Ljuba	11 PI 26.9	2827 Vellamo	15 PI 07.2
1231 Auricula	11 PI 49.4	1340 Yvette	15 PI 19.0
214 Aschera	11 PI 51.7	134 Sophrosyne	15 PI 50.8
1050 Meta	11 PI 58.8	2036 Sheragul	15 PI 51.7
624 Hektor	12 PI 05.2	1942 Jablunka	15 PI 55.1
2231 Durrell	12 PI 12.0	1077 Campanula	15 PI 55.7
2683 Brian	12 PI 14.0	935 Clivia	16 PI 05.7
605 Juvisia	12 PI 30.5	1145 Robelmonte	16 PI 22.6
573 Recha	12 PI 39.8	399 Persephone	16 PI 26.5
3506 French	12 PI 46.4	366 Vincentina	16 PI 34.3
1538 Detre	12 PI 47.8	1863 Antinous	16 PI 55.7
1049 Gotho	12 PI 52.0	3233 Krisbarons	17 PI 14.8
1859 Kovalevskaya	12 PI 53.6	2644 Victor Jara	17 PI 16.4
192 Nausikaa	12 PI 58.9	2487 Juhani	17 PI 17.4
101 Helena	13 PI 02.5	177 Irma	17 PI 20.6
603 Timandra	13 PI 04.9	804 Hispania	17 PI 25.6
2552 Remek	13 PI 05.9	2374 Vladvysotskij	17 PI 27.7
3217 Seidelmann	13 PI 05.9	1159 Granada	17 PI 29.8
917 Lyka	13 PI 09.5	1524 Joensuu	17 PI 41.9
958 Asplinda	13 PI 15.0	996 Hilaritas	17 PI 42.2
2509 Chukotka	13 PI 24.3	203 Pompeja	17 PI 43.3
672 Astarte	13 PI 25.5	2274 Ehrsson	18 PI 01.7
135 Hertha	13 PI 30.4	2547 Hubei	18 PI 03.3
1002 Olbersia	13 PI 38.5	973 Aralia	18 PI 08.2
303 Josephina	13 PI 55.7	2836 Sobolev	18 PI 09.6
2735 Ellen	14 PI 02.8	2917 Sawyer Hogg	18 PI 26.2
2015 Kachuevskaya	14 PI 06.5	3245 Jensch	18 PI 31.3
717 Wisibada	14 PI 11.2	117 Lomia	18 PI 33.3
2753 Duncan	14 PI 18.3	1887 Virton	18 PI 33.5
1158 Luda	14 PI 24.5	844 Leontina	18 PI 34.6
2358 Bahner	14 PI 28.2	1005 Arago	19 PI 11.2

Name	Position	Name	Position
2157 Ashbrook	19 PI 13.9	902 Probitas	22 PI 35.0
2645 Daphne Plane	19 PI 20.2	1921 Pala	22 PI 36.4
575 Renate	19 PI 21.4	224 Oceana	22 PI 39.0
193 Ambrosia	19 PI 39.5	1416 Renauxa	22 PI 39.2
2130 Evdokiya	19 PI 42.5	2079 Jacchia	22 PI 43.0
1242 Zambesia	19 PI 45.7	1382 Gerti	22 PI 44.6
2939 Coconino	19 PI 49.8	1997 Leverrier	22 PI 48.0
1924 Horus	19 PI 53.6	833 Monica	22 PI 50.2
3150 Tosa	19 PI 54.5	2599 Veseli	23 PI 13.7
2748 Patrick Gene	20 PI 08.2	1884 Skip	23 PI 18.1
2756 Dzhangar	20 PI 10.3	3350 Scobee	23 PI 20.2
2923 Schuyler	20 PI 12.6	35 Leukothea	23 PI 26.9
108 Hecuba	20 PI 15.8	98 Ianthe	23 PI 40.5
659 Nestor	20 PI 24.8	990 Yerkes	23 PI 44.6
1135 Colchis	20 PI 30.2	333 Badenia	23 PI 45.5
3043 San Diego	20 PI 34.2	1481 Tubingia	23 PI 49.9
2101 Adonis	20 PI 36.9	1308 Halleria	24 PI 07.6
231 Vindobona	20 PI 40.7	1488 Aura	24 PI 10.6
2159 Kukkamaki	20 PI 59.7	388 Charybdis	24 PI 15.9
658 Asteria	21 PI 00.1	841 Arabella	24 PI 18.7
2888 Hodgson	21 PI 02.2	1906 Naef	24 PI 24.6
436 Patricia	21 PI 02.9	169 Zelia	24 PI 25.3
1678 Hveen	21 PI 12.8	613 Ginevra	24 PI 33.4
955 Alstede	21 PI 31.7	327 Columbia	24 PI 34.6
1381 Danubia	21 PI 35.2	1418 Fayeta	24 PI 41.2
1998 Titius	21 PI 35.8	356 Liguria	24 PI 51.8
693 Zerbinetta	21 PI 36.4	337 Devosa	25 PI 01.1
355 Gabriella	21 PI 43.2	3012 Minsk	25 PI 01.4
1268 Libya	21 PI 50.2	708 Raphaela	25 PI 02.2
2646 Abetti	22 PI 08.6	1557 Roehla	25 PI 03.7
1454 Kalevala	22 PI 11.3	1640 Nemo	25 PI 05.6
328 Gudrun	22 PI 13.0	637 Chrysothemis	25 PI 16.1
829 Academia	22 PI 13.1	1184 Gaea	25 PI 31.0
2915 Moskvina	22 PI 21.4	2726 Kotelnikov	25 PI 37.9
1877 Marsden	22 PI 29.6	2637 Bobrovnikoff	25 PI 41.9
2760 Kacha	22 PI 33.4	2626 Belnika	25 PI 49.6
1923 Osiris	22 PI 34.0	616 Elly	25 PI 51.2

Name	Position	Name	Position
29 Amphitrite	25 PI 57.1	3402 Wisdom	27 PI 36.6
1710 Gotard	26 PI 13.6	1318 Nerina	27 PI 53.5
2793 Valdaj	26 PI 18.8	1392 Pierre	27 PI 56.4
309 Fraternitas	26 PI 21.7	2252 CERGA	28 PI 04.1
2890 Vilyujsk	26 PI 24.4	3190 Aposhanskij	28 PI 09.6
1116 Catriona	26 PI 27.1	36 Atalante	28 PI 15.1
1919 Clemence	26 PI 28.4	2128 Wetherill	28 PI 16.3
1700 Zvezdara	26 PI 31.1	2514 Taiyuan	28 PI 21.2
1981 Midas	26 PI 32.9	1913 Sekanina	28 PI 29.9
2001 Einstein	26 PI 33.4	1470 Carla	28 PI 58.0
1608 Munoz	26 PI 33.8	3314 Beals	29 PI 03.1
632 Pyrrha	26 PI 44.4	1396 Outeniqua	29 PI 08.6
711 Marmulla	26 PI 47.5	75 Eurydike	29 PI 11.8
649 Josefa	26 PI 52.4	1018 Arnolda	29 PI 13.7
2414 Vibeke	26 PI 53.2	2930 Euripides	29 PI 16.5
2809 Vernadskij	26 PI 56.7	2088 Sahlia	29 PI 18.5
2869 Nepryadva	27 PI 02.0	1610 Mirnaya	29 PI 18.7
501 Urhixidur	27 PI 08.3	2607 Yakutia	29 PI 27.8
948 Jucunda	27 PI 12.7	2486 Metsahovi	29 PI 33.8
1310 Villigera	27 PI 16.0	2174 Asmodeus	29 PI 39.6
493 Griseldis	27 PI 19.7	620 Drakonia	29 PI 45.2
1587 Kahrstedt	27 PI 28.4	247 Eukrate	29 PI 52.0
1121 Natascha	27 PI 35.6	1380 Volodia	29 PI 57.5

Ephemeris	80			433			1221			1489			114			3018			2041		
0 Hours ET	Sappho			Eros			Amor			Attila			Kassandra			Godiva			Lancelot		
Mo Dy Year	h long	g long	g lat	h long	g long	g lat	h long	g long	g lat	h long	g long	g lat	h long	g long	g lat	h long	g long	g lat	h long	g long	g lat
1/ 2/1920	26 9	8 25	1 59	29 56	0pi39	7 6	0 55	12 8	1 51	1 30	15 18	2 8	4 34	22 8	0 42	15 46	7 17	2 17	17 31	15 52	1 4
1/12/1920	28 36	12 55	2 18	4 48	6 50	7 10	2 35	15 18	2 5	2 55	16 37	2 4	6 17	25 35	0 34	18 23	11 15	2 13	19 5	19 25	1 1
1/22/1920	1sa 5	17 24	2 38	9 51	13 14	7 13	4 13	18 37	2 17	4 20	18 20	2 1	8 0	29 8	0 26	21 4	15 2	2 9	20 41	22 59	0 57
2/ 1/1920	3 37	21 49	3 0	15 8	19 52	7 13	5 50	22 4	2 29	5 46	20 21	1 58	9 43	2 47	0 19	23 47	18 36	2 3	22 17	26 33	0 54
2/11/1920	6 11	26 12	3 24	20 39	26 43	7 11	7 27	25 37	2 41	7 12	22 40	1 55	11 27	6 28	0 12	26 33	21 54	1 57	23 54	0aq 7	0 51
2/21/1920	8 49	0cp29	3 50	26 26	3 47	7 6	9 3	29 15	2 52	8 39	25 13	1 53	13 10	10 12	0 5	29 22	24 52	1 49	25 33	3 40	0 48
3/ 2/1920	11 30	4 40	4 19	2ta29	11 5	6 58	10 39	2 57	3 4	10 7	27 58	1 51	14 53	13 58	0s 2	2 15	27 25	1 39	27 12	7 9	0 45
3/12/1920	14 14	8 43	4 51	8 49	18 35	6 46	12 15	6 41	3 15	11 35	0ta53	1 49	16 37	17 44	0 9	5 10	29 30	1 28	28 53	10 34	0 42
3/22/1920	17 2	12 36	5 26	15 29	26 18	6 30	13 50	10 27	3 27	13 4	3 56	1 48	18 21	21 29	0 15	8 9	0 59	1 13	0aq35	13 54	0 38
4/ 1/1920	19 53	16 17	6 6	22 27	4 15	6 10	15 26	14 14	3 40	14 33	7 6	1 47	20 5	25 13	0 22	11 11	1 48	0 56	2 18	17 8	0 35
4/11/1920	22 48	19 44	6 50	29 44	12 23	5 45	17 1	18 2	3 53	16 4	10 22	1 47	21 49	28 55	0 29	14 17	1 51R	0 34	4 2	20 13	0 32
4/21/1920	25 48	22 51	7 39	7 20	20 44	5 14	18 37	21 49	4 7	17 35	13 42	1 47	23 34	2 34	0 37	17 26	1 6	0 9	5 48	23 8	0 28
5/ 1/1920	28 51	25 36	8 34	15 14	29 16	4 39	20 13	25 35	4 22	19 7	17 5	1 47	25 19	6 8	0 44	20 39	29sc35	0 20	7 35	25 51	0 24
5/11/1920	1 58	27 54	9 34	23 25	7 58	3 58	21 49	29 19	4 38	20 39	20 30	1 47	27 4	9 38	0 53	23 55	27 28	0 51	9 23	28 20	0 19
5/21/1920	5 9	29 38	10 41	1cn50	16 49	3 12	23 26	3 1	4 55	22 13	23 57	1 48	28 50	13 1	1 1	27 15	25 3	1 23	11 13	0pi32	0 14
5/31/1920	8 25	0 42	11 53	10 27	25 48	2 23	25 4	6 40	5 13	23 47	27 23	1 49	0ar36	16 16	1 11	0sa38	22 40	1 54	13 3	2 24	0 8
6/10/1920	11 46	1 2	13 8	19 13	4 53	1 30	26 42	10 15	5 34	25 22	0gm50	1 50	2 23	19 23	1 21	4 5	20 43	2 21	14 56	3 53	0 2
6/20/1920	15 11	0 31	14 22	28 4	14 2	0 35	28 22	13 46	5 56	26 58	4 15	1 52	4 10	22 18	1 32	7 36	19 27	2 45	16 50	4 56	0 5
6/30/1920	18 41	29 12	15 29	6 56	23 13	0s20	0ta 2	17 10	6 21	28 35	7 37	1 54	5 57	25 1	1 44	11 10	19 3	3 3	18 45	5 28	0 13
7/10/1920	22 15	27 12	16 22	15 46	2le25	1 15	1 44	20 28	6 48	0gm13	10 57	1 56	7 45	27 28	1 57	14 48	19 31	3 17	20 41	5 28R	0 22
7/20/1920	25 54	24 47	16 54	24 30	11 35	2 7	3 28	23 38	7 19	1 52	14 12	1 59	9 34	29 38	2 11	18 28	20 49	3 28	22 39	4 55	0 31
7/30/1920	29 37	22 21	17 1	3vi 4	20 41	2 56	5 13	26 36	7 53	3 32	17 21	2 3	11 24	1 26	2 27	22 12	22 53	3 36	24 39	3 48	0 41
8/ 9/1920	3 25	20 17	16 42	11 25	29 42	3 41	6 59	29 23	8 31	5 13	20 23	2 6	13 14	2 50	2 45	25 59	25 35	3 42	26 40	2 14	0 52
8/19/1920	7 18	18 57	16 3	19 31	8 37	4 20	8 48	1 53	9 15	6 55	23 16	2 10	15 4	3 45	3 4	29 49	28 52	3 46	28 43	0 21	1 2
8/29/1920	11 15	18 32	15 9	27 20	17 23	4 54	10 39	4 4	10 3	8 38	25 59	2 15	16 56	4 8	3 24	3 41	2 37	3 48	0pi47	28 20	1 12
9/ 8/1920	15 15	19 4	14 8	4 52	26 0	5 22	12 32	5 52	10 58	10 22	28 28	2 20	18 48	3 55	3 46	7 35	6 46	3 50	2 53	26 25	1 20
9/18/1920	19 20	20 31	13 3	12 4	4 26	5 44	14 28	7 10	11 59	12 7	0cn42	2 26	20 41	3 5	4 8	11 32	11 16	3 50	5 1	24 50	1 27
9/28/1920	23 28	22 47	11 59	18 58	12 42	6 1	16 27	7 52	13 7	13 53	2 36	2 33	22 35	1 41	4 29	15 30	16 3	3 50	7 10	23 43	1 33
10/ 8/1920	27 39	25 45	10 56	25 33	20 46	6 12	18 29	7 52R	14 21	15 40	4 8	2 40	24 30	29ar47	4 47	19 29	21 4	3 49	9 21	23 12	1 37
10/18/1920	1pi53	29 20	9 57	1sc50	28 39	6 19	20 34	7 3	15 38	17 28	5 14	2 47	26 26	27 33	5 2	23 29	26 18	3 47	11 33	23 18D	1 40
10/28/1920	6 10	3 25	9 1	7 50	6 21	6 22	22 43	5 21	16 54	19 18	5 50	2 54	28 23	25 15	5 12	27 30	1cp42	3 45	13 47	24 2	1 42
11/ 7/1920	10 28	7 55	8 8	13 33	13 51	6 21	24 57	2 46	18 2	21 9	5 52R	3 2	0ta20	23 6	5 17	1aq31	7 14	3 42	16 2	25 20	1 44
11/17/1920	14 47	12 46	7 18	19 2	21 11	6 17	27 15	29ta29	18 55	23 0	5 20	3 8	2 19	21 21	5 17	5 32	12 54	3 39	18 19	27 9	1 45
11/27/1920	19 8	17 54	6 32	24 16	28 19	6 10	29 38	25 48	19 24	24 54	4 14	3 14	4 19	20 9	5 14	9 33	18 39	3 35	20 38	29 27	1 46
12/ 7/1920	23 29	23 16	5 48	29 18	5 18	6 2	2 6	22 13	19 25	26 48	2 37	3 17	6 20	19 36	5 7	13 32	24 28	3 31	22 58	2 8	1 46
12/17/1920	27 50	28 50	5 7	4 7	12 7	5 51	4 41	19 8	19 1	28 44	0 40	3 18	8 21	19 43D	4 58	17 31	0aq20	3 27	25 20	5 10	1 47
12/27/1920	2ar10	4 32	4 27	8 46	18 46	5 39	7 23	16 55	18 17	0cn40	28 34	3 15	10 25	20 28	4 49	21 27	6 14	3 23	27 43	8 31	1 47
1/ 6/1921	6 29	10 20	3 51	13 16	25 16	5 25	10 11	15 43	17 21	2 39	26 33	3 9	12 29	21 47	4 40	25 22	12 9	3 18	0ar 8	12 6	1 48
1/16/1921	10 46	16 15	3 16	17 36	1cp38	5 11	13 9	15 35D	16 19	4 38	24 52	3 1	14 35	23 38	4 31	29 15	18 5	3 13	2 34	15 54	1 49
1/26/1921	15 2	22 12	2 42	21 49	7 51	4 54	16 15	16 25	15 17	6 39	23 40	2 51	16 42	25 55	4 22	3 4	23 59	3 8	5 1	19 53	1 49

Ephemeris 0 Hours ET	80 Sappho			433 Eros			1221 Amor			1489 Attila			114 Kassandra			3018 Godiva			2041 Lancelot		
Mo Dy Year	h long	g long	g lat	h long	g long	g lat	h long	g long	g lat	h long	g long	g lat	h long	g long	g lat	h long	g long	g lat	h long	g long	g lat
2/ 5/1921	19 15	28 12	2 11	25 55	13 55	4 37	19 31	18 8	14 17	8 41	23 3	2 40	18 50	28 34	4 15	6 52	29 52	3 3	7 30	24 2	1 50
2/15/1921	23 26	4 13	1 41	29 55	19 52	4 19	22 59	20 37	13 20	10 44	23 4D	2 28	20 59	1 33	4 8	10 36	5 43	2 57	10 0	28 18	1 51
2/25/1921	27 33	10 15	1 12	3 49	25 40	3 59	26 39	23 47	12 26	12 48	23 42	2 17	23 10	4 49	4 2	14 17	11 31	2 52	12 31	2 39	1 52
3/ 7/1921	1ta37	16 15	0 45	7 39	1aq20	3 38	0cn33	27 32	11 35	14 54	24 53	2 6	25 23	8 18	3 56	17 55	17 15	2 46	15 3	7 6	1 53
3/17/1921	5 38	22 14	0 19	11 25	6 53	3 15	4 43	1gm48	10 46	17 1	26 35	1 56	27 36	11 59	3 51	21 29	22 55	2 40	17 37	11 37	1 55
3/27/1921	9 35	28 10	0s 6	15 7	12 16	2 51	9 11	6 33	9 59	19 10	28 43	1 47	29 52	15 50	3 47	25 0	28 30	2 35	20 11	16 11	1 56
4/ 6/1921	13 28	4 4	0 30	18 47	17 31	2 24	13 57	11 45	9 12	21 19	1cn14	1 38	2 9	19 49	3 43	28 27	4 0	2 29	22 46	20 47	1 58
4/16/1921	17 17	9 55	0 52	22 25	22 37	1 55	19 6	17 22	8 25	23 30	4 4	1 30	4 27	23 56	3 40	1 50	9 25	2 22	25 22	25 24	2 0
4/26/1921	21 2	15 42	1 15	26 1	27 33	1 22	24 38	23 24	7 37	25 43	7 11	1 23	6 47	28 8	3 38	5 10	14 43	2 16	27 59	0ta 1	2 2
5/ 6/1921	24 42	21 25	1 36	29 37	2pi18	0 47	0le38	29 49	6 47	27 56	10 31	1 16	9 8	2 25	3 35	8 26	19 55	2 10	0ta36	4 38	2 4
5/16/1921	28 19	27 4	1 57	3 12	6 51	0 6	7 6	6 39	5 54	0le11	14 4	1 10	11 31	6 46	3 34	11 38	25 0	2 3	3 14	9 14	2 6
5/26/1921	1 51	2gm38	2 17	6 47	11 10	0 39	14 6	13 54	4 57	2 27	17 46	1 4	13 56	11 11	3 32	14 47	29 56	1 57	5 52	13 49	2 8
6/ 5/1921	5 19	8 7	2 37	10 23	15 12	1 31	21 39	21 33	3 56	4 44	21 37	0 59	16 22	15 38	3 31	17 53	4 44	1 50	8 30	18 21	2 11
6/15/1921	8 43	13 31	2 56	14 1	18 55	2 31	29 47	29 36	2 51	7 2	25 36	0 53	18 50	20 8	3 30	20 54	9 23	1 43	11 9	22 49	2 14
6/25/1921	12 3	18 49	3 16	17 40	22 14	3 41	8 29	8 5	1 41	9 21	29 40	0 48	21 20	24 39	3 30	23 53	13 51	1 36	13 47	27 14	2 17
7/ 5/1921	15 18	24 2	3 35	21 22	25 3	5 2	17 44	16 57	0 27	11 41	3 50	0 44	23 51	29 11	3 29	26 48	18 8	1 28	16 26	1gm33	2 20
7/15/1921	18 30	29 8	3 55	25 7	27 15	6 37	27 27	26 11	0 50	14 3	8 4	0 39	26 24	3 44	3 29	29 39	22 12	1 20	19 4	5 46	2 23
7/25/1921	21 38	4 8	4 15	28 56	28 40	8 27	7 31	5 46	2 7	16 25	12 21	0 35	28 59	8 16	3 30	2 28	26 2	1 11	21 43	9 52	2 27
8/ 4/1921	24 43	9 0	4 35	2 49	29 8	10 34	17 47	15 37	3 22	18 48	16 41	0 30	1 35	12 49	3 30	5 13	29 35	1 2	24 20	13 49	2 31
8/14/1921	27 43	13 45	4 56	6 48	28 26	12 57	28 5	25 40	4 31	21 12	21 3	0 26	4 13	17 19	3 31	7 56	2 49	0 52	26 57	17 35	2 36
8/24/1921	0cn40	18 21	5 18	10 52	26 26	15 28	8 12	5 50	5 34	23 37	25 26	0 22	6 52	21 48	3 32	10 35	5 40	0 41	29 34	21 9	2 41
9/ 3/1921	3 34	22 47	5 41	15 4	23 9	17 56	17 59	16 1	6 25	26 2	29 51	0 18	9 33	26 14	3 33	13 12	8 6	0 29	2 10	24 27	2 46
9/13/1921	6 24	27 2	6 5	19 23	18 55	19 59	27 18	26 6	7 6	28 29	4 15	0 13	12 16	0le35	3 34	15 46	10 1	0 16	4 45	27 28	2 51
9/23/1921	9 11	1le 5	6 31	23 50	14 23	21 22	6 2	6 0	7 34	0vi55	8 38	0 9	15 0	4 51	3 35	18 18	11 21	0 1	7 19	0cn 7	2 57
10/ 3/1921	11 55	4 53	6 59	28 27	10 23	21 56	14 10	15 36	7 50	3 23	12 59	0 4	17 45	9 0	3 37	20 47	12 2	0 15	9 53	2 22	3 4
10/13/1921	14 36	8 25	7 29	3 15	7 34	21 48	21 41	24 52	7 56	5 51	17 18	0n 0	20 32	13 1	3 38	23 14	11 58R	0 33	12 25	4 7	3 10
10/23/1921	17 14	11 37	8 1	8 15	6 16	21 10	28 36	3sa45	7 53	8 19	21 33	0 5	23 21	16 50	3 40	25 38	11 9	0 53	14 56	5 18	3 17
11/ 2/1921	19 50	14 25	8 37	13 27	6 28D	20 14	4 57	12 14	7 44	10 47	25 43	0 11	26 11	20 26	3 42	28 0	9 35	1 13	17 26	5 52	3 23
11/12/1921	22 22	16 46	9 15	18 53	8 4	19 10	10 48	20 18	7 30	13 16	29 47	0 16	29 2	23 44	3 43	0gm20	7 24	1 34	19 55	5 45R	3 27
11/22/1921	24 53	18 36	9 57	24 35	10 51	18 3	16 11	28 0	7 12	15 45	3 42	0 22	1 54	26 42	3 44	2 39	4 51	1 53	22 22	4 58	3 30
12/ 2/1921	27 20	19 48	10 41	0ta33	14 38	16 54	21 8	5 20	6 53	18 14	7 28	0 28	4 47	29 15	3 45	4 55	2 12	2 10	24 48	3 33	3 31
12/12/1921	29 46	20 18	11 27	6 48	19 16	15 45	25 44	12 19	6 32	20 43	11 1	0 35	7 41	1 17	3 45	7 10	29ta48	2 23	27 13	1 41	3 27
12/22/1921	2 9	20 1	12 12	13 21	24 36	14 34	0aq 0	19 0	6 10	23 12	14 19	0 42	10 37	2 43	3 44	9 22	27 53	2 34	29 36	29gm36	3 21
1/ 1/1922	4 30	18 57	12 53	20 13	0ar34	13 22	3 59	25 24	5 49	25 41	17 20	0 51	13 33	3 27	3 40	11 34	26 39	2 41	1 58	27 32	3 10
1/11/1922	6 50	17 8	13 27	27 24	7 6	12 8	7 42	1aq32	5 28	28 10	19 59	1 0	16 30	3 25R	3 33	13 43	26 8	2 45	4 18	25 46	2 57
1/21/1922	9 7	14 44	13 49	4 54	14 6	10 50	11 12	7 25	5 7	0li38	22 14	1 10	19 27	2 37	3 23	15 52	26 20	2 48	6 37	24 29	2 42
1/31/1922	11 22	12 1	13 55	12 42	21 32	9 27	14 29	13 5	4 48	3 6	23 59	1 20	22 25	1 6	3 8	17 59	27 11	2 49	8 54	23 50	2 27
2/10/1922	13 36	9 18	13 45	20 48	29 21	8 1	17 35	18 33	4 28	5 33	25 10	1 32	25 24	29le 2	2 48	20 4	28 37	2 49	11 9	23 49D	2 12
2/20/1922	15 48	6 55	13 20	29 8	7 31	6 29	20 32	23 48	4 10	8 0	25 45	1 44	28 23	26 44	2 23	22 9	0gm31	2 49	13 23	24 25	1 58
3/ 2/1922	17 58	5 6	12 44	7 42	15 59	4 54	23 20	28 52	3 52	10 26	25 41R	1 57	1vi22	24 31	1 55	24 12	2 50	2 49	15 35	25 35	1 44

Ephemeris 0 Hours ET Mo Dy Year	80 Sappho h long	 g long	 g lat	433 Eros h long	 g long	 g lat	1221 Amor h long	 g long	 g lat	1489 Attila h long	 g long	 g lat	114 Kassandra h long	 g long	 g lat	3018 Godiva h long	 g long	 g lat	2041 Lancelot h long	 g long	 g lat
3/12/1922	20 7	3 59	12 2	16 25	24 43	3 15	26 0	3 44	3 35	12 51	24 57	2 10	4 21	22 43	1 27	26 14	5 29	2 48	17 46	27 15	1 32
3/22/1922	22 15	3 37	11 17	25 15	3gm40	1 35	28 33	8 25	3 18	15 16	23 38	2 22	7 20	21 34	0 58	28 15	8 24	2 48	19 55	29 20	1 21
4/ 1/1922	24 21	3 56	10 32	4le 7	12 48	0s 4	1pi 0	12 55	3 2	17 40	21 53	2 33	10 19	21 10	0 32	0cn15	11 33	2 48	22 2	1 47	1 11
4/11/1922	26 26	4 53	9 50	12 58	22 3	1 41	3 22	17 14	2 46	20 3	19 54	2 41	13 18	21 32	0 8	2 15	14 53	2 48	24 8	4 30	1 2
4/21/1922	28 30	6 22	9 10	21 44	1cn22	3 11	5 38	21 21	2 29	22 25	17 56	2 46	16 17	22 37	0 13	4 13	18 22	2 49	26 12	7 29	0 54
5/ 1/1922	0vi33	8 20	8 34	0vi21	10 44	4 34	7 49	25 15	2 13	24 47	16 14	2 48	19 15	24 21	0 31	6 11	21 59	2 49	28 15	10 38	0 47
5/11/1922	2 35	10 40	8 1	8 47	20 5	5 47	9 57	28 55	1 56	27 7	14 59	2 48	22 13	26 37	0 47	8 8	25 41	2 50	0le16	13 58	0 40
5/21/1922	4 36	13 19	7 31	16 58	29 22	6 49	12 0	2 21	1 39	29 26	14 19	2 46	25 10	29 21	1 1	10 5	29 27	2 51	2 16	17 24	0 34
5/31/1922	6 36	16 15	7 4	24 53	8 32	7 38	14 0	5 31	1 21	1 44	14 15D	2 42	28 6	2 28	1 13	12 1	3 17	2 53	4 14	20 57	0 28
6/10/1922	8 36	19 23	6 40	2li30	17 34	8 15	15 57	8 23	1 2	4 1	14 47	2 38	1li 2	5 55	1 24	13 56	7 10	2 55	6 11	24 34	0 23
6/20/1922	10 34	22 43	6 18	9 48	26 26	8 40	17 52	10 54	0 41	6 16	15 52	2 33	3 56	9 36	1 34	15 51	11 5	2 57	8 6	28 14	0 18
6/30/1922	12 32	26 11	5 58	16 48	5 5	8 54	19 43	13 2	0 19	8 31	17 26	2 28	6 50	13 31	1 43	17 46	15 1	2 59	9 59	1 57	0 13
7/10/1922	14 30	29 47	5 40	23 29	13 31	8 59	21 33	14 42	0s 6	10 44	19 27	2 23	9 42	17 37	1 51	19 40	18 57	3 2	11 52	5 42	0 9
7/20/1922	16 27	3 29	5 23	29 51	21 44	8 56	23 20	15 51	0 33	12 56	21 49	2 18	12 33	21 51	1 58	21 35	22 53	3 6	13 43	9 26	0 4
7/30/1922	18 23	7 17	5 8	5 57	29 44	8 45	25 5	16 25	1 3	15 7	24 30	2 14	15 23	26 13	2 5	23 29	26 49	3 10	15 32	13 11	0 0
8/ 9/1922	20 20	11 8	4 54	11 45	7 30	8 30	26 49	16 18R	1 36	17 16	27 26	2 10	18 12	0li41	2 11	25 22	0le43	3 14	17 20	16 55	0 4
8/19/1922	22 16	15 3	4 42	17 18	15 3	8 10	28 31	15 28	2 11	19 24	0sc36	2 7	20 59	5 13	2 17	27 16	4 35	3 19	19 7	20 37	0 8
8/29/1922	24 11	19 1	4 30	22 37	22 25	7 47	0ar12	13 54	2 49	21 31	3 56	2 3	23 45	9 50	2 22	29 10	8 24	3 24	20 53	24 16	0 13
9/ 8/1922	26 7	23 1	4 19	27 43	29 35	7 22	1 51	11 38	3 27	23 37	7 26	2 1	26 30	14 29	2 28	1 3	12 10	3 31	22 37	27 53	0 17
9/18/1922	28 2	27 2	4 9	2 36	6 35	6 55	3 30	8 51	4 5	25 41	11 3	1 58	29 13	19 11	2 33	2 57	15 51	3 38	24 20	1vi25	0 21
9/28/1922	29 57	1li 4	3 59	7 19	13 27	6 27	5 8	5 44	4 38	27 44	14 47	1 56	1 54	23 54	2 38	4 51	19 26	3 45	26 2	4 51	0 26
10/ 8/1922	1 53	5 5	3 50	11 51	20 10	6 0	6 45	2 39	5 6	29 46	18 35	1 54	4 34	28 38	2 43	6 45	22 54	3 54	27 43	8 11	0 30
10/18/1922	3 48	9 6	3 41	16 14	26 46	5 31	8 21	29pi52	5 27	1 46	22 27	1 52	7 13	3 22	2 48	8 39	26 12	4 4	29 22	11 24	0 35
10/28/1922	5 44	13 6	3 33	20 30	3 15	5 4	9 57	27 38	5 42	3 45	26 21	1 51	9 49	8 6	2 54	10 34	29 20	4 15	1 1	14 27	0 40
11/ 7/1922	7 40	17 3	3 25	24 38	9 39	4 36	11 33	26 6	5 51	5 43	0sa18	1 50	12 24	12 48	2 59	12 29	2 14	4 28	2 38	17 18	0 46
11/17/1922	9 36	20 57	3 17	28 39	15 58	4 9	13 9	25 18	5 56	7 39	4 15	1 49	14 58	17 29	3 5	14 24	4 52	4 41	4 14	19 57	0 52
11/27/1922	11 33	24 46	3 10	2 35	22 12	3 42	14 44	25 13D	5 57	9 35	8 12	1 49	17 30	22 7	3 11	16 20	7 10	4 57	5 50	22 19	0 58
12/ 7/1922	13 30	28 29	3 2	6 27	28 22	3 15	16 19	25 48	5 57	11 29	12 7	1 48	20 0	26 42	3 17	18 16	9 4	5 14	7 24	24 24	1 5
12/17/1922	15 27	2 6	2 54	10 14	4 29	2 49	17 55	26 58	5 56	13 21	16 1	1 48	22 29	1sa12	3 24	20 13	10 30	5 32	8 57	26 7	1 13
12/27/1922	17 25	5 33	2 46	13 57	10 33	2 23	19 31	28 37	5 55	15 13	19 52	1 49	24 56	5 37	3 31	22 11	11 24	5 51	10 30	27 26	1 21
1/ 6/1923	19 24	8 49	2 38	17 38	16 34	1 57	21 7	0ar43	5 54	17 3	23 38	1 50	27 21	9 56	3 39	24 9	11 40	6 11	12 1	28 18	1 30
1/16/1923	21 24	11 51	2 29	21 16	22 33	1 31	22 43	3 9	5 53	18 52	27 20	1 51	29 45	14 7	3 47	26 8	11 16	6 31	13 32	28 40	1 39
1/26/1923	23 24	14 38	2 18	24 53	28 30	1 5	24 21	5 54	5 53	20 40	0cp54	1 52	2 7	18 10	3 56	28 8	10 9	6 49	15 2	28 29	1 49
2/ 5/1923	25 25	17 4	2 7	28 29	4 25	0 39	25 59	8 54	5 54	22 27	4 21	1 54	4 27	22 1	4 6	0vi 9	8 24	7 3	16 30	27 47	1 59
2/15/1923	27 27	19 8	1 54	2 4	10 18	0 13	27 38	12 7	5 55	24 13	7 39	1 56	6 46	25 41	4 17	2 11	6 7	7 10	17 58	26 35	2 8
2/25/1923	29 31	20 43	1 39	5 39	16 11	0 15	29 18	15 31	5 57	25 57	10 46	1 58	9 4	29 7	4 29	4 14	3 33	7 11	19 26	24 57	2 16
3/ 7/1923	1 35	21 46	1 22	9 15	22 2	0 42	0 59	19 4	6 1	27 41	13 40	2 1	11 20	2 16	4 43	6 18	1 0	7 3	20 52	23 4	2 23
3/17/1923	3 41	22 12	1 1	12 52	27 53	1 11	2 42	22 44	6 5	29 23	16 20	2 4	13 34	5 7	4 57	8 23	28 43	6 48	22 18	21 4	2 28
3/27/1923	5 48	21 58	0 38	16 30	3 44	1 41	4 26	26 31	6 10	1 5	18 42	2 7	15 47	7 35	5 12	10 29	26 58	6 28	23 43	19 9	2 32
4/ 6/1923	7 56	21 0	0 10	20 11	9 35	2 12	6 12	0ta24	6 16	2 45	20 45	2 11	17 59	9 38	5 29	12 37	25 53	6 4	25 8	17 30	2 33

Ephemeris	80			433			1221			1489			114			3018			2041		
0 Hours ET	Sappho			Eros			Amor			Attila			Kassandra			Godiva			Lancelot		
Mo Dy Year	h long	g long	g lat	h long	g long	g lat	h long	g long	g lat	h long	g long	g lat	h long	g long	g lat	h long	g long	g lat	h long	g long	g lat
4/16/1923	10 7	19 21	0 21	23 55	15 27	2 45	7 59	4 21	6 23	4 25	22 27	2 15	20 9	11 12	5 47	14 46	25 30	5 39	26 31	16 14	2 33
4/26/1923	12 18	17 8	0 54	27 43	21 20	3 19	9 49	8 23	6 31	6 3	23 43	2 19	22 18	12 13	6 5	16 56	25 49	5 14	27 55	15 25	2 31
5/ 6/1923	14 32	14 34	1 28	1pi35	27 15	3 55	11 41	12 29	6 40	7 41	24 32	2 24	24 26	12 39	6 23	19 8	26 46	4 50	29 17	15 6	2 29
5/16/1923	16 47	11 56	2 2	5 31	3 12	4 34	13 36	16 37	6 50	9 17	24 52	2 28	26 32	12 26	6 41	21 22	28 18	4 27	0li39	15 16	2 26
5/26/1923	19 4	9 31	2 32	9 34	9 13	5 14	15 33	20 49	7 2	10 53	24 40	2 32	28 37	11 35	6 56	23 37	0vi20	4 6	2 0	15 53	2 23
6/ 5/1923	21 23	7 37	2 59	13 43	15 16	5 57	17 33	25 4	7 15	12 28	23 57	2 35	0cp41	10 9	7 7	25 54	2 47	3 47	3 21	16 55	2 19
6/15/1923	23 45	6 22	3 22	18 0	21 25	6 43	19 37	29 21	7 29	14 1	22 46	2 36	2 43	8 15	7 13	28 13	5 37	3 29	4 41	18 18	2 16
6/25/1923	26 9	5 51	3 41	22 24	27 39	7 31	21 44	3 41	7 45	15 35	21 11	2 36	4 45	6 4	7 12	0li34	8 45	3 13	6 1	20 1	2 14
7/ 5/1923	28 35	6 6	3 56	26 58	4 0	8 22	23 56	8 3	8 3	17 7	19 20	2 35	6 45	3 50	7 4	2 57	12 10	2 58	7 20	22 0	2 11
7/15/1923	1sa 4	7 1	4 8	1ar43	10 28	9 15	26 11	12 27	8 22	18 38	17 24	2 31	8 45	1 48	6 49	5 23	15 50	2 44	8 39	24 13	2 9
7/25/1923	3 35	8 35	4 18	6 39	17 4	10 11	28 32	16 53	8 43	20 9	15 34	2 25	10 43	0 10	6 30	7 50	19 41	2 31	9 57	26 37	2 7
8/ 4/1923	6 10	10 42	4 27	11 47	23 51	11 9	0gm58	21 22	9 7	21 39	13 58	2 18	12 40	29 4	6 7	10 20	23 44	2 19	11 15	29 12	2 6
8/14/1923	8 47	13 18	4 34	17 9	0gm49	12 8	3 29	25 53	9 33	23 8	12 45	2 10	14 36	28 33	5 43	12 52	27 57	2 8	12 33	1 55	2 5
8/24/1923	11 28	16 19	4 40	22 45	8 0	13 7	6 8	0cn25	10 2	24 37	12 0	2 2	16 32	28 38D	5 19	15 27	2 19	1 57	13 50	4 44	2 4
9/ 3/1923	14 12	19 42	4 46	28 38	15 25	14 4	8 53	5 0	10 34	26 5	11 44	1 53	18 26	29 18	4 56	18 5	6 48	1 46	15 6	7 39	2 4
9/13/1923	16 59	23 25	4 51	4 47	23 5	14 57	11 46	9 36	11 10	27 32	11 57	1 45	20 20	0cp28	4 34	20 45	11 25	1 36	16 23	10 38	2 4
9/23/1923	19 50	27 25	4 56	11 15	1cn 2	15 44	14 47	14 14	11 49	28 58	12 39	1 37	22 13	2 6	4 13	23 28	16 9	1 26	17 39	13 41	2 4
10/ 3/1923	22 45	1sa39	5 0	18 1	9 16	16 20	17 59	18 53	12 33	0aq24	13 46	1 29	24 4	4 8	3 55	26 14	20 58	1 16	18 55	16 45	2 5
10/13/1923	25 43	6 8	5 5	25 6	17 47	16 41	21 21	23 34	13 21	1 50	15 16	1 22	25 56	6 31	3 38	29 3	25 53	1 6	20 10	19 50	2 6
10/23/1923	28 46	10 48	5 9	2gm30	26 34	16 42	24 55	28 14	14 14	3 14	17 5	1 16	27 46	9 11	3 23	1 55	0sc54	0 57	21 25	22 56	2 8
11/ 2/1923	1 53	15 40	5 14	10 12	5 36	16 19	28 42	2 55	15 13	4 39	19 12	1 10	29 36	12 6	3 9	4 51	5 59	0 47	22 40	25 59	2 10
11/12/1923	5 3	20 42	5 18	18 13	14 49	15 24	2 44	7 35	16 17	6 2	21 34	1 5	1 25	15 13	2 56	7 49	11 8	0 37	23 54	29 1	2 12
11/22/1923	8 19	25 52	5 22	26 29	24 8	13 54	7 2	12 13	17 28	7 25	24 8	1 0	3 14	18 30	2 45	10 51	16 21	0 27	25 9	1 59	2 15
12/ 2/1923	11 38	1cp12	5 26	4 59	3vi28	11 44	11 39	16 50	18 43	8 48	26 52	0 56	5 2	21 56	2 34	13 57	21 37	0 16	26 23	4 51	2 18
12/12/1923	15 3	6 39	5 30	13 39	12 42	8 52	16 37	21 23	20 3	10 10	29 45	0 52	6 49	25 29	2 25	17 6	26 56	0 5	27 37	7 38	2 22
12/22/1923	18 31	12 13	5 33	22 28	21 41	5 21	21 57	25 53	21 24	11 32	2 45	0 48	8 36	29 8	2 16	20 18	2sa17	0 6	28 50	10 15	2 27
1/ 1/1924	22 5	17 53	5 37	1le19	0li18	1 13	27 43	0vi19	22 43	12 53	5 50	0 45	10 22	2 51	2 8	23 35	7 39	0 18	0sc 4	12 43	2 32
1/11/1924	25 43	23 39	5 40	10 11	8 23	3 25	3 56	4 43	23 51	14 14	8 59	0 42	12 8	6 37	2 1	26 54	13 2	0 31	1 17	14 59	2 38
1/21/1924	29 25	29 31	5 42	18 59	15 46	8 24	10 40	9 8	24 37	15 34	12 11	0 39	13 54	10 25	1 54	0sa18	18 26	0 44	2 31	17 0	2 45
1/31/1924	3 12	5 26	5 44	27 39	22 17	13 35	17 56	13 43	24 39	16 54	15 24	0 36	15 39	14 13	1 47	3 44	23 49	0 59	3 44	18 44	2 52
2/10/1924	7 4	11 26	5 46	6 9	27 41	18 49	25 47	18 43	23 22	18 14	18 37	0 34	17 24	18 2	1 41	7 15	29 10	1 14	4 57	20 10	3 0
2/20/1924	10 59	17 29	5 47	14 25	1sc44	23 55	4vi12	24 39	19 41	19 33	21 48	0 32	19 8	21 50	1 36	10 49	4 29	1 30	6 10	21 13	3 9
3/ 1/1924	14 59	23 34	5 47	22 25	4 11	28 42	13 11	2li16	11 47	20 52	24 58	0 29	20 52	25 35	1 30	14 26	9 44	1 48	7 23	21 51	3 18
3/11/1924	19 2	29 41	5 46	0li 8	4 44	32 52	22 40	12 32	2n50	22 11	28 5	0 27	22 36	29 18	1 25	18 7	14 54	2 7	8 35	22 3	3 27
3/21/1924	23 9	5 49	5 45	7 32	3 21	36 5	2li35	26 13	23 18	23 29	1pi 6	0 25	24 20	2 57	1 19	21 51	19 57	2 28	9 48	21 46	3 36
3/31/1924	27 19	11 58	5 42	14 38	0 19	37 58	12 46	12 25	41 4	24 47	4 2	0 23	26 3	6 31	1 14	25 37	24 52	2 51	11 1	21 2	3 45
4/10/1924	1pi32	18 7	5 38	21 24	26 23	38 14	23 4	27 50	50 55	26 5	6 51	0 20	27 46	9 58	1 9	29 27	29 37	3 15	12 13	19 52	3 52
4/20/1924	5 47	24 14	5 32	27 53	22 39	36 57	3sc18	9 28	55 2	27 22	9 32	0 18	29 30	13 19	1 4	3 19	4 8	3 42	13 26	18 20	3 58
4/30/1924	10 4	0ar19	5 26	4 3	19 55	34 29	13 16	16 35	56 9	28 40	12 2	0 16	1 13	16 31	0 58	7 13	8 23	4 12	14 39	16 34	4 1
5/10/1924	14 23	6 22	5 17	9 57	18 36	31 20	22 49	19 54	55 34	29 57	14 21	0 13	2 56	19 32	0 52	11 9	12 19	4 43	15 52	14 41	4 1

Ephemeris	80			433			1221			1489			114			3018			2041		
0 Hours ET	Sappho			Eros			Amor			Attila			Kassandra			Godiva			Lancelot		
Mo Dy Year	h long	g long	g lat	h long	g long	g lat	h long	g long	g lat	h long	g long	g lat	h long	g long	g lat	h long	g long	g lat	h long	g long	g lat
5/20/1924	18 42	12 20	5 7	15 35	18 42D	27 56	1sa51	20 40	53 43	1 13	16 25	0 11	4 39	22 22	0 46	15 7	15 49	5 18	17 4	12 52	3 59
5/30/1924	23 2	18 14	4 55	20 58	19 58	24 37	10 17	20 3	50 46	2 30	18 14	0 8	6 22	24 58	0 40	19 6	18 50	5 56	18 17	11 15	3 55
6/ 9/1924	27 22	24 1	4 42	26 8	22 11	21 31	18 6	19 1	46 52	3 46	19 45	0 5	8 5	27 17	0 33	23 7	21 15	6 36	19 30	9 57	3 48
6/19/1924	1ar42	29 42	4 26	1sa 5	25 7	18 44	25 19	18 17	42 17	5 3	20 56	0 2	9 48	29 17	0 25	27 8	22 57	7 18	20 43	9 4	3 40
6/29/1924	6 0	5 13	4 8	5 51	28 36	16 15	1cp56	18 11D	37 24	6 19	21 43	0 2	11 31	0 55	0 17	1aq 9	23 51	8 0	21 56	8 37	3 32
7/ 9/1924	10 17	10 34	3 48	10 27	2 30	14 4	8 2	18 50	32 34	7 35	22 5	0 6	13 14	2 8	0 7	5 10	23 52R	8 40	23 9	8 37D	3 23
7/19/1924	14 33	15 43	3 25	14 53	6 43	12 8	13 38	20 13	28 3	8 51	22 1R	0 10	14 57	2 52	0s 3	9 10	23 0	9 15	24 23	9 4	3 14
7/29/1924	18 45	20 37	2 59	19 11	11 10	10 27	18 48	22 12	24 2	10 7	21 28	0 14	16 41	3 4	0 15	13 10	21 24	9 39	25 36	9 55	3 5
8/ 8/1924	22 56	25 14	2 31	23 21	15 49	8 58	23 35	24 42	20 32	11 22	20 29	0 19	18 25	2 42	0 27	17 8	19 18	9 49	26 50	11 9	2 57
8/18/1924	27 3	29 30	1 59	27 24	20 38	7 39	28 1	27 35	17 32	12 38	19 6	0 24	20 9	1 45	0 41	21 4	17 6	9 44	28 4	12 42	2 49
8/28/1924	1ta 7	3 22	1 22	1cp22	25 33	6 29	2 8	0cp48	15 0	13 54	17 25	0 28	21 53	0 16	0 55	24 59	15 11	9 23	29 18	14 33	2 42
9/ 7/1924	5 8	6 44	0 41	5 15	0sa35	5 26	5 58	4 16	12 51	15 9	15 33	0 33	23 37	28 20	1 10	28 52	13 55	8 51	0sa32	16 39	2 36
9/17/1924	9 5	9 31	0s 5	9 3	5 43	4 30	9 34	7 55	11 2	16 25	13 40	0 37	25 22	26 9	1 24	2 42	13 27	8 12	1 46	18 58	2 30
9/27/1924	12 58	11 37	0 57	12 48	10 56	3 39	12 58	11 44	9 30	17 40	11 56	0 40	27 8	23 55	1 37	6 29	13 50	7 30	3 1	21 29	2 25
10/ 7/1924	16 47	12 53	1 55	16 30	16 13	2 53	16 9	15 41	8 12	18 56	10 30	0 43	28 53	21 51	1 48	10 13	15 2	6 49	4 16	24 9	2 21
10/17/1924	20 32	13 16	2 59	20 9	21 34	2 12	19 11	19 43	7 5	20 11	9 26	0 46	0ar40	20 11	1 57	13 54	16 58	6 9	5 31	26 57	2 17
10/27/1924	24 13	12 40	4 8	23 46	26 59	1 33	22 3	23 50	6 7	21 27	8 50	0 48	2 26	19 1	2 4	17 32	19 31	5 32	6 47	29 52	2 14
11/ 6/1924	27 50	11 7	5 18	27 22	2cp28	0 57	24 47	28 2	5 17	22 42	8 43	0 50	4 13	18 27	2 10	21 6	22 34	4 58	8 3	2 53	2 11
11/16/1924	1gm22	8 50	6 24	0aq57	8 1	0 24	27 24	2 15	4 33	23 58	9 4	0 51	6 1	18 30D	2 14	24 37	26 3	4 28	9 19	5 58	2 8
11/26/1924	4 51	6 7	7 21	4 32	13 38	0n 7	29 54	6 31	3 55	25 14	9 52	0 52	7 49	19 8	2 17	28 5	29 52	4 0	10 36	9 5	2 7
12/ 6/1924	8 15	3 26	8 4	8 8	19 18	0 36	2 18	10 49	3 21	26 30	11 3	0 53	9 38	20 19	2 19	1ar28	3 57	3 35	11 53	12 15	2 5
12/16/1924	11 36	1 11	8 32	11 44	25 1	1 4	4 37	15 7	2 51	27 46	12 37	0 53	11 27	21 58	2 21	4 48	8 15	3 13	13 10	15 25	2 4
12/26/1924	14 52	29ta40	8 46	15 22	0aq49	1 30	6 51	19 26	2 24	29 2	14 29	0 54	13 18	24 3	2 22	8 4	12 44	2 52	14 28	18 34	2 3
1/ 5/1925	18 5	29 1	8 50	19 2	6 40	1 56	9 0	23 45	1 59	0ar18	16 37	0 55	15 8	26 30	2 24	11 17	17 20	2 34	15 46	21 42	2 3
1/15/1925	21 13	29 15	8 45	22 45	12 36	2 21	11 5	28 3	1 36	1 34	19 0	0 56	17 0	29 14	2 25	14 26	22 2	2 17	17 4	24 47	2 3
1/20/1925	22gm46	29ta40	8s41	24aq38	15aq35	2n33	12pi 7	0pi11	1n26	2ar13	20pi15	0s56	17ar56	0ar43	2s26	15ar59	24pi25	2n 9	17sa44	26sa17	2n 3
1/30/1925	25 49	1 3	8 31	28 26	21 37	2 57	14 7	4 28	1 6	3 29	22 54	0 57	19 49	3 50	2 27	19 3	29 14	1 54	19 3	29 15	2 4
2/ 9/1925	28 49	3 4	8 18	2 19	27 43	3 20	16 4	8 42	0 47	4 46	25 42	0 58	21 43	7 10	2 29	22 3	4 5	1 40	20 23	2 5	2 5
2/19/1925	1 45	5 36	8 6	6 17	3 54	3 43	17 58	12 54	0 29	6 3	28 36	1 0	23 37	10 40	2 31	25 0	8 57	1 27	21 43	4 47	2 6
3/ 1/1925	4 37	8 33	7 53	10 21	10 11	4 5	19 50	17 4	0 12	7 20	1 36	1 1	25 33	14 18	2 33	27 53	13 50	1 15	23 4	7 19	2 8
3/11/1925	7 27	11 50	7 40	14 31	16 33	4 27	21 39	21 10	0s 5	8 38	4 40	1 2	27 29	18 3	2 36	0ta43	18 43	1 4	24 25	9 39	2 10
3/21/1925	10 13	15 23	7 29	18 49	23 2	4 49	23 27	25 13	0 21	9 56	7 47	1 4	29 26	21 54	2 39	3 31	23 34	0 54	25 47	11 44	2 12
3/31/1925	12 56	19 8	7 18	23 15	29 37	5 9	25 12	29 12	0 38	11 14	10 55	1 6	1 25	25 49	2 42	6 15	28 24	0 43	27 10	13 33	2 15
4/10/1925	15 36	23 4	7 8	27 51	6 20	5 29	26 56	3 6	0 54	12 32	14 4	1 8	3 24	29 48	2 45	8 56	3 13	0 34	28 33	15 1	2 18
4/20/1925	18 13	27 7	6 59	2 38	13 11	5 48	28 38	6 55	1 11	13 51	17 12	1 10	5 25	3 50	2 49	11 34	7 58	0 24	29 56	16 8	2 21
4/30/1925	20 48	1cn16	6 51	7 35	20 12	6 5	0ar19	10 38	1 28	15 10	20 19	1 13	7 26	7 53	2 53	14 10	12 41	0 15	1 21	16 49	2 24
5/10/1925	23 20	5 30	6 44	12 46	27 21	6 20	1 59	14 15	1 47	16 29	23 23	1 16	9 29	11 58	2 58	16 43	17 20	0 6	2 46	17 2	2 27
5/20/1925	25 50	9 47	6 37	18 10	4 41	6 33	3 38	17 44	2 6	17 49	26 23	1 19	11 33	16 4	3 3	19 13	21 55	0s 2	4 11	16 47	2 29
5/30/1925	28 17	14 7	6 31	23 50	12 12	6 43	5 15	21 4	2 26	19 9	29 19	1 23	13 38	20 9	3 9	21 41	26 27	0 11	5 38	16 3	2 31

Ephemeris 0 Hours ET	80 Sappho			433 Eros			1221 Amor			1489 Attila			114 Kassandra			3018 Godiva			2041 Lancelot		
Mo Dy Year	h long	g long	g lat	h long	g long	g lat	h long	g long	g lat	h long	g long	g lat	h long	g long	g lat	h long	g long	g lat	h long	g long	g lat
6/ 9/1925	0le42	18 28	6 26	29 45	19 55	6 50	6 53	24 14	2 48	20 29	2 8	1 26	15 45	24 13	3 15	24 7	0gm53	0 20	7 5	14 51	2 31
6/19/1925	3 4	22 50	6 22	5 58	27 50	6 51	8 29	27 13	3 11	21 50	4 50	1 31	17 53	28 16	3 22	26 31	5 15	0 29	8 33	13 16	2 29
6/29/1925	5 25	27 13	6 18	12 29	5 56	6 48	10 5	29 59	3 37	23 11	7 23	1 36	20 2	2 17	3 29	28 52	9 30	0 38	10 2	11 26	2 26
7/ 9/1925	7 44	1le35	6 15	19 18	14 15	6 39	11 41	2 29	4 6	24 33	9 45	1 41	22 13	6 15	3 37	1 11	13 39	0 47	11 31	9 30	2 21
7/19/1925	10 1	5 57	6 13	26 27	22 46	6 23	13 17	4 41	4 37	25 55	11 54	1 47	24 25	10 8	3 46	3 29	17 41	0 57	13 2	7 38	2 14
7/29/1925	12 16	10 17	6 11	3 55	1cn28	6 0	14 52	6 31	5 12	27 18	13 47	1 53	26 38	13 57	3 56	5 44	21 35	1 7	14 33	6 2	2 5
8/ 8/1925	14 29	14 36	6 11	11 41	10 21	5 29	16 27	7 56	5 50	28 41	15 23	2 0	28 53	17 39	4 6	7 58	25 19	1 18	16 5	4 49	1 56
8/18/1925	16 40	18 53	6 10	19 44	19 22	4 50	18 3	8 50	6 33	0ta 5	16 39	2 8	1 10	21 13	4 18	10 10	28 53	1 30	17 38	4 3	1 46
8/28/1925	18 51	23 8	6 11	28 3	28 30	4 4	19 39	9 9	7 19	1 29	17 31	2 16	3 28	24 37	4 31	12 21	2 14	1 42	19 12	3 49	1 36
9/ 7/1925	20 59	27 19	6 12	6 36	7 43	3 12	21 15	8 49	8 9	2 54	17 56	2 25	5 47	27 48	4 46	14 30	5 21	1 56	20 47	4 6	1 26
9/17/1925	23 7	1vi26	6 14	15 18	17 0	2 13	22 52	7 45	9 1	4 19	17 53R	2 33	8 8	0cn45	5 2	16 38	8 10	2 10	22 22	4 52	1 17
9/27/1925	25 13	5 29	6 17	24 7	26 18	1 11	24 30	5 57	9 52	5 45	17 19	2 42	10 31	3 23	5 19	18 44	10 39	2 26	23 59	6 6	1 9
10/ 7/1925	27 18	9 26	6 21	3le 0	5 35	0 6	26 8	3 29	10 39	7 12	16 16	2 50	12 56	5 39	5 38	20 49	12 45	2 44	25 37	7 45	1 1
10/17/1925	29 21	13 16	6 26	11 51	14 48	1 1	27 47	0 32	11 19	8 39	14 47	2 57	15 22	7 28	5 59	22 53	14 23	3 3	27 16	9 46	0 54
10/27/1925	1 24	16 59	6 32	20 38	23 55	2 6	29 27	27 20	11 46	10 7	12 58	3 2	17 49	8 44	6 20	24 56	15 28	3 24	28 56	12 7	0 47
11/ 6/1925	3 26	20 32	6 40	29 17	2li55	3 10	1 9	24 14	12 0	11 35	10 59	3 6	20 19	9 24	6 42	26 58	15 57	3 46	0aq37	14 45	0 40
11/16/1925	5 27	23 54	6 48	7 45	11 45	4 11	2 52	21 33	12 1	13 5	9 1	3 6	22 50	9 22R	7 4	28 59	15 46	4 9	2 20	17 37	0 35
11/26/1925	7 27	27 2	6 58	15 58	20 24	5 7	4 36	19 31	11 52	14 35	7 15	3 4	25 22	8 37	7 23	0cn59	14 52	4 33	4 3	20 43	0 29
12/ 6/1925	9 26	29 54	7 9	23 55	28 49	5 59	6 22	18 15	11 35	16 5	5 51	3 1	27 56	7 10	7 38	2 58	13 17	4 55	5 48	23 59	0 24
12/16/1925	11 25	2 27	7 21	1li35	7 0	6 46	8 10	17 46	11 13	17 37	4 55	2 55	0cn32	5 9	7 46	4 56	11 9	5 14	7 34	27 25	0 19
12/26/1925	13 23	4 36	7 35	8 56	14 56	7 28	10 0	18 3	10 51	19 9	4 32	2 49	3 10	2 48	7 44	6 53	8 39	5 28	9 21	0aq59	0 14
1/ 5/1926	15 20	6 19	7 50	15 58	22 35	8 6	11 53	19 1	10 28	20 42	4 41	2 43	5 49	0 27	7 33	8 50	6 3	5 37	11 9	4 40	0 10
1/15/1926	17 17	7 31	8 5	22 41	29 57	8 39	13 47	20 36	10 7	22 16	5 21	2 36	8 30	28 23	7 13	10 47	3 39	5 39	12 59	8 25	0 5
1/25/1926	19 14	8 6	8 21	29 6	7 0	9 9	15 45	22 42	9 47	23 51	6 31	2 30	11 13	26 54	6 47	12 43	1 42	5 36	14 50	12 16	0 1
2/ 4/1926	21 10	8 2R	8 34	5 14	13 45	9 35	17 46	25 14	9 29	25 26	8 6	2 24	13 56	26 9	6 17	14 38	0 22	5 29	16 42	16 9	0 3
2/14/1926	23 6	7 15	8 45	11 5	20 10	9 59	19 50	28 10	9 14	27 3	10 3	2 18	16 42	26 9D	5 46	16 33	29 42	5 19	18 36	20 4	0 8
2/24/1926	25vi 2	5li48R	8s51	16sc40	26sa14	10s20	21ta58	1ta25	9s 0	28ta40	12ta20	2s13	19cn29	26gm55	5s15	18cn28	29gm44	5s 8	20aq31	24aq 1	0s12
3/ 6/1926	26 57	3 44	8 50	22 0	1cp56	10 39	24 10	4 58	8 48	0gm19	14 54	2 9	22 17	28 22	4 46	20 22	0cn23	4 56	22 27	27 58	0 17
3/16/1926	28 53	1 16	8 40	27 8	7 14	10 56	26 26	8 46	8 38	1 58	17 42	2 5	25 7	0cn24	4 18	22 16	1 35	4 45	24 25	1pi55	0 22
3/26/1926	0li48	28 38	8 20	2sa 3	12 5	11 12	28 47	12 48	8 29	3 38	20 42	2 1	27 58	2 58	3 53	24 10	3 17	4 34	26 25	5 50	0 27
4/ 5/1926	2 44	26 8	7 51	6 46	16 26	11 27	1gm13	17 3	8 22	5 20	23 51	1 58	0le50	5 57	3 30	26 4	5 24	4 24	28 25	9 42	0 32
4/15/1926	4 39	24 1	7 16	11 20	20 12	11 40	3 46	21 29	8 16	7 2	27 10	1 55	3 43	9 18	3 9	27 57	7 51	4 15	0pi28	13 31	0 38
4/25/1926	6 35	22 27	6 38	15 44	23 19	11 52	6 24	26 6	8 10	8 45	0gm35	1 53	6 37	12 57	2 49	29 51	10 36	4 7	2 31	17 16	0 44
5/ 5/1926	8 31	21 34	5 58	20 1	25 38	12 0	9 10	0gm53	8 5	10 30	4 6	1 51	9 32	16 51	2 31	1 45	13 36	4 0	4 37	20 54	0 50
5/15/1926	10 27	21 21	5 19	24 10	27 2	12 4	12 4	5 52	8 1	12 16	7 41	1 49	12 29	20 58	2 14	3 39	16 47	3 54	6 43	24 26	0 57
5/25/1926	12 24	21 47	4 43	28 12	27 20	12 0	15 7	11 1	7 57	14 2	11 21	1 48	15 25	25 15	1 59	5 33	20 9	3 48	8 51	27 49	1 5
6/ 4/1926	14 22	22 48	4 9	2 9	26 24	11 44	18 19	16 21	7 52	15 50	15 3	1 47	18 23	29 41	1 44	7 27	23 39	3 44	11 1	1ar 1	1 13
6/14/1926	16 19	24 20	3 37	6 1	24 10	11 10	21 43	21 53	7 48	17 39	18 47	1 46	21 21	4 15	1 30	9 21	27 16	3 39	13 12	4 0	1 23
6/24/1926	18 18	26 19	3 9	9 48	20 48	10 15	25 18	27 37	7 42	19 29	22 32	1 46	24 20	8 55	1 17	11 16	1le 0	3 36	15 25	6 45	1 33
7/ 4/1926	20 17	28 41	2 43	13 32	16 39	8 55	29 6	3 34	7 35	21 20	26 18	1 45	27 19	13 40	1 5	13 11	4 48	3 33	17 39	9 11	1 44

Ephemeris	80			433			1221			1489			114			3018			2041		
0 Hours ET	Sappho			Eros			Amor			Attila			Kassandra			Godiva			Lancelot		
Mo Dy Year	h long	g long	g lat	h long	g long	g lat	h long	g long	g lat	h long	g long	g lat	h long	g long	g lat	h long	g long	g lat	h long	g long	g lat
7/14/1926	22 17	1li22	2 19	17 13	12 22	7 17	3 10	9 46	7 27	23 13	0cn 3	1 45	0vi18	18 29	0 53	15 7	8 40	3 31	19 55	11 16	1 56
7/24/1926	24 17	4 21	1 57	20 52	8 36	5 29	7 30	16 13	7 17	25 7	3 47	1 45	3 17	23 22	0 41	17 3	12 35	3 29	22 12	12 55	2 10
8/ 3/1926	26 19	7 34	1 37	24 29	5 50	3 43	12 9	22 57	7 3	27 2	7 29	1 46	6 17	28 18	0 30	18 59	16 33	3 28	24 30	14 5	2 24
8/13/1926	28 22	10 59	1 19	28 5	4 16	2 7	17 9	29 59	6 46	28 58	11 9	1 47	9 16	3 16	0 19	20 57	20 33	3 27	26 50	14 42	2 40
8/23/1926	0sc25	14 35	1 1	1aq40	3 55	0 43	22 32	7 20	6 25	0cn56	14 44	1 47	12 15	8 16	0 8	22 55	24 35	3 27	29 11	14 43R	2 56
9/ 2/1926	2 30	18 21	0 45	5 15	4 37	0n28	28 20	15 2	5 58	2 54	18 14	1 49	15 14	13 16	0n 3	24 53	28 37	3 27	1 34	14 7	3 12
9/12/1926	4 36	22 15	0 29	8 51	6 14	1 27	4 36	23 7	5 25	4 54	21 37	1 50	18 12	18 17	0 13	26 53	2 39	3 28	3 58	12 55	3 27
9/22/1926	6 44	26 17	0 14	12 27	8 38	2 16	11 23	1vi34	4 46	6 56	24 51	1 52	21 10	23 18	0 24	28 53	6 41	3 29	6 23	11 15	3 40
10/ 2/1926	8 53	0sc25	0n 1	16 6	11 38	2 58	18 43	10 24	3 58	8 58	27 54	1 53	24 7	28 18	0 35	0vi54	10 41	3 31	8 49	9 17	3 50
10/12/1926	11 4	4 39	0 16	19 46	15 10	3 33	26 37	19 36	3 3	11 2	0le44	1 56	27 4	3 17	0 46	2 56	14 39	3 33	11 17	7 17	3 55
10/22/1926	13 16	8 57	0 30	23 30	19 8	4 3	5 5	29 9	2 1	13 7	3 18	1 58	29 59	8 13	0 57	4 59	18 34	3 36	13 45	5 29	3 57
11/ 1/1926	15 30	13 21	0 44	27 17	23 28	4 29	14 7	8 59	0 53	15 14	5 33	2 0	2 54	13 7	1 8	7 4	22 24	3 40	16 15	4 9	3 54
11/11/1926	17 46	17 47	0 59	1pi 8	28 8	4 52	23 39	19 4	0n19	17 22	7 23	2 3	5 48	17 56	1 20	9 9	26 9	3 44	18 45	3 23	3 49
11/21/1926	20 4	22 18	1 14	5 4	3 5	5 12	3li35	29 17	1 33	19 31	8 46	2 5	8 41	22 41	1 32	11 16	29 46	3 48	21 17	3 18D	3 42
12/ 1/1926	22 24	26 50	1 29	9 6	8 18	5 29	13 48	9 33	2 45	21 41	9 38	2 8	11 32	27 20	1 45	13 24	3 14	3 54	23 49	3 53	3 33
12/11/1926	24 47	1sa25	1 45	13 14	13 44	5 45	24 5	19 46	3 52	23 53	9 53	2 9	14 22	1sc51	1 59	15 34	6 30	4 0	26 22	5 6	3 25
12/21/1926	27 11	6 1	2 1	17 30	19 24	5 58	4sc17	29 48	4 53	26 6	9 31	2 10	17 11	6 13	2 13	17 45	9 31	4 7	28 55	6 53	3 16
12/31/1926	29 39	10 38	2 19	21 53	25 16	6 11	14 13	9 35	5 46	28 20	8 31	2 9	19 59	10 25	2 29	19 57	12 16	4 15	1 29	9 10	3 7
1/10/1927	2 8	15 15	2 38	26 26	1pi21	6 21	23 43	19 2	6 29	0le35	6 58	2 6	22 45	14 25	2 45	22 11	14 38	4 23	4 4	11 52	3 0
1/20/1927	4 41	19 51	2 58	1ar 9	7 37	6 30	2sa41	28 4	7 4	2 51	5 2	2 1	25 30	18 9	3 3	24 27	16 35	4 32	6 38	14 57	2 52
1/30/1927	7 17	24 25	3 19	6 3	14 5	6 36	11 4	6 41	7 30	5 9	2 56	1 54	28 13	21 36	3 23	26 45	18 0	4 41	9 13	18 19	2 45
2/ 9/1927	9 55	28 58	3 42	11 9	20 45	6 41	18 49	14 50	7 49	7 27	0 56	1 45	0sc55	24 42	3 44	29 5	18 50	4 50	11 49	21 57	2 39
2/19/1927	12 37	3 26	4 7	16 29	27 37	6 44	25 58	22 31	8 3	9 47	29cn18	1 34	3 35	27 24	4 8	1 26	18 59R	4 58	14 24	25 47	2 33
3/ 1/1927	15 23	7 51	4 35	22 4	4 40	6 44	2cp32	29 46	8 12	12 8	28 12	1 23	6 14	29 37	4 33	3 50	18 23	5 3	16 59	29 47	2 28
3/11/1927	18 11	12 9	5 5	27 54	11 56	6 40	8 35	6 33	8 17	14 30	27 43	1 11	8 51	1 18	5 0	6 16	17 3	5 5	19 34	3 56	2 23
3/21/1927	21 4	16 20	5 39	4 1	19 23	6 34	14 8	12 53	8 19	16 52	27 54	1 0	11 26	2 21	5 28	8 44	15 4	5 2	22 9	8 11	2 19
3/31/1927	24sa 0	20cp22	6n16	10ta26	27ar 3	6n24	19cp16	18aq48	8n19	19le16	28cn43	0s50	14sc 0	2sa44	5n58	11li15	12li37R	4s52	24ta43	12ta31	2s15
4/10/1927	27 0	24 13	6 57	17 9	4 56	6 9	24 0	24 17	8 18	21 40	0le 7	0 41	16 32	2 24	6 27	13 48	10 0	4 36	27 17	16 55	2 11
4/20/1927	0cp 4	27 48	7 43	24 11	13 0	5 50	28 24	29 20	8 16	24 5	2 2	0 32	19 3	1 22	6 54	16 24	7 33	4 14	29 51	21 22	2 8
4/30/1927	3 13	1aq 7	8 33	1gm33	21 16	5 25	2 29	3 56	8 13	26 31	4 22	0 24	21 31	29sc43	7 17	19 2	5 34	3 47	2 23	25 50	2 5
5/10/1927	6 26	4 3	9 30	9 13	29 44	4 56	6 18	8 5	8 10	28 58	7 6	0 17	23 59	27 38	7 33	21 43	4 16	3 19	4 55	0gm19	2 2
5/20/1927	9 43	6 33	10 32	17 11	8 21	4 21	9 53	11 44	8 7	1 25	10 8	0 10	26 24	25 22	7 41	24 27	3 44	2 50	7 27	4 49	1 59
5/30/1927	13 4	8 30	11 39	25 25	17 9	3 41	13 15	14 52	8 3	3 52	13 27	0 4	28 48	23 12	7 40	27 14	3 59	2 23	9 57	9 18	1 57
6/ 9/1927	16 31	9 48	12 51	3cn53	26 4	2 56	16 26	17 23	7 59	6 20	16 59	0n 1	1sa11	21 23	7 31	0sc 4	4 59	1 57	12 27	13 46	1 54
6/19/1927	20 1	10 21	14 5	12 32	5 6	2 7	19 26	19 15	7 53	8 48	20 42	0 7	3 32	20 6	7 17	2 57	6 39	1 33	14 55	18 12	1 52
6/29/1927	23 37	10 4	15 18	21 19	14 13	1 15	22 18	20 22	7 46	11 17	24 35	0 12	5 51	19 27	6 59	5 53	8 54	1 12	17 22	22 35	1 50
7/ 9/1927	27 17	8 57	16 23	0le11	23 23	0 22	25 1	20 39	7 36	13 45	28 36	0 16	8 9	19 26D	6 38	8 53	11 40	0 52	19 49	26 55	1 48
7/19/1927	1aq 2	7 6	17 13	9 3	2le34	0 32	27 37	20 2	7 21	16 14	2 44	0 21	10 25	20 4	6 18	11 56	14 52	0 34	22 14	1cn10	1 47
7/29/1927	4 51	4 47	17 40	17 51	11 45	1 25	0pi 7	18 29	7 0	18 43	6 57	0 25	12 40	21 15	5 57	15 2	18 27	0 17	24 37	5 21	1 45
8/ 8/1927	8 44	2 23	17 39	26 33	20 53	2 16	2 30	16 5	6 32	21 12	11 15	0 29	14 53	22 56	5 38	18 12	22 22	0 2	27 0	9 27	1 43

Ephemeris	80			433			1221			1489			114			3018			2041		
0 Hours ET	Sappho			Eros			Amor			Attila			Kassandra			Godiva			Lancelot		
Mo Dy Year	h long	g long	g lat	h long	g long	g lat	h long	g long	g lat	h long	g long	g lat	h long	g long	g lat	h long	g long	g lat	h long	g long	g lat
8/18/1927	12 42	0 20	17 11	5 4	29 57	3 3	4 49	13 0	5 55	23 41	15 36	0 33	17 5	25 3	5 20	21 25	26 34	0 13	29 21	13 25	1 42
8/28/1927	16 43	28 58	16 21	13 22	8 55	3 46	7 2	9 33	5 10	26 9	20 0	0 37	19 16	27 31	5 3	24 42	1sc 2	0 26	1 41	17 15	1 41
9/ 7/1927	20 48	28 31	15 16	21 25	17 45	4 24	9 11	6 8	4 20	28 37	24 26	0 41	21 25	0sa18	4 48	28 3	5 43	0 38	3 59	20 56	1 39
9/17/1927	24 57	29 2	14 3	29 10	26 26	4 57	11 16	3 6	3 28	1li 5	28 54	0 45	23 33	3 21	4 34	1sa27	10 35	0 50	6 16	24 26	1 38
9/27/1927	29 9	0aq29	12 48	6 37	4 58	5 24	13 18	0 45	2 37	3 33	3 22	0 49	25 39	6 36	4 22	4 55	15 39	1 1	8 31	27 42	1 37
10/ 7/1927	3 23	2 45	11 34	13 46	13 19	5 45	15 16	29 12	1 51	6 0	7 50	0 53	27 45	10 3	4 10	8 26	20 52	1 12	10 45	0le43	1 36
10/17/1927	7 40	5 45	10 23	20 35	21 28	6 2	17 12	28 29	1 10	8 26	12 17	0 57	29 49	13 38	4 0	12 1	26 14	1 22	12 58	3 25	1 35
10/27/1927	11 58	9 21	9 16	27 6	29 26	6 14	19 4	28 31D	0 34	10 52	16 42	1 2	1 52	17 21	3 51	15 39	1sa44	1 32	15 9	5 46	1 33
11/ 6/1927	16 17	13 26	8 14	3 19	7 13	6 21	20 55	29 15	0 4	13 16	21 4	1 6	3 54	21 10	3 42	19 20	7 21	1 42	17 18	7 42	1 32
11/16/1927	20 38	17 57	7 16	9 16	14 47	6 25	22 43	0pi34	0 23	15 41	25 23	1 11	5 54	25 4	3 35	23 4	13 4	1 51	19 26	9 10	1 30
11/26/1927	24 58	22 48	6 23	14 56	22 11	6 25	24 29	2 21	0 46	18 4	29 37	1 16	7 54	29 1	3 28	26 52	18 53	2 0	21 33	10 4	1 28
12/ 6/1927	29 18	27 55	5 33	20 21	29 23	6 22	26 13	4 34	1 5	20 26	3 44	1 21	9 52	3 1	3 22	0cp42	24 46	2 8	23 38	10 23	1 25
12/16/1927	3 38	3 15	4 47	25 33	6 24	6 17	27 56	7 6	1 23	22 48	7 44	1 27	11 50	7 3	3 17	4 34	0cp43	2 16	25 41	10 4	1 21
12/26/1927	7 56	8 45	4 4	0sa31	13 14	6 10	29 38	9 55	1 39	25 9	11 36	1 33	13 46	11 5	3 12	8 29	6 43	2 24	27 43	9 7	1 16
1/ 5/1928	12 13	14 24	3 24	5 19	19 54	6 1	1 18	12 57	1 53	27 28	15 16	1 39	15 42	15 7	3 8	12 25	12 45	2 32	29 44	7 38	1 10
1/15/1928	16 28	20 8	2 46	9 56	26 24	5 50	2 57	16 11	2 6	29 46	18 43	1 47	17 37	19 8	3 5	16 23	18 49	2 39	1 43	5 44	1 2
1/25/1928	20 40	25 57	2 11	14 23	2cp44	5 38	4 35	19 33	2 19	2 4	21 55	1 54	19 30	23 7	3 2	20 23	24 54	2 47	3 41	3 38	0 54
2/ 4/1928	24 49	1ar49	1 38	18 42	8 55	5 24	6 13	23 3	2 31	4 20	24 49	2 3	21 23	27 2	2 59	24 23	0aq59	2 53	5 37	1 34	0 44
2/14/1928	28 55	7 43	1 8	22 53	14 55	5 9	7 50	26 38	2 42	6 35	27 21	2 12	23 16	0aq54	2 57	28 24	7 2	3 0	7 32	29cn47	0 35
2/24/1928	2 58	13 37	0 39	26 58	20 47	4 53	9 26	0ar17	2 54	8 49	29 30	2 22	25 7	4 41	2 55	2 25	13 4	3 6	9 26	28 25	0 25
3/ 5/1928	6 58	19 30	0 12	0cp56	26 28	4 35	11 2	4 0	3 5	11 1	1 11	2 33	26 58	8 21	2 54	6 25	19 4	3 12	11 18	27 36	0 17
3/15/1928	10 53	25 23	0 14	4 50	2aq 0	4 15	12 37	7 45	3 17	13 12	2 20	2 44	28 48	11 54	2 53	10 26	25 0	3 18	13 9	27 23	0 8
3/25/1928	14 45	1ta14	0 38	8 39	7 22	3 54	14 13	11 32	3 29	15 23	2 54	2 55	0aq37	15 19	2 52	14 25	0pi51	3 24	14 58	27 43	0 1
4/ 4/1928	18 32	7 2	1 1	12 24	12 33	3 30	15 48	15 19	3 42	17 31	2 53R	3 7	2 26	18 33	2 52	18 22	6 37	3 29	16 47	28 34	0 5
4/14/1928	22 16	12 47	1 23	16 6	17 32	3 4	17 24	19 6	3 56	19 39	2 14	3 18	4 14	21 35	2 52	22 18	12 17	3 34	18 34	29 54	0 11
4/24/1928	25 55	18 29	1 44	19 45	22 19	2 35	18 59	22 53	4 10	21 45	1 2	3 27	6 2	24 24	2 52	26 13	17 50	3 38	20 19	1 37	0 16
5/ 4/1928	29ta30	24ta 7	2s 5	23cp23	26aq52	2s 3	20ar35	26ar38	4s25	23sc50	29sc23R	3n34	7aq49	26aq57	2n52	0pi 4	23pi14	3n43	22le 4	3le41	0n21
5/14/1928	3 1	29 41	2 24	26 59	1pi10	1 26	22 12	0ta22	4 41	25 53	27 27	3 38	9 35	29 11	2 53	3 54	28 28	3 47	23 47	6 2	0 25
5/24/1928	6 28	5 11	2 43	0aq34	5 9	0 43	23 49	4 3	4 59	27 56	25 28	3 38	11 21	1 4	2 53	7 40	3 32	3 51	25 29	8 37	0 29
6/ 3/1928	9 51	10 36	3 1	4 9	8 46	0n 5	25 27	7 40	5 18	29 57	23 39	3 35	13 7	2 32	2 53	11 23	8 23	3 55	27 11	11 24	0 33
6/13/1928	13 9	15 57	3 19	7 45	11 58	1 2	27 5	11 14	5 39	1 56	22 10	3 30	14 52	3 33	2 53	15 4	13 1	3 58	28 51	14 21	0 36
6/23/1928	16 24	21 12	3 37	11 21	14 37	2 9	28 45	14 42	6 2	3 55	21 10	3 22	16 37	4 2	2 53	18 40	17 21	4 2	0vi29	17 25	0 39
7/ 3/1928	19 35	26 22	3 55	14 59	16 37	3 27	0ta26	18 4	6 27	5 52	20 43	3 13	18 22	3 59R	2 51	22 14	21 23	4 5	2 7	20 36	0 42
7/13/1928	22 42	1cn27	4 13	18 39	17 49	4 59	2 8	21 19	6 56	7 48	20 49D	3 4	20 6	3 20	2 48	25 44	25 2	4 8	3 44	23 52	0 45
7/23/1928	25 45	6 25	4 31	22 21	18 1R	6 46	3 51	24 24	7 27	9 42	21 28	2 54	21 50	2 8	2 42	29 10	28 16	4 10	5 20	27 12	0 48
8/ 2/1928	28 45	11 16	4 49	26 7	17 4	8 48	5 36	27 18	8 3	11 36	22 35	2 45	23 34	0 27	2 35	2 33	1ta 0	4 13	6 55	0vi34	0 51
8/12/1928	1 41	16 1	5 8	29 57	14 51	10 58	7 23	29 59	8 42	13 28	24 9	2 36	25 18	28 25	2 25	5 52	3 8	4 14	8 29	3 58	0 54
8/22/1928	4 33	20 37	5 28	3 52	11 26	13 9	9 12	2 22	9 27	15 19	26 5	2 28	27 1	26 12	2 13	9 7	4 36	4 14	10 2	7 24	0 57
9/ 1/1928	7 23	25 4	5 48	7 52	7 12	15 3	11 3	4 25	10 17	17 9	28 21	2 21	28 44	24 3	1 59	12 19	5 18	4 12	11 34	10 49	1 1
9/11/1928	10 9	29 22	6 10	11 58	2 46	16 27	12 56	6 2	11 14	18 57	0sa53	2 14	0pi27	22 10	1 44	15 27	5 10R	4 7	13 5	14 13	1 4

Ephemeris	80			433			1221			1489			114			3018			2041		
0 Hours ET	Sappho			Eros			Amor			Attila			Kassandra			Godiva			Lancelot		
Mo Dy Year	h long	g long	g lat	h long	g long	g lat	h long	g long	g lat	h long	g long	g lat	h long	g long	g lat	h long	g long	g lat	h long	g long	g lat
9/21/1928	12 52	3 28	6 34	16 11	28aq54	17 13	14 52	7 8	12 17	20 45	3 39	2 8	2 11	20 43	1 28	18 31	4 11	3 58	14 36	17 36	1 7
10/ 1/1928	15 33	7 21	6 59	20 32	26 12	17 27	16 51	7 35	13 26	22 31	6 37	2 3	3 54	19 48	1 13	21 33	2 25	3 44	16 5	20 55	1 11
10/11/1928	18 10	11 0	7 26	25 2	24 54	17 15	18 54	7 18	14 41	24 16	9 45	1 58	5 37	19 28	0 58	24 30	0 4	3 26	17 34	24 11	1 15
10/21/1928	20 45	14 20	7 55	29 41	25 0D	16 49	20 59	6 10	15 57	26 1	13 1	1 53	7 20	19 44	0 44	27 24	27 27	3 2	19 2	27 22	1 20
10/31/1928	23 17	17 20	8 27	4 32	26 24	16 14	23 8	4 9	17 11	27 44	16 23	1 49	9 3	20 32	0 32	0ta16	24 55	2 36	20 29	0li26	1 24
11/10/1928	25 46	19 55	9 2	9 34	28 53	15 35	25 22	1 18	18 14	29 26	19 51	1 46	10 46	21 50	0 21	3 3	22 50	2 8	21 56	3 23	1 30
11/20/1928	28 14	22 1	9 40	14 50	2 18	14 55	27 40	27 51	18 58	1 7	23 23	1 43	12 29	23 34	0 11	5 48	21 24	1 40	23 22	6 10	1 35
11/30/1928	0le39	23 34	10 21	20 20	6 30	14 13	0gm 4	24 10	19 17	2 47	26 57	1 40	14 13	25 41	0 1	8 30	20 45	1 15	24 47	8 46	1 41
12/10/1928	3 1	24 28	11 3	26 5	11 22	13 31	2 32	20 43	19 10	4 26	0cp33	1 38	15 56	28 8	0 7	11 9	20 52D	0 52	26 12	11 8	1 48
12/20/1928	5 22	24 38R	11 46	2ta 7	16 50	12 47	5 7	17 56	18 38	6 4	4 10	1 36	17 40	0pi52	0 15	13 46	21 44	0 31	27 36	13 15	1 55
12/30/1928	7 41	24 2	12 28	8 26	22 48	12 2	7 49	16 4	17 50	7 41	7 46	1 34	19 24	3 50	0 22	16 19	23 13	0 14	28 59	15 2	2 3
1/ 9/1929	9 57	22 39	13 4	15 4	29 14	11 14	10 38	15 14	16 53	9 17	11 21	1 33	21 9	7 0	0 29	18 50	25 15	0s 1	0li22	16 28	2 12
1/19/1929	12 12	20 35	13 31	22 0	6 6	10 23	13 35	15 25D	15 52	10 52	14 53	1 32	22 53	10 19	0 35	21 19	27 44	0 15	1 44	17 30	2 22
1/29/1929	14 26	18 2	13 44	29 16	13 20	9 28	16 42	16 33	14 51	12 27	18 20	1 31	24 38	13 47	0 42	23 45	0ta35	0 26	3 5	18 5	2 32
2/ 8/1929	16 37	15 18	13 42	6 50	20 55	8 29	19 59	18 31	13 53	14 0	21 44	1 30	26 23	17 22	0 48	26 9	3 44	0 36	4 26	18 11R	2 42
2/18/1929	18 48	12 41	13 24	14 43	28 50	7 24	23 26	21 13	12 58	15 33	25 0	1 30	28 9	21 1	0 54	28 31	7 7	0 46	5 47	17 46	2 52
2/28/1929	20 56	10 30	12 53	22 52	7 3	6 15	27 7	24 33	12 6	17 5	28 10	1 30	29 55	24 45	1 0	0gm51	10 42	0 54	7 7	16 52	3 2
3/10/1929	23 4	8 56	12 14	1cn16	15 31	5 1	1cn 1	28 27	11 17	18 36	1aq10	1 30	1 42	28 32	1 6	3 9	14 26	1 1	8 27	15 31	3 11
3/20/1929	25 10	8 5	11 30	9 52	24 14	3 43	5 11	2 50	10 30	20 7	4 0	1 30	3 29	2 20	1 12	5 25	18 18	1 8	9 46	13 50	3 18
3/30/1929	27 15	7 58D	10 44	18 38	3gm 8	2 22	9 39	7 41	9 44	21 36	6 37	1 31	5 17	6 10	1 18	7 39	22 15	1 15	11 5	11 56	3 23
4/ 9/1929	29 18	8 31	10 0	27 28	12 11	0 59	14 26	12 58	8 59	23 5	9 1	1 32	7 5	10 0	1 25	9 52	26 16	1 21	12 23	10 0	3 25
4/19/1929	1 21	9 39	9 18	6 20	21 22	0s24	19 35	18 39	8 13	24 33	11 8	1 33	8 54	13 49	1 32	12 2	0gm21	1 27	13 41	8 13	3 25
4/29/1929	3 23	11 18	8 40	15 10	0cn38	1 44	25 7	24 44	7 26	26 1	12 56	1 34	10 43	17 37	1 39	14 12	4 27	1 32	14 58	6 42	3 23
5/ 9/1929	5 24	13 22	8 5	23 54	9 56	2 59	1le 7	1cn14	6 37	27 28	14 24	1 35	12 33	21 22	1 47	16 20	8 35	1 38	16 16	5 34	3 19
5/19/1929	7 24	15 49	7 33	2vi29	19 14	4 8	7 35	8 7	5 45	28 54	15 27	1 36	14 24	25 5	1 56	18 26	12 43	1 44	17 32	4 53	3 15
5/29/1929	9 23	18 33	7 4	10 51	28 30	5 9	14 35	15 23	4 49	0aq20	16 5	1 37	16 16	28 43	2 5	20 32	16 51	1 50	18 49	4 41	3 9
6/ 8/1929	11vi22	21le32	6s38	18vi59	7le40	6s 1	22le 8	23cn 5	3s49	1aq45	16aq14	1n38	18ar 8	2ta16	2s14	22gm36	20gm59	1s55	20li 5	4li55	3n 3
6/18/1929	13 20	24 44	6 15	26 49	16 44	6 43	0vi16	1le11	2 44	3 10	15 54	1 39	20 1	5 43	2 25	24 39	25 5	2 2	21 21	5 36	2 57
6/28/1929	15 17	28 6	5 53	4 22	25 38	7 15	8 58	9 40	1 35	4 33	15 5	1 38	21 55	9 2	2 36	26 41	29 10	2 8	22 37	6 40	2 51
7/ 8/1929	17 14	1vi36	5 34	11 36	4 23	7 38	18 13	18 34	0 23	5 57	13 51	1 37	23 50	12 12	2 49	28 41	3 12	2 14	23 52	8 5	2 46
7/18/1929	19 10	5 14	5 16	18 31	12 55	7 52	27 55	27 50	0 53	7 20	12 14	1 35	25 46	15 11	3 3	0cn41	7 11	2 21	25 7	9 48	2 41
7/28/1929	21 7	8 58	5 0	25 8	21 15	7 57	7 59	7 25	2 8	8 42	10 25	1 32	27 43	17 57	3 18	2 41	11 7	2 29	26 22	11 46	2 37
8/ 7/1929	23 3	12 46	4 45	1sc26	29 23	7 56	18 15	17 16	3 22	10 4	8 31	1 27	29 40	20 26	3 35	4 39	14 58	2 37	27 37	13 58	2 33
8/17/1929	24 58	16 39	4 31	7 27	7 18	7 49	28 32	27 20	4 30	11 26	6 42	1 22	1 39	22 37	3 53	6 36	18 44	2 45	28 51	16 22	2 29
8/27/1929	26 54	20 35	4 18	13 12	15 0	7 37	8 38	7 29	5 31	12 47	5 9	1 16	3 39	24 24	4 13	8 33	22 23	2 54	0sc 6	18 56	2 27
9/ 6/1929	28 49	24 34	4 7	18 42	22 31	7 22	18 25	17 39	6 21	14 7	3 58	1 9	5 40	25 45	4 36	10 30	25 55	3 4	1 20	21 37	2 24
9/16/1929	0li45	28 35	3 55	23 58	29 51	7 3	27 43	27 43	7 0	15 28	3 13	1 3	7 42	26 35	4 59	12 25	29 18	3 15	2 34	24 26	2 22
9/26/1929	2 40	2 36	3 45	29 0	7 0	6 43	6 27	7 34	7 28	16 48	2 57	0 56	9 45	26 50	5 24	14 21	2 30	3 28	3 48	27 19	2 21
10/ 6/1929	4 36	6 39	3 35	3 51	14 0	6 20	14 34	17 9	7 43	18 7	3 9	0 50	11 50	26 26	5 50	16 16	5 30	3 41	5 2	0sc17	2 20
10/16/1929	6 32	10 41	3 25	8 31	20 51	5 57	22 4	26 23	7 49	19 26	3 49	0 45	13 56	25 23	6 14	18 10	8 14	3 56	6 15	3 18	2 19

Ephemeris	80			433			1221			1489			114			3018			2041		
0 Hours ET	Sappho			Eros			Amor			Attila			Kassandra			Godiva			Lancelot		
Mo Dy Year	h long	g long	g lat	h long	g long	g lat	h long	g long	g lat	h long	g long	g lat	h long	g long	g lat	h long	g long	g lat	h long	g long	g lat
10/26/1929	8 28	14 42	3 16	13 1	27 34	5 33	28 59	5 14	7 47	20 45	4 53	0 39	16 3	23 45	6 36	20 4	10 40	4 12	7 29	6 21	2 19
11/ 5/1929	10 24	18 42	3 7	17 23	4 10	5 9	5 20	13 41	7 38	22 3	6 20	0 34	18 11	21 39	6 52	21 58	12 43	4 30	8 42	9 25	2 20
11/15/1929	12 21	22 39	2 58	21 36	10 40	4 44	11 11	21 44	7 24	23 22	8 6	0 30	20 21	19 18	7 2	23 52	14 21	4 50	9 56	12 28	2 21
11/25/1929	14 18	26 33	2 50	25 43	17 3	4 20	16 33	29 24	7 6	24 39	10 10	0 26	22 32	17 0	7 5	25 46	15 28	5 11	11 9	15 30	2 22
12/ 5/1929	16 16	0sc21	2 41	29 43	23 22	3 55	21 31	6 43	6 47	25 57	12 27	0 22	24 44	14 59	7 0	27 39	16 1	5 34	12 23	18 29	2 24
12/15/1929	18 15	4 4	2 32	3 38	29 36	3 31	26 7	13 41	6 27	27 14	14 57	0 19	26 58	13 31	6 49	29 33	15 55R	5 57	13 36	21 24	2 26
12/25/1929	20 14	7 38	2 23	7 28	5 45	3 6	0aq23	20 21	6 6	28 32	17 37	0 15	29 14	12 42	6 34	1 26	15 8	6 20	14 50	24 13	2 29
1/ 4/1930	22 14	11 4	2 13	11 14	11 51	2 41	4 21	26 44	5 45	29 49	20 26	0 12	1 31	12 35D	6 17	3 20	13 40	6 40	16 3	26 55	2 33
1/14/1930	24 14	14 17	2 2	14 57	17 53	2 16	8 4	2aq51	5 24	1 5	23 21	0 9	3 49	13 10	5 58	5 13	11 38	6 57	17 16	29 28	2 37
1/24/1930	26 16	17 16	1 50	18 37	23 52	1 51	11 34	8 44	5 4	2 22	26 21	0 7	6 9	14 23	5 40	7 7	9 12	7 7	18 30	1 50	2 42
2/ 3/1930	28 19	19 58	1 37	22 15	29 48	1 26	14 51	14 23	4 44	3 38	29 25	0 4	8 31	16 10	5 23	9 1	6 37	7 10	19 44	3 59	2 47
2/13/1930	0sc22	22 19	1 23	25 52	5 41	1 0	17 57	19 49	4 26	4 54	2 31	0 2	10 54	18 27	5 7	10 56	4 11	7 5	20 57	5 53	2 53
2/23/1930	2 27	24 15	1 6	29 27	11 32	0 33	20 54	25 3	4 7	6 10	5 39	0s 1	13 19	21 9	4 52	12 51	2 8	6 54	22 11	7 29	3 0
3/ 5/1930	4 34	25 42	0 47	3 2	17 21	0 6	23 42	0pi 5	3 50	7 26	8 46	0 3	15 45	24 13	4 38	14 46	0 40	6 38	23 25	8 44	3 7
3/15/1930	6 41	26 35	0 25	6 38	23 8	0 23	26 22	4 56	3 33	8 42	11 52	0 6	18 13	27 35	4 25	16 42	29cn51	6 20	24 39	9 36	3 15
3/25/1930	8 50	26 50	0n 1	10 14	28 53	0 52	28 56	9 36	3 16	9 58	14 55	0 9	20 43	1gm12	4 14	18 38	29 44D	6 0	25 53	10 3	3 22
4/ 4/1930	11 1	26 22	0 30	13 51	4 37	1 24	1 23	14 4	3 0	11 13	17 55	0 11	23 14	5 2	4 3	20 35	0le15	5 40	27 7	10 3R	3 30
4/14/1930	13 13	25 12	1 3	17 30	10 21	1 56	3 44	18 20	2 43	12 29	20 51	0 14	25 47	9 3	3 53	22 32	1 22	5 21	28 22	9 35	3 38
4/24/1930	15 28	23 21	1 39	21 12	16 4	2 31	6 0	22 24	2 27	13 45	23 40	0 17	28 22	13 13	3 44	24 31	2 59	5 3	29 36	8 39	3 44
5/ 4/1930	17 44	21 0	2 15	24 57	21 47	3 9	8 12	26 15	2 11	15 0	26 22	0 20	0cn58	17 31	3 36	26 30	5 2	4 46	0 51	7 19	3 49
5/14/1930	20 2	18 22	2 51	28 45	27 30	3 48	10 19	29 52	1 54	16 15	28 55	0 23	3 36	21 55	3 28	28 30	7 28	4 31	2 6	5 40	3 52
5/24/1930	22 22	15 46	3 24	2 38	3 13	4 31	12 23	3 14	1 36	17 31	1 19	0 27	6 15	26 25	3 21	0vi30	10 13	4 18	3 21	3 50	3 52
6/ 3/1930	24 44	13 29	3 53	6 37	8 58	5 17	14 23	6 19	1 18	18 46	3 29	0 31	8 56	1cn 0	3 14	2 32	13 15	4 5	4 37	1 57	3 50
6/13/1930	27 9	11 48	4 16	10 41	14 45	6 7	16 20	9 5	0 58	20 2	5 26	0 35	11 39	5 39	3 7	4 35	16 30	3 54	5 53	0 12	3 45
6/23/1930	29 36	10 49	4 34	14 51	20 34	7 1	18 14	11 30	0 37	21 17	7 7	0 39	14 23	10 22	3 1	6 39	19 57	3 44	7 9	28 43	3 38
7/ 3/1930	2 6	10 36	4 48	19 10	26 25	7 59	20 6	13 30	0 14	22 33	8 28	0 44	17 8	15 7	2 55	8 44	23 34	3 34	8 25	27 36	3 30
7/13/1930	4sa39	11sc 7	4n59	23pi37	2ta20	9n 3	21pi55	15ar 1	0s11	23pi49	9ar29	0s49	19cn55	19cn54	2s49	10vi50	27le20	3s26	9sa42	26sc55R	3n20
7/23/1930	7 15	12 19	5 7	28 13	8 19	10 11	23 43	15 59	0 39	25 4	10 5	0 55	22 43	24 44	2 43	12 58	1vi14	3 18	10 59	26 42	3 10
8/ 2/1930	9 53	14 9	5 13	3 0	14 22	11 26	25 28	16 21	1 10	26 20	10 16	1 1	25 33	29 34	2 37	15 7	5 15	3 11	12 16	26 57	3 0
8/12/1930	12 35	16 31	5 17	7 59	20 31	12 46	27 12	16 1	1 44	27 36	9 59	1 7	28 24	4 25	2 31	17 17	9 21	3 4	13 34	27 38	2 51
8/22/1930	15 21	19 21	5 21	13 10	26 46	14 12	28 54	14 56	2 20	28 52	9 14	1 13	1le15	9 16	2 25	19 29	13 33	2 58	14 52	28 43	2 41
9/ 1/1930	18 9	22 35	5 24	18 35	3gm 8	15 44	0ar35	13 8	2 58	0ar 8	8 3	1 19	4 8	14 6	2 18	21 43	17 49	2 52	16 11	0sa10	2 33
9/11/1930	21 2	26 11	5 27	24 16	9 37	17 21	2 15	10 41	3 37	1 25	6 31	1 25	7 2	18 55	2 12	23 59	22 10	2 46	17 30	1 56	2 25
9/21/1930	23 58	0sa 6	5 29	0ta12	16 15	19 2	3 54	7 45	4 13	2 41	4 43	1 30	9 57	23 42	2 6	26 16	26 33	2 41	18 49	4 0	2 18
10/ 1/1930	26 58	4 18	5 31	6 26	23 1	20 43	5 32	4 37	4 45	3 58	2 48	1 34	12 53	28 27	1 59	28 35	1li 0	2 36	20 9	6 18	2 11
10/11/1930	0cp 2	8 44	5 34	12 58	29 56	22 23	7 9	1 35	5 11	5 15	0 57	1 37	15 49	3 7	1 52	0li56	5 29	2 31	21 29	8 50	2 5
10/21/1930	3 11	13 24	5 35	19 48	7 0	23 56	8 45	28pi57	5 30	6 32	29pi19	1 39	18 46	7 42	1 44	3 20	10 0	2 27	22 50	11 32	2 0
10/31/1930	6 24	18 16	5 37	26 58	14 12	25 16	10 21	26 56	5 43	7 50	28 2	1 39	21 43	12 10	1 36	5 45	14 33	2 22	24 12	14 24	1 55
11/10/1930	9 41	23 19	5 39	4 26	21 31	26 14	11 57	25 38	5 51	9 7	27 11	1 39	24 41	16 30	1 27	8 13	19 6	2 17	25 33	17 24	1 50
11/20/1930	13 2	28 31	5 41	12 14	28 54	26 39	13 33	25 4	5 54	10 25	26 50	1 38	27 39	20 40	1 18	10 43	23 38	2 13	26 56	20 31	1 46

Ephemeris	80			433			1221			1489			114			3018			2041		
0 Hours ET	Sappho			Eros			Amor			Attila			Kassandra			Godiva			Lancelot		
Mo Dy Year	h long	g long	g lat	h long	g long	g lat	h long	g long	g lat	h long	g long	g lat	h long	g long	g lat	h long	g long	g lat	h long	g long	g lat
11/30/1930	16 28	3 53	5 42	20 18	6 14	26 17	15 8	25 13D	5 56	11 44	26 57	1 37	0vi38	24 37	1 7	13 16	28 10	2 8	28 19	23 43	1 43
12/10/1930	19 59	9 23	5 43	28 38	13 24	24 50	16 44	26 0	5 55	13 2	27 34	1 36	3 36	28 18	0 56	15 51	2 40	2 3	29 43	26 59	1 40
12/20/1930	23 34	15 0	5 44	7 10	20 13	21 54	18 19	27 20	5 54	14 21	28 36	1 34	6 35	1 40	0 43	18 29	7 8	1 58	1 7	0cp18	1 37
12/30/1930	27 14	20 44	5 45	15 53	26 23	17 2	19 55	29 8	5 53	15 40	0ar 2	1 33	9 33	4 39	0 28	21 9	11 31	1 52	2 32	3 38	1 34
1/ 9/1931	0aq59	26 34	5 45	24 43	1vi39	9 50	21 32	1 21	5 52	17 0	1 48	1 32	12 31	7 11	0 12	23 53	15 49	1 46	3 57	7 0	1 32
1/19/1931	4 48	2aq30	5 45	3le35	5 39	0 15	23 8	3 54	5 52	18 20	3 52	1 31	15 29	9 10	0n 7	26 39	19 59	1 39	5 24	10 20	1 30
1/29/1931	8 41	8 30	5 44	12 26	8 5	10 49	24 46	6 44	5 52	19 40	6 12	1 30	18 27	10 31	0 28	29 29	24 0	1 31	6 50	13 39	1 28
2/ 8/1931	12 38	14 34	5 42	21 13	8 57	21 30	26 24	9 49	5 53	21 1	8 44	1 29	21 23	11 9	0 52	2 21	27 50	1 22	8 18	16 55	1 27
2/18/1931	16 40	20 41	5 40	29 51	8 30	29 56	28 3	13 5	5 55	22 22	11 28	1 29	24 19	11 1R	1 19	5 17	1sa26	1 11	9 47	20 7	1 25
2/28/1931	20 45	26 50	5 37	8 18	7 26	35 18	29 44	16 32	5 57	23 43	14 20	1 29	27 15	10 6	1 47	8 17	4 44	0 59	11 16	23 13	1 24
3/10/1931	24 53	3pi 2	5 33	16 30	6 37	37 52	1 25	20 7	6 1	25 5	17 20	1 29	0li 9	8 31	2 16	11 19	7 40	0 44	12 46	26 12	1 23
3/20/1931	29 4	9 14	5 27	24 26	6 37D	38 17	3 8	23 50	6 5	26 28	20 25	1 29	3 3	6 26	2 43	14 25	10 10	0 27	14 17	29 2	1 22
3/30/1931	3 18	15 27	5 21	2li 5	7 45	37 13	4 52	27 38	6 10	27 51	23 35	1 30	5 56	4 10	3 8	17 35	12 8	0 7	15 48	1 41	1 21
4/ 9/1931	7 34	21 40	5 13	9 25	9 56	35 16	6 39	1ta32	6 17	29 14	26 48	1 31	8 47	2 0	3 28	20 48	13 28	0 17	17 21	4 8	1 20
4/19/1931	11 52	27 51	5 4	16 26	12 59	32 49	8 26	5 31	6 24	0ta38	0ta 3	1 32	11 38	0 16	3 43	24 5	14 4	0 44	18 54	6 20	1 19
4/29/1931	16 11	4 0	4 53	23 9	16 45	30 7	10 17	9 33	6 32	2 3	3 19	1 33	14 27	29 9	3 53	27 26	13 53R	1 16	20 29	8 14	1 18
5/ 9/1931	20 31	10 5	4 41	29 33	21 0	27 22	12 9	13 39	6 42	3 28	6 35	1 35	17 15	28 45	3 59	0sa50	12 53	1 51	22 4	9 48	1 17
5/19/1931	24 51	16 8	4 28	5 39	25 37	24 42	14 4	17 48	6 53	4 54	9 51	1 37	20 2	29 4	4 2	4 18	11 10	2 29	23 40	10 59	1 15
5/29/1931	29 11	22 5	4 12	11 29	0li31	22 10	16 1	22 1	7 5	6 20	13 5	1 40	22 47	0li 4	4 2	7 49	8 57	3 5	25 18	11 43	1 13
6/ 8/1931	3 30	27 56	3 55	17 4	5 35	19 49	18 2	26 15	7 18	7 47	16 16	1 43	25 31	1 40	4 1	11 24	6 35	3 39	26 56	11 58	1 11
6/18/1931	7 48	3 41	3 37	22 24	10 46	17 39	20 6	0gm32	7 33	9 15	19 23	1 46	28 13	3 46	3 59	15 3	4 27	4 6	28 36	11 42	1 8
6/28/1931	12 4	9 17	3 16	27 31	16 3	15 40	22 14	4 52	7 49	10 44	22 25	1 49	0sc54	6 19	3 57	18 44	2 53	4 27	0aq16	10 55	1 4
7/ 8/1931	16 18	14 44	2 53	2sa25	21 24	13 53	24 26	9 13	8 8	12 13	25 20	1 53	3 33	9 13	3 54	22 29	2 8	4 42	1 58	9 39	0 59
7/18/1931	20 30	20 0	2 28	7 8	26 47	12 16	26 42	13 37	8 28	13 42	28 8	1 58	6 11	12 25	3 51	26 17	2 15D	4 50	3 40	8 0	0 52
7/28/1931	24 39	25 3	2 0	11 42	2sc12	10 48	29 3	18 2	8 50	15 13	0gm47	2 3	8 47	15 52	3 49	0cp 8	3 16	4 54	5 24	6 5	0 45
8/ 7/1931	28 45	29 51	1 30	16 6	7 38	9 29	1 30	22 30	9 15	16 44	3 14	2 9	11 22	19 32	3 46	4 1	5 5	4 55	7 9	4 7	0 37
8/17/1931	2ta48	4gm21	0n57	20sa22	13sc 6	8s17	4gm 2	26gm59	9s42	18ta16	5gm27	2s15	13sc55	23li22	3n44	7cp56	7sa36	4n53	8aq56	2aq16R	0n29
8/27/1931	6 47	8 29	0 20	24 30	18 35	7 12	6 41	1cn30	10 12	19 49	7 24	2 22	16 26	27 21	3 41	11 54	10 44	4 50	10 43	0 44	0 20
9/ 6/1931	10 42	12 13	0s21	28 33	24 5	6 13	9 27	6 3	10 45	21 23	9 2	2 30	18 56	1sc27	3 39	15 53	14 22	4 46	12 32	29cp40	0 12
9/16/1931	14 34	15 27	1 6	2 29	29 36	5 20	12 20	10 37	11 22	22 57	10 17	2 38	21 24	5 38	3 37	19 53	18 27	4 40	14 22	29 8	0 4
9/26/1931	18 21	18 5	1 57	6 21	5 8	4 31	15 23	15 11	12 3	24 33	11 6	2 47	23 51	9 54	3 36	23 55	22 52	4 34	16 13	29 9D	0s 4
10/ 6/1931	22 4	20 2	2 52	10 8	10 41	3 46	18 35	19 46	12 49	26 9	11 26	2 56	26 16	14 14	3 35	27 57	27 36	4 28	18 6	29 45	0 10
10/16/1931	25 44	21 10	3 54	13 52	16 16	3 4	21 59	24 22	13 40	27 47	11 15	3 5	28 39	18 36	3 34	1aq59	2 34	4 21	20 0	0 52	0 16
10/26/1931	29 19	21 24R	5 0	17 33	21 52	2 26	25 34	28 56	14 36	29 25	10 31	3 13	1sa 1	23 1	3 33	6 1	7 45	4 14	21 56	2 28	0 22
11/ 5/1931	2 50	20 40	6 9	21 12	27 31	1 50	29 22	3 29	15 38	1 4	9 16	3 20	3 22	27 26	3 33	10 2	13 6	4 7	23 53	4 30	0 27
11/15/1931	6 16	19 2	7 17	24 49	3 11	1 16	3 26	7 59	16 46	2 44	7 35	3 26	5 41	1sa52	3 33	14 2	18 35	3 59	25 51	6 55	0 32
11/25/1931	9 39	16 40	8 18	28 25	8 53	0 45	7 46	12 25	18 0	4 25	5 37	3 29	7 58	6 17	3 34	18 1	24 10	3 51	27 50	9 40	0 36
12/ 5/1931	12 58	13 55	9 8	2 0	14 37	0 15	12 25	16 46	19 21	6 7	3 33	3 28	10 14	10 41	3 35	21 59	29 50	3 44	29 52	12 41	0 40
12/15/1931	16 12	11 12	9 41	5 35	20 23	0n14	17 25	20 59	20 46	7 51	1 36	3 25	12 28	15 3	3 36	25 54	5 34	3 36	1 54	15 58	0 44
12/25/1931	19 23	8 56	9 59	9 11	26 12	0 42	22 47	25 2	22 12	9 35	29ta59	3 20	14 41	19 22	3 38	29 47	11 20	3 28	3 58	19 27	0 47

Ephemeris	80			433			1221			1489			114			3018			2041		
0 Hours ET	Sappho			Eros			Amor			Attila			Kassandra			Godiva			Lancelot		
Mo Dy Year	h long	g long	g lat	h long	g long	g lat	h long	g long	g lat	h long	g long	g lat	h long	g long	g lat	h long	g long	g lat	h long	g long	g lat
1/ 4/1932	22 30	7 24	10 2	12 48	2aq 3	1 8	28 36	28 55	23 36	11 20	28 49	3 12	16 53	23 37	3 40	3 37	17 8	3 20	6 4	23 7	0 51
1/14/1932	25 34	6 43	9 55	16 27	7 58	1 34	4 52	2 34	24 49	13 7	28 14	3 3	19 3	27 48	3 43	7 25	22 56	3 13	8 11	26 56	0 55
1/24/1932	28 33	6 54D	9 41	20 8	13 55	2 0	11 39	6 2	25 35	14 55	28 13D	2 53	21 12	1cp53	3 47	11 9	28 43	3 5	10 19	0pi53	0 58
2/ 3/1932	1 30	7 52	9 24	23 52	19 55	2 24	18 59	9 25	25 32	16 44	28 47	2 44	23 20	5 51	3 51	14 50	4 30	2 57	12 29	4 57	1 2
2/13/1932	4 22	9 30	9 5	27 39	25 59	2 49	26 53	12 54	24 0	18 34	29 53	2 35	25 27	9 41	3 55	18 28	10 14	2 49	14 41	9 6	1 5
2/23/1932	7 12	11 43	8 47	1pi31	2pi 7	3 13	5 21	16 56	19 47	20 25	1 28	2 26	27 32	13 22	4 1	22 3	15 55	2 42	16 54	13 18	1 9
3/ 4/1932	9 58	14 22	8 29	5 27	8 19	3 38	14 23	22 16	10 50	22 17	3 28	2 18	29 36	16 51	4 7	25 33	21 34	2 34	19 8	17 34	1 13
3/14/1932	12 42	17 24	8 12	9 30	14 36	4 2	23 55	29 59	5n35	24 11	5 49	2 11	1 39	20 8	4 14	29 0	27 8	2 27	21 24	21 53	1 17
3/24/1932	15 22	20 44	7 56	13 38	20 59	4 25	3li52	11 23	28 4	26 6	8 29	2 4	3 40	23 10	4 22	2 24	2ar38	2 19	23 41	26 12	1 21
4/ 3/1932	18 0	24 18	7 41	17 54	27 27	4 49	14 4	26 37	46 45	28 2	11 25	1 58	5 41	25 54	4 30	5 43	8 4	2 12	26 0	0ar33	1 26
4/13/1932	20 35	28 4	7 28	22 19	4 1	5 12	24 23	12 56	56 20	29 59	14 35	1 52	7 40	28 18	4 39	8 59	13 24	2 4	28 20	4 53	1 30
4/23/1932	23 7	1cn58	7 16	26 52	10 43	5 34	4sc35	26 14	59 35	1 58	17 56	1 47	9 39	0aq20	4 49	12 12	18 39	1 57	0ar41	9 12	1 35
5/ 3/1932	25 37	5 59	7 5	1ar36	17 33	5 56	14 31	4 45	59 35	3 58	21 27	1 42	11 36	1 55	5 0	15 20	23 47	1 49	3 4	13 30	1 40
5/13/1932	28 4	10 6	6 56	6 31	24 32	6 16	24 2	9 14	57 48	6 0	25 5	1 38	13 33	3 0	5 11	18 25	28 49	1 42	5 27	17 45	1 46
5/23/1932	0le29	14 16	6 47	11 38	1ta41	6 35	3sa 0	11 16	54 47	8 2	28 51	1 34	15 29	3 33	5 21	21 27	3 44	1 34	7 53	21 56	1 52
6/ 2/1932	2 52	18 30	6 39	16 59	9 0	6 52	11 23	12 4	50 51	10 6	2 42	1 30	17 23	3 30R	5 31	24 25	8 31	1 26	10 19	26 3	1 58
6/12/1932	5 13	22 46	6 32	22 35	16 30	7 5	19 8	12 32	46 10	12 11	6 39	1 27	19 17	2 51	5 39	27 19	13 10	1 18	12 46	0ta 4	2 5
6/22/1932	7 32	27 3	6 26	28 26	24 12	7 15	26 17	13 15	41 7	14 18	10 39	1 24	21 10	1 37	5 44	0ta10	17 40	1 10	15 15	3 59	2 13
7/ 2/1932	9 49	1le21	6 21	4 35	2gm 7	7 21	2cp51	14 25	36 1	16 26	14 42	1 21	23 3	29cp53	5 45	2 59	22 0	1 2	17 44	7 45	2 21
7/12/1932	12 4	5 39	6 17	11 1	10 15	7 21	8 54	16 7	31 11	18 35	18 48	1 18	24 54	27 48	5 42	5 44	26 9	0 53	20 14	11 21	2 29
7/22/1932	14 18	9 57	6 13	17 45	18 35	7 14	14 27	18 20	26 47	20 46	22 56	1 15	26 45	25 35	5 32	8 26	0gm 6	0 44	22 46	14 44	2 39
8/ 1/1932	16 30	14 14	6 10	24 49	27 7	7 1	19 35	20 58	22 54	22 57	27 5	1 12	28 35	23 28	5 18	11 5	3 48	0 34	25 18	17 53	2 49
8/11/1932	18 40	18 30	6 8	2gm12	5 51	6 39	24 19	23 58	19 34	25 10	1le14	1 10	0aq25	21 38	5 0	13 41	7 14	0 23	27 50	20 44	3 1
8/21/1932	20 49	22 45	6 7	9 53	14 46	6 9	28 43	27 15	16 44	27 25	5 23	1 7	2 14	20 16	4 40	16 15	10 20	0 12	0ta23	23 14	3 13
8/31/1932	22 56	26 57	6 6	17 53	23 49	5 30	2 48	0cp46	14 20	29 40	9 30	1 5	4 2	19 27	4 18	18 46	13 5	0s 1	2 57	25 20	3 26
9/10/1932	25 3	1vi 6	6 6	26 8	3le 0	4 42	6 38	4 28	12 18	1 57	13 36	1 2	5 50	19 13	3 56	21 15	15 25	0 14	5 31	26 55	3 40
9/20/1932	27le 8	5vi11	6s 7	4cn37	12le17	3n47	10aq12	8cp20	10n34	4le14	17le38	1s 0	7aq37	19cp34	3n35	23ta41	17gm14	0s29	8ta 5	27ta57	3s54
9/30/1932	29 12	9 12	6 9	13 17	21 36	2 44	13 34	12 18	9 7	6 33	21 36	0 57	9 24	20 27	3 16	26 5	18 29	0 46	10 40	28 22	4 8
10/10/1932	1 14	13 8	6 12	22 4	0vi56	1 36	16 45	16 22	7 52	8 53	25 29	0 55	11 10	21 49	2 57	28 27	19 5	1 4	13 14	28 6	4 22
10/20/1932	3 16	16 57	6 16	0le56	10 15	0 24	19 45	20 31	6 48	11 15	29 14	0 52	12 56	23 37	2 41	0gm47	18 57R	1 24	15 49	27 11	4 33
10/30/1932	5 17	20 39	6 20	9 48	19 29	0 50	22 37	24 43	5 53	13 37	2 51	0 49	14 41	25 47	2 25	3 4	18 5	1 45	18 24	25 41	4 41
11/ 9/1932	7 17	24 11	6 26	18 36	28 36	2 4	25 20	28 58	5 5	16 0	6 16	0 46	16 27	28 16	2 12	5 20	16 29	2 6	20 58	23 47	4 45
11/19/1932	9 17	27 32	6 33	27 17	7 35	3 17	27 56	3 16	4 23	18 24	9 27	0 42	18 11	1aq 1	1 59	7 35	14 18	2 27	23 32	21 44	4 44
11/29/1932	11 16	0li40	6 41	5 48	16 23	4 27	0pi26	7 35	3 46	20 48	12 21	0 38	19 56	3 59	1 48	9 47	11 44	2 45	26 6	19 48	4 37
12/ 9/1932	13 14	3 32	6 51	14 5	24 58	5 33	2 49	11 54	3 13	23 14	14 54	0 34	21 40	7 9	1 37	11 58	9 6	3 0	28 39	18 15	4 26
12/19/1932	15 11	6 6	7 1	22 6	3sc18	6 35	5 7	16 15	2 44	25 40	17 3	0 29	23 24	10 29	1 28	14 7	6 42	3 11	1gm12	17 15	4 11
12/29/1932	17 8	8 16	7 13	29 50	11 21	7 32	7 21	20 35	2 17	28 7	18 42	0 23	25 8	13 56	1 19	16 15	4 48	3 18	3 44	16 55	3 55
1/ 8/1933	19 5	10 0	7 25	7 16	19 7	8 24	9 30	24 55	1 53	0vi35	19 48	0 17	26 52	17 29	1 10	18 21	3 34	3 21	6 15	17 16	3 39
1/18/1933	21 1	11 13	7 38	14 23	26 35	9 11	11 35	29 14	1 31	3 3	20 17	0 9	28 35	21 7	1 3	20 27	3 2	3 22	8 45	18 15	3 23
1/28/1933	22 57	11 51	7 51	21 11	3 42	9 55	13 37	3 31	1 11	5 31	20 6	0 1	0pi19	24 49	0 55	22 31	3 12	3 21	11 15	19 48	3 8

Ephemeris 0 Hours ET Mo Dy Year	80 Sappho h long	 g long	 g lat	433 Eros h long	 g long	 g lat	1221 Amor h long	 g long	 g lat	1489 Attila h long	 g long	 g lat	114 Kassandra h long	 g long	 g lat	3018 Godiva h long	 g long	 g lat	2041 Lancelot h long	 g long	 g lat
2/ 7/1933	24 53	11 49R	8 2	27 41	10 27	10 35	15 35	7 47	0 52	8 0	19 16	0 9	2 2	28 34	0 48	24 34	4 1	3 20	13 44	21 52	2 53
2/17/1933	26 48	11 5	8 10	3 52	16 50	11 12	17 30	12 1	0 33	10 29	17 50	0 19	3 46	2 19	0 41	26 35	5 24	3 17	16 11	24 21	2 40
2/27/1933	28 44	9 40	8 14	9 47	22 47	11 47	19 23	16 11	0 16	12 59	15 59	0 29	5 29	6 5	0 35	28 36	7 15	3 15	18 38	27 11	2 29
3/ 9/1933	0li39	7 39	8 11	15 27	28 18	12 20	21 13	20 19	0s 1	15 28	13 57	0 39	7 12	9 51	0 28	0cn36	9 31	3 12	21 3	0gm19	2 18
3/19/1933	2 35	5 13	7 59	20 51	3 19	12 52	23 1	24 23	0 17	17 58	12 0	0 49	8 56	13 35	0 22	2 35	12 7	3 10	23 27	3 41	2 8
3/29/1933	4 30	2 35	7 38	26 1	7 47	13 22	24 48	28 23	0 34	20 27	10 23	0 57	10 39	17 17	0 15	4 33	15 0	3 8	25 50	7 15	1 59
4/ 8/1933	6 26	0 4	7 10	0sa59	11 36	13 52	26 32	2 19	0 50	22 57	9 17	1 3	12 22	20 55	0 8	6 31	18 6	3 7	28 11	10 58	1 51
4/18/1933	8 22	27 55	6 35	5 46	14 40	14 20	28 15	6 9	1 7	25 26	8 49	1 9	14 6	24 30	0 1	8 28	21 24	3 5	0cn32	14 48	1 43
4/28/1933	10 18	26 19	5 58	10 22	16 51	14 46	29 56	9 54	1 24	27 55	8 59	1 13	15 50	27 58	0 6	10 24	24 50	3 5	2 51	18 44	1 36
5/ 8/1933	12 15	25 23	5 19	14 49	17 59	15 7	1 37	13 32	1 42	0li23	9 48	1 16	17 34	1ar21	0 13	12 20	28 24	3 4	5 8	22 44	1 30
5/18/1933	14 12	25 8	4 42	19 7	17 55R	15 20	3 16	17 2	2 1	2 51	11 10	1 19	19 18	4 35	0 21	14 15	2 4	3 4	7 24	26 48	1 24
5/28/1933	16 10	25 31	4 6	23 18	16 32	15 18	4 54	20 25	2 21	5 19	13 2	1 21	21 3	7 41	0 30	16 10	5 49	3 4	9 39	0cn53	1 19
6/ 7/1933	18 8	26 31	3 33	27 22	13 49	14 54	6 31	23 37	2 43	7 46	15 20	1 23	22 48	10 35	0 39	18 4	9 37	3 5	11 52	5 0	1 13
6/17/1933	20 7	28 1	3 3	1cp20	10 3	14 1	8 8	26 38	3 6	10 12	17 59	1 24	24 33	13 16	0 49	19 58	13 29	3 6	14 3	9 7	1 9
6/27/1933	22 7	29 59	2 35	5 13	5 43	12 36	9 45	29 27	3 31	12 37	20 58	1 25	26 19	15 43	1 0	21 52	17 22	3 8	16 14	13 14	1 4
7/ 7/1933	24 7	2 20	2 10	9 2	1 32	10 47	11 20	2 0	3 59	15 2	24 11	1 26	28 4	17 51	1 12	23 46	21 17	3 10	18 22	17 20	0 59
7/17/1933	26 8	5 2	1 47	12 47	28 8	8 47	12 56	4 16	4 30	17 26	27 38	1 27	29 51	19 38	1 26	25 39	25 12	3 12	20 30	21 24	0 55
7/27/1933	28 11	8 0	1 26	16 29	25 52	6 47	14 32	6 10	5 4	19 49	1li15	1 27	1 38	21 1	1 40	27 33	29 8	3 15	22 35	25 26	0 51
8/ 6/1933	0sc14	11 14	1 7	20 8	24 50	4 57	16 7	7 39	5 42	22 11	5 2	1 28	3 25	21 55	1 56	29 26	3 3	3 18	24 40	29 25	0 47
8/16/1933	2 19	14 40	0 49	23 46	24 57D	3 21	17 43	8 39	6 23	24 32	8 55	1 29	5 13	22 18	2 14	1 20	6 57	3 22	26 43	3 20	0 43
8/26/1933	4 25	18 18	0 32	27 22	26 5	1 58	19 18	9 5	7 9	26 52	12 55	1 30	7 1	22 7	2 32	3 13	10 49	3 26	28 44	7 10	0 39
9/ 5/1933	6 32	22 6	0 15	0aq57	28 2	0 48	20 55	8 52	7 58	29 11	16 59	1 31	8 50	21 19	2 51	5 7	14 38	3 31	0le44	10 54	0 35
9/15/1933	8 41	26 2	0 0	4 32	0cp40	0n11	22 31	7 57	8 49	1 29	21 8	1 32	10 40	19 57	3 10	7 1	18 24	3 37	2 42	14 32	0 31
9/25/1933	10 51	0sc 6	0 15	8 8	3 52	1 2	24 8	6 17	9 41	3 45	25 19	1 33	12 30	18 6	3 28	8 55	22 5	3 44	4 40	18 2	0 27
10/ 5/1933	13 3	4 16	0 30	11 45	7 32	1 45	25 46	3 56	10 29	6 1	29 32	1 34	14 21	15 56	3 43	10 49	25 41	3 51	6 35	21 22	0 23
10/15/1933	15 16	8 33	0 44	15 23	11 34	2 22	27 25	1 3	11 10	8 15	3 47	1 35	16 13	13 39	3 55	12 44	29 9	3 59	8 30	24 31	0 19
10/25/1933	17sc32	12sc55	0n58	19aq 3	15cp57	2n55	29ar 5	27ar53R	11s39	10sc28	8sc 1	1n37	18ar 5	11ar29R	4s 3	14le40	2vi28	4s 9	10le22	27le27	0s14
11/ 4/1933	19 49	17 21	1 13	22 46	20 37	3 24	0ta46	24 45	11 56	12 39	12 16	1 38	19 58	9 42	4 8	16 35	5 35	4 19	12 14	0vi 7	0 9
11/14/1933	22 9	21 52	1 27	26 32	25 32	3 50	2 28	21 58	11 59	14 50	16 28	1 40	21 53	8 26	4 9	18 31	8 30	4 31	14 4	2 29	0 4
11/24/1933	24 31	26 27	1 42	0pi23	0aq40	4 13	4 12	19 48	11 51	16 59	20 39	1 42	23 47	7 47	4 7	20 28	11 7	4 44	15 53	4 30	0n 2
12/ 4/1933	26 55	1sa 4	1 57	4 18	6 1	4 33	5 57	18 23	11 36	19 7	24 46	1 44	25 43	7 47D	4 4	22 26	13 25	4 59	17 41	6 6	0 8
12/14/1933	29 21	5 43	2 13	8 18	11 34	4 52	7 44	17 46	11 15	21 13	28 50	1 47	27 40	8 24	3 59	24 24	15 19	5 15	19 28	7 14	0 15
12/24/1933	1 51	10 25	2 30	12 25	17 17	5 9	9 33	17 54D	10 53	23 18	2 48	1 50	29 38	9 35	3 55	26 23	16 44	5 33	21 13	7 51	0 22
1/ 3/1934	4 23	15 7	2 48	16 39	23 12	5 25	11 25	18 45	10 30	25 22	6 39	1 53	1 37	11 17	3 50	28 23	17 36	5 51	22 57	7 54R	0 30
1/13/1934	6 58	19 51	3 6	21 1	29 16	5 39	13 19	20 12	10 9	27 25	10 23	1 57	3 37	13 25	3 45	0vi24	17 50	6 10	24 40	7 22	0 38
1/23/1934	9 35	24 34	3 26	25 31	5 31	5 52	15 15	22 12	9 49	29 26	13 57	2 1	5 38	15 57	3 41	2 25	17 23	6 28	26 22	6 17	0 47
2/ 2/1934	12 16	29 16	3 48	0ar12	11 56	6 3	17 14	24 40	9 31	1 26	17 20	2 5	7 40	18 48	3 37	4 28	16 14	6 44	28 3	4 43	0 55
2/12/1934	15 1	3 56	4 11	5 4	18 31	6 12	19 17	27 31	9 15	3 24	20 31	2 10	9 43	21 55	3 34	6 32	14 25	6 55	29 42	2 49	1 3
2/22/1934	17 49	8 34	4 36	10 7	25 17	6 19	21 23	0ta42	9 1	5 21	23 26	2 15	11 47	25 16	3 31	8 37	12 6	7 1	1 21	0 46	1 11
3/ 4/1934	20 40	13 8	5 4	15 24	2ar14	6 24	23 33	4 11	8 49	7 17	26 4	2 21	13 53	28 50	3 29	10 44	9 32	6 59	2 59	28 47	1 17

Ephemeris	80			433			1221			1489			114			3018			2041		
0 Hours ET	Sappho			Eros			Amor			Attila			Kassandra			Godiva			Lancelot		
Mo Dy Year	h long	g long	g lat	h long	g long	g lat	h long	g long	g lat	h long	g long	g lat	h long	g long	g lat	h long	g long	g lat	h long	g long	g lat
3/14/1934	23 35	17 38	5 34	20 55	9 22	6 27	25 48	7 56	8 39	9 12	28 22	2 28	16 0	2 33	3 27	12 51	6 58	6 49	4 35	27 2	1 22
3/24/1934	26 35	22 2	6 7	26 42	16 41	6 26	28 6	11 54	8 30	11 6	0cp17	2 34	18 8	6 24	3 26	15 0	4 44	6 31	6 11	25 42	1 25
4/ 3/1934	29 38	26 17	6 43	2ta45	24 12	6 23	0gm31	16 5	8 22	12 58	1 46	2 42	20 18	10 22	3 25	17 11	3 3	6 9	7 45	24 52	1 28
4/13/1934	2 45	0aq24	7 23	9 6	1ta54	6 15	3 0	20 28	8 16	14 49	2 46	2 49	22 29	14 26	3 25	19 23	2 2	5 44	9 19	24 33	1 30
4/23/1934	5 57	4 17	8 7	15 45	9 49	6 3	5 36	25 2	8 10	16 39	3 15	2 57	24 41	18 35	3 25	21 37	1 45	5 18	10 52	24 46	1 31
5/ 3/1934	9 13	7 56	8 55	22 43	17 55	5 46	8 19	29 46	8 5	18 28	3 10R	3 4	26 55	22 48	3 26	23 52	2 10	4 52	12 24	25 27	1 31
5/13/1934	12 33	11 16	9 48	0gm 0	26 12	5 25	11 9	4 41	8 1	20 15	2 32	3 11	29 10	27 4	3 26	26 10	3 13	4 28	13 55	26 35	1 32
5/23/1934	15 58	14 13	10 47	7 36	4 41	4 58	14 8	9 46	7 57	22 1	1 23	3 16	1 27	1gm22	3 28	28 29	4 51	4 5	15 25	28 5	1 32
6/ 2/1934	19 28	16 41	11 51	15 30	13 20	4 25	17 16	15 2	7 53	23 47	29sa49	3 19	3 45	5 42	3 29	0li50	6 59	3 43	16 54	29 55	1 32
6/12/1934	23 2	18 35	12 59	23 40	22 8	3 47	20 35	20 30	7 49	25 31	27 57	3 19	6 4	10 4	3 31	3 13	9 33	3 24	18 23	2 2	1 32
6/22/1934	26 41	19 47	14 10	2cn 6	1cn 4	3 4	24 4	26 9	7 44	27 14	25 59	3 17	8 26	14 26	3 34	5 39	12 29	3 6	19 51	4 23	1 32
7/ 2/1934	0aq24	20 11	15 21	10 42	10 6	2 17	27 47	2cn 1	7 38	28 56	24 7	3 12	10 49	18 48	3 37	8 7	15 44	2 49	21 18	6 55	1 33
7/12/1934	4 12	19 44	16 27	19 28	19 14	1 26	1cn44	8 7	7 31	0cp37	22 31	3 5	13 13	23 9	3 40	10 37	19 16	2 34	22 45	9 38	1 33
7/22/1934	8 4	18 27	17 20	28 19	28 24	0 33	5 57	14 28	7 22	2 17	21 19	2 57	15 39	27 30	3 43	13 10	23 3	2 19	24 10	12 29	1 34
8/ 1/1934	12 1	16 29	17 53	7 11	7 36	0s20	10 27	21 5	7 10	3 56	20 37	2 47	18 7	1cn48	3 47	15 45	27 2	2 6	25 35	15 26	1 35
8/11/1934	16 1	14 10	17 59	16 1	16 47	1 13	15 18	28 0	6 54	5 34	20 25	2 37	20 36	6 4	3 52	18 23	1li12	1 53	27 0	18 29	1 36
8/21/1934	20 6	11 54	17 35	24 44	25 55	2 5	20 30	5 14	6 35	7 11	20 44	2 27	23 7	10 16	3 57	21 3	5 33	1 41	28 24	21 36	1 37
8/31/1934	24 13	10 7	16 45	3vi18	4 58	2 53	26 7	12 48	6 11	8 47	21 32	2 18	25 40	14 23	4 2	23 47	10 3	1 29	29 47	24 45	1 38
9/10/1934	28 24	9 7	15 35	11 39	13 56	3 37	2le11	20 44	5 41	10 22	22 45	2 9	28 14	18 23	4 9	26 33	14 41	1 18	1 10	27 57	1 40
9/20/1934	2 38	9 4D	14 15	19 45	22 46	4 17	8 45	29 2	5 4	11 57	24 22	2 1	0cn49	22 16	4 15	29 23	19 26	1 7	2 32	1li10	1 42
9/30/1934	6 53	10 0	12 50	27 34	1li26	4 51	15 50	7 44	4 19	13 30	26 19	1 53	3 27	25 59	4 23	2 15	24 19	0 57	3 53	4 23	1 44
10/10/1934	11 11	11 49	11 26	5 6	9 57	5 20	23 29	16 48	3 27	15 3	28 32	1 47	6 6	29 29	4 30	5 11	29 18	0 46	5 14	7 35	1 47
10/20/1934	15 30	14 25	10 6	12 18	18 17	5 44	1vi43	26 14	2 27	16 35	1cp 1	1 40	8 46	2 44	4 39	8 10	4 23	0 36	6 35	10 45	1 50
10/30/1934	19 50	17 41	8 51	19 12	26 25	6 3	10 32	5 59	1 21	18 6	3 42	1 35	11 28	5 40	4 48	11 13	9 33	0 25	7 55	13 51	1 53
11/ 9/1934	24 11	21 30	7 42	25 47	4 21	6 17	19 52	16 0	0 10	19 36	6 34	1 30	14 12	8 13	4 58	14 19	14 49	0 15	9 15	16 53	1 57
11/19/1934	28 31	25 46	6 38	2sc 4	12 5	6 27	29 39	26 11	1 4	21 6	9 34	1 25	16 57	10 17	5 8	17 29	20 9	0 4	10 34	19 50	2 1
11/29/1934	2ar50	0pi24	5n40	8sc 4	19sc38	6s34	9li46	6sc28	2n18	22cp35	12cp41	1n21	19cn43	11le48	5s18	20sc42	25sc32	0n 7	11li53	22li39	2n 6
12/ 9/1934	7 9	5 20	4 47	13 48	26 58	6 37	20 3	16 44	3 28	24 3	15 54	1 18	22 31	12 40	5 27	23 58	1sa 0	0 18	13 11	25 19	2 11
12/19/1934	11 26	10 30	3 58	19 16	4 6	6 37	0sc20	26 52	4 32	25 31	19 10	1 14	25 20	12 48R	5 33	27 19	6 30	0 30	14 29	27 48	2 17
12/29/1934	15 41	15 51	3 13	24 31	11 3	6 34	10 25	6 46	5 28	26 57	22 30	1 11	28 10	12 10	5 37	0sa43	12 2	0 42	15 47	0sc 3	2 24
1/ 8/1935	19 54	21 21	2 32	29 33	17 48	6 29	20 8	16 21	6 16	28 24	25 50	1 9	1le 2	10 48	5 35	4 10	17 36	0 55	17 4	2 4	2 31
1/18/1935	24 4	26 57	1 54	4 23	24 22	6 23	29 20	25 34	6 54	29 49	29 12	1 6	3 54	8 51	5 26	7 41	23 12	1 8	18 21	3 46	2 39
1/28/1935	28 11	2ar37	1 20	9 2	0cp45	6 14	7 59	4 20	7 24	1 14	2 32	1 4	6 48	6 33	5 10	11 16	28 47	1 22	19 37	5 8	2 48
2/ 7/1935	2 15	8 21	0 47	13 31	6 57	6 4	16 0	12 40	7 46	2 39	5 51	1 2	9 42	4 14	4 47	14 54	4 23	1 37	20 54	6 6	2 57
2/17/1935	6 15	14 6	0 17	17 52	12 57	5 52	23 25	20 31	8 2	4 3	9 6	1 0	12 37	2 15	4 19	18 35	9 56	1 52	22 10	6 39	3 7
2/27/1935	10 12	19 53	0s10	22 5	18 47	5 39	0cp14	27 55	8 12	5 26	12 17	0 58	15 33	0 51	3 48	22 20	15 28	2 9	23 25	6 43R	3 17
3/ 9/1935	14 4	25 39	0 36	26 11	24 25	5 24	6 29	4 51	8 18	6 49	15 23	0 57	18 30	0 12	3 16	26 7	20 57	2 26	24 41	6 19	3 27
3/19/1935	17 53	1ta23	1 0	0cp11	29 51	5 7	12 15	11 20	8 22	8 11	18 22	0 56	21 27	0 20D	2 45	29 57	26 20	2 44	25 56	5 26	3 36
3/29/1935	21 38	7 6	1 23	4 5	5 5	4 49	17 33	17 23	8 23	9 33	21 13	0 54	24 25	1 12	2 16	3 50	1aq38	3 4	27 11	4 8	3 44
4/ 8/1935	25 18	12 47	1 44	7 55	10 6	4 28	22 27	23 0	8 22	10 55	23 54	0 53	27 23	2 46	1 49	7 45	6 48	3 25	28 26	2 30	3 50

Ephemeris	80			433			1221			1489			114			3018			2041		
0 Hours ET	Sappho			Eros			Amor			Attila			Kassandra			Godiva			Lancelot		
Mo Dy Year	h long	g long	g lat	h long	g long	g lat	h long	g long	g lat	h long	g long	g lat	h long	g long	g lat	h long	g long	g lat	h long	g long	g lat
4/18/1935	28 55	18 26	2 5	11 41	14 53	4 4	26 59	28 10	8 21	12 16	26 24	0 52	0vi21	4 54	1 25	11 42	11 48	3 48	29 41	0 39	3 54
4/28/1935	2 27	24 1	2 24	15 24	19 23	3 38	1aq12	2 54	8 18	13 36	28 41	0 51	3 19	7 32	1 4	15 40	16 36	4 12	0 55	28 46	3 55
5/ 8/1935	5 55	29 32	2 42	19 4	23 35	3 7	5 8	7 10	8 16	14 56	0pi42	0 50	6 18	10 35	0 44	19 40	21 10	4 37	2 10	26 59	3 53
5/18/1935	9 19	5 0	3 0	22 42	27 26	2 32	8 48	10 56	8 12	16 16	2 27	0 49	9 16	13 59	0 27	23 41	25 26	5 5	3 24	25 28	3 49
5/28/1935	12 38	10 24	3 17	26 19	0pi53	1 51	12 15	14 11	8 9	17 36	3 51	0 47	12 14	17 40	0 11	27 42	29 20	5 34	4 38	24 18	3 43
6/ 7/1935	15 54	15 43	3 33	29 54	3 49	1 3	15 31	16 51	8 5	18 55	4 53	0 46	15 12	21 35	0n 3	1aq44	2 47	6 5	5 52	23 34	3 36
6/17/1935	19 6	20 58	3 50	3 29	6 10	0 6	18 35	18 51	7 59	20 13	5 31	0 44	18 9	25 42	0 17	5 45	5 42	6 38	7 6	23 18	3 29
6/27/1935	22 14	26 8	4 6	7 5	7 46	1 2	21 30	20 7	7 53	21 32	5 42	0 42	21 6	29 59	0 29	9 46	7 59	7 13	8 19	23 28	3 21
7/ 7/1935	25 18	1cn13	4 22	10 41	8 28	2 22	24 17	20 34	7 43	22 50	5 26	0 40	24 2	4 23	0 40	13 46	9 31	7 47	9 33	24 4	3 13
7/17/1935	28 19	6 12	4 38	14 19	8 6	3 56	26 56	20 7	7 29	24 8	4 41	0 38	26 58	8 55	0 51	17 44	10 12	8 21	10 47	25 4	3 5
7/27/1935	1cn16	11 6	4 54	17 58	6 32	5 43	29 28	18 45	7 10	25 25	3 31	0 35	29 52	13 31	1 1	21 41	10 0R	8 50	12 0	26 25	2 58
8/ 6/1935	4 10	15 53	5 11	21 40	3 44	7 38	1 54	16 28	6 42	26 43	1 59	0 31	2 46	18 12	1 10	25 36	8 54	9 12	13 14	28 4	2 52
8/16/1935	7 0	20 33	5 28	25 25	29aq55	9 32	4 15	13 29	6 6	28 0	0 12	0 27	5 39	22 57	1 19	29 29	7 6	9 22	14 28	29 60	2 46
8/26/1935	9 47	25 6	5 46	29 14	25 33	11 13	6 30	10 3	5 22	29 17	28 19	0 22	8 31	27 44	1 28	3 19	4 51	9 18	15 41	2 9	2 41
9/ 5/1935	12 31	29 30	6 5	3 8	21 22	12 29	8 41	6 35	4 32	0pi33	26 29	0 18	11 22	2 34	1 36	7 6	2 35	9 0	16 55	4 31	2 36
9/15/1935	15 13	3 45	6 25	7 7	17 59	13 17	10 47	3 28	3 40	1 50	24 52	0 13	14 11	7 25	1 45	10 50	0 40	8 29	18 8	7 3	2 32
9/25/1935	17 51	7 48	6 47	11 12	15 50	13 39	12 50	0 59	2 49	3 6	23 36	0 9	16 59	12 16	1 53	14 31	29 25	7 50	19 22	9 44	2 28
10/ 5/1935	20 26	11 39	7 10	15 23	15 2	13 44	14 50	29aq17	2 1	4 22	22 45	0 4	19 46	17 8	2 1	18 9	28 58	7 6	20 36	12 32	2 25
10/15/1935	22 59	15 15	7 35	19 43	15 32	13 36	16 47	28 25	1 19	5 38	22 22	0 0	22 32	22 0	2 9	21 43	29 22	6 22	21 50	15 26	2 23
10/25/1935	25 29	18 35	8 3	24 11	17 10	13 22	18 40	28 21D	0 42	6 54	22 27D	0 3	25 16	26 50	2 17	25 13	0pi34	5 40	23 4	18 25	2 21
11/ 4/1935	27 57	21 33	8 33	28 48	19 47	13 3	20 32	28 58	0 10	8 10	23 0	0 7	27 58	1sc38	2 25	28 40	2 27	5 2	24 18	21 27	2 19
11/14/1935	0le23	24 8	9 6	3 36	23 13	12 42	22 21	0pi11	0 17	9 26	23 58	0 10	0sc40	6 24	2 34	2 4	4 56	4 26	25 32	24 31	2 18
11/24/1935	2 46	26 15	9 42	8 36	27 21	12 20	24 8	1 55	0 41	10 41	25 18	0 13	3 19	11 7	2 42	5 23	7 54	3 55	26 46	27 36	2 18
12/ 4/1935	5 7	27 49	10 20	13 49	2pi 4	11 57	25 53	4 3	1 1	11 57	26 59	0 15	5 57	15 45	2 51	8 39	11 16	3 26	28 1	0sa42	2 17
12/14/1935	7 27	28 45	11 0	19 16	7 19	11 32	27 36	6 33	1 19	13 12	28 58	0 18	8 34	20 18	3 1	11 51	14 58	3 1	29 15	3 46	2 18
12/24/1935	9 44	28 59	11 40	24 58	13 2	11 6	29 18	9 19	1 35	14 28	1 12	0 20	11 9	24 45	3 11	15 0	18 54	2 38	0sa30	6 47	2 19
1/ 3/1936	12le 0	28le26R	12s19	0ta56	19pi 9	10n37	0ar59	12pi20	1s50	15pi43	3pi38	0s22	13sc42	29sc 4	3n22	18ar 5	23pi 3	2n18	1sa45	9sa44	2n20
1/13/1936	14 13	27 8	12 54	7 12	25 40	10 6	2 39	15 32	2 3	16 59	6 15	0 24	16 14	3 14	3 34	21 7	27 22	2 0	3 0	12 36	2 22
1/23/1936	16 25	25 9	13 19	13 45	2ar32	9 32	4 17	18 53	2 16	18 14	9 0	0 26	18 44	7 14	3 46	24 5	1ar48	1 43	4 15	15 21	2 24
2/ 2/1936	18 36	22 39	13 31	20 38	9 44	8 54	5 55	22 22	2 28	19 30	11 53	0 28	21 13	11 2	4 0	27 0	6 20	1 28	5 31	17 58	2 27
2/12/1936	20 45	19 55	13 29	27 49	17 15	8 12	7 32	25 56	2 40	20 45	14 51	0 30	23 40	14 35	4 15	29 51	10 56	1 14	6 47	20 24	2 30
2/22/1936	22 53	17 17	13 11	5 20	25 3	7 25	9 9	29 35	2 51	22 1	17 53	0 32	26 6	17 51	4 31	2 39	15 34	1 2	8 3	22 37	2 34
3/ 3/1936	24 59	15 3	12 40	13 9	3ta 8	6 33	10 45	3 18	3 3	23 16	20 57	0 34	28 29	20 47	4 48	5 25	20 15	0 50	9 20	24 35	2 38
3/13/1936	27 4	13 24	12 1	21 15	11 28	5 36	12 20	7 3	3 15	24 32	24 3	0 37	0sa52	23 21	5 7	8 7	24 56	0 39	10 36	26 16	2 43
3/23/1936	29 8	12 28	11 17	29 36	20 2	4 34	13 56	10 49	3 27	25 48	27 9	0 39	3 13	25 29	5 28	10 46	29 37	0 29	11 54	27 38	2 48
4/ 2/1936	1 11	12 14	10 31	8 10	28 48	3 28	15 32	14 36	3 39	27 4	0ar15	0 42	5 32	27 6	5 50	13 23	4 18	0 19	13 11	28 36	2 54
4/12/1936	3 13	12 41	9 46	16 53	7 44	2 18	17 7	18 24	3 53	28 20	3 18	0 44	7 50	28 10	6 12	15 57	8 57	0 10	14 29	29 10	2 59
4/22/1936	5 14	13 44	9 4	25 43	16 48	1 6	18 43	22 11	4 7	29 36	6 19	0 47	10 6	28 36	6 35	18 28	13 35	0 1	15 47	29 16R	3 5
5/ 2/1936	7 15	15 18	8 26	4 35	25 59	0s 6	20 19	25 57	4 22	0 52	9 15	0 50	12 21	28 22	6 57	20 57	18 11	0 8	17 5	28 54	3 10
5/12/1936	9 14	17 18	7 50	13 26	5 13	1 17	21 55	29 40	4 38	2 9	12 6	0 54	14 34	27 29	7 17	23 23	22 43	0 17	18 24	28 4	3 14

Ephemeris	80			433			1221			1489			114			3018			2041		
0 Hours ET	Sappho			Eros			Amor			Attila			Kassandra			Godiva			Lancelot		
Mo Dy Year	h long	g long	g lat	h long	g long	g lat	h long	g long	g lat	h long	g long	g lat	h long	g long	g lat	h long	g long	g lat	h long	g long	g lat
5/22/1936	11 13	19 41	7 18	22 12	14 29	2 25	23 32	3 22	4 55	3 26	14 50	0 57	16 46	25 59	7 33	25 47	27 13	0 25	19 44	26 49	3 17
6/ 1/1936	13 11	22 22	6 49	0vi49	23 44	3 28	25 10	7 0	5 14	4 43	17 27	1 1	18 57	24 2	7 42	28 10	1gm39	0 34	21 3	25 12	3 18
6/11/1936	15 8	25 18	6 22	9 14	2le56	4 24	26 48	10 34	5 34	6 0	19 53	1 6	21 6	21 49	7 45	0gm30	6 1	0 42	22 24	23 23	3 16
6/21/1936	17 5	28 27	5 58	17 25	12 4	5 13	28 28	14 4	5 57	7 17	22 8	1 10	23 14	19 36	7 39	2 48	10 18	0 51	23 44	21 29	3 13
7/ 1/1936	19 2	1 47	5 37	25 19	21 5	5 54	0ta 8	17 27	6 22	8 35	24 10	1 16	25 21	17 37	7 26	5 4	14 30	1 0	25 5	19 42	3 7
7/11/1936	20 58	5 15	5 17	2li56	29 57	6 26	1 50	20 43	6 50	9 53	25 56	1 21	27 26	16 4	7 7	7 18	18 37	1 9	26 27	18 9	2 59
7/21/1936	22 54	8 52	4 59	10 14	8 39	6 50	3 33	23 51	7 20	11 11	27 24	1 27	29 30	15 5	6 45	9 31	22 37	1 18	27 49	16 58	2 50
7/31/1936	24 50	12 34	4 42	17 13	17 10	7 6	5 18	26 47	7 55	12 30	28 31	1 34	1 33	14 43	6 21	11 42	26 29	1 28	29 12	16 14	2 40
8/10/1936	26 45	16 22	4 27	23 54	25 30	7 15	7 4	29 30	8 34	13 48	29 15	1 41	3 35	14 57	5 57	13 51	0cn13	1 39	0cp35	15 59	2 30
8/20/1936	28 41	20 14	4 13	0sc16	3li38	7 17	8 52	1 57	9 17	15 8	29 32	1 49	5 36	15 46	5 33	15 59	3 47	1 50	1 59	16 13	2 20
8/30/1936	0li36	24 10	4 0	6 21	11 33	7 14	10 43	4 4	10 7	16 27	29 21	1 57	7 36	17 5	5 11	18 6	7 9	2 3	3 24	16 54	2 10
9/ 9/1936	2 31	28 8	3 47	12 9	19 17	7 7	12 36	5 47	11 2	17 47	28 42	2 5	9 34	18 52	4 50	20 12	10 18	2 16	4 49	18 1	2 1
9/19/1936	4 27	2 9	3 36	17 42	26 49	6 56	14 31	6 59	12 4	19 7	27 35	2 12	11 32	21 1	4 31	22 16	13 11	2 30	6 15	19 32	1 52
9/29/1936	6 23	6 11	3 25	23 1	4 11	6 41	16 29	7 34	13 12	20 28	26 4	2 19	13 29	23 31	4 14	24 19	15 46	2 46	7 41	21 23	1 44
10/ 9/1936	8 18	10 14	3 14	28 6	11 22	6 25	18 31	7 26R	14 25	21 49	24 15	2 24	15 25	26 17	3 59	26 21	17 58	3 4	9 8	23 32	1 37
10/19/1936	10 15	14 17	3 4	2 59	18 24	6 6	20 35	6 28	15 41	23 11	22 19	2 27	17 19	29 18	3 44	28 22	19 44	3 23	10 36	25 57	1 30
10/29/1936	12 11	18 20	2 54	7 41	25 17	5 46	22 44	4 37	16 56	24 33	20 24	2 29	19 13	2 31	3 32	0cn22	21 0	3 44	12 4	28 36	1 24
11/ 8/1936	14 8	22 20	2 44	12 14	2sa 1	5 26	24 56	1 55	18 1	25 55	18 42	2 29	21 7	5 53	3 20	2 21	21 41	4 6	13 34	1 27	1 18
11/18/1936	16 6	26 19	2 34	16 37	8 39	5 4	27 13	28ta34	18 50	27 18	17 21	2 28	22 59	9 24	3 10	4 20	21 44R	4 29	15 4	4 29	1 13
11/28/1936	18 4	0sc14	2 25	20 52	15 10	4 42	29 35	24 53	19 14	28 42	16 28	2 25	24 50	13 2	3 0	6 17	21 5	4 53	16 35	7 39	1 8
12/ 8/1936	20 3	4 5	2 15	25 0	21 34	4 20	2 3	21 22	19 12	0ta 6	16 5	2 22	26 41	16 45	2 52	8 14	19 44	5 15	18 6	10 56	1 3
12/18/1936	22 2	7 50	2 4	29 2	27 53	3 57	4 36	18 25	18 44	1 31	16 12	2 18	28 31	20 32	2 44	10 11	17 47	5 35	19 39	14 19	0 59
12/28/1936	24 3	11 27	1 54	2 58	4 6	3 34	7 16	16 21	17 59	2 56	16 50	2 14	0aq21	24 22	2 37	12 7	15 23	5 51	21 12	17 46	0 55
1/ 7/1937	26 4	14 56	1 42	6 49	10 15	3 11	10 3	15 19	17 3	4 22	17 55	2 10	2 10	28 14	2 30	14 2	12 48	6 0	22 47	21 17	0 51
1/17/1937	28 6	18 13	1 29	10 36	16 19	2 47	12 59	15 19D	16 2	5 48	19 24	2 6	3 58	2 6	2 24	15 57	10 19	6 3	24 22	24 51	0 48
1/27/1937	0sc10	21 16	1 16	14 20	22 19	2 23	16 3	16 17	15 2	7 15	21 16	2 2	5 46	5 59	2 18	17 52	8 11	6 0	25 58	28 26	0 44
2/ 6/1937	2sc14	24sc 3	1s 1	18cp 0	28cp15	1s59	19gm17	18ta 5	14s 3	8ta43	23ar26	1s59	7aq33	9aq50	2n13	19cn46	6cn36R	5s52	27cp35	2aq 1	0n41
2/16/1937	4 20	26 29	0 43	21 39	4 7	1 33	22 42	20 39	13 7	10 11	25 51	1 57	9 20	13 40	2 9	21 40	5 41	5 41	29 13	5 35	0 37
2/26/1937	6 27	28 32	0 24	25 16	9 56	1 7	26 19	23 51	12 15	11 40	28 31	1 54	11 6	17 26	2 4	23 34	5 27	5 28	0 52	9 7	0 34
3/ 8/1937	8 36	0sa 7	0 1	28 51	15 42	0 40	0cn10	27 38	11 26	13 10	1ta22	1 52	12 52	21 8	2 0	25 27	5 51	5 15	2 32	12 37	0 30
3/18/1937	10 46	1 8	0 24	2 27	21 24	0 12	4 16	1gm56	10 39	14 41	4 22	1 51	14 37	24 46	1 56	27 21	6 51	5 2	4 14	16 2	0 27
3/28/1937	12 58	1 31	0 53	6 2	27 4	0 18	8 39	6 41	9 53	16 12	7 31	1 49	16 22	28 17	1 52	29 14	8 22	4 49	5 56	19 21	0 23
4/ 7/1937	15 11	1 13	1 27	9 38	2pi42	0 50	13 22	11 52	9 8	17 44	10 45	1 48	18 7	1pi40	1 48	1 8	10 20	4 38	7 39	22 34	0 19
4/17/1937	17 27	0 10	2 3	13 15	8 17	1 24	18 25	17 28	8 23	19 17	14 5	1 48	19 51	4 55	1 45	3 2	12 39	4 27	9 24	25 38	0 15
4/27/1937	19 44	28 27	2 42	16 54	13 50	2 0	23 51	23 28	7 36	20 51	17 29	1 48	21 35	8 0	1 41	4 55	15 18	4 17	11 10	28 32	0 10
5/ 7/1937	22 4	26 10	3 22	20 35	19 21	2 39	29 44	29 52	6 48	22 25	20 55	1 48	23 19	10 53	1 37	6 49	18 12	4 9	12 57	1pi14	0 5
5/17/1937	24 25	23 33	3 59	24 19	24 50	3 21	6 5	6 40	5 57	24 0	24 24	1 48	25 2	13 31	1 33	8 44	21 19	4 1	14 45	3 41	0s 0
5/27/1937	26 50	20 55	4 33	28 7	0ar17	4 6	12 57	13 51	5 2	25 37	27 54	1 49	26 46	15 54	1 29	10 38	24 38	3 55	16 34	5 51	0 6
6/ 6/1937	29 16	18 35	5 1	1 59	5 43	4 56	20 22	21 27	4 3	27 14	1gm24	1 50	28 29	17 57	1 24	12 33	28 5	3 49	18 25	7 40	0 13
6/16/1937	1 45	16 47	5 23	5 56	11 8	5 52	28 21	29 27	3 0	28 52	4 53	1 51	0pi12	19 38	1 19	14 28	1le41	3 44	20 16	9 7	0 20

Ephemeris	80			433			1221			1489			114			3018			2041		
0 Hours ET	Sappho			Eros			Amor			Attila			Kassandra			Godiva			Lancelot		
Mo Dy Year	h long	g long	g lat	h long	g long	g lat	h long	g long	g lat	h long	g long	g lat	h long	g long	g lat	h long	g long	g lat	h long	g long	g lat
6/26/1937	4 17	15 41	5 39	9 59	16 31	6 52	6 55	7 51	1 52	0gm31	8 21	1 53	1 56	20 55	1 13	16 24	5 23	3 39	22 10	10 6	0 29
7/ 6/1937	6 52	15 22	5 50	14 8	21 52	8 0	16 2	16 39	0 40	2 11	11 46	1 55	3 39	21 42	1 6	18 20	9 10	3 35	24 4	10 36	0 38
7/16/1937	9 30	15 49	5 57	18 25	27 11	9 14	25 38	25 50	0n34	3 53	15 9	1 57	5 22	21 59	0 59	20 17	13 2	3 32	26 0	10 33R	0 48
7/26/1937	12 11	16 59	6 1	22 50	2ta28	10 38	5 38	5 21	1 50	5 35	18 26	2 0	7 5	21 41	0 50	22 14	16 58	3 29	27 57	9 57	0 58
8/ 5/1937	14 56	18 46	6 3	27 25	7 42	12 11	15 51	15 9	3 4	7 18	21 38	2 3	8 48	20 49	0 40	24 13	20 58	3 27	29 56	8 48	1 9
8/15/1937	17 43	21 8	6 5	2ar10	12 51	13 56	26 9	25 9	4 14	9 2	24 42	2 7	10 31	19 24	0 29	26 12	24 59	3 26	1 55	7 13	1 20
8/25/1937	20 35	23 58	6 5	7 6	17 54	15 53	6 19	5 18	5 16	10 48	27 37	2 11	12 14	17 33	0 16	28 11	29 3	3 24	3 57	5 19	1 30
9/ 4/1937	23 30	27 14	6 5	12 14	22 49	18 4	16 11	15 28	6 9	12 34	0cn22	2 15	13 57	15 24	0 3	0vi12	3 8	3 24	5 59	3 18	1 39
9/14/1937	26 29	0sa53	6 4	17 37	27 32	20 29	25 37	25 34	6 51	14 22	2 52	2 20	15 41	13 11	0 10	2 14	7 14	3 23	8 3	1 25	1 47
9/24/1937	29 33	4 51	6 4	23 14	1gm58	23 10	4 29	5 28	7 21	16 10	5 7	2 26	17 25	11 6	0 23	4 16	11 19	3 23	10 9	29aq52	1 52
10/ 4/1937	2 40	9 7	6 3	29 7	6 2	26 5	12 46	15 7	7 40	18 0	7 1	2 32	19 9	9 21	0 35	6 20	15 24	3 24	12 16	28 48	1 57
10/14/1937	5 52	13 38	6 3	5 17	9 33	29 11	20 25	24 26	7 48	19 51	8 33	2 39	20 53	8 6	0 46	8 25	19 28	3 25	14 24	28 19	1 59
10/24/1937	9 8	18 22	6 2	11 45	12 21	32 21	27 28	3sa23	7 47	21 44	9 37	2 46	22 37	7 26	0 56	10 31	23 29	3 27	16 34	28 28D	2 0
11/ 3/1937	12 28	23 19	6 1	18 32	14 12	35 26	3 57	11 55	7 39	23 37	10 11	2 53	24 22	7 22D	1 4	12 39	27 26	3 29	18 45	29 12	2 1
11/13/1937	15 54	28 28	6 0	25 37	14 50	38 5	9 55	20 4	7 27	25 32	10 10R	3 0	26 7	7 52	1 11	14 48	1li19	3 31	20 57	0pi31	2 1
11/23/1937	19 23	3 46	5 59	3gm 2	14 10	39 52	15 24	27 50	7 10	27 28	9 34	3 6	27 53	8 55	1 17	16 58	5 5	3 34	23 11	2 21	2 1
12/ 3/1937	22 58	9 14	5 57	10 45	12 19	40 14	20 28	5 13	6 51	29 25	8 23	3 10	29 39	10 27	1 23	19 10	8 44	3 38	25 26	4 38	2 0
12/13/1937	26 37	14 50	5 56	18 46	9 55	38 35	25 8	12 17	6 31	1 24	6 43	3 13	1 25	12 25	1 27	21 24	12 12	3 42	27 43	7 18	2 0
12/23/1937	0aq20	20 33	5 54	27 3	7 58	34 33	29 29	19 1	6 10	3 24	4 44	3 12	3 12	14 44	1 32	23 39	15 28	3 47	0ar 1	10 19	1 59
1/ 2/1938	4 8	26 23	5 51	5 34	7 20	28 9	3 32	25 27	5 49	5 25	2 37	3 8	5 0	17 22	1 36	25 56	18 29	3 52	2 20	13 37	1 59
1/12/1938	8 1	2aq19	5 48	14 15	8 34	19 54	7 18	1aq38	5 28	7 27	0 39	3 2	6 48	20 17	1 40	28 15	21 11	3 57	4 41	17 10	1 58
1/22/1938	11 57	8 20	5 44	23 3	11 43	10 46	10 51	7 34	5 8	9 31	29 2	2 53	8 36	23 24	1 44	0li36	23 30	4 3	7 2	20 55	1 58
2/ 1/1938	15 58	14 26	5 40	1le55	16 29	1 50	14 11	13 16	4 49	11 36	27 56	2 42	10 26	26 43	1 48	2 59	25 21	4 9	9 25	24 50	1 58
2/11/1938	20 2	20 35	5 35	10 47	22 35	6 1	17 20	18 45	4 30	13 42	27 27	2 30	12 15	0ar11	1 52	5 24	26 39	4 15	11 49	28 55	1 58
2/21/1938	24 10	26 47	5 29	19 35	29 38	12 19	20 19	24 2	4 11	15 50	27 36D	2 18	14 6	3 47	1 56	7 51	27 19	4 19	14 14	3 6	1 59
3/ 3/1938	28 21	3pi 1	5 22	28 16	7 19	16 59	23 9	29 7	3 54	17 59	28 22	2 7	15 57	7 29	2 0	10 21	27 16R	4 22	16 40	7 23	1 59
3/13/1938	2pi35	9pi16	5n14	6vi45	15cn24	20s10	25aq51	4pi 1	3n37	20cn 9	29gm42	1s56	17ar49	11ar17	2s 4	12li53	26li28R	4s22	19ar 7	11ar45	2s 0
3/23/1938	6 51	15 32	5 5	15 1	23 38	22 7	28 26	8 43	3 20	22 20	1 31	1 46	19 42	15 8	2 9	15 28	24 56	4 18	21 35	16 11	2 1
4/ 2/1938	11 8	21 48	4 54	23 0	1le53	23 4	0pi55	13 13	3 3	24 33	3 46	1 37	21 36	19 2	2 13	18 5	22 46	4 8	24 4	20 39	2 2
4/12/1938	15 28	28 2	4 42	0li43	10 2	23 13	3 17	17 32	2 47	26 47	6 23	1 28	23 31	22 59	2 18	20 45	20 15	3 52	26 33	25 9	2 3
4/22/1938	19 48	4 15	4 29	8 7	18 0	22 48	5 35	21 39	2 31	29 2	9 19	1 20	25 26	26 56	2 24	23 28	17 40	3 30	29 3	29 39	2 4
5/ 2/1938	24 8	10 25	4 15	15 12	25 45	21 58	7 48	25 33	2 14	1 18	12 32	1 13	27 23	0ta54	2 30	26 14	15 22	3 4	1 34	4 10	2 6
5/12/1938	28 28	16 32	3 59	21 58	3vi17	20 51	9 56	29 13	1 57	3 36	15 58	1 6	29 20	4 52	2 36	29 3	13 38	2 36	4 5	8 41	2 8
5/22/1938	2 48	22 34	3 42	28 26	10 35	19 34	12 0	2 38	1 40	5 54	19 35	1 0	1 18	8 49	2 43	1 55	12 39	2 7	6 37	13 10	2 10
6/ 1/1938	7 7	28 31	3 23	4 36	17 40	18 11	14 1	5 47	1 22	8 14	23 22	0 54	3 18	12 44	2 51	4 50	12 28D	1 40	9 9	17 37	2 12
6/11/1938	11 24	4 22	3 3	10 30	24 33	16 46	15 59	8 37	1 2	10 35	27 17	0 48	5 19	16 37	2 59	7 49	13 5	1 14	11 41	22 1	2 14
6/21/1938	15 39	10 6	2 41	16 7	1li15	15 22	17 54	11 6	0 41	12 56	1le20	0 43	7 20	20 26	3 7	10 51	14 27	0 50	14 14	26 22	2 17
7/ 1/1938	19 51	15 42	2 17	21 30	7 48	14 0	19 46	13 11	0 19	15 19	5 28	0 38	9 23	24 11	3 17	13 56	16 27	0 28	16 46	0gm38	2 20
7/11/1938	24 1	21 8	1 52	26 39	14 12	12 41	21 36	14 48	0s 6	17 42	9 41	0 33	11 27	27 51	3 28	17 5	19 2	0 9	19 19	4 49	2 23
7/21/1938	28 8	26 23	1 24	1sa36	20 29	11 28	23 24	15 54	0 33	20 7	13 58	0 29	13 33	1gm24	3 39	20 18	22 7	0 9	21 51	8 53	2 26

Ephemeris 0 Hours ET	80 Sappho			433 Eros			1221 Amor			1489 Attila			114 Kassandra			3018 Godiva			2041 Lancelot		
Mo Dy Year	h long	g long	g lat	h long	g long	g lat	h long	g long	g lat	h long	g long	g lat	h long	g long	g lat	h long	g long	g lat	h long	g long	g lat
7/31/1938	2 12	1gm26	0 54	6 21	26 39	10 18	25 10	16 23	1 4	22 32	18 18	0 24	15 39	4 49	3 52	23 34	25 37	0 25	24 23	12 49	2 30
8/10/1938	6 12	6 14	0 22	10 56	2sc44	9 14	26 54	16 11	1 37	24 58	22 41	0 20	17 47	8 4	4 6	26 53	29 29	0 40	26 55	16 35	2 34
8/20/1938	10 8	10 45	0s13	15 22	8 45	8 13	28 37	15 16	2 13	27 24	27 6	0 15	19 57	11 7	4 21	0sa17	3 40	0 53	29 27	20 10	2 38
8/30/1938	14 1	14 57	0 51	19 39	14 42	7 17	0ar18	13 36	2 51	29 51	1vi32	0 11	22 8	13 54	4 39	3 44	8 8	1 5	1 58	23 31	2 43
9/ 9/1938	17 49	18 44	1 33	23 49	20 36	6 26	1 58	11 16	3 29	2 19	5 59	0 6	24 20	16 24	4 57	7 14	12 50	1 17	4 28	26 35	2 48
9/19/1938	21 34	22 4	2 20	27 53	26 28	5 37	3 37	8 24	4 6	4 47	10 25	0 2	26 33	18 31	5 18	10 48	17 46	1 28	6 58	29 21	2 54
9/29/1938	25 14	24 51	3 11	1cp50	2sa18	4 53	5 15	5 17	4 39	7 15	14 50	0 3	28 48	20 13	5 41	14 26	22 53	1 38	9 27	1 43	3 0
10/ 9/1938	28 50	26 59	4 7	5 43	8 7	4 11	6 52	2 13	5 6	9 44	19 13	0 8	1gm 5	21 23	6 5	18 6	28 10	1 47	11 56	3 39	3 6
10/19/1938	2 22	28 21	5 8	9 31	13 55	3 32	8 29	29pi29	5 27	12 12	23 33	0 12	3 23	21 58	6 30	21 50	3 36	1 56	14 23	5 3	3 12
10/29/1938	5 50	28 51	6 13	13 15	19 43	2 55	10 5	27 20	5 41	14 41	27 49	0 18	5 43	21 53R	6 55	25 37	9 11	2 4	16 50	5 51	3 18
11/ 8/1938	9 14	28 25	7 21	16 57	25 30	2 21	11 41	25 53	5 49	17 10	1li59	0 23	8 4	21 6	7 19	29 27	14 52	2 12	19 15	6 1R	3 23
11/18/1938	12 34	27 4	8 28	20 36	1cp18	1 48	13 17	25 11	5 54	19 39	6 3	0 29	10 27	19 39	7 39	3 19	20 40	2 20	21 40	5 31	3 26
11/28/1938	15 49	24 54	9 29	24 13	7 6	1 17	14 52	25 11D	5 55	22 8	9 59	0 35	12 51	17 39	7 53	7 14	26 33	2 26	24 3	4 22	3 28
12/ 8/1938	19 1	22 14	10 18	27 49	12 56	0 48	16 27	25 51	5 55	24 37	13 44	0 42	15 17	15 19	7 59	11 11	2cp30	2 33	26 25	2 42	3 26
12/18/1938	22 9	19 26	10 51	1aq24	18 46	0 19	18 3	27 4	5 54	27 5	17 17	0 49	17 45	12 57	7 55	15 9	8 31	2 39	28 46	0 41	3 21
12/28/1938	25 13	16 55	11 6	4 59	24 38	0n 9	19 39	28 48	5 53	29 33	20 34	0 57	20 15	10 50	7 42	19 9	14 35	2 45	1cn 6	28 35	3 12
1/ 7/1939	28 14	15 2	11 5	8 35	0aq31	0 35	21 15	0ar56	5 52	2 1	23 33	1 5	22 46	9 16	7 23	23 10	20 40	2 50	3 25	26 40	3 0
1/17/1939	1cn11	13 58	10 53	12 11	6 25	1 2	22 51	3 25	5 52	4 28	26 11	1 15	25 19	8 22	6 58	27 11	26 47	2 55	5 42	25 9	2 46
1/27/1939	4 4	13 45	10 34	15 50	12 22	1 28	24 29	6 12	5 52	6 55	28 24	1 25	27 53	8 13D	6 32	1aq13	2aq54	2 59	7 58	24 12	2 31
2/ 6/1939	6 55	14 21	10 11	19 30	18 21	1 53	26 7	9 14	5 53	9 21	0sc 7	1 36	0cn29	8 49	6 5	5 15	8 59	3 3	10 12	23 53	2 16
2/16/1939	9 42	15 39	9 47	23 13	24 22	2 19	27 46	12 28	5 54	11 46	1 17	1 48	3 7	10 5	5 38	9 16	15 4	3 7	12 26	24 13	2 1
2/26/1939	12 26	17 34	9 23	26 59	0pi27	2 45	29 26	15 52	5 56	14 10	1 50	2 1	5 46	11 58	5 13	13 16	21 6	3 10	14 37	25 7	1 48
3/ 8/1939	15 7	19 58	9 0	0pi50	6 35	3 10	1 7	19 26	6 0	16 34	1 44R	2 14	8 27	14 22	4 50	17 15	27 5	3 13	16 48	26 34	1 35
3/18/1939	17 45	22 47	8 39	4 45	12 46	3 36	2 49	23 7	6 4	18 57	0 59	2 26	11 10	17 12	4 29	21 13	3 0	3 16	18 57	28 27	1 24
3/28/1939	20 20	25 55	8 19	8 46	19 2	4 2	4 33	26 54	6 9	21 18	29li40	2 38	13 54	20 24	4 9	25 8	8 51	3 18	21 4	0cn44	1 14
4/ 7/1939	22 53	29 19	8 1	12 54	25 23	4 28	6 18	0ta47	6 15	23 39	27 54	2 48	16 39	23 56	3 51	29 1	14 35	3 20	23 10	3 20	1 4
4/17/1939	25cn23	2cn55	7s45	17pi 8	1ar49	4n54	8ta 6	4ta45	6s22	25li59	25li55R	2n55	19cn27	27gm43	3s34	2pi52	20pi14	3n21	25cn15	6cn11	0s56
4/27/1939	27 51	6 41	7 30	21 30	8 22	5 20	9 55	8 46	6 30	28 18	23 57	3 0	22 15	1cn43	3 19	6 40	25 45	3 23	27 19	9 16	0 48
5/ 7/1939	0le17	10 34	7 17	26 2	15 2	5 45	11 47	12 51	6 40	0sc35	22 15	3 1	25 5	5 55	3 4	10 25	1ar 7	3 24	29 20	12 31	0 41
5/17/1939	2 40	14 34	7 5	0ar43	21 50	6 11	13 41	17 0	6 50	2 52	20 59	2 59	27 56	10 16	2 50	14 6	6 21	3 24	1 21	15 55	0 35
5/27/1939	5 1	18 39	6 54	5 36	28 47	6 35	15 38	21 11	7 2	5 7	20 17	2 56	0le48	14 45	2 38	17 45	11 24	3 24	3 20	19 25	0 29
6/ 6/1939	7 21	22 47	6 45	10 40	5 54	6 58	17 38	25 25	7 15	7 21	20 11D	2 51	3 41	19 20	2 26	21 20	16 16	3 25	5 18	23 1	0 23
6/16/1939	9 38	26 58	6 36	15 58	13 11	7 18	19 41	29 41	7 30	9 34	20 40	2 45	6 36	24 1	2 14	24 51	20 54	3 24	7 14	26 40	0 18
6/26/1939	11 53	1le11	6 29	21 30	20 40	7 36	21 47	4 0	7 46	11 46	21 42	2 40	9 31	28 47	2 3	28 19	25 18	3 24	9 10	0le23	0 13
7/ 6/1939	14 7	5 25	6 22	27 18	28 21	7 51	23 58	8 20	8 4	13 56	23 13	2 34	12 28	3 37	1 52	1 43	29 25	3 23	11 3	4 7	0 9
7/16/1939	16 19	9 39	6 16	3 22	6 16	8 1	26 13	12 43	8 24	16 5	25 9	2 28	15 25	8 30	1 41	5 3	3 11	3 22	12 56	7 52	0 4
7/26/1939	18 30	13 54	6 11	9 45	14 23	8 5	28 33	17 8	8 45	18 13	27 28	2 23	18 22	13 26	1 31	8 20	6 36	3 20	14 47	11 38	0n 0
8/ 5/1939	20 39	18 8	6 7	16 25	22 44	8 2	0gm58	21 35	9 10	20 19	0sc 5	2 18	21 21	18 24	1 21	11 33	9 33	3 18	16 37	15 24	0 5
8/15/1939	22 47	22 22	6 4	23 25	1cn19	7 52	3 28	26 3	9 36	22 25	2 57	2 13	24 19	23 24	1 10	14 42	12 0	3 15	18 26	19 8	0 9
8/25/1939	24 53	26 33	6 2	0gm43	10 5	7 32	6 5	0cn33	10 6	24 28	6 3	2 9	27 19	28 24	1 0	17 48	13 50	3 11	20 13	22 50	0 13

Ephemeris	80			433			1221			1489			114			3018			2041		
0 Hours ET	Sappho			Eros			Amor			Attila			Kassandra			Godiva			Lancelot		
Mo Dy Year	h long	g long	g lat	h long	g long	g lat	h long	g long	g lat	h long	g long	g lat	h long	g long	g lat	h long	g long	g lat	h long	g long	g lat
9/ 4/1939	26 58	0vi43	6 0	8 20	19 2	7 3	8 49	5 4	10 39	26 31	9 20	2 6	0vi18	3 25	0 50	20 50	15 0	3 5	21 59	26 29	0 17
9/14/1939	29 2	4 50	5 59	16 15	28 10	6 23	11 41	9 37	11 15	28 32	12 45	2 2	3 17	8 26	0 40	23 49	15 23	2 57	23 44	0vi 5	0 22
9/24/1939	1 5	8 54	5 59	24 27	7 24	5 32	14 41	14 11	11 56	0sa32	16 19	1 59	6 17	13 26	0 30	26 44	14 57	2 47	25 28	3 35	0 26
10/ 4/1939	3 7	12 54	6 0	2cn53	16 44	4 32	17 51	18 45	12 41	2 31	19 58	1 57	9 16	18 24	0 19	29 36	13 42	2 33	27 11	7 0	0 31
10/14/1939	5 8	16 48	6 1	11 31	26 7	3 23	21 11	23 19	13 31	4 28	23 42	1 55	12 16	23 20	0 8	2 25	11 43	2 16	28 53	10 18	0 36
10/24/1939	7 9	20 37	6 4	20 17	5 29	2 7	24 43	27 52	14 26	6 25	27 30	1 53	15 14	28 13	0n 3	5 11	9 15	1 55	0vi34	13 28	0 41
11/ 3/1939	9 8	24 18	6 7	29 8	14 49	0 45	28 28	2 23	15 27	8 20	1sa20	1 51	18 13	3 2	0 15	7 53	6 35	1 32	2 13	16 27	0 47
11/13/1939	11 7	27 50	6 12	8 0	24 4	0s40	2 28	6 52	16 35	10 13	5 13	1 50	21 11	7 45	0 27	10 33	4 6	1 8	3 52	19 14	0 53
11/23/1939	13 5	1li11	6 17	16 49	3li11	2 6	6 44	11 16	17 49	12 6	9 6	1 49	24 8	12 22	0 41	13 10	2 7	0 44	5 29	21 46	0 59
12/ 3/1939	15 3	4 19	6 24	25 32	12 7	3 32	11 18	15 34	19 10	13 57	12 58	1 48	27 5	16 51	0 54	15 44	0 48	0 22	7 6	24 2	1 6
12/13/1939	17 0	7 11	6 31	4vi 5	20 50	4 55	16 12	19 45	20 35	15 47	16 49	1 48	0li 0	21 11	1 9	18 16	0 16	0 3	8 42	25 58	1 14
12/23/1939	18 56	9 45	6 40	12 25	29 18	6 15	21 29	23 45	22 4	17 36	20 38	1 48	2 55	25 18	1 25	20 45	0 29	0 14	10 16	27 31	1 22
1/ 2/1940	20 53	11 56	6 49	20 30	7 29	7 30	27 11	27 33	23 31	19 23	24 24	1 48	5 49	29 11	1 43	23 12	1 24	0 29	11 50	28 38	1 31
1/12/1940	22 48	13 42	6 59	28 18	15 21	8 41	3 21	1vi 6	24 50	21 10	28 5	1 49	8 42	2 47	2 2	25 36	2 56	0 41	13 23	29 16	1 40
1/22/1940	24 44	14 57	7 10	5 48	22 52	9 48	10 0	4 25	25 46	22 56	1cp41	1 49	11 33	6 3	2 22	27 59	4 58	0 52	14 55	29 23R	1 50
2/ 1/1940	26 40	15 37	7 20	12 59	29 60	10 50	17 13	7 34	26 0	24 40	5 10	1 51	14 23	8 55	2 45	0gm19	7 26	1 1	16 26	28 58	2 0
2/11/1940	28 35	15 37R	7 28	19 51	6 43	11 49	24 59	10 42	24 54	26 23	8 31	1 52	17 12	11 18	3 10	2 37	10 15	1 9	17 57	28 0	2 10
2/21/1940	0li31	14 56	7 34	26 25	12 58	12 44	3vi21	14 11	21 27	28 5	11 43	1 54	20 0	13 9	3 37	4 53	13 22	1 16	19 27	26 35	2 19
3/ 2/1940	2 26	13 34	7 35	2sc41	18 44	13 38	12 16	18 45	13 46	29 47	14 43	1 56	22 46	14 22	4 7	7 8	16 42	1 22	20 55	24 49	2 27
3/12/1940	4 22	11 36	7 30	8 39	23 56	14 30	21 43	25 23	1n 1	1 27	17 29	1 59	25 31	14 53	4 38	9 21	20 13	1 28	22 24	22 51	2 34
3/22/1940	6 17	9 10	7 17	14 22	28 30	15 22	1li37	5li34	23 20	3 6	20 1	2 2	28 14	14 39	5 9	11 32	23 53	1 33	23 51	20 52	2 38
4/ 1/1940	8 13	6 33	6 56	19 49	2 21	16 12	11 48	20 10	44 19	4 44	22 15	2 5	0sc56	13 42	5 40	13 41	27 40	1 39	25 18	19 5	2 40
4/11/1940	10 9	4 1	6 27	25 3	5 20	17 2	22 8	7sc17	56 0	6 22	24 10	2 8	3 36	12 6	6 7	15 50	1gm33	1 43	26 44	17 37	2 40
4/21/1940	12 6	1 50	5 53	0sa 3	7 18	17 50	2sc26	22 21	60 12	7 58	25 41	2 12	6 15	10 2	6 29	17 56	5 29	1 48	28 9	16 35	2 39
5/ 1/1940	14 3	0 12	5 16	4 52	8 6	18 32	12 28	2sa32	60 32	9 33	26 48	2 16	8 52	7 46	6 44	20 2	9 29	1 53	29 34	16 2	2 36
5/11/1940	16 1	29 13	4 39	9 31	7 33	19 3	22 7	8 7	58 53	11 8	27 26	2 20	11 27	5 35	6 51	22 6	13 31	1 58	0 58	15 59D	2 33
5/21/1940	17li59	28vi55R	4s 2	13sa59	5cp35R	19s13	1sa15	10sa46	55n58	12cp42	27cp35	2n23	14sc 1	3sc47R	6n50	24gm 9	17gm34	2s 2	2li22	16vi24	2n30
5/31/1940	19 58	29 16	3 28	18 19	2 19	18 54	9 47	11 54	52 5	14 15	27 13	2 27	16 33	2 33	6 42	26 11	21 38	2 7	3 45	17 15	2 27
6/10/1940	21 57	0li14	2 56	22 32	28sa 8	17 55	17 42	12 28	47 27	15 47	26 20	2 29	19 3	1 58	6 30	28 12	25 42	2 12	5 7	18 30	2 23
6/20/1940	23 58	1 43	2 27	26 37	23 44	16 18	25 0	13 11	42 22	17 18	25 1	2 30	21 32	2 4D	6 16	0cn12	29 45	2 17	6 29	20 6	2 20
6/30/1940	25 59	3 40	2 1	0cp37	19 51	14 13	1cp43	14 18	37 12	18 49	23 21	2 29	23 59	2 48	6 1	2 12	3 47	2 23	7 51	21 58	2 17
7/10/1940	28 1	6 1	1 37	4 31	16 59	11 55	7 53	15 55	32 15	20 19	21 28	2 27	26 25	4 7	5 45	4 10	7 47	2 29	9 12	24 6	2 15
7/20/1940	0sc 5	8 43	1 15	8 20	15 22	9 41	13 33	18 3	27 42	21 48	19 33	2 22	28 49	5 56	5 30	6 8	11 44	2 35	10 32	26 27	2 13
7/30/1940	2 9	11 42	0 54	12 6	14 59	7 38	18 46	20 36	23 42	23 16	17 46	2 16	1 11	8 11	5 15	8 5	15 39	2 42	11 52	28 59	2 11
8/ 9/1940	4 15	14 57	0 36	15 48	15 41	5 49	23 35	23 32	20 14	24 44	16 17	2 9	3 32	10 48	5 2	10 1	19 30	2 49	13 12	1 40	2 9
8/19/1940	6 22	18 25	0 18	19 28	17 17	4 16	28 3	26 46	17 17	26 11	15 11	2 1	5 51	13 43	4 50	11 57	23 16	2 57	14 31	4 28	2 8
8/29/1940	8 30	22 4	0 2	23 6	19 37	2 57	2 13	0cp15	14 47	27 37	14 34	1 53	8 9	16 54	4 38	13 52	26 57	3 6	15 50	7 23	2 8
9/ 8/1940	10 40	25 54	0 14	26 42	22 34	1 49	6 5	3 55	12 40	29 3	14 26	1 45	10 25	20 17	4 28	15 47	0le30	3 15	17 8	10 22	2 7
9/18/1940	12 52	29 52	0 29	0aq17	26 0	0 50	9 43	7 44	10 53	0aq28	14 47	1 37	12 40	23 51	4 19	17 42	3 56	3 25	18 26	13 25	2 7
9/28/1940	15 6	3 59	0 44	3 53	29 50	0 0	13 7	11 41	9 23	1 53	15 35	1 29	14 53	27 34	4 10	19 36	7 12	3 37	19 44	16 31	2 8

Ephemeris	80			433			1221			1489			114			3018			2041		
0 Hours ET	Sappho			Eros			Amor			Attila			Kassandra			Godiva			Lancelot		
Mo Dy Year	h long	g long	g lat	h long	g long	g lat	h long	g long	g lat	h long	g long	g lat	h long	g long	g lat	h long	g long	g lat	h long	g long	g lat
10/ 8/1940	17 21	8 13	0 58	7 28	4 0	0 44	16 20	15 44	8 5	3 17	16 48	1 22	17 5	1sa25	4 2	21 30	10 16	3 50	21 1	19 38	2 9
10/18/1940	19 38	12 33	1 12	11 4	8 27	1 23	19 23	19 52	6 59	4 41	18 23	1 15	19 15	5 21	3 56	23 23	13 6	4 4	22 18	22 46	2 10
10/28/1940	21 57	16 59	1 26	14 42	13 9	1 57	22 16	24 3	6 2	6 4	20 16	1 10	21 24	9 22	3 50	25 17	15 38	4 19	23 35	25 53	2 12
11/ 7/1940	24 19	21 29	1 41	18 21	18 4	2 28	25 1	28 18	5 13	7 26	22 27	1 4	23 32	13 26	3 44	27 11	17 51	4 36	24 52	28 58	2 14
11/17/1940	26 43	26 5	1 55	22 4	23 11	2 55	27 38	2 35	4 29	8 48	24 51	0 59	25 38	17 34	3 39	29 4	19 39	4 55	26 8	2 1	2 16
11/27/1940	29 9	0sa44	2 10	25 49	28 28	3 21	0pi 9	6 53	3 51	10 10	27 27	0 55	27 44	21 43	3 35	0 57	20 58	5 16	27 24	4 59	2 19
12/ 7/1940	1 38	5 26	2 25	29 38	3 56	3 44	2 34	11 13	3 18	11 31	0aq13	0 51	29 48	25 53	3 32	2 51	21 45	5 38	28 40	7 51	2 23
12/17/1940	4 10	10 12	2 41	3 32	9 33	4 5	4 53	15 33	2 48	12 52	3 6	0 47	1 50	0cp 2	3 29	4 45	21 54R	6 0	29 55	10 36	2 27
12/27/1940	6 45	14 59	2 58	7 31	15 19	4 25	7 7	19 53	2 21	14 12	6 7	0 43	3 52	4 11	3 26	6 39	21 22	6 23	1 10	13 13	2 32
1/ 6/1941	9 22	19 48	3 15	11 36	21 15	4 43	9 17	24 13	1 57	15 32	9 12	0 40	5 52	8 18	3 24	8 33	20 9	6 44	2 26	15 38	2 38
1/16/1941	12 3	24 39	3 34	15 48	27 19	5 0	11 23	28 32	1 34	16 52	12 20	0 37	7 52	12 23	3 23	10 27	18 19	7 1	3 41	17 50	2 44
1/26/1941	14 47	29 29	3 54	20 8	3 31	5 16	13 25	2 50	1 13	18 11	15 31	0 35	9 50	16 23	3 22	12 22	16 0	7 13	4 55	19 47	2 51
2/ 5/1941	17 34	4 20	4 15	24 37	9 53	5 30	15 24	7 6	0 54	19 30	18 43	0 32	11 47	20 20	3 22	14 17	13 27	7 17	6 10	21 26	2 58
2/15/1941	20 25	9 10	4 38	29 15	16 24	5 43	17 20	11 20	0 36	20 48	21 54	0 29	13 44	24 10	3 22	16 13	10 55	7 14	7 25	22 45	3 6
2/25/1941	23 20	13 58	5 2	4 4	23 4	5 55	19 13	15 31	0 18	22 7	25 4	0 27	15 39	27 54	3 22	18 9	8 42	7 4	8 39	23 40	3 15
3/ 7/1941	26 19	18 43	5 29	9 4	29 54	6 4	21 3	19 40	0 1	23 25	28 12	0 25	17 34	1aq29	3 23	20 6	7 0	6 48	9 53	24 10	3 24
3/17/1941	29 22	23 25	5 58	14 18	6 54	6 12	22 52	23 44	0 15	24 42	1pi15	0 22	19 27	4 55	3 25	22 3	5 57	6 28	11 8	24 13R	3 34
3/27/1941	2 29	28 2	6 30	19 46	14 5	6 17	24 38	27 45	0 32	26 0	4 14	0 20	21 20	8 11	3 27	24 1	5 35	6 7	12 22	23 47	3 42
4/ 6/1941	5 40	2 33	7 4	25 29	21 26	6 19	26 23	1ar42	0 48	27 17	7 7	0 18	23 12	11 13	3 30	26 0	5 53	5 46	13 36	22 53	3 51
4/16/1941	8 55	6 56	7 42	1ta28	28 59	6 17	28 6	5 33	1 5	28 34	9 52	0 16	25 4	14 0	3 33	28 0	6 48	5 25	14 50	21 35	3 57
4/26/1941	12 15	11 9	8 23	7 44	6 43	6 12	29 48	9 19	1 22	29 51	12 28	0 13	26 54	16 30	3 36	0vi 0	8 15	5 6	16 4	19 57	4 2
5/ 6/1941	15 39	15 9	9 7	14 19	14 39	6 3	1 28	12 58	1 40	1 8	14 53	0 11	28 44	18 40	3 40	2 2	10 11	4 48	17 18	18 8	4 4
5/16/1941	19 8	18 54	9 56	21 13	22 47	5 49	3 8	16 31	1 58	2 24	17 6	0 8	0aq33	20 28	3 44	4 5	12 31	4 31	18 33	16 15	4 3
5/26/1941	22 42	22 19	10 50	28 26	1gm 5	5 29	4 46	19 55	2 18	3 40	19 4	0 5	2 22	21 49	3 48	6 8	15 12	4 16	19 47	14 30	4 0
6/ 5/1941	26 20	25 20	11 48	5 57	9 35	5 4	6 23	23 9	2 39	4 56	20 46	0 2	4 10	22 40	3 51	8 13	18 10	4 2	21 1	12 59	3 54
6/15/1941	0aq 3	27 50	12 49	13 47	18 15	4 33	8 0	26 13	3 2	6 12	22 9	0s 1	5 57	23 0	3 54	10 19	21 23	3 50	22 15	11 50	3 47
6/25/1941	3aq50	29aq44	13n54	21gm54	27gm 4	3n57	9ar37	29ar 5	3s27	7pi28	23pi11	0s 5	7aq44	22aq45R	3n56	12vi27	24le49	3s38	23sc29	11sc 7R	3n38
7/ 5/1941	7 42	0 54	14 59	0cn16	6 1	3 15	11 13	1 41	3 55	8 44	23 49	0 8	9 30	21 55	3 56	14 36	28 26	3 28	24 44	10 51	3 29
7/15/1941	11 38	1 15	16 1	8 51	15 5	2 28	12 48	4 1	4 25	10 0	24 1	0 13	11 16	20 32	3 54	16 46	2 13	3 18	25 58	11 2	3 19
7/25/1941	15 38	0 43	16 54	17 35	24 13	1 38	14 24	6 0	4 58	11 15	23 45	0 17	13 2	18 42	3 48	18 58	6 9	3 9	27 13	11 38	3 10
8/ 4/1941	19 41	29aq21	17 31	26 25	3le24	0 46	15 59	7 35	5 35	12 31	23 2	0 22	14 47	16 35	3 40	21 12	10 12	3 1	28 28	12 39	3 1
8/14/1941	23 48	27 20	17 43	5 17	12 36	0s 9	17 35	8 42	6 16	13 47	21 53	0 26	16 32	14 22	3 27	23 27	14 21	2 53	29 42	14 1	2 53
8/24/1941	27 59	25 2	17 27	14 7	21 48	1 3	19 10	9 16	7 1	15 2	20 22	0 31	18 16	12 18	3 12	25 44	18 36	2 46	0 57	15 42	2 46
9/ 3/1941	2 11	22 54	16 41	22 53	0vi56	1 55	20 46	9 12R	7 49	16 18	18 36	0 36	20 0	10 33	2 55	28 3	22 57	2 39	2 13	17 39	2 39
9/13/1941	6 27	21 20	15 31	1vi29	10 0	2 45	22 23	8 25	8 40	17 33	16 43	0 40	21 44	9 17	2 37	0li24	27 22	2 32	3 28	19 52	2 33
9/23/1941	10 44	20 37	14 6	9 53	18 57	3 31	24 0	6 55	9 32	18 49	14 52	0 44	23 27	8 35	2 19	2 47	1li52	2 26	4 44	22 16	2 27
10/ 3/1941	15 3	20 52	12 33	18 3	27 47	4 13	25 38	4 42	10 21	20 4	13 14	0 47	25 11	8 28D	2 2	5 12	6 25	2 19	5 59	24 52	2 22
10/13/1941	19 22	22 3	11 1	25 56	6 27	4 50	27 16	1 55	11 4	21 20	11 56	0 50	26 54	8 55	1 46	7 39	11 1	2 13	7 15	27 36	2 18
10/23/1941	23 42	24 6	9 32	3li32	14 57	5 22	28 56	28ar47	11 37	22 36	11 2	0 52	28 37	9 53	1 31	10 9	15 39	2 7	8 32	0sa28	2 14
11/ 2/1941	28 2	26 54	8 10	10 49	23 15	5 48	0ta37	25 36	11 56	23 51	10 38	0 54	0pi20	11 20	1 17	12 42	20 20	2 1	9 48	3 26	2 10

Ephemeris	80			433			1221			1489			114			3018			2041		
0 Hours ET	Sappho			Eros			Amor			Attila			Kassandra			Godiva			Lancelot		
Mo Dy Year	h long	g long	g lat	h long	g long	g lat	h long	g long	g lat	h long	g long	g lat	h long	g long	g lat	h long	g long	g lat	h long	g long	g lat
11/12/1941	2 22	0pi19	6 55	17 47	1sc21	6 10	2 19	22 44	12 2	25 7	10 42D	0 55	2 3	13 12	1 4	15 16	25 3	1 55	11 5	6 29	2 7
11/22/1941	6 40	4 15	5 47	24 26	9 15	6 27	4 2	20 25	11 56	26 23	11 13	0 56	3 46	15 25	0 53	17 54	29 46	1 49	12 22	9 36	2 5
12/ 2/1941	10 57	8 35	4 45	0sc47	16 56	6 40	5 47	18 50	11 41	27 40	12 11	0 57	5 29	17 57	0 43	20 34	4 29	1 42	13 40	12 46	2 3
12/12/1941	15 12	13 15	3 50	6 51	24 25	6 50	7 34	18 2	11 22	28 56	13 32	0 57	7 12	20 45	0 33	23 17	9 12	1 35	14 57	15 56	2 1
12/22/1941	19 25	18 10	3 0	12 38	1sa41	6 55	9 23	18 1D	10 59	0ar12	15 13	0 58	8 54	23 46	0 24	26 3	13 53	1 28	16 16	19 7	2 0
1/ 1/1942	23 35	23 18	2 15	18 10	8 44	6 58	11 14	18 43	10 36	1 29	17 12	0 58	10 37	26 57	0 16	28 52	18 32	1 20	17 34	22 17	1 59
1/11/1942	27 43	28 36	1 34	23 28	15 34	6 59	13 7	20 3	10 14	2 46	19 27	0 59	12 21	0pi18	0 9	1 45	23 7	1 11	18 53	25 25	1 59
1/21/1942	1ta47	4 0	0 57	28 32	22 12	6 57	15 3	21 56	9 54	4 3	21 54	1 0	14 4	3 47	0 1	4 40	27 37	1 1	20 12	28 29	1 59
1/31/1942	5 47	9 30	0 23	3 24	28 37	6 53	17 2	24 18	9 36	5 20	24 33	1 1	15 47	7 21	0 6	7 39	2sa 1	0 51	21 32	1cp27	1 59
2/10/1942	9 44	15 3	0s 7	8 6	4 50	6 47	19 4	27 5	9 19	6 37	27 20	1 1	17 31	11 0	0 12	10 41	6 16	0 39	22 52	4 20	2 0
2/20/1942	13 37	20 39	0 36	12 37	10 50	6 39	21 10	0ta12	9 5	7 55	0ar15	1 3	19 15	14 43	0 19	13 47	10 21	0 25	24 13	7 4	2 1
3/ 2/1942	17 26	26 15	1 1	17 0	16 37	6 30	23 19	3 38	8 52	9 13	3 15	1 4	20 59	18 28	0 26	16 56	14 12	0 9	25 34	9 38	2 2
3/12/1942	21 11	1ta52	1 25	21 15	22 10	6 19	25 33	7 19	8 41	10 31	6 20	1 5	22 43	22 14	0 32	20 9	17 47	0n 8	26 55	12 1	2 4
3/22/1942	24 52	7 28	1 47	25 22	27 30	6 6	27 51	11 15	8 32	11 50	9 27	1 7	24 28	26 1	0 39	23 25	21 1	0 29	28 18	14 9	2 6
4/ 1/1942	28 29	13 2	2 7	29 23	2 34	5 51	0gm14	15 24	8 24	13 9	12 36	1 9	26 13	29 48	0 46	26 45	23 51	0 53	29 40	16 1	2 8
4/11/1942	2 1	18 35	2 27	3 19	7 22	5 34	2 43	19 44	8 17	14 28	15 46	1 11	27 58	3 33	0 53	0sa 9	26 10	1 21	1 3	17 34	2 10
4/21/1942	5 30	24 5	2 45	7 10	11 52	5 14	5 18	24 16	8 12	15 48	18 56	1 13	29 44	7 16	1 0	3 36	27 53	1 53	2 27	18 44	2 13
5/ 1/1942	8 54	29 32	3 2	10 57	16 1	4 51	8 0	28 59	8 7	17 8	22 4	1 16	1 31	10 56	1 8	7 7	28 54	2 29	3 51	19 31	2 15
5/11/1942	12 14	4 56	3 18	14 40	19 47	4 24	10 49	3 51	8 2	18 28	25 10	1 18	3 18	14 32	1 16	10 42	29 7	3 9	5 16	19 50	2 18
5/21/1942	15 31	10 17	3 33	18 20	23 4	3 52	13 46	8 55	7 58	19 49	28 13	1 21	5 5	18 3	1 25	14 19	28 31	3 52	6 42	19 40	2 19
5/31/1942	18 43	15 34	3 48	21 59	25 49	3 13	16 53	14 9	7 54	21 10	1ta11	1 25	6 53	21 27	1 34	18 1	27 8	4 35	8 8	19 1	2 20
6/10/1942	21 51	20 47	4 3	25 35	27 52	2 27	20 10	19 35	7 49	22 32	4 3	1 29	8 42	24 45	1 45	21 45	25 9	5 16	9 35	17 55	2 20
6/20/1942	24 56	25 56	4 17	29 11	29 7	1 31	23 38	25 12	7 45	23 54	6 48	1 33	10 31	27 53	1 56	25 32	22 53	5 50	11 3	16 24	2 18
6/30/1942	27 57	1cn 0	4 31	2 46	29 23	0 24	27 19	1cn 2	7 39	25 16	9 24	1 38	12 21	0ta51	2 8	29 23	20 44	6 15	12 31	14 36	2 15
7/10/1942	0cn55	5 59	4 45	6 21	28 32	0 57	1cn13	7 6	7 32	26 39	11 50	1 43	14 11	3 36	2 21	3 15	19 5	6 30	14 0	12 41	2 10
7/20/1942	3 49	10 53	4 59	9 57	26 27	2 30	5 23	13 24	7 22	28 3	14 3	1 49	16 3	6 6	2 36	7 11	18 11	6 36	15 30	10 48	2 3
7/30/1942	6cn40	15cn42	5s14	13aq34	23aq14R	4n12	9cn51	19cn59	7s11	29ar27	16ta 2	1s55	17ar55	8ta18	2s52	11cp 8	18sa 9	6n34	17cp 1	9cp	1n55
8/ 9/1942	9 27	20 25	5 29	17 13	19 10	5 55	14 39	26 51	6 56	0 52	17 43	2 2	19 48	10 8	3 9	15 6	19 1	6 27	18 32	7 50	1 46
8/19/1942	12 12	25 1	5 44	20 55	14 50	7 28	19 48	4 1	6 37	2 17	19 4	2 10	21 41	11 34	3 29	19 7	20 43	6 17	20 5	7 0	1 36
8/29/1942	14 53	29 29	6 1	24 39	10 56	8 44	25 21	11 32	6 14	3 43	20 2	2 18	23 36	12 30	3 50	23 8	23 9	6 4	21 38	6 39	1 27
9/ 8/1942	17 32	3 50	6 18	28 27	8 0	9 38	1le21	19 24	5 44	5 9	20 34	2 27	25 32	12 54	4 12	27 9	26 12	5 50	23 12	6 49	1 17
9/18/1942	20 8	8 1	6 36	2 19	6 19	10 12	7 50	27 39	5 8	6 36	20 38R	2 36	27 28	12 41	4 36	1aq11	29 46	5 36	24 47	7 30	1 9
9/28/1942	22 41	12 2	6 56	6 16	5 56	10 32	14 50	6 17	4 25	8 4	20 11	2 45	29 26	11 51	4 59	5 13	3 46	5 22	26 23	8 38	1 0
10/ 8/1942	25 12	15 50	7 18	10 20	6 43	10 41	22 24	15 17	3 34	9 32	19 14	2 53	1 24	10 25	5 21	9 15	8 8	5 8	28 0	10 12	0 53
10/18/1942	27 40	19 24	7 41	14 29	8 33	10 43	0vi33	24 40	2 36	11 2	17 49	3 1	3 24	8 28	5 40	13 15	12 47	4 54	29 38	12 9	0 46
10/28/1942	0le 6	22 41	8 7	18 47	11 15	10 41	9 16	4li22	1 31	12 31	16 3	3 6	5 25	6 13	5 55	17 14	17 41	4 40	1 16	14 25	0 39
11/ 7/1942	2 30	25 39	8 35	23 12	14 41	10 36	18 31	14 20	0 20	14 2	14 4	3 10	7 27	3 53	6 3	21 12	22 46	4 27	2 56	16 59	0 33
11/17/1942	4 51	28 13	9 6	27 47	18 44	10 29	28 15	24 30	0 53	15 33	12 4	3 11	9 30	1 45	6 5	25 7	28 0	4 14	4 37	19 48	0 27
11/27/1942	7 11	0vi20	9 39	2 32	23 20	10 20	8 20	4sc46	2 7	17 5	10 14	3 9	11 34	0 2	6 2	29 0	3 22	4 1	6 19	22 50	0 22
12/ 7/1942	9 28	1 55	10 15	7 29	28 23	10 10	18 36	15 2	3 18	18 38	8 46	3 5	13 40	28 54	5 54	2 51	8 49	3 49	8 2	26 3	0 17

Ephemeris	80			433			1221			1489			114			3018			2041		
0 Hours ET	Sappho			Eros			Amor			Attila			Kassandra			Godiva			Lancelot		
Mo Dy Year	h long	g long	g lat	h long	g long	g lat	h long	g long	g lat	h long	g long	g lat	h long	g long	g lat	h long	g long	g lat	h long	g long	g lat
12/17/1942	11 44	2 52	10 52	12 38	3 52	9 58	28 54	25 12	4 23	20 12	7 45	3 0	15 47	28 26	5 43	6 39	14 21	3 37	9 47	29 26	0 12
12/27/1942	13 58	3 8	11 30	18 2	9 43	9 44	9 1	5 8	5 20	21 47	7 16	2 53	17 55	28 39	5 31	10 24	19 55	3 26	11 32	2 57	0 8
1/ 6/1943	16 10	2 39	12 7	23 39	15 55	9 28	18 48	14 47	6 9	23 22	7 20D	2 47	20 5	29 30	5 18	14 5	25 30	3 15	13 19	6 34	0 3
1/16/1943	18 21	1 24	12 39	29 33	22 26	9 10	28 5	24 3	6 48	24 59	7 57	2 40	22 16	0 57	5 6	17 43	1pi 7	3 4	15 6	10 17	0s 1
1/26/1943	20 30	29le29	13 2	5 44	29 16	8 48	6 49	2cp54	7 19	26 36	9 3	2 33	24 28	2 54	4 54	21 18	6 43	2 54	16 56	14 5	0 5
2/ 5/1943	22 38	27 1	13 14	12 13	6 24	8 23	14 56	11 18	7 42	28 14	10 35	2 26	26 42	5 19	4 42	24 49	12 18	2 44	18 46	17 55	0 10
2/15/1943	24 44	24 19	13 11	19 1	13 48	7 54	22 26	19 15	7 58	29 53	12 32	2 20	28 57	8 6	4 32	28 17	17 51	2 34	20 37	21 48	0 14
2/25/1943	26 50	21 41	12 53	26 7	21 29	7 21	29 20	26 43	8 9	1 34	14 48	2 15	1 14	11 12	4 23	1ar41	23 22	2 25	22 30	25 42	0 18
3/ 7/1943	28 54	19 24	12 23	3gm33	29 25	6 43	5 41	3 44	8 16	3 15	17 21	2 10	3 33	14 35	4 14	5 1	28 50	2 15	24 24	29 37	0 23
3/17/1943	0vi57	17 41	11 43	11 17	7 36	6 0	11 31	10 18	8 19	4 57	20 9	2 6	5 53	18 11	4 6	8 17	4 15	2 7	26 20	3 30	0 28
3/27/1943	2 59	16 40	10 59	19 19	16 1	5 12	16 53	16 25	8 21	6 40	23 10	2 2	8 15	22 0	4 0	11 30	9 36	1 58	28 17	7 23	0 33
4/ 6/1943	5 0	16 21	10 14	27 36	24 37	4 18	21 50	22 7	8 20	8 25	26 21	1 58	10 38	25 58	3 53	14 39	14 52	1 49	0pi15	11 12	0 38
4/16/1943	7 0	16 43	9 29	6 7	3gm25	3 21	26 25	27 22	8 18	10 10	29 41	1 55	13 3	0gm 4	3 48	17 45	20 4	1 41	2 14	14 59	0 43
4/26/1943	9 0	17 41	8 47	14 49	12 22	2 19	0aq41	2pi11	8 16	11 57	3 9	1 53	15 30	4 18	3 43	20 47	25 10	1 32	4 15	18 40	0 49
5/ 6/1943	10 59	19 10	8 8	23 37	21 26	1 15	4 39	6 33	8 13	13 45	6 42	1 50	17 58	8 37	3 38	23 45	0ta11	1 24	6 17	22 16	0 56
5/16/1943	12 57	21 7	7 33	2le29	0cn35	0 10	8 22	10 25	8 10	15 33	10 21	1 49	20 28	13 2	3 34	26 40	5 6	1 16	8 21	25 44	1 3
5/26/1943	14 54	23 26	7 0	11 21	9 48	0 54	11 51	13 47	8 6	17 24	14 3	1 47	22 59	17 31	3 30	29 32	9 54	1 7	10 26	29 4	1 11
6/ 5/1943	16 51	26 3	6 31	20 9	19 2	1 56	15 8	16 34	8 2	19 15	17 48	1 45	25 33	22 4	3 27	2 21	14 36	0 59	12 33	2 13	1 19
6/15/1943	18 48	28 57	6 5	28 48	28 16	2 55	18 14	18 43	7 57	21 7	21 36	1 44	28 8	26 39	3 24	5 6	19 10	0 50	14 41	5 10	1 28
6/25/1943	20 44	2 4	5 41	7 17	7 26	3 48	21 11	20 9	7 50	23 1	25 26	1 43	0cn45	1cn17	3 21	7 49	23 35	0 41	16 50	7 51	1 38
7/ 5/1943	22 40	5 22	5 19	15 32	16 31	4 35	23 59	20 47	7 41	24 56	29 16	1 43	3 23	5 57	3 18	10 28	27 52	0 32	19 1	10 14	1 49
7/15/1943	24 35	8 49	4 59	23 30	25 30	5 15	26 39	20 32	7 28	26 52	3 6	1 42	6 3	10 38	3 15	13 5	1gm58	0 23	21 13	12 15	2 2
7/25/1943	26 31	12 24	4 40	1li12	4vi21	5 48	29 12	19 22	7 9	28 50	6 55	1 42	8 45	15 19	3 13	15 39	5 53	0 13	23 26	13 52	2 15
8/ 4/1943	28 26	16 5	4 24	8 35	13 2	6 13	1 39	17 17	6 44	0cn48	10 43	1 42	11 28	20 2	3 10	18 11	9 35	0 2	25 41	14 59	2 29
8/14/1943	0li21	19 53	4 8	15 39	21 32	6 32	4 0	14 26	6 10	2 48	14 29	1 42	14 12	24 43	3 8	20 40	13 2	0 9	27 57	15 34	2 45
8/24/1943	2 17	23 44	3 54	22 24	29 51	6 43	6 16	11 4	5 27	4 50	18 11	1 43	16 59	29 24	3 6	23 6	16 12	0 21	0ar15	15 32R	3 0
9/ 3/1943	4li12	27vi40	3s40	28li51	7li59	6s49	8pi28	7pi35R	4n38	6cn52	21cn48	1s43	19cn46	4le 3	3s 3	25ta31	19gm 2	0s34	2ar34	14ar54R	3s16
9/13/1943	6 8	1li39	3 27	5 0	15 55	6 49	10 35	4 21	3 46	8 56	25 19	1 44	22 35	8 39	3 1	27 53	21 28	0 49	4 54	13 41	3 31
9/23/1943	8 3	5 40	3 15	10 52	23 40	6 45	12 38	1 43	2 55	11 1	28 43	1 45	25 26	13 12	2 59	0gm13	23 28	1 5	7 15	12 1	3 43
10/ 3/1943	10 0	9 42	3 4	16 29	1sc13	6 38	14 39	29aq50	2 7	13 8	1 57	1 46	28 18	17 40	2 56	2 31	24 55	1 22	9 38	10 3	3 52
10/13/1943	11 56	13 46	2 52	21 51	8 35	6 27	16 36	28 47	1 24	15 16	4 58	1 47	1le11	22 1	2 53	4 47	25 46	1 41	12 1	8 2	3 58
10/23/1943	13 53	17 50	2 42	27 0	15 48	6 14	18 30	28 33	0 46	17 25	7 45	1 49	4 5	26 14	2 50	7 2	25 56R	2 1	14 26	6 15	3 59
11/ 2/1943	15 50	21 54	2 31	1sa56	22 51	5 58	20 21	29 1	0 13	19 35	10 14	1 50	7 0	0vi18	2 47	9 15	25 22	2 23	16 52	4 53	3 56
11/12/1943	17 48	25 56	2 20	6 41	29 44	5 42	22 11	0pi 7	0 15	21 47	12 21	1 52	9 56	4 9	2 43	11 26	24 5	2 45	19 18	4 7	3 50
11/22/1943	19 47	29 57	2 10	11 15	6 30	5 23	23 58	1 44	0 39	24 0	14 2	1 53	12 52	7 44	2 39	13 35	22 9	3 6	21 46	4 0D	3 43
12/ 2/1943	21 46	3 54	1 59	15 40	13 8	5 5	25 43	3 47	1 0	26 14	15 13	1 55	15 50	11 1	2 34	15 43	19 44	3 25	24 14	4 33	3 34
12/12/1943	23 46	7 47	1 48	19 57	19 38	4 45	27 27	6 12	1 18	28 29	15 50	1 55	18 48	13 54	2 27	17 50	17 6	3 40	26 44	5 43	3 25
12/22/1943	25 47	11 35	1 36	24 7	26 2	4 24	29 9	8 55	1 34	0le46	15 49R	1 55	21 47	16 19	2 20	19 56	14 34	3 50	29 13	7 28	3 16
1/ 1/1944	27 49	15 16	1 24	28 10	2cp19	4 3	0ar50	11 53	1 49	3 4	15 9	1 54	24 46	18 10	2 10	22 0	12 25	3 56	1 44	9 42	3 8
1/11/1944	29 52	18 47	1 11	2 7	8 31	3 42	2 30	15 3	2 3	5 23	13 53	1 51	27 46	19 23	1 58	24 3	10 50	3 58	4 15	12 22	3 0

Ephemeris	80			433			1221			1489			114			3018			2041		
0 Hours ET	Sappho			Eros			Amor			Attila			Kassandra			Godiva			Lancelot		
Mo Dy Year	h long	g long	g lat	h long	g long	g lat	h long	g long	g lat	h long	g long	g lat	h long	g long	g lat	h long	g long	g lat	h long	g long	g lat
1/21/1944	1 57	22 9	0 57	6 0	14 37	3 20	4 9	18 22	2 15	7 42	12 8	1 46	0vi46	19 52	1 44	26 5	9 57	3 57	6 46	15 23	2 52
1/31/1944	4 2	25 17	0 41	9 48	20 38	2 57	5 47	21 49	2 27	10 3	10 5	1 39	3 46	19 34	1 26	28 6	9 46	3 54	9 18	18 43	2 46
2/10/1944	6 9	28 9	0 24	13 32	26 33	2 34	7 24	25 22	2 39	12 25	8 1	1 30	6 45	18 29	1 5	0cn 6	10 15	3 50	11 50	22 18	2 39
2/20/1944	8 17	0sa42	0 4	17 13	2aq24	2 9	9 0	29 0	2 51	14 48	6 11	1 19	9 45	16 45	0 41	2 5	11 20	3 46	14 22	26 5	2 34
3/ 1/1944	10 27	2 51	0 18	20 52	8 11	1 44	10 36	2 42	3 2	17 12	4 49	1 8	12 44	14 34	0 15	4 4	12 57	3 41	16 55	0ta 3	2 28
3/11/1944	12 38	4 34	0 44	24 29	13 53	1 17	12 12	6 26	3 14	19 37	4 3	0 57	15 44	12 17	0n12	6 2	14 59	3 36	19 27	4 9	2 24
3/21/1944	14 51	5 44	1 13	28 5	19 30	0 49	13 48	10 13	3 26	22 2	3 56D	0 46	18 42	10 13	0 38	7 59	17 24	3 32	21 59	8 22	2 19
3/31/1944	17 6	6 16	1 46	1aq40	25 4	0 20	15 23	14 0	3 38	24 28	4 28	0 36	21 40	8 39	1 2	9 55	20 7	3 28	24 31	12 40	2 15
4/10/1944	19 23	6 8R	2 22	5 16	0pi33	0n12	16 59	17 47	3 51	26 55	5 37	0 27	24 38	7 46	1 22	11 51	23 5	3 25	27 3	17 1	2 12
4/20/1944	21 42	5 16	3 3	8 51	5 59	0 46	18 34	21 34	4 5	29 22	7 18	0 18	27 34	7 38D	1 40	13 46	26 15	3 22	29 34	21 26	2 8
4/30/1944	24 4	3 40	3 45	12 28	11 20	1 23	20 10	25 21	4 20	1 50	9 28	0 11	0li30	8 15	1 54	15 41	29 36	3 20	2 5	25 53	2 5
5/10/1944	26 27	1 29	4 28	16 6	16 37	2 3	21 47	29 5	4 36	4 18	12 2	0 4	3 25	9 33	2 6	17 36	3 4	3 18	4 36	0gm20	2 2
5/20/1944	28 53	28sc55	5 9	19 47	21 50	2 47	23 24	2 47	4 53	6 47	14 57	0n 2	6 18	11 26	2 16	19 30	6 40	3 17	7 5	4 48	2 0
5/30/1944	1 22	26 15	5 44	23 30	26 58	3 36	25 1	6 26	5 11	9 15	18 9	0 8	9 11	13 49	2 24	21 24	10 22	3 16	9 34	9 16	1 57
6/ 9/1944	3 53	23 50	6 12	27 17	2ar 0	4 30	26 40	10 1	5 31	11 44	21 35	0 13	12 2	16 37	2 30	23 18	14 7	3 16	12 3	13 42	1 55
6/19/1944	6 28	21 56	6 32	1pi 7	6 58	5 30	28 19	13 32	5 53	14 13	25 14	0 18	14 52	19 46	2 36	25 12	17 56	3 16	14 30	18 7	1 53
6/29/1944	9 5	20 43	6 45	5 3	11 48	6 38	29 59	16 57	6 18	16 43	29 3	0 23	17 41	23 13	2 40	27 5	21 48	3 16	16 57	22 29	1 51
7/ 9/1944	11 46	20 18	6 53	9 4	16 29	7 55	1 41	20 15	6 45	19 12	3 0	0 27	20 28	26 54	2 44	28 59	25 42	3 17	19 22	26 48	1 49
7/19/1944	14 29	20 40	6 56	13 12	21 1	9 22	3 23	23 25	7 15	21 41	7 5	0 31	23 14	0li47	2 48	0le52	29 37	3 19	21 47	1cn 3	1 47
7/29/1944	17 17	21 46	6 56	17 27	25 20	11 1	5 8	26 24	7 49	24 9	11 16	0 35	25 59	4 50	2 51	2 46	3 33	3 21	24 10	5 13	1 46
8/ 8/1944	20 7	23 32	6 55	21 50	29 22	12 55	6 54	29 10	8 27	26 38	15 31	0 39	28 41	9 1	2 54	4 40	7 29	3 23	26 33	9 17	1 44
8/18/1944	23 2	25 53	6 52	26 21	3 3	15 6	8 42	1 41	9 9	29 6	19 50	0 43	1 23	13 18	2 56	6 34	11 25	3 26	28 54	13 15	1 43
8/28/1944	26 0	28 44	6 49	1ar 3	6 15	17 36	10 32	3 53	9 57	1 34	24 13	0 47	4 3	17 41	2 59	8 28	15 19	3 30	1 14	17 4	1 41
9/ 7/1944	29 3	2 2	6 45	5 56	8 49	20 25	12 25	5 42	10 51	4 1	28 37	0 50	6 41	22 8	3 1	10 22	19 11	3 34	3 32	20 44	1 40
9/17/1944	2 10	5 44	6 41	11 2	10 33	23 35	14 20	7 1	11 52	6 27	3 3	0 54	9 17	26 38	3 4	12 17	23 0	3 39	5 50	24 13	1 39
9/27/1944	5 21	9 45	6 37	16 20	11 11	26 59	16 17	7 45	12 59	8 53	7 29	0 57	11 52	1sc11	3 6	14 12	26 45	3 44	8 6	27 28	1 38
10/ 7/1944	8cp36	14sa 5	6n33	21ar53	10ta29R	30n25	18ta18	7gm48R	14s11	11li18	11li56	1n 1	14sc25	5sc46	3n 8	16le 8	0vi25	3s50	10cn20	0le28	1s37
10/17/1944	11 56	18 41	6 29	27 42	8 17	33 32	20 22	7 2	15 27	13 43	16 22	1 5	16 57	10 21	3 11	18 4	3 59	3 58	12 34	3 9	1 35
10/27/1944	15 20	23 31	6 25	3 48	4 43	35 49	22 30	5 22	16 43	16 6	20 46	1 9	19 27	14 58	3 14	20 1	7 24	4 6	14 46	5 28	1 34
11/ 6/1944	18 49	28 33	6 21	10 11	0 26	36 47	24 42	2 51	17 51	18 29	25 7	1 13	21 55	19 33	3 17	21 59	10 40	4 15	16 56	7 21	1 33
11/16/1944	22 23	3 47	6 17	16 52	26 27	36 13	26 59	29ta36	18 44	20 51	29 25	1 17	24 22	24 8	3 20	23 57	13 44	4 25	19 5	8 46	1 31
11/26/1944	26 1	9 12	6 13	23 53	23 48	34 13	29 20	25 58	19 14	23 12	3 38	1 22	26 47	28 41	3 24	25 56	16 32	4 37	21 13	9 37	1 28
12/ 6/1944	29 44	14 45	6 8	1gm12	23 1	31 12	1 46	22 21	19 18	25 31	7 46	1 27	29 10	3 11	3 28	27 56	19 3	4 50	23 20	9 52	1 25
12/16/1944	3 31	20 27	6 3	8 51	24 12	27 30	4 19	19 15	18 55	27 50	11 46	1 32	1 32	7 38	3 32	29 56	21 11	5 4	25 24	9 29	1 21
12/26/1944	7 23	26 17	5 58	16 47	27 9	23 25	6 58	16 58	18 13	0sc 7	15 38	1 38	3 52	12 0	3 37	1 58	22 53	5 20	27 28	8 29	1 16
1/ 5/1945	11 19	2aq13	5 52	25 0	1ta34	19 8	9 44	15 42	17 18	2 23	19 19	1 44	6 11	16 18	3 42	4 1	24 4	5 36	29 30	6 57	1 9
1/15/1945	15 20	8 14	5 46	3cn26	7 13	14 45	12 38	15 28D	16 18	4 39	22 49	1 51	8 28	20 29	3 48	6 4	24 39	5 54	1 31	5 1	1 1
1/25/1945	19 24	14 21	5 38	12 5	13 52	10 22	15 41	16 14	15 16	6 52	26 3	1 58	10 44	24 32	3 55	8 9	24 34R	6 10	3 30	2 55	0 52
2/ 4/1945	23 31	20 31	5 30	20 51	21 18	6 2	18 54	17 52	14 17	9 5	29 1	2 6	12 58	28 27	4 2	10 16	23 47	6 26	5 28	0 53	0 43
2/14/1945	27 42	26 45	5 22	29 43	29 23	1 53	22 17	20 17	13 20	11 17	1 39	2 14	15 11	2 11	4 11	12 23	22 17	6 37	7 25	29 9	0 33

Ephemeris	80			433			1221			1489			114			3018			2041		
0 Hours ET	Sappho			Eros			Amor			Attila			Kassandra			Godiva			Lancelot		
Mo Dy Year	h long	g long	g lat	h long	g long	g lat	h long	g long	g lat	h long	g long	g lat	h long	g long	g lat	h long	g long	g lat	h long	g long	g lat
2/24/1945	1pi56	3pi 1	5 12	8 35	7 58	2 0	25 53	23 22	12 27	13 27	3 55	2 24	17 22	5 44	4 20	14 32	20 12	6 44	9 20	27 52	0 23
3/ 6/1945	6 12	9 18	5 1	17 24	16 56	5 29	29 42	27 2	11 36	15 36	5 43	2 34	19 32	9 2	4 30	16 42	17 43	6 43	11 14	27 8	0 15
3/16/1945	10 30	15 36	4 49	26 6	26 8	8 29	3 46	1gm14	10 48	17 43	7 3	2 44	21 41	12 4	4 40	18 54	15 6	6 35	13 7	27 0D	0 6
3/26/1945	14 49	21 55	4 36	4 39	5 30	10 57	8 7	5 54	10 2	19 49	7 49	2 55	23 48	14 47	4 52	21 8	12 40	6 19	14 58	27 25	0n 1
4/ 5/1945	19 9	28 12	4 22	12 58	14 53	12 51	12 47	11 1	9 17	21 54	8 0	3 6	25 54	17 9	5 5	23 23	10 42	5 57	16 49	28 22	0 7
4/15/1945	23 30	4 28	4 7	21 2	24 13	14 12	17 47	16 33	8 31	23 58	7 34	3 17	27 59	19 6	5 19	25 40	9 22	5 32	18 38	29 46	0 13
4/25/1945	27 51	10 41	3 51	28 49	3le24	15 2	23 11	22 29	7 44	26 0	6 34	3 26	0cp 3	20 34	5 34	27 59	8 46	5 5	20 25	1 34	0 18
5/ 5/1945	2 11	16 51	3 33	6 18	12 22	15 26	29 0	28 49	6 56	28 1	5 5	3 33	2 5	21 31	5 49	0li20	8 54D	4 38	22 12	3 42	0 23
5/15/1945	6 31	22 57	3 14	13 28	21 6	15 28	5 18	5 32	6 5	0sa 1	3 14	3 38	4 6	21 52	6 4	2 42	9 42	4 12	23 57	6 7	0 27
5/25/1945	10 48	28 59	2 54	20 19	29 32	15 11	12 7	12 40	5 10	2 0	1 15	3 39	6 6	21 36	6 18	5 7	11 9	3 48	25 41	8 46	0 31
6/ 4/1945	15 4	4 54	2 33	26 52	7 42	14 41	19 28	20 13	4 12	3 57	29sc20	3 37	8 5	20 43	6 29	7 35	13 7	3 25	27 24	11 36	0 35
6/14/1945	19 18	10 44	2 10	3 7	15 35	14 1	27 24	28 9	3 8	5 53	27 41	3 32	10 3	19 16	6 37	10 4	15 34	3 4	29 6	14 36	0 38
6/24/1945	23 29	16 26	1 45	9 4	23 12	13 15	5 54	6 31	2 1	7 48	26 28	3 24	12 1	17 22	6 40	12 36	18 26	2 45	0vi47	17 43	0 41
7/ 4/1945	27 37	22 0	1 20	14 46	0li34	12 25	14 59	15 16	0 49	9 41	25 45	3 15	13 57	15 12	6 36	15 11	21 38	2 28	2 27	20 57	0 44
7/14/1945	1ta42	27 25	0 52	20 12	7 42	11 32	24 33	24 24	0n26	11 33	25 35	3 6	15 52	12 59	6 27	17 48	25 9	2 11	4 6	24 16	0 47
7/24/1945	5 43	2 38	0 23	25 25	14 39	10 39	4li31	3vi53	1 42	13 24	25 58	2 56	17 46	10 57	6 11	20 28	28 55	1 56	5 44	27 38	0 50
8/ 3/1945	9 40	7 40	0s 8	0sa25	21 25	9 47	14 45	13 39	2 57	15 14	26 50	2 47	19 39	9 18	5 52	23 11	2 56	1 42	7 20	1vi 3	0 53
8/13/1945	13 34	12 28	0 41	5 13	28 1	8 56	25 5	23 40	4 8	17 3	28 10	2 38	21 32	8 10	5 30	25 56	7 9	1 28	8 56	4 30	0 57
8/23/1945	17 23	16 59	1 17	9 51	4 29	8 7	5 18	3li49	5 11	18 50	29 54	2 29	23 24	7 38	5 6	28 45	11 33	1 15	10 31	7 57	1 0
9/ 2/1945	21 9	21 11	1 56	14 19	10 51	7 20	15 14	14 0	6 6	20 37	1 58	2 21	25 15	7 41D	4 43	1 37	16 6	1 3	12 5	11 25	1 3
9/12/1945	24 50	25 2	2 38	18 39	17 6	6 36	24 44	24 8	6 49	22 22	4 20	2 14	27 5	8 18	4 21	4 32	20 49	0 51	13 39	14 51	1 6
9/22/1945	28 27	28 25	3 24	22 50	23 17	5 53	3sa41	4sc 5	7 21	24 6	6 57	2 8	28 55	9 26	4 0	7 30	25 40	0 40	15 11	18 16	1 10
10/ 2/1945	2 0	1cn18	4 14	26 55	29 24	5 13	12 3	13 47	7 40	25 49	9 47	2 2	0aq44	11 1	3 40	10 32	0sc39	0 28	16 42	21 38	1 14
10/12/1945	5 29	3 35	5 10	0cp54	5 27	4 35	19 47	23 10	7 49	27 31	12 48	1 57	2 32	13 0	3 23	13 37	5 45	0 17	18 13	24 56	1 18
10/22/1945	8 53	5 8	6 9	4 48	11 27	3 58	26 54	2sa10	7 49	29 12	15 57	1 52	4 20	15 20	3 7	16 45	10 57	0 6	19 43	28 9	1 22
11/ 1/1945	12 14	5 52	7 14	8 37	17 26	3 24	3 27	10 46	7 42	0 52	19 14	1 48	6 7	17 58	2 52	19 57	16 15	0n 5	21 12	1li15	1 27
11/11/1945	15gm30	5cn42R	8s20	12cp23	23sa22	2s51	9cp28	18sa58	7n29	2cp31	22sa36	1n44	7aq54	20cp50	2n39	23sc13	21sc38	0n16	22vi40	4li14	1n32
11/21/1945	18 43	4 37	9 26	16 5	29 18	2 20	15 1	26 47	7 13	4 9	26 3	1 41	9 40	23 55	2 27	26 32	27 6	0 27	24 8	7 3	1 38
12/ 1/1945	21 52	2 41	10 25	19 45	5 13	1 50	20 7	4 14	6 54	5 47	29 33	1 38	11 26	27 10	2 16	29 55	2sa39	0 38	25 35	9 41	1 44
12/11/1945	24 56	0 7	11 14	23 22	11 7	1 21	24 50	11 19	6 34	7 23	3 5	1 36	13 11	0aq34	2 6	3 21	8 15	0 49	27 1	12 6	1 51
12/21/1945	27 58	27 18	11 46	26 58	17 1	0 53	29 13	18 6	6 12	8 58	6 38	1 34	14 56	4 5	1 57	6 51	13 55	1 1	28 27	14 14	1 59
12/31/1945	0cn55	24 38	12 0	0aq34	22 55	0 25	3 17	24 35	5 51	10 33	10 11	1 32	16 41	7 41	1 48	10 24	19 37	1 13	29 52	16 4	2 7
1/10/1946	3 49	22 28	11 58	4 9	28 50	0n 2	7 5	0aq48	5 30	12 6	13 42	1 30	18 25	11 22	1 40	14 1	25 21	1 26	1 16	17 32	2 16
1/20/1946	6 40	21 3	11 42	7 44	4 45	0 28	10 39	6 45	5 10	13 39	17 12	1 29	20 9	15 6	1 33	17 41	1cp 7	1 38	2 40	18 35	2 26
1/30/1946	9 28	20 29	11 19	11 20	10 41	0 55	14 1	12 29	4 50	15 11	20 37	1 28	21 53	18 52	1 26	21 24	6 53	1 52	4 3	19 12	2 36
2/ 9/1946	12 12	20 43	10 51	14 58	16 39	1 21	17 11	18 0	4 31	16 42	23 58	1 27	23 36	22 38	1 19	25 11	12 39	2 5	5 26	19 19R	2 46
2/19/1946	14 54	21 43	10 22	18 38	22 37	1 47	20 11	23 19	4 12	18 13	27 13	1 27	25 20	26 25	1 13	29 0	18 24	2 20	6 48	18 56	2 57
3/ 1/1946	17 32	23 20	9 53	22 20	28 38	2 14	23 2	28 25	3 54	19 42	0aq21	1 26	27 3	0pi11	1 7	2 51	24 7	2 34	8 10	18 3	3 7
3/11/1946	20 8	25 29	9 26	26 5	4 42	2 41	25 44	3 21	3 37	21 11	3 21	1 26	28 46	3 54	1 1	6 45	29 47	2 50	9 31	16 44	3 16
3/21/1946	22 41	28 4	9 1	29 55	10 47	3 8	28 20	8 4	3 20	22 40	6 10	1 26	0pi29	7 35	0 55	10 41	5 23	3 6	10 52	15 3	3 23

Ephemeris	80			433			1221			1489			114			3018			2041		
0 Hours ET	Sappho			Eros			Amor			Attila			Kassandra			Godiva			Lancelot		
Mo Dy Year	h long	g long	g lat	h long	g long	g lat	h long	g long	g lat	h long	g long	g lat	h long	g long	g lat	h long	g long	g lat	h long	g long	g lat
3/31/1946	25 12	1cn 1	8 37	3 49	16 57	3 36	0pi49	12 37	3 3	24 7	8 47	1 27	2 12	11 12	0 49	14 39	10 53	3 23	12 12	13 10	3 28
4/10/1946	27 40	4 14	8 16	7 48	23 10	4 4	3 12	16 58	2 47	25 34	11 11	1 27	3 55	14 44	0 43	18 38	16 17	3 40	13 32	11 14	3 30
4/20/1946	0le 6	7 41	7 57	11 54	29 27	4 33	5 30	21 7	2 31	27 1	13 19	1 28	5 38	18 10	0 37	22 38	21 32	3 58	14 51	9 26	3 30
4/30/1946	2 29	11 19	7 40	16 6	5 50	5 2	7 43	25 3	2 14	28 26	15 9	1 29	7 21	21 29	0 31	26 39	26 37	4 17	16 10	7 55	3 28
5/10/1946	4 51	15 5	7 25	20 26	12 19	5 32	9 52	28 46	1 57	29 52	16 38	1 30	9 4	24 39	0 24	0aq40	1pi29	4 37	17 28	6 47	3 24
5/20/1946	7 10	18 58	7 11	24 55	18 55	6 2	11 57	2 15	1 40	1 16	17 44	1 31	10 47	27 38	0 17	4 41	6 7	4 58	18 47	6 6	3 19
5/30/1946	9 28	22 57	6 58	29 34	25 39	6 31	13 58	5 27	1 22	2 40	18 25	1 31	12 30	0ar26	0 9	8 42	10 25	5 20	20 4	5 53	3 13
6/ 9/1946	11 43	27 0	6 47	4 23	2ta32	7 1	15 56	8 21	1 3	4 4	18 38	1 32	14 13	3 0	0 1	12 42	14 22	5 44	21 22	6 8	3 7
6/19/1946	13 57	1le 6	6 37	9 24	9 35	7 29	17 51	10 55	0 42	5 27	18 23	1 32	15 56	5 17	0 8	16 40	17 52	6 8	22 39	6 49	3 0
6/29/1946	16 9	5 15	6 28	14 39	16 49	7 56	19 44	13 6	0 20	6 49	17 39	1 32	17 40	7 15	0 18	20 37	20 51	6 33	23 56	7 53	2 55
7/ 9/1946	18 20	9 25	6 20	20 7	24 15	8 20	21 34	14 50	0s 5	8 11	16 28	1 30	19 24	8 51	0 29	24 32	23 12	6 58	25 12	9 18	2 49
7/19/1946	20 29	13 36	6 13	25 51	1gm53	8 42	23 22	16 3	0 32	9 33	14 55	1 28	21 8	10 0	0 40	28 25	24 49	7 23	26 29	11 2	2 44
7/29/1946	22 37	17 47	6 7	1ta51	9 45	8 58	25 8	16 41	1 1	10 54	13 7	1 25	22 53	10 41	0 54	2 16	25 36	7 47	27 45	13 1	2 39
8/ 8/1946	24 43	21 59	6 2	8 8	17 52	9 9	26 52	16 39R	1 34	12 15	11 13	1 21	24 38	10 48R	1 8	6 3	25 29R	8 6	29 1	15 14	2 35
8/18/1946	26 49	26 9	5 58	14 44	26 12	9 12	28 35	15 54	2 10	13 35	9 24	1 15	26 23	10 22	1 23	9 48	24 30	8 19	0sc16	17 39	2 32
8/28/1946	28 53	0vi19	5 54	21 38	4 47	9 7	0ar17	14 24	2 47	14 55	7 47	1 10	28 9	9 20	1 39	13 30	22 43	8 21	1 32	20 13	2 28
9/ 7/1946	0vi56	4 27	5 52	28 52	13 35	8 51	1 57	12 12	3 26	16 15	6 32	1 3	29 55	7 46	1 56	17 8	20 27	8 12	2 47	22 56	2 26
9/17/1946	2 58	8 32	5 50	6 24	22 36	8 24	3 36	9 27	4 4	17 34	5 42	0 57	1 41	5 47	2 11	20 43	18 2	7 50	4 2	25 45	2 24
9/27/1946	4 59	12 35	5 48	14 15	1le46	7 43	5 14	6 22	4 38	18 53	5 21	0 51	3 28	3 34	2 25	24 15	15 53	7 18	5 17	28 40	2 22
10/ 7/1946	6 59	16 33	5 48	22 22	11 5	6 50	6 51	3 15	5 6	20 11	5 28	0 45	5 16	1 19	2 37	27 43	14 19	6 39	6 31	1 38	2 21
10/17/1946	8 58	20 26	5 49	0cn45	20 29	5 44	8 28	0 25	5 28	21 30	6 3	0 40	7 4	29pi18	2 46	1ar 7	13 32	5 56	7 46	4 40	2 20
10/27/1946	10 57	24 14	5 50	9 20	29 55	4 26	10 4	28 7	5 43	22 48	7 2	0 35	8 53	27 41	2 53	4 27	13 35D	5 14	9 0	7 44	2 20
11/ 6/1946	12 55	27 54	5 52	18 4	9 21	2 58	11 40	26 30	5 53	24 5	8 25	0 30	10 42	26 38	2 57	7 44	14 26	4 35	10 14	10 49	2 20
11/16/1946	14 53	1li26	5 55	26 54	18 43	1 21	13 16	25 37	5 58	25 23	10 7	0 26	12 32	26 11	3 0	10 57	16 0	3 59	11 29	13 53	2 21
11/26/1946	16 50	4 47	5 59	5 46	27 58	0s21	14 51	25 28D	5 59	26 40	12 7	0 22	14 23	26 21	3 0	14 7	18 10	3 26	12 43	16 56	2 22
12/ 6/1946	18 46	7 55	6 4	14 37	7 2	2 7	16 27	25 59	5 59	27 57	14 22	0 18	16 15	27 7	3 0	17 13	20 51	2 57	13 57	19 56	2 24
12/16/1946	20vi42	10li47	6s10	23le22	15li54	3s53	18ar 2	27pi 6	5s58	29aq14	16aq50	0n15	18ar 7	28pi26	3s 0	20ar15	23pi57	2n31	15sc11	22sc52	2n26
12/26/1946	22 38	13 22	6 17	1vi57	24 30	5 38	19 38	28 43	5 57	0pi30	19 28	0 12	20 0	0ar13	2 59	23 15	27 23	2 9	16 25	25 42	2 29
1/ 5/1947	24 34	15 35	6 24	10 21	2sc47	7 20	21 14	0ar45	5 56	1 47	22 14	0 9	21 54	2 25	2 58	26 10	1ar 6	1 49	17 39	28 25	2 32
1/15/1947	26 29	17 21	6 32	18 30	10 42	8 58	22 51	3 10	5 55	3 3	25 8	0 6	23 49	4 58	2 57	29 3	5 1	1 31	18 53	0sa59	2 36
1/25/1947	28 25	18 38	6 40	26 23	18 14	10 33	24 28	5 53	5 54	4 19	28 7	0 3	25 45	7 49	2 57	1 52	9 7	1 14	20 6	3 22	2 41
2/ 4/1947	0li20	19 20	6 47	3 57	25 19	12 3	26 6	8 52	5 55	5 35	1pi10	0 1	27 42	10 56	2 56	4 38	13 20	1 0	21 20	5 33	2 46
2/14/1947	2 15	19 23R	6 53	11 13	1sa53	13 30	27 45	12 4	5 56	6 51	4 15	0 2	29 39	14 16	2 57	7 21	17 40	0 47	22 34	7 28	2 52
2/24/1947	4 11	18 45	6 56	18 10	7 54	14 55	29 25	15 26	5 58	8 7	7 22	0 5	1 38	17 46	2 57	10 1	22 5	0 35	23 48	9 5	2 58
3/ 6/1947	6 6	17 26	6 55	24 49	13 15	16 18	1 6	18 58	6 1	9 23	10 29	0 7	3 38	21 26	2 58	12 39	26 32	0 23	25 2	10 22	3 5
3/16/1947	8 2	15 29	6 47	1sc 9	17 53	17 40	2 48	22 38	6 5	10 38	13 35	0 10	5 39	25 14	2 59	15 13	1ta 2	0 13	26 17	11 17	3 13
3/26/1947	9 58	13 6	6 33	7 12	21 40	19 1	4 32	26 24	6 10	11 54	16 38	0 12	7 41	29 7	3 0	17 45	5 32	0 3	27 31	11 46	3 20
4/ 5/1947	11 55	10 29	6 11	12 59	24 25	20 21	6 18	0ta16	6 16	13 10	19 39	0 15	9 45	3 6	3 2	20 15	10 4	0 6	28 45	11 48R	3 28
4/15/1947	13 51	7 56	5 42	18 30	25 58	21 37	8 5	4 13	6 22	14 25	22 35	0 18	11 49	7 9	3 4	22 42	14 34	0 15	29 60	11 22	3 35
4/25/1947	15 49	5 43	5 8	23 47	26 9R	22 45	9 54	8 14	6 30	15 41	25 25	0 21	13 55	11 15	3 7	25 7	19 4	0 23	1 14	10 29	3 41

Ephemeris	80			433			1221			1489			114			3018			2041		
0 Hours ET	Sappho			Eros			Amor			Attila			Kassandra			Godiva			Lancelot		
Mo Dy Year	h long	g long	g lat	h long	g long	g lat	h long	g long	g lat	h long	g long	g lat	h long	g long	g lat	h long	g long	g lat	h long	g long	g lat
5/ 5/1947	17 47	4 2	4 32	28 51	24 48	23 33	11 46	12 19	6 39	16 56	28 8	0 24	16 3	15 24	3 10	27 30	23 33	0 32	2 29	9 11	3 46
5/15/1947	19 45	3 1	3 56	3 43	21 59	23 51	13 40	16 27	6 49	18 12	0ar43	0 27	18 11	19 35	3 13	29 51	27 59	0 40	3 44	7 33	3 49
5/25/1947	21 45	2 40	3 21	8 24	18 0	23 23	15 37	20 38	7 1	19 28	3 8	0 31	20 21	23 47	3 17	2 9	2 23	0 48	4 59	5 44	3 49
6/ 4/1947	23 45	3 0	2 48	12 55	13 32	22 4	17 36	24 52	7 13	20 43	5 21	0 35	22 32	27 59	3 22	4 26	6 44	0 56	6 14	3 52	3 47
6/14/1947	25 46	3 55	2 18	17 17	9 22	20 1	19 39	29 8	7 28	21 59	7 20	0 39	24 45	2 11	3 26	6 41	11 2	1 4	7 30	2 6	3 42
6/24/1947	27 48	5 24	1 50	21 31	6 9	17 31	21 46	3 27	7 43	23 15	9 4	0 43	26 59	6 23	3 32	8 54	15 16	1 12	8 45	0 36	3 35
7/ 4/1947	29 51	7 20	1 25	25 38	4 13	14 53	23 56	7 48	8 1	24 31	10 29	0 48	29 15	10 33	3 37	11 6	19 25	1 21	10 1	29 27	3 26
7/14/1947	1 56	9 41	1 2	29 39	3 34	12 24	26 10	12 11	8 20	25 47	11 33	0 54	1 32	14 41	3 44	13 16	23 29	1 29	11 17	28 44	3 17
7/24/1947	4 1	12 23	0 41	3 35	4 3	10 8	28 30	16 37	8 41	27 3	12 14	0 59	3 51	18 46	3 51	15 25	27 27	1 39	12 34	28 29	3 7
8/ 3/1947	6 8	15 24	0 22	7 25	5 31	8 10	0gm54	21 4	9 4	28 19	12 28	1 6	6 11	22 47	3 59	17 32	1cn19	1 48	13 50	28 41	2 57
8/13/1947	8 16	18 39	0 4	11 12	7 44	6 28	3 25	25 33	9 30	29 36	12 16	1 12	8 33	26 42	4 7	19 38	5 2	1 59	15 7	29 19	2 47
8/23/1947	10 26	22 9	0 13	14 55	10 35	5 1	6 1	0cn 4	9 59	0 52	11 36	1 18	10 56	0cn32	4 17	21 42	8 37	2 10	16 25	0sa22	2 38
9/ 2/1947	12 38	25 51	0 29	18 35	13 55	3 46	8 44	4 37	10 31	2 9	10 28	1 25	13 22	4 13	4 27	23 46	12 1	2 22	17 42	1 46	2 29
9/12/1947	14 51	29 43	0 44	22 13	17 40	2 41	11 35	9 12	11 6	3 26	8 58	1 31	15 48	7 44	4 39	25 48	15 13	2 35	19 0	3 31	2 22
9/22/1947	17 6	3 44	0 59	25 50	21 44	1 44	14 35	13 47	11 45	4 44	7 12	1 36	18 17	11 3	4 51	27 50	18 9	2 49	20 19	5 32	2 14
10/ 2/1947	19 23	7 54	1 13	29 25	26 5	0 54	17 44	18 24	12 28	6 1	5 17	1 40	20 47	14 6	5 5	29 50	20 49	3 5	21 38	7 48	2 8
10/12/1947	21 42	12 11	1 27	3 1	0cp41	0 11	21 3	23 1	13 16	7 19	3 25	1 43	23 19	16 50	5 20	1 50	23 8	3 22	22 57	10 18	2 2
10/22/1947	24 3	16 35	1 40	6 36	5 28	0 28	24 34	27 38	14 9	8 37	1 44	1 45	25 52	19 10	5 37	3 49	25 3	3 40	24 16	12 58	1 56
11/ 1/1947	26 27	21 5	1 54	10 12	10 27	1 4	28 18	2le14	15 7	9 55	0 24	1 45	28 27	21 3	5 54	5 47	26 29	4 1	25 36	15 48	1 51
11/11/1947	28 53	25 40	2 8	13 49	15 35	1 36	2 16	6 49	16 12	11 14	29 29	1 45	1cn 4	22 23	6 12	7 44	27 23	4 23	26 57	18 46	1 47
11/21/1947	1 22	0sa20	2 22	17 28	20 52	2 5	6 30	11 21	17 22	12 33	29 4	1 44	3 43	23 4	6 30	9 41	27 39	4 46	28 18	21 50	1 43
12/ 1/1947	3 53	5 5	2 37	21 9	26 17	2 32	11 2	15 49	18 38	13 52	29 8D	1 42	6 23	23 2R	6 46	11 37	27 15	5 10	29 39	25 0	1 39
12/11/1947	6 28	9 53	2 52	24 53	1aq51	2 57	15 54	20 13	20 0	15 12	29 41	1 41	9 4	22 15	7 0	13 32	26 9	5 33	1 1	28 14	1 36
12/21/1947	9 5	14 44	3 8	28 41	7 32	3 20	21 9	24 29	21 23	16 32	0 41	1 39	11 47	20 45	7 8	15 28	24 25	5 54	2 24	1cp30	1 33
12/31/1947	11 45	19 38	3 24	2 34	13 20	3 42	26 48	28 37	22 46	17 52	2 4	1 37	14 32	18 42	7 8	17 22	22 10	6 11	3 47	4 48	1 30
1/10/1948	14 29	24 34	3 41	6 31	19 16	4 3	2le54	2 38	24 0	19 13	3 49	1 36	17 18	16 21	7 0	19 17	19 37	6 21	5 11	8 7	1 28
1/20/1948	17sa16	29sa32	4n 0	10pi35	25aq19	4n22	9le30	6vi30	24s55	20ar34	5ar52	1s35	20cn 6	14cn 2R	6s42	21cn11	17cn 4R	6s25	6cp35	11cp25	1n26
1/30/1948	20 7	4 31	4 19	14 45	1pi29	4 41	16 39	10 22	25 10	21 56	8 11	1 34	22 55	12 4	6 17	23 5	14 46	6 23	8 0	14 41	1 24
2/ 9/1948	23 1	9 31	4 40	19 2	7 48	4 58	24 21	14 23	24 14	23 18	10 44	1 33	25 45	10 42	5 47	24 59	12 58	6 14	9 25	17 54	1 23
2/19/1948	26 0	14 30	5 2	23 28	14 14	5 14	2vi38	18 55	21 9	24 40	13 27	1 32	28 37	10 6	5 14	26 53	11 48	6 2	10 51	21 2	1 21
2/29/1948	29 2	19 28	5 25	28 4	20 49	5 29	11 29	24 42	14 11	26 4	16 20	1 32	1 29	10 16D	4 41	28 46	11 18	5 48	12 18	24 5	1 20
3/10/1948	2 9	24 25	5 51	2 50	27 32	5 43	20 52	2li41	0 41	27 27	19 20	1 32	4 23	11 11	4 10	0le40	11 29	5 33	13 46	27 1	1 18
3/20/1948	5 19	29 19	6 18	7 47	4 24	5 54	0li41	14 3	20 11	28 51	22 27	1 32	7 18	12 47	3 40	2 34	12 16	5 18	15 14	29 47	1 17
3/30/1948	8 34	4 9	6 48	12 57	11 26	6 4	10 50	29 15	40 39	0ta16	25 38	1 32	10 14	14 57	3 13	4 28	13 36	5 4	16 43	2 23	1 16
4/ 9/1948	11 54	8 55	7 20	18 20	18 38	6 12	21 8	15 54	52 42	1 41	28 53	1 33	13 10	17 38	2 48	6 22	15 24	4 50	18 13	4 45	1 15
4/19/1948	15 18	13 34	7 54	23 59	26 1	6 16	1sc25	29 51	57 34	3 7	2 10	1 34	16 8	20 44	2 25	8 16	17 36	4 38	19 44	6 53	1 14
4/29/1948	18 47	18 6	8 31	29 54	3ta34	6 18	11 28	9 4	58 37	4 33	5 28	1 36	19 6	24 10	2 5	10 10	20 8	4 27	21 15	8 42	1 13
5/ 9/1948	22 20	22 27	9 11	6 6	11 20	6 15	21 9	13 58	57 42	6 0	8 47	1 37	22 4	27 54	1 46	12 5	22 57	4 17	22 47	10 12	1 11
5/19/1948	25 58	26 35	9 54	12 35	19 16	6 9	0sa20	15 53	55 27	7 28	12 5	1 39	25 3	1le52	1 28	14 0	26 0	4 8	24 20	11 17	1 10

Ephemeris	80			433			1221			1489			114			3018			2041		
0 Hours ET	Sappho			Eros			Amor			Attila			Kassandra			Godiva			Lancelot		
Mo Dy Year	h long	g long	g lat	h long	g long	g lat	h long	g long	g lat	h long	g long	g lat	h long	g long	g lat	h long	g long	g lat	h long	g long	g lat
5/29/1948	29 40	0pi27	10 41	19 24	27 24	5 57	8 55	16 14	52 9	8 57	15 22	1 42	28 2	6 2	1 12	15 56	29 15	4 0	25 54	11 56	1 7
6/ 8/1948	3 27	3 58	11 31	26 31	5 44	5 39	16 54	15 59	47 58	10 26	18 36	1 44	1vi 1	10 22	0 57	17 52	2 41	3 53	27 29	12 7	1 5
6/18/1948	7 18	7 4	12 24	3 58	14 15	5 16	24 15	15 54D	43 10	11 55	21 47	1 47	4 1	14 51	0 43	19 49	6 15	3 46	29 5	11 47	1 1
6/28/1948	11 14	9 38	13 19	11 43	22 55	4 47	1cp 2	16 20	38 7	13 26	24 53	1 51	7 0	19 26	0 30	21 47	9 56	3 41	0aq42	10 56	0 57
7/ 8/1948	15 13	11 34	14 15	19 46	1cn46	4 12	7 15	17 22	33 10	14 57	27 54	1 55	10 0	24 7	0 18	23 45	13 44	3 36	2 20	9 37	0 52
7/18/1948	19 17	12 43	15 8	28 4	10 44	3 31	12 58	19 3	28 33	16 29	0gm47	1 59	12 59	28 53	0 6	25 44	17 36	3 32	3 59	7 56	0 46
7/28/1948	23 23	13 1	15 56	6 35	19 49	2 45	18 14	21 15	24 26	18 2	3 31	2 4	15 57	3 43	0 6	27 43	21 34	3 28	5 39	6 2	0 39
8/ 7/1948	27 33	12 25	16 32	15 17	28 58	1 55	23 6	23 54	20 52	19 36	6 5	2 9	18 55	8 35	0 17	29 44	25 34	3 25	7 20	4 4	0 31
8/17/1948	1pi46	10 59	16 48	24 6	8 10	1 2	27 37	26 55	17 48	21 10	8 25	2 16	21 53	13 30	0 27	1 45	29 39	3 22	9 2	2 15	0 23
8/27/1948	6 1	8 56	16 38	2le58	17 24	0 6	1aq48	0cp14	15 13	22 46	10 30	2 22	24 50	18 27	0 38	3 48	3 45	3 20	10 45	0 45	0 15
9/ 6/1948	10 18	6 40	15 59	11 50	26 36	0 50	5 42	3 46	13 2	24 22	12 17	2 30	27 46	23 25	0 48	5 52	7 54	3 18	12 29	29cp42	0 7
9/16/1948	14 37	4 38	14 52	20 37	5 46	1 44	9 22	7 29	11 11	25 59	13 42	2 38	0li41	28 23	0 58	7 56	12 3	3 16	14 15	29 11	0s 0
9/26/1948	18 57	3 14	13 27	29 16	14 50	2 37	12 48	11 21	9 37	27 37	14 42	2 46	3 36	3 22	1 8	10 2	16 14	3 15	16 2	29 13D	0 7
10/ 6/1948	23 17	2 43	11 50	7 44	23 48	3 26	16 2	15 20	8 17	29 16	15 13	2 55	6 29	8 19	1 18	12 9	20 24	3 15	17 49	29 48	0 13
10/16/1948	27 37	3 10	10 12	15 58	2li37	4 11	19 6	19 25	7 9	0 56	15 13R	3 4	9 21	13 16	1 28	14 18	24 34	3 14	19 39	0 55	0 19
10/26/1948	1ar56	4 32	8 38	23 56	11 17	4 52	22 1	23 33	6 10	2 37	14 40	3 13	12 12	18 10	1 38	16 28	28 42	3 14	21 29	2 29	0 24
11/ 5/1948	6 15	6 44	7 11	1li37	19 45	5 27	24 46	27 46	5 19	4 19	13 35	3 20	15 2	23 2	1 49	18 40	2 47	3 15	23 21	4 28	0 29
11/15/1948	10 32	9 38	5 51	8 59	28 2	5 58	27 25	2aq 1	4 35	6 2	12 1	3 26	17 51	27 50	2 0	20 53	6 49	3 16	25 14	6 50	0 33
11/25/1948	14 47	13 6	4 41	16 2	6 6	6 23	29 56	6 18	3 56	7 47	10 7	3 30	20 38	2 33	2 11	23 8	10 46	3 17	27 8	9 31	0 37
12/ 5/1948	19 1	17 3	3 38	22 46	13 57	6 44	2 22	10 36	3 22	9 32	8 2	3 30	23 23	7 11	2 22	25 24	14 37	3 18	29 4	12 29	0 41
12/15/1948	23 11	21 23	2 42	29 12	21 34	7 1	4 42	14 56	2 51	11 18	6 2	3 27	26 7	11 42	2 35	27 43	18 19	3 20	1 1	15 42	0 45
12/25/1948	27 19	26 1	1 53	5 21	28 58	7 14	6 57	19 15	2 24	13 6	4 18	3 22	28 50	16 5	2 48	0li 3	21 51	3 22	2 59	19 6	0 48
1/ 4/1949	1ta23	0ar53	1 9	11 12	6 8	7 24	9 7	23 35	1 59	14 54	3 0	3 14	1 31	20 18	3 2	2 26	25 9	3 25	4 59	22 42	0 51
1/14/1949	5 24	5 55	0 30	16 48	13 3	7 30	11 14	27 54	1 36	16 44	2 15	3 5	4 11	24 20	3 17	4 51	28 12	3 27	7 1	26 26	0 55
1/24/1949	9 21	11 6	0s 5	22 10	19 45	7 34	13 16	2 11	1 15	18 35	2 6	2 55	6 49	28 8	3 33	7 17	0sc55	3 29	9 3	0pi18	0 58
2/ 3/1949	13 14	16 23	0 36	27 17	26 13	7 36	15 16	6 28	0 56	20 27	2 32	2 45	9 25	1sa40	3 51	9 47	3 14	3 32	11 7	4 16	1 2
2/13/1949	17 4	21 45	1 4	2sa13	2cp26	7 36	17 12	10 42	0 37	22 20	3 31	2 35	12 0	4 54	4 10	12 18	5 4	3 33	13 13	8 19	1 5
2/23/1949	20ta49	27ar 9	1s30	6sa57	8cp25	7s33	19pi 6	14pi53	0n20	24gm15	5gm 0	2s26	14sc33	7sa46	4n31	14li52	6sc20	3s34	15pi20	12pi26	1s 9
3/ 5/1949	24 30	2ta35	1 53	11 31	14 8	7 29	20 57	19 2	0 3	26 11	6 56	2 17	17 4	10 13	4 54	17 29	6 56	3 33	17 28	16 36	1 13
3/15/1949	28 7	8 2	2 14	15 56	19 35	7 23	22 46	23 8	0 14	28 8	9 15	2 10	19 34	12 11	5 18	20 9	6 48R	3 30	19 38	20 48	1 17
3/25/1949	1gm40	13 28	2 33	20 13	24 46	7 16	24 32	27 10	0 30	0cn 6	11 53	2 2	22 2	13 35	5 44	22 51	5 54	3 22	21 50	25 1	1 21
4/ 4/1949	5 9	18 54	2 51	24 22	29 38	7 6	26 17	1ar 7	0 47	2 5	14 48	1 56	24 29	14 22	6 10	25 36	4 15	3 11	24 2	29 15	1 25
4/14/1949	8 34	24 18	3 7	28 25	4 9	6 54	28 1	4 59	1 4	4 6	17 57	1 50	26 54	14 28R	6 37	28 24	2 2	2 54	26 17	3 29	1 30
4/24/1949	11 54	29 39	3 23	2 22	8 18	6 39	29 43	8 47	1 21	6 8	21 17	1 44	29 17	13 53	7 3	1sc16	29li29	2 32	28 32	7 41	1 35
5/ 4/1949	15 11	4 59	3 37	6 14	12 0	6 21	1 23	12 28	1 38	8 12	24 49	1 39	1 39	12 37	7 25	4 10	26 58	2 7	0ar49	11 51	1 40
5/14/1949	18 24	10 15	3 51	10 2	15 10	5 59	3 3	16 1	1 57	10 16	28 29	1 34	3 59	10 50	7 42	7 8	24 49	1 39	3 7	15 58	1 46
5/24/1949	21 32	15 29	4 4	13 46	17 43	5 32	4 42	19 27	2 16	12 22	2 16	1 30	6 18	8 40	7 52	10 9	23 17	1 10	5 27	20 2	1 52
6/ 3/1949	24 37	20 39	4 17	17 27	19 32	4 58	6 19	22 44	2 37	14 30	6 9	1 26	8 35	6 25	7 53	13 13	22 34	0 42	7 48	24 1	1 59
6/13/1949	27 39	25 45	4 29	21 6	20 27	4 14	7 56	25 51	3 0	16 38	10 7	1 22	10 51	4 19	7 46	16 21	22 40D	0 17	10 10	27 53	2 6
6/23/1949	0cn37	0cn48	4 41	24 43	20 19R	3 20	9 33	28 46	3 24	18 48	14 10	1 18	13 5	2 36	7 33	19 33	23 35	0n 6	12 34	1ta38	2 14

Ephemeris	80			433			1221			1489			114			3018			2041		
0 Hours ET	Sappho			Eros			Amor			Attila			Kassandra			Godiva			Lancelot		
Mo Dy Year	h long	g long	g lat	h long	g long	g lat	h long	g long	g lat	h long	g long	g lat	h long	g long	g lat	h long	g long	g lat	h long	g long	g lat
7/ 3/1949	3 31	5 46	4 53	28 18	19 0	2 13	11 9	1 26	3 51	20 59	18 16	1 15	15 18	1 27	7 14	22 48	25 14	0 27	14 58	5 14	2 22
7/13/1949	6 22	10 40	5 6	1 54	16 28	0 52	12 45	3 50	4 21	23 11	22 24	1 12	17 29	0 54	6 52	26 7	27 32	0 45	17 24	8 39	2 32
7/23/1949	9 10	15 29	5 18	5 29	12 53	0n40	14 20	5 54	4 53	25 24	26 35	1 9	19 39	1 0D	6 29	29 29	0sc23	1 2	19 50	11 51	2 42
8/ 2/1949	11 55	20 13	5 31	9 4	8 40	2 16	15 56	7 35	5 30	27 39	0le47	1 6	21 48	1 42	6 6	2 55	3 44	1 16	22 18	14 47	2 53
8/12/1949	14 37	24 51	5 44	12 41	4 28	3 48	17 32	8 49	6 10	29 55	5 0	1 3	23 55	2 56	5 44	6 25	7 29	1 29	24 47	17 23	3 5
8/22/1949	17 16	29 23	5 58	16 19	0 53	5 7	19 7	9 31	6 54	2 12	9 12	1 0	26 2	4 39	5 24	9 58	11 36	1 41	27 16	19 37	3 18
9/ 1/1949	19 52	3 48	6 12	19 59	28 23	6 10	20 44	9 36R	7 42	4 30	13 24	0 57	28 6	6 46	5 5	13 34	16 2	1 52	29 46	21 24	3 31
9/11/1949	22 25	8 5	6 28	23 43	27 9	6 58	22 20	9 0	8 33	6 49	17 34	0 54	0cp10	9 14	4 47	17 14	20 44	2 1	2 17	22 40	3 46
9/21/1949	24 56	12 13	6 45	27 30	27 7D	7 32	23 57	7 40	9 25	9 9	21 42	0 51	2 13	12 0	4 32	20 57	25 39	2 10	4 49	23 20	4 1
10/ 1/1949	27 25	16 11	7 3	1pi20	28 11	7 56	25 35	5 36	10 15	11 30	25 45	0 48	4 14	15 1	4 17	24 43	0sa48	2 18	7 21	23 21R	4 15
10/11/1949	29 51	19 57	7 23	5 16	0aq12	8 14	27 14	2 56	11 0	13 52	29 44	0 45	6 14	18 15	4 4	28 32	6 6	2 25	9 54	22 44	4 29
10/21/1949	2 15	23 29	7 45	9 18	2 59	8 26	28 53	29ar51	11 35	16 15	3 36	0 41	8 13	21 39	3 52	2 24	11 35	2 31	12 27	21 29	4 39
10/31/1949	4 36	26 44	8 8	13 25	6 26	8 34	0ta34	26 39	11 58	18 39	7 19	0 38	10 11	25 12	3 42	6 18	17 11	2 37	15 0	19 44	4 46
11/10/1949	6 56	29 41	8 34	17 40	10 27	8 39	2 16	23 40	12 6	21 4	10 53	0 34	12 9	28 52	3 32	10 14	22 55	2 43	17 33	17 44	4 48
11/20/1949	9 14	2 14	9 3	22 3	14 56	8 42	3 59	21 12	12 3	23 29	14 13	0 30	14 5	2 37	3 24	14 12	28 45	2 48	20 7	15 43	4 44
11/30/1949	11 30	4 21	9 34	26 35	19 51	8 43	5 44	19 26	11 49	25 55	17 18	0 25	16 0	6 28	3 16	18 11	4 40	2 52	22 40	13 58	4 35
12/10/1949	13 44	5 56	10 7	1ar18	25 8	8 42	7 31	18 28	11 30	28 21	20 4	0 20	17 54	10 21	3 9	22 12	10 38	2 56	25 14	12 42	4 23
12/20/1949	15 57	6 56	10 43	6 11	0pi45	8 40	9 20	18 17D	11 8	0vi48	22 27	0 14	19 48	14 17	3 3	26 13	16 40	2 59	27 47	12 4	4 8
12/30/1949	18 8	7 14	11 18	11 16	6 41	8 35	11 11	18 49	10 45	3 16	24 24	0 8	21 41	18 14	2 57	0aq14	22 44	3 2	0gm19	12 6D	3 52
1/ 9/1950	20 17	6 48	11 52	16 35	12 55	8 29	13 4	20 2	10 22	5 44	25 49	0 0	23 33	22 11	2 52	4 16	28 50	3 4	2 52	12 49	3 36
1/19/1950	22 25	5 36	12 22	22 9	19 25	8 20	15 0	21 48	10 1	8 12	26 38	0 8	25 24	26 8	2 47	8 17	4 55	3 6	5 24	14 9	3 21
1/29/1950	24 32	3 44	12 44	27 58	26 12	8 9	16 59	24 4	9 42	10 40	26 48	0 17	27 14	0aq 3	2 43	12 17	11 1	3 7	7 55	16 1	3 6
2/ 8/1950	26 37	1 20	12 54	4 4	3ar14	7 54	19 0	26 46	9 25	13 9	26 19	0 27	29 4	3 56	2 39	16 16	17 5	3 8	10 26	18 21	2 52
2/18/1950	28 42	28le39	12 50	10 27	10 32	7 36	21 6	29 49	9 10	15 37	25 12	0 38	0aq53	7 45	2 36	20 14	23 7	3 9	12 56	21 4	2 40
2/28/1950	0vi45	26 0	12 32	17 9	18 4	7 14	23 15	3 11	8 57	18 6	23 33	0 48	2 42	11 30	2 33	24 9	29 6	3 9	15 25	24 7	2 29
3/10/1950	2 47	23 41	12 2	24 11	25 51	6 47	25 28	6 50	8 45	20 34	21 36	0 59	4 30	15 9	2 31	28 3	5 1	3 9	17 54	27 26	2 18
3/20/1950	4 48	21 55	11 23	1gm31	3ta53	6 15	27 46	10 43	8 35	23 2	19 34	1 8	6 17	18 42	2 28	1pi54	10 52	3 8	20 21	0gm58	2 9
3/30/1950	6 49	20 49	10 40	9 10	12 7	5 39	0gm 9	14 50	8 27	25 30	17 45	1 16	8 4	22 6	2 26	5 42	16 38	3 8	22 48	4 41	2 0
4/ 9/1950	8 49	20 26	9 54	17 6	20 34	4 57	2 37	19 8	8 20	27 58	16 21	1 22	9 50	25 21	2 24	9 28	22 18	3 6	25 13	8 33	1 52
4/19/1950	10 47	20 42	9 10	25 19	29 12	4 11	5 12	23 38	8 14	0li25	15 31	1 27	11 36	28 25	2 23	13 10	27 52	3 5	27 37	12 31	1 45
4/29/1950	12 46	21 36	8 28	3cn47	8 0	3 19	7 53	28 19	8 8	2 52	15 20	1 31	13 22	1pi16	2 21	16 49	3 19	3 3	0cn 0	16 34	1 38
5/ 9/1950	14 43	23 1	7 49	12 26	16 57	2 24	10 42	3 11	8 3	5 18	15 46	1 33	15 7	3 52	2 19	20 25	8 37	3 1	2 22	20 41	1 32
5/19/1950	16 41	24 54	7 14	21 13	26 1	1 26	13 39	8 13	7 59	7 44	16 48	1 34	16 52	6 11	2 18	23 57	13 47	2 59	4 43	24 51	1 26
5/29/1950	18 37	27 10	6 41	0le 4	5 10	0 26	16 45	13 26	7 54	10 9	18 22	1 35	18 36	8 9	2 16	27 25	18 47	2 56	7 2	29 3	1 21
6/ 8/1950	20 34	29 45	6 12	8 56	14 21	0 33	20 1	18 50	7 50	12 33	20 24	1 36	20 20	9 45	2 14	0ar50	23 36	2 54	9 20	3 16	1 16
6/18/1950	22 30	2 36	5 45	17 46	23 34	1 32	23 28	24 26	7 45	14 56	22 49	1 36	22 4	10 54	2 11	4 11	28 13	2 51	11 36	7 29	1 11
6/28/1950	24 25	5 41	5 21	26 28	2le46	2 28	27 7	0cn15	7 39	17 19	25 35	1 36	23 48	11 34	2 8	7 29	2 37	2 47	13 52	11 41	1 6
7/ 8/1950	26 21	8 58	4 59	5 0	11 55	3 19	1cn 1	6 17	7 31	19 40	28 37	1 36	25 31	11 41R	2 3	10 42	6 45	2 43	16 5	15 52	1 2
7/18/1950	28 16	12 24	4 39	13 19	20 59	4 5	5 10	12 34	7 22	22 1	1 54	1 36	27 14	11 14	1 58	13 52	10 35	2 39	18 18	20 2	0 57
7/28/1950	0li12	15 58	4 21	21 23	29 57	4 45	9 37	19 7	7 11	24 20	5 22	1 36	28 57	10 13	1 51	16 59	14 5	2 34	20 29	24 9	0 53

Ephemeris	80			433			1221			1489			114			3018			2041		
0 Hours ET	Sappho			Eros			Amor			Attila			Kassandra			Godiva			Lancelot		
Mo Dy Year	h long	g long	g lat	h long	g long	g lat	h long	g long	g lat	h long	g long	g lat	h long	g long	g lat	h long	g long	g lat	h long	g long	g lat
8/ 7/1950	2 7	19 39	4 4	29 10	8 46	5 19	14 22	25 57	6 56	26 39	9 0	1 36	0pi40	8 41	1 42	20 2	17 12	2 29	22 38	28 12	0 49
8/17/1950	4 3	23 26	3 48	6 38	17 26	5 45	19 30	3le 5	6 37	28 56	12 46	1 36	2 23	6 45	1 31	23 1	19 50	2 22	24 46	2 12	0 45
8/27/1950	5 58	27 18	3 33	13 48	25 56	6 6	25 1	10 34	6 14	1 13	16 39	1 36	4 6	4 34	1 19	25 57	21 57	2 15	26 53	6 6	0 41
9/ 6/1950	7 54	1li14	3 19	20 38	4 16	6 20	0le59	18 24	5 45	3 28	20 37	1 37	5 49	2 23	1 6	28 50	23 27	2 6	28 58	9 55	0 37
9/16/1950	9 50	5 13	3 6	27 11	12 23	6 28	7 26	26 37	5 10	5 42	24 40	1 37	7 32	0 23	0 51	1 39	24 15	1 56	1 2	13 37	0 33
9/26/1950	11 47	9 15	2 54	3 25	20 20	6 32	14 25	5 12	4 27	7 55	28 45	1 38	9 15	28 46	0 37	4 26	24 17R	1 43	3 4	17 10	0 28
10/ 6/1950	13 44	13 19	2 42	9 22	28 5	6 31	21 56	14 10	3 37	10 6	2 53	1 38	10 58	27 40	0 24	7 9	23 30	1 29	5 5	20 34	0 24
10/16/1950	15 41	17 24	2 30	15 3	5 39	6 26	0vi 3	23 30	2 39	12 17	7 3	1 39	12 41	27 8	0 11	9 50	21 56	1 11	7 4	23 46	0 19
10/26/1950	17 39	21 30	2 18	20 29	13 2	6 19	8 44	3li10	1 35	14 26	11 13	1 40	14 24	27 12D	0s 0	12 27	19 44	0 51	9 3	26 45	0 15
11/ 5/1950	19 38	25 35	2 7	25 41	20 14	6 9	17 57	13 7	0 25	16 34	15 23	1 41	16 8	27 50	0 11	15 2	17 9	0 30	10 59	29 27	0 9
11/15/1950	21 37	29 40	1 56	0sa41	27 18	5 56	27 39	23 16	0 48	18 40	19 32	1 43	17 51	28 59	0 20	17 35	14 30	0 9	12 55	1 51	0 4
11/25/1950	23 37	3 43	1 44	5 29	4 11	5 42	7 43	3sc32	2 1	20 46	23 39	1 44	19 35	0pi36	0 28	20 4	12 9	0 12	14 49	3 52	0n 2
12/ 5/1950	25 38	7 43	1 32	10 6	10 56	5 27	18 0	13 48	3 12	22 50	27 43	1 46	21 19	2 36	0 36	22 32	10 20	0 30	16 41	5 28	0 9
12/15/1950	27 40	11 39	1 20	14 34	17 33	5 10	28 18	23 58	4 17	24 52	1sa44	1 48	23 3	4 58	0 43	24 57	9 14	0 46	18 33	6 35	0 16
12/25/1950	29 43	15 30	1 7	18 53	24 2	4 52	8 27	3sa56	5 15	26 54	5 39	1 51	24 48	7 38	0 49	27 20	8 53	1 0	20 23	7 10	0 24
1/ 4/1951	1 47	19 15	0 54	23 5	0cp24	4 34	18 16	13 37	6 3	28 54	9 28	1 54	26 33	10 32	0 55	29 41	9 16	1 11	22 12	7 10R	0 32
1/14/1951	3 53	22 51	0 39	27 10	6 39	4 14	27 36	22 56	6 43	0sa53	13 10	1 57	28 18	13 40	1 1	1 59	10 19	1 21	23 59	6 35	0 41
1/24/1951	6 0	26 17	0 23	1cp 9	12 47	3 54	6 22	1cp50	7 14	2 50	16 42	2 0	0ar 4	16 57	1 6	4 16	11 56	1 29	25 46	5 25	0 50
2/ 3/1951	8 8	29 31	0 5	5 2	18 48	3 33	14 32	10 17	7 37	4 47	20 4	2 4	1 50	20 24	1 12	6 31	14 2	1 35	27 31	3 48	0 59
2/13/1951	10 17	2 30	0 14	8 51	24 44	3 11	22 5	18 16	7 54	6 42	23 14	2 9	3 37	23 58	1 17	8 45	16 32	1 41	29 15	1 51	1 7
2/23/1951	12 29	5 10	0 36	12 36	0aq33	2 48	29 1	25 48	8 5	8 36	26 9	2 14	5 24	27 37	1 22	10 56	19 22	1 46	0 58	29le47	1 15
3/ 5/1951	14 42	7 29	1 1	16 18	6 17	2 24	5 24	2aq53	8 11	10 28	28 48	2 19	7 12	1ar21	1 28	13 7	22 28	1 50	2 40	27 49	1 21
3/15/1951	16 56	9 21	1 30	19 58	11 55	1 58	11 16	9 30	8 15	12 20	1 7	2 25	9 0	5 9	1 33	15 15	25 47	1 54	4 20	26 8	1 26
3/25/1951	19 13	10 42	2 2	23 35	17 27	1 31	16 40	15 42	8 16	14 10	3 5	2 32	10 49	8 59	1 39	17 22	29 17	1 58	6 0	24 53	1 30
4/ 4/1951	21 32	11 27	2 39	27 12	22 53	1 2	21 39	21 27	8 15	15 59	4 37	2 38	12 38	12 50	1 45	19 28	2 54	2 1	7 38	24 9	1 32
4/14/1951	23 53	11 31R	3 20	0aq47	28 14	0 30	26 16	26 47	8 13	17 47	5 42	2 45	14 29	16 42	1 51	21 33	6 39	2 5	9 16	23 57	1 34
4/24/1951	26 16	10 51	4 4	4 22	3 28	0n 4	0aq33	1pi40	8 10	19 34	6 16	2 53	16 19	20 34	1 58	23 37	10 29	2 8	10 53	24 17	1 35
5/ 4/1951	28 42	9 27	4 51	7 57	8 36	0 41	4 32	6 7	8 7	21 19	6 17R	3 0	18 11	24 25	2 5	25 39	14 22	2 12	12 28	25 6	1 35
5/14/1951	1sa10	7 24	5 37	11 33	13 36	1 22	8 16	10 6	8 3	23 4	5 46	3 6	20 4	28 14	2 12	27 41	18 19	2 16	14 3	26 20	1 36
5/24/1951	3 41	4 54	6 21	15 11	18 30	2 8	11 46	13 34	7 59	24 48	4 43	3 11	21 57	2 1	2 20	29 41	22 18	2 19	15 37	27 57	1 36
6/ 3/1951	6 15	2 14	6 58	18 51	23 14	2 59	15 4	16 29	7 55	26 30	3 13	3 14	23 51	5 44	2 29	1 41	26 17	2 23	17 10	29 54	1 36
6/13/1951	8 52	29sc44	7 27	22 33	27 49	3 57	18 11	18 47	7 49	28 12	1 25	3 15	25 46	9 23	2 39	3 39	0cn18	2 28	18 42	2 6	1 36
6/23/1951	11 32	27 42	7 46	26 19	2 11	5 2	21 8	20 23	7 43	29 52	29sa28	3 13	27 42	12 57	2 49	5 38	4 19	2 32	20 13	4 33	1 36
7/ 3/1951	14 15	26 21	7 57	0pi 8	6 20	6 17	23 56	21 13	7 34	1 32	27 34	3 8	29 39	16 24	3 0	7 35	8 18	2 37	21 43	7 11	1 36
7/13/1951	17 2	25 47	8 1	4 3	10 11	7 43	26 37	21 12R	7 21	3 10	25 53	3 2	1 37	19 42	3 13	9 32	12 17	2 42	23 13	9 58	1 37
7/23/1951	19 52	26 2	8 0	8 2	13 39	9 22	29 10	20 15	7 4	4 48	24 35	2 53	3 36	22 51	3 26	11 28	16 14	2 48	24 41	12 53	1 37
8/ 2/1951	22 46	27 2	7 56	12 8	16 40	11 17	1 38	18 24	6 40	6 24	23 45	2 44	5 36	25 48	3 41	13 23	20 8	2 54	26 9	15 55	1 38
8/12/1951	25 44	28 45	7 49	16 20	19 3	13 30	3 59	15 43	6 8	8 0	23 25	2 34	7 37	28 30	3 58	15 19	24 0	3 0	27 37	19 2	1 39
8/22/1951	28 46	1sa 4	7 42	20 41	20 39	16 3	6 16	12 28	5 28	9 35	23 36	2 25	9 40	0gm55	4 16	17 13	27 47	3 7	29 3	22 12	1 40
9/ 1/1951	1 52	3 55	7 34	25 10	21 15	18 55	8 27	8 59	4 40	11 9	24 15	2 15	11 43	2 59	4 36	19 8	1le29	3 15	0li29	25 26	1 42

Ephemeris	80			433			1221			1489			114			3018			2041		
0 Hours ET	Sappho			Eros			Amor			Attila			Kassandra			Godiva			Lancelot		
Mo Dy Year	h long	g long	g lat	h long	g long	g lat	h long	g long	g lat	h long	g long	g lat	h long	g long	g lat	h long	g long	g lat	h long	g long	g lat
9/11/1951	5 2	7 14	7 26	29 49	20 35	22 0	10 35	5 39	3 49	12 42	25 21	2 7	13 48	4 38	4 58	21 2	5 5	3 24	1 55	28 41	1 43
9/21/1951	8 16	10 57	7 18	4 39	18 32	25 6	12 39	2 51	2 58	14 15	26 51	1 59	15 54	5 48	5 22	22 56	8 34	3 33	3 19	1 58	1 45
10/ 1/1951	11 35	15 2	7 10	9 40	15 5	27 49	14 39	0 46	2 10	15 46	28 41	1 51	18 1	6 25	5 47	24 50	11 54	3 44	4 43	5 14	1 47
10/11/1951	14 59	19 25	7 2	14 55	10 41	29 43	16 36	29aq30	1 26	17 17	0cp50	1 44	20 10	6 23R	6 13	26 44	15 3	3 56	6 7	8 29	1 50
10/21/1951	18 27	24 5	6 55	20 24	6 10	30 28	18 31	29 4	0 47	18 47	3 13	1 38	22 20	5 43	6 39	28 37	17 59	4 9	7 29	11 41	1 53
10/31/1951	22 0	29 0	6 47	26 8	2 28	30 4	20 23	29 21	0 14	20 17	5 50	1 32	24 31	4 23	7 2	0le31	20 39	4 24	8 52	14 51	1 56
11/10/1951	25 37	4 7	6 40	2ta 8	0 18	28 44	22 12	0pi18	0 14	21 46	8 38	1 27	26 44	2 29	7 20	2 25	23 0	4 40	10 14	17 56	2 0
11/20/1951	29 19	9 27	6 32	8 26	29 54D	26 50	24 0	1 47	0 39	23 14	11 34	1 23	28 58	0 13	7 32	4 19	24 59	4 58	11 35	20 55	2 4
11/30/1951	3 5	14 56	6 24	15 2	1 13	24 37	25 45	3 44	1 0	24 41	14 38	1 19	1 14	27 51	7 36	6 13	26 31	5 17	12 56	23 47	2 9
12/10/1951	6 56	20 35	6 16	21 57	4 1	22 16	27 29	6 4	1 19	26 8	17 48	1 15	3 31	25 38	7 31	8 7	27 32	5 38	14 16	26 29	2 15
12/20/1951	10 51	26 22	6 7	29 11	8 5	19 51	29 12	8 43	1 35	27 34	21 1	1 12	5 50	23 53	7 19	10 1	27 56	6 0	15 36	29 1	2 21
12/30/1951	14 50	2aq17	5 58	6 44	13 11	17 23	0 53	11 37	1 50	28 59	24 19	1 9	8 11	22 44	7 2	11 56	27 42	6 22	16 55	1 19	2 27
1/ 9/1952	18 53	8 17	5 48	14 35	19 10	14 54	2 33	14 43	2 3	0aq24	27 37	1 6	10 33	22 18	6 42	13 52	26 45	6 43	18 14	3 22	2 35
1/19/1952	23 0	14 23	5 38	22 43	25 53	12 21	4 11	18 0	2 16	1 49	0aq57	1 3	12 56	22 36	6 20	15 47	25 9	7 1	19 33	5 7	2 43
1/29/1952	27 10	20 34	5 27	1cn 6	3ta12	9 46	5 49	21 25	2 28	3 13	4 16	1 1	15 22	23 34	5 58	17 44	23 0	7 14	20 51	6 32	2 52
2/ 8/1952	1pi23	26 47	5 16	9 41	11 4	7 9	7 27	24 57	2 40	4 36	7 33	0 59	17 49	25 9	5 37	19 40	20 30	7 21	22 9	7 32	3 1
2/18/1952	5 38	3pi 4	5 3	18 26	19 21	4 31	9 3	28 34	2 51	5 59	10 47	0 57	20 17	27 16	5 17	21 38	17 56	7 19	23 26	8 7	3 11
2/28/1952	9 55	9 22	4 50	27 16	28 1	1 55	10 40	2 15	3 3	7 21	13 58	0 55	22 47	29 51	4 59	23 36	15 33	7 9	24 43	8 14R	3 21
3/ 9/1952	14 14	15 41	4 35	6 8	6 59	0s36	12 15	5 59	3 14	8 43	17 3	0 54	25 19	2 49	4 42	25 35	13 38	6 54	26 0	7 51	3 31
3/19/1952	18 34	22 0	4 20	14 58	16 10	2 58	13 51	9 45	3 26	10 5	20 1	0 52	27 53	6 7	4 27	27 34	12 20	6 34	27 17	7 0	3 40
3/29/1952	22 54	28 19	4 4	23 43	25 31	5 8	15 27	13 32	3 38	11 26	22 52	0 51	0cn28	9 42	4 12	29 35	11 42	6 12	28 33	5 43	3 49
4/ 8/1952	27 14	4 36	3 47	2vi19	4 57	7 2	17 2	17 19	3 51	12 46	25 34	0 50	3 5	13 30	3 59	1 37	11 46D	5 49	29 49	4 6	3 55
4/18/1952	1ar35	10 51	3 29	10 42	14 24	8 37	18 38	21 7	4 5	14 7	28 4	0 48	5 43	17 31	3 47	3 39	12 29	5 27	1 4	2 15	3 58
4/28/1952	5 54	17 3	3 9	18 51	23 48	9 52	20 14	24 53	4 19	15 26	0pi21	0 47	8 24	21 41	3 36	5 43	13 46	5 6	2 20	0 21	3 59
5/ 8/1952	10 11	23 11	2 49	26 43	3le 5	10 48	21 50	28 38	4 34	16 46	2 24	0 46	11 5	26 0	3 25	7 47	15 33	4 46	3 35	28 34	3 57
5/18/1952	14 27	29 15	2 28	4 17	12 12	11 25	23 27	2 21	4 51	18 5	4 10	0 44	13 49	0cn27	3 15	9 53	17 47	4 28	4 50	27 1	3 53
5/28/1952	18 41	5 13	2 6	11 32	21 6	11 44	25 5	6 1	5 9	19 24	5 36	0 43	16 33	4 59	3 6	12 1	20 23	4 12	6 5	25 50	3 47
6/ 7/1952	22 52	11 6	1 43	18 29	29 46	11 49	26 43	9 38	5 29	20 43	6 41	0 41	19 20	9 37	2 57	14 9	23 17	3 56	7 20	25 5	3 40
6/17/1952	27 0	16 53	1 18	25 7	8 11	11 41	28 22	13 10	5 51	22 1	7 21	0 40	22 8	14 19	2 48	16 19	26 28	3 43	8 34	24 47	3 32
6/27/1952	1ta 5	22 32	0 53	1sc27	16 21	11 23	0ta 2	16 36	6 14	23 19	7 35	0 38	24 57	19 5	2 40	18 31	29 53	3 30	9 49	24 57	3 23
7/ 7/1952	5 7	28 2	0 26	7 29	24 15	10 58	1 44	19 57	6 41	24 37	7 22	0 35	27 47	23 53	2 31	20 44	3 31	3 18	11 3	25 32	3 15
7/17/1952	9 5	3 23	0s 2	13 15	1li54	10 27	3 27	23 9	7 10	25 54	6 40	0 32	0le39	28 45	2 23	22 59	7 18	3 7	12 17	26 31	3 7
7/27/1952	12 59	8 34	0 32	18 46	9 20	9 52	5 11	26 11	7 43	27 11	5 33	0 29	3 32	3 38	2 15	25 16	11 15	2 57	13 31	27 51	3 0
8/ 6/1952	16 49	13 33	1 3	24 3	16 33	9 15	6 57	29 1	8 20	28 28	4 3	0 25	6 26	8 33	2 7	27 34	15 20	2 48	14 45	29 30	2 53
8/16/1952	20 35	18 19	1 37	29 6	23 34	8 37	8 45	1 37	9 1	29 45	2 17	0 21	9 21	13 29	1 59	29 55	19 33	2 39	15 59	1 25	2 47
8/26/1952	24 17	22 49	2 13	3 58	0sc25	7 59	10 35	3 54	9 48	1 2	0 24	0 17	12 16	18 25	1 51	2 17	23 52	2 31	17 13	3 35	2 42
9/ 5/1952	27 55	27 0	2 51	8 38	7 8	7 20	12 27	5 49	10 40	2 19	28 34	0 12	15 13	23 21	1 43	4 42	28 17	2 22	18 27	5 56	2 37
9/15/1952	1gm28	0cn51	3 33	13 9	13 42	6 43	14 22	7 17	11 39	3 35	26 55	0 8	18 11	28 16	1 35	7 9	2 47	2 15	19 41	8 28	2 32
9/25/1952	4 58	4 17	4 18	17 31	20 9	6 6	16 19	8 11	12 45	4 51	25 36	0 3	21 9	3 10	1 26	9 38	7 22	2 7	20 55	11 9	2 29
10/ 5/1952	8 23	7 13	5 7	21 45	26 31	5 30	18 20	8 25	13 56	6 7	24 42	0s 1	24 7	8 1	1 17	12 10	12 1	2 0	22 9	13 57	2 25

Ephemeris	80			433			1221			1489			114			3018			2041		
0 Hours ET	Sappho			Eros			Amor			Attila			Kassandra			Godiva			Lancelot		
Mo Dy Year	h long	g long	g lat	h long	g long	g lat	h long	g long	g lat	h long	g long	g lat	h long	g long	g lat	h long	g long	g lat	h long	g long	g lat
10/15/1952	11 44	9 35	6 1	25 52	2sa47	4 56	20 23	7 52	15 12	7 23	24 16	0 5	27 6	12 48	1 8	14 44	16 44	1 52	23 22	16 51	2 23
10/25/1952	15 1	11 16	6 59	29 53	8 59	4 23	22 31	6 27	16 28	8 39	24 18D	0 8	0vi 6	17 31	0 58	17 20	21 31	1 45	24 36	19 49	2 21
11/ 4/1952	18 15	12 11	8 1	3 48	15 8	3 51	24 42	4 8	17 39	9 55	24 48	0 11	3 5	22 9	0 47	20 0	26 20	1 37	25 50	22 51	2 19
11/14/1952	21 24	12 13R	9 6	7 38	21 13	3 20	26 58	1 3	18 37	11 11	25 43	0 14	6 5	26 39	0 36	22 42	1sc11	1 30	27 4	25 55	2 18
11/24/1952	24 29	11 21	10 9	11 25	27 16	2 50	29 19	27 28	19 13	12 27	27 2	0 17	9 5	1li 0	0 24	25 27	6 5	1 22	28 18	29 0	2 17
12/ 4/1952	27 31	9 37	11 8	15 8	3 17	2 22	1 44	23 49	19 23	13 43	28 41	0 19	12 4	5 10	0 11	28 15	10 59	1 14	29 33	2 5	2 17
12/14/1952	0cn29	7 11	11 56	18 48	9 16	1 53	4 16	20 32	19 6	14 59	0pi38	0 21	15 4	9 6	0n 3	1sc 7	15 53	1 5	0 47	5 9	2 17
12/24/1952	3 24	4 24	12 28	22 26	15 14	1 26	6 54	18 1	18 27	16 14	2 50	0 24	18 2	12 46	0 19	4 1	20 48	0 56	2 1	8 10	2 18
1/ 3/1953	6 16	1 38	12 41	26 3	21 10	0 59	9 39	16 30	17 34	17 30	5 16	0 26	21 1	16 6	0 37	6 59	25 41	0 46	3 16	11 8	2 19
1/13/1953	9 4	29gm16	12 38	29 38	27 6	0 32	12 32	16 1	16 34	18 46	7 52	0 28	23 59	19 2	0 56	10 0	0sa32	0 35	4 31	13 59	2 20
1/23/1953	11 49	27 35	12 20	3 13	3 2	0 6	15 34	16 33	15 32	20 2	10 37	0 29	26 56	21 29	1 18	13 4	5 20	0 24	5 46	16 44	2 22
2/ 2/1953	14 31	26 43	11 53	6 49	8 58	0 21	18 45	17 59	14 31	21 18	13 29	0 31	29 52	23 23	1 42	16 12	10 3	0 11	7 1	19 21	2 25
2/12/1953	17 10	26 40D	11 22	10 25	14 54	0 47	22 6	20 13	13 33	22 34	16 27	0 33	2 48	24 39	2 9	19 23	14 40	0n 3	8 16	21 46	2 28
2/22/1953	19 46	27 22	10 49	14 2	20 50	1 14	25 40	23 9	12 38	23 50	19 29	0 35	5 43	25 11	2 38	22 38	19 10	0 19	9 31	23 59	2 32
3/ 4/1953	22 20	28 45	10 16	17 41	26 47	1 42	29 26	26 41	11 47	25 6	22 34	0 38	8 36	24 57	3 8	25 57	23 30	0 37	10 47	25 58	2 36
3/14/1953	24 51	0cn40	9 46	21 22	2pi46	2 10	3 28	0gm46	10 58	26 23	25 41	0 40	11 29	23 58	3 39	29 19	27 37	0 57	12 3	27 39	2 40
3/24/1953	27 20	3 4	9 17	25 7	8 46	2 38	7 45	5 20	10 11	27 39	28 48	0 42	14 20	22 19	4 9	2 45	1cp30	1 20	13 20	29 0	2 45
4/ 3/1953	29 46	5 50	8 51	28 55	14 48	3 8	12 22	10 21	9 24	28 56	1 55	0 45	17 10	20 13	4 36	6 14	5 3	1 46	14 36	29 59	2 50
4/13/1953	2 10	8 54	8 27	2 47	20 53	3 38	17 18	15 48	8 38	0ar13	4 59	0 47	19 58	17 57	4 57	9 47	8 13	2 16	15 53	0 33	2 56
4/23/1953	4 32	12 12	8 6	6 45	27 2	4 10	22 37	21 39	7 52	1 30	8 1	0 50	22 46	15 49	5 11	13 23	10 55	2 49	17 10	0 40R	3 1
5/ 3/1953	6 52	15 43	7 46	10 49	3 14	4 42	28 21	27 53	7 3	2 47	10 59	0 53	25 31	14 7	5 18	17 3	13 2	3 27	18 28	0 19	3 6
5/13/1953	9 9	19 23	7 29	14 59	9 31	5 15	4 33	4 32	6 12	4 4	13 53	0 57	28 16	13 2	5 20	20 45	14 29	4 9	19 46	29 30	3 10
5/23/1953	11 25	23 10	7 14	19 17	15 54	5 50	11 15	11 35	5 18	5 22	16 39	1 0	0sc59	12 39	5 18	24 31	15 10	4 54	21 4	28 15	3 12
6/ 2/1953	13 40	27 4	7 0	23 43	22 24	6 25	18 30	19 2	4 20	6 40	19 18	1 4	3 40	12 58	5 13	28 20	15 0R	5 42	22 23	26 40	3 13
6/12/1953	15 52	1le 2	6 47	28 19	29 1	7 1	26 18	26 53	3 18	7 58	21 48	1 9	6 20	13 57	5 6	2 11	14 1	6 29	23 42	24 51	3 11
6/22/1953	18 3	5 4	6 36	3 5	5 47	7 37	4 41	5 9	2 11	9 16	24 6	1 14	8 58	15 30	4 58	6 5	12 19	7 11	25 1	22 57	3 8
7/ 2/1953	20 13	9 9	6 26	8 2	12 43	8 13	13 37	13 49	1 0	10 35	26 11	1 19	11 34	17 34	4 50	10 0	10 13	7 44	26 21	21 9	3 2
7/12/1953	22 21	13 16	6 17	13 13	19 50	8 49	23 5	22 52	0n15	11 54	28 2	1 25	14 9	20 3	4 42	13 58	8 3	8 5	27 41	19 36	2 54
7/22/1953	24 28	17 24	6 9	18 37	27 9	9 22	2li57	2vi16	1 30	13 14	29 34	1 31	16 42	22 54	4 34	17 56	6 14	8 13	29 2	18 24	2 45
8/ 1/1953	26 33	21 33	6 2	24 16	4 41	9 54	13 7	11 58	2 45	14 33	0 46	1 38	19 14	26 2	4 26	21 56	5 6	8 9	0cp24	17 38	2 35
8/11/1953	28 37	25 42	5 55	0ta11	12 28	10 21	23 24	21 55	3 57	15 54	1 35	1 45	21 44	29 25	4 19	25 57	4 48	7 56	1 45	17 20	2 25
8/21/1953	0vi41	29 50	5 50	6 23	20 29	10 42	3sc38	2li 2	5 2	17 14	1 58	1 53	24 12	3 1	4 13	29 58	5 23	7 37	3 8	17 32	2 15
8/31/1953	2 43	3 58	5 46	12 53	28 47	10 56	13 36	12 12	5 58	18 35	1 53R	2 1	26 39	6 47	4 7	3 59	6 49	7 14	4 31	18 11	2 6
9/10/1953	4 44	8 4	5 42	19 42	7 19	11 0	23 10	22 20	6 43	19 56	1 19	2 9	29 4	10 41	4 2	8 0	9 0	6 50	5 54	19 16	1 57
9/20/1953	6 45	12 9	5 39	26 50	16 7	10 51	2sa13	2sc19	7 16	21 18	0 16	2 17	1 27	14 42	3 57	12 0	11 50	6 26	7 19	20 44	1 48
9/30/1953	8 44	16 10	5 37	4 18	25 8	10 29	10 41	12 4	7 38	22 40	28 49	2 24	3 49	18 49	3 53	15 58	15 12	6 3	8 43	22 33	1 41
10/10/1953	10 43	20 7	5 36	12 3	4le21	9 51	18 31	21 30	7 48	24 3	27 2	2 29	6 9	23 0	3 49	19 55	19 1	5 40	10 9	24 40	1 33
10/20/1953	12 41	24 0	5 35	20 6	13 43	8 55	25 45	0sa34	7 49	25 26	25 6	2 33	8 28	27 15	3 46	23 51	23 12	5 18	11 35	27 3	1 27
10/30/1953	14 39	27 47	5 35	28 24	23 11	7 41	2cp25	9 15	7 43	26 50	23 10	2 35	10 45	1sa32	3 43	27 44	27 40	4 58	13 2	29 39	1 21
11/ 9/1953	16 36	1li27	5 36	6 56	2vi41	6 10	8 32	17 32	7 31	28 14	21 24	2 35	13 1	5 51	3 40	1pi35	2aq23	4 39	14 29	2 28	1 15

Ephemeris	80			433			1221			1489			114			3018			2041		
0 Hours ET	Sappho			Eros			Amor			Attila			Kassandra			Godiva			Lancelot		
Mo Dy Year	h long	g long	g lat	h long	g long	g lat	h long	g long	g lat	h long	g long	g lat	h long	g long	g lat	h long	g long	g lat	h long	g long	g lat
11/19/1953	18 33	4 59	5 38	15 38	12 9	4 23	14 9	25 25	7 15	29 39	20 0	2 34	15 15	10 11	3 39	5 23	7 17	4 20	15 57	5 27	1 10
11/29/1953	20 29	8 20	5 41	24 27	21 32	2 22	19 20	2cp56	6 56	1 4	19 2	2 31	17 27	14 31	3 37	9 8	12 20	4 3	17 26	8 35	1 5
12/ 9/1953	22 25	11 29	5 44	3le19	0li45	0 10	24 8	10 5	6 36	2 30	18 34	2 27	19 39	18 50	3 36	12 50	17 30	3 47	18 56	11 50	1 1
12/19/1953	24 21	14 22	5 48	12 11	9 45	2 8	28 35	16 56	6 15	3 57	18 38D	2 23	21 49	23 8	3 36	16 29	22 45	3 32	20 27	15 10	0 56
12/29/1953	26 17	16 58	5 53	20 58	18 27	4 30	2 43	23 28	5 54	5 24	19 12	2 18	23 57	27 24	3 36	20 5	28 4	3 17	21 58	18 35	0 53
1/ 8/1954	28 12	19 12	5 58	29 36	26 47	6 53	6 34	29 44	5 32	6 52	20 14	2 14	26 5	1cp36	3 36	23 37	3 26	3 4	23 30	22 4	0 49
1/18/1954	0li 7	21 0	6 4	8 4	4 42	9 15	10 11	5 45	5 12	8 20	21 42	2 10	28 10	5 44	3 37	27 5	8 49	2 51	25 4	25 35	0 45
1/28/1954	2 3	22 19	6 9	16 18	12 8	11 35	13 35	11 32	4 52	9 49	23 32	2 6	0cp15	9 46	3 39	0ar30	14 12	2 39	26 38	29 7	0 42
2/ 7/1954	3 58	23 4	6 14	24 15	19 0	13 52	16 47	17 5	4 32	11 19	25 41	2 3	2 19	13 43	3 41	3 51	19 35	2 27	28 13	2 40	0 38
2/17/1954	5 54	23 10R	6 17	1li55	25 12	16 6	19 49	22 26	4 14	12 50	28 7	1 59	4 21	17 32	3 44	7 8	24 57	2 16	29 48	6 11	0 35
2/27/1954	7 50	22 35	6 17	9 17	0sa38	18 18	22 42	27 36	3 55	14 21	0ta46	1 57	6 22	21 12	3 47	10 22	0ar16	2 5	1 25	9 40	0 31
3/ 9/1954	9 46	21 19	6 13	16 19	5 10	20 28	25 27	2 33	3 38	15 53	3 38	1 55	8 22	24 42	3 51	13 32	5 34	1 55	3 3	13 7	0 28
3/19/1954	11 42	19 25	6 4	23 3	8 37	22 37	28 4	7 20	3 21	17 26	6 39	1 53	10 21	28 0	3 56	16 39	10 49	1 45	4 42	16 28	0 24
3/29/1954	13 39	17 3	5 48	29 29	10 49	24 41	0pi34	11 55	3 4	19 0	9 49	1 51	12 19	1aq 5	4 1	19 42	16 0	1 35	6 22	19 45	0 21
4/ 8/1954	15 37	14 26	5 25	5 36	11 31	26 35	2 59	16 18	2 48	20 35	13 6	1 50	14 16	3 53	4 7	22 41	21 7	1 26	8 3	22 54	0 17
4/18/1954	17 35	11 52	4 56	11 28	10 34	28 8	5 18	20 30	2 31	22 10	16 28	1 49	16 12	6 23	4 13	25 37	26 11	1 17	9 45	25 54	0 12
4/28/1954	19 33	9 38	4 23	17 3	7 59	29 5	7 32	24 30	2 15	23 47	19 54	1 49	18 7	8 32	4 20	28 30	1ta 9	1 8	11 29	28 44	0 8
5/ 8/1954	21 33	7 54	3 48	22 24	4 4	29 7	9 42	28 16	1 58	25 24	23 23	1 49	20 1	10 16	4 28	1ta20	6 3	0 59	13 13	1 21	0 3
5/18/1954	23 33	6 50	3 12	27 32	29sc35	28 4	11 48	1 48	1 40	27 2	26 55	1 49	21 55	11 33	4 36	4 7	10 52	0 51	14 59	3 44	0s 3
5/28/1954	25 34	6 27	2 39	2 27	25 23	26 1	13 50	5 4	1 22	28 41	0gm28	1 49	23 47	12 19	4 43	6 50	15 34	0 42	16 45	5 49	0 9
6/ 7/1954	27 36	6 45	2 7	7 11	22 11	23 18	15 49	8 3	1 3	0gm21	4 1	1 50	25 39	12 30	4 50	9 31	20 10	0 33	18 34	7 33	0 15
6/17/1954	29 39	7 39	1 38	11 45	20 21	20 19	17 45	10 42	0 43	2 2	7 35	1 51	27 30	12 6	4 56	12 9	24 40	0 24	20 23	8 54	0 23
6/27/1954	1 43	9 7	1 12	16 9	19 50	17 24	19 38	12 58	0 20	3 44	11 7	1 52	29 20	11 7	4 59	14 44	29 2	0 15	22 13	9 48	0 31
7/ 7/1954	3 48	11 3	0 48	20 26	20 30	14 43	21 29	14 49	0s 4	5 27	14 37	1 54	1 10	9 36	4 59	17 16	3 15	0 5	24 5	10 12	0 40
7/17/1954	5 55	13 24	0 26	24 35	22 7	12 20	23 17	16 9	0 30	7 12	18 5	1 56	2 59	7 39	4 55	19 47	7 20	0s 4	25 58	10 3	0 50
7/27/1954	8 3	16 7	0 7	28 37	24 30	10 16	25 4	16 56	1 0	8 57	21 28	1 58	4 47	5 29	4 47	22 14	11 13	0 15	27 53	9 21	1 0
8/ 6/1954	10 13	19 8	0 12	2 34	27 29	8 28	26 49	17 3R	1 32	10 43	24 46	2 1	6 35	3 17	4 34	24 40	14 55	0 26	29 49	8 7	1 11
8/16/1954	12 24	22 26	0 29	6 26	0sa57	6 55	28 32	16 29	2 7	12 31	27 57	2 4	8 22	1 17	4 17	27 3	18 23	0 37	1 46	6 28	1 21
8/26/1954	14 37	25 57	0 45	10 14	4 47	5 35	0ar14	15 9	2 45	14 19	1cn 1	2 8	10 8	29cp41	3 59	29 24	21 36	0 50	3 45	4 32	1 31
9/ 5/1954	16 52	29 41	1 0	13 58	8 55	4 25	1 54	13 7	3 24	16 9	3 53	2 12	11 55	28 35	3 38	1 43	24 30	1 4	5 45	2 32	1 40
9/15/1954	19 9	3 36	1 14	17 39	13 19	3 24	3 34	10 28	4 2	18 0	6 34	2 17	13 40	28 4	3 18	4 0	27 2	1 19	7 46	0 41	1 47
9/25/1954	21 28	7 41	1 28	21 18	17 54	2 30	5 12	7 26	4 37	19 52	8 59	2 22	15 26	28 8D	2 58	6 15	29 9	1 35	9 49	29 10	1 52
10/ 5/1954	23 49	11 54	1 42	24 55	22 41	1 41	6 50	4 18	5 7	21 45	11 6	2 27	17 10	28 46	2 40	8 29	0 47	1 53	11 54	28 10	1 56
10/15/1954	26 12	16 15	1 55	28 30	27 38	0 58	8 27	1 23	5 30	23 40	12 51	2 33	18 55	29 54	2 23	10 41	1 51	2 12	14 0	27 45	1 58
10/25/1954	28 38	20 43	2 9	2 6	2 42	0 19	10 4	28pi56	5 47	25 36	14 10	2 40	20 39	1 29	2 7	12 51	2 16	2 33	16 7	27 56	1 59
11/ 4/1954	1sa 7	25 17	2 22	5 41	7 54	0n16	11 40	27 9	5 57	27 33	14 59	2 46	22 23	3 28	1 52	15 0	1 59	2 55	18 16	28 44	2 0
11/14/1954	3 38	29 58	2 36	9 16	13 14	0 49	13 16	26 7	6 2	29 31	15 16	2 53	24 7	5 48	1 39	17 8	0 58	3 18	20 26	0pi 6	2 0
11/24/1954	6 12	4 43	2 50	12 53	18 39	1 19	14 51	25 48	6 4	1 30	14 56	2 58	25 50	8 24	1 27	19 14	29gm17	3 40	22 37	1 57	1 59
12/ 4/1954	8 49	9 33	3 4	16 31	24 12	1 47	16 27	26 10	6 4	3 31	14 1	3 3	27 34	11 16	1 17	21 19	27 3	3 59	24 50	4 15	1 59
12/14/1954	11 29	14 27	3 18	20 12	29 50	2 13	18 3	27 9	6 3	5 33	12 33	3 5	29 17	14 20	1 7	23 23	24 29	4 15	27 5	6 56	1 58

Ephemeris	80			433			1221			1489			114			3018			2041		
0 Hours ET	Sappho			Eros			Amor			Attila			Kassandra			Godiva			Lancelot		
Mo Dy Year	h long	g long	g lat	h long	g long	g lat	h long	g long	g lat	h long	g long	g lat	h long	g long	g lat	h long	g long	g lat	h long	g long	g lat
12/24/1954	14 12	19 25	3 34	23 55	5 34	2 38	19 39	28 39	6 1	7 37	10 40	3 5	1 0	17 35	0 57	25 25	21 53	4 26	29 21	9 57	1 58
1/ 3/1955	16 59	24 26	3 50	27 42	11 24	3 2	21 15	0ar37	5 59	9 41	8 34	3 2	2 43	20 58	0 49	27 27	19 33	4 32	1 38	13 15	1 57
1/13/1955	19 49	29 29	4 6	1pi33	17 20	3 24	22 52	2 57	5 58	11 47	6 31	2 56	4 25	24 27	0 41	29 28	17 43	4 33	3 57	16 48	1 57
1/23/1955	22 43	4 35	4 24	5 29	23 22	3 46	24 29	5 36	5 58	13 54	4 45	2 47	6 8	28 3	0 33	1 27	16 31	4 30	6 17	20 33	1 57
2/ 2/1955	25 41	9 42	4 42	9 31	29 31	4 6	26 7	8 32	5 58	16 2	3 26	2 36	7 51	1pi42	0 26	3 26	16 2	4 25	8 38	24 28	1 57
2/12/1955	28 43	14 50	5 2	13 39	5 45	4 26	27 46	11 41	5 58	18 12	2 43	2 25	9 34	5 25	0 19	5 25	16 14	4 19	11 0	28 32	1 57
2/22/1955	1 49	19 59	5 23	17 54	12 7	4 45	29 26	15 1	6 0	20 23	2 39D	2 13	11 17	9 9	0 12	7 22	17 3	4 12	13 24	2 43	1 58
3/ 4/1955	5 0	25 8	5 45	22 18	18 36	5 3	1 7	18 32	6 3	22 35	3 13	2 1	13 0	12 54	0 6	9 19	18 25	4 5	15 49	7 0	1 58
3/14/1955	8 14	0aq16	6 9	26 50	25 13	5 20	2 49	22 10	6 6	24 48	4 21	1 50	14 43	16 40	0s 1	11 15	20 15	3 58	18 15	11 21	1 59
3/24/1955	11 33	5 22	6 34	1ar33	1ar57	5 36	4 33	25 55	6 11	27 3	6 1	1 39	16 26	20 24	0 8	13 11	22 30	3 52	20 42	15 46	2 0
4/ 3/1955	14 57	10 26	7 0	6 27	8 51	5 50	6 18	29 46	6 16	29 18	8 9	1 30	18 9	24 7	0 15	15 7	25 4	3 46	23 10	20 14	2 1
4/13/1955	18 25	15 26	7 29	11 33	15 53	6 2	8 6	3 42	6 23	1 35	10 40	1 21	19 53	27 46	0 22	17 2	27 54	3 41	25 39	24 43	2 3
4/23/1955	21 58	20 21	7 59	16 52	23 6	6 12	9 55	7 42	6 30	3 53	13 31	1 13	21 37	1ar22	0 29	18 56	0cn58	3 37	28 9	29 14	2 4
5/ 3/1955	25 36	25 10	8 32	22 26	0ta29	6 20	11 46	11 47	6 39	6 11	16 40	1 5	23 21	4 54	0 37	20 51	4 14	3 33	0ta39	3 45	2 6
5/13/1955	29 18	29 50	9 6	28 16	8 3	6 24	13 40	15 54	6 48	8 31	20 3	0 58	25 5	8 19	0 45	22 45	7 38	3 30	3 10	8 15	2 8
5/23/1955	3 4	4 20	9 42	4 23	15 48	6 24	15 36	20 5	6 59	10 52	23 38	0 52	26 50	11 38	0 54	24 39	11 11	3 28	5 42	12 44	2 10
6/ 2/1955	6 55	8 36	10 21	10 47	23 45	6 20	17 36	24 19	7 12	13 14	27 23	0 46	28 35	14 48	1 3	26 33	14 49	3 26	8 15	17 11	2 12
6/12/1955	10 50	12 35	11 2	17 30	1gm53	6 11	19 38	28 35	7 25	15 36	1le17	0 40	0ar21	17 48	1 13	28 26	18 33	3 24	10 48	21 36	2 15
6/22/1955	14 50	16 13	11 44	24 31	10 14	5 56	21 44	2 54	7 40	18 0	5 19	0 35	2 7	20 36	1 24	0le20	22 21	3 23	13 21	25 57	2 17
7/ 2/1955	18 53	19 23	12 28	1gm52	18 45	5 35	23 54	7 15	7 57	20 24	9 26	0 29	3 54	23 10	1 36	2 14	26 12	3 23	15 54	0gm13	2 20
7/12/1955	23 0	22 1	13 12	9 32	27 27	5 8	26 8	11 38	8 16	22 48	13 39	0 24	5 41	25 28	1 50	4 8	0le 5	3 23	18 28	4 24	2 24
7/22/1955	27 10	23 57	13 54	17 29	6 18	4 34	28 27	16 4	8 36	25 14	17 56	0 20	7 28	27 26	2 4	6 2	4 1	3 24	21 2	8 28	2 27
8/ 1/1955	1pi22	25 5	14 32	25 43	15 18	3 53	0gm51	20 32	8 59	27 40	22 16	0 15	9 16	29 1	2 20	7 56	7 57	3 25	23 35	12 25	2 31
8/11/1955	5 37	25 20	15 0	4cn11	24 24	3 7	3 20	25 1	9 24	0vi 6	26 39	0 10	11 5	0ta10	2 38	9 51	11 54	3 26	26 9	16 11	2 35
8/21/1955	9 55	24 39	15 14	12 50	3le35	2 16	5 56	29 33	9 52	2 33	1vi 4	0 6	12 54	0 48	2 57	11 46	15 51	3 28	28 42	19 46	2 40
8/31/1955	14 13	23 8	15 7	21 37	12 49	1 21	8 38	4 7	10 22	5 0	5 30	0 1	14 44	0 53R	3 17	13 41	19 48	3 31	1gm15	23 7	2 45
9/10/1955	18 33	21 3	14 34	0le28	22 4	0 23	11 28	8 42	10 56	7 28	9 57	0 3	16 35	0 21	3 38	15 37	23 42	3 34	3 48	26 12	2 50
9/20/1955	22 53	18 47	13 35	9 21	1vi17	0 35	14 26	13 19	11 34	9 55	14 24	0 8	18 27	29 14	3 59	17 33	27 35	3 38	6 20	28 56	2 56
9/30/1955	27 14	16 49	12 13	18 9	10 28	1 33	17 34	17 57	12 15	12 23	18 49	0 12	20 19	27 34	4 18	19 30	1vi24	3 42	8 52	1 17	3 2
10/10/1955	1ar34	15 32	10 38	26 51	19 33	2 30	20 52	22 36	13 1	14 51	23 13	0 17	22 12	25 29	4 35	21 27	5 9	3 48	11 22	3 11	3 8
10/20/1955	5 53	15 8	8 58	5 23	28 31	3 23	24 21	27 16	13 52	17 19	27 34	0 22	24 6	23 12	4 48	23 25	8 48	3 54	13 53	4 33	3 14
10/30/1955	10 10	15 42	7 22	13 42	7 20	4 13	28 3	1le55	14 48	19 47	1li51	0 27	26 1	20 57	4 57	25 24	12 21	4 0	16 22	5 19	3 20
11/ 9/1955	14 26	17 9	5 53	21 45	15 59	4 58	1cn58	6 34	15 50	22 14	6 3	0 32	27 57	18 58	5 0	27 23	15 44	4 8	18 50	5 25R	3 25
11/19/1955	18 39	19 25	4 32	29 30	24 26	5 38	6 10	11 11	16 57	24 41	10 8	0 38	29 54	17 28	4 59	29 24	18 57	4 17	21 18	4 51	3 29
11/29/1955	22 50	22 21	3 22	6 58	2sc40	6 14	10 39	15 45	18 10	27 8	14 6	0 44	1 52	16 34	4 55	1 25	21 56	4 27	23 44	3 38	3 30
12/ 9/1955	26 58	25 49	2 20	14 7	10 41	6 44	15 28	20 16	19 27	29 35	17 54	0 51	3 50	16 19	4 49	3 27	24 39	4 38	26 10	1 55	3 27
12/19/1955	1ta 3	29 44	1 26	20 57	18 28	7 10	20 39	24 43	20 47	2 1	21 30	0 58	5 50	16 43	4 41	5 31	27 1	4 50	28 34	29gm53	3 21
12/29/1955	5 4	4 0	0 39	27 28	25 59	7 32	26 14	29 4	22 6	4 26	24 51	1 6	7 51	17 43	4 33	7 35	29 0	5 4	0cn57	27 47	3 12
1/ 8/1956	9 2	8 33	0s 2	3 42	3sa16	7 50	2le16	3 20	23 17	6 51	27 56	1 14	9 54	19 16	4 25	9 41	0li30	5 18	3 19	25 55	2 59
1/18/1956	12 56	13 18	0 38	9 38	10 18	8 4	8 47	7 33	24 11	9 15	0sc41	1 23	11 57	21 17	4 17	11 48	1 26	5 33	5 39	24 29	2 45

Ephemeris	80			433			1221			1489			114			3018			2041		
0 Hours ET	Sappho			Eros			Amor			Attila			Kassandra			Godiva			Lancelot		
Mo Dy Year	h long	g long	g lat	h long	g long	g lat	h long	g long	g lat	h long	g long	g lat	h long	g long	g lat	h long	g long	g lat	h long	g long	g lat
1/28/1956	16 45	18 14	1 9	15 19	17 3	8 16	15 50	11 49	24 29	11 39	3 1	1 33	14 2	23 43	4 10	13 56	1 43	5 48	7 59	23 38	2 29
2/ 7/1956	20 31	23 17	1 37	20 44	23 33	8 24	23 27	16 20	23 43	14 1	4 55	1 44	16 8	26 30	4 4	16 5	1 19	6 2	10 16	23 26	2 14
2/17/1956	24 13	28 25	2 2	25 56	29 47	8 31	1vi38	21 25	20 59	16 23	6 17	1 56	18 15	29 35	3 58	18 17	0 11	6 13	12 33	23 53	1 59
2/27/1956	27 50	3 37	2 24	0sa55	5 43	8 36	10 23	27 46	14 46	18 44	7 3	2 9	20 24	2 55	3 53	20 29	28 23	6 20	14 48	24 54	1 45
3/ 8/1956	1gm24	8 52	2 43	5 43	11 21	8 38	19 40	6 17	2 41	21 4	7 12R	2 21	22 34	6 27	3 49	22 44	26 4	6 20	17 2	26 27	1 33
3/18/1956	4 53	14 8	3 1	10 20	16 40	8 39	29 25	17 59	16 32	23 23	6 42	2 34	24 45	10 11	3 45	25 0	23 28	6 13	19 14	28 27	1 21
3/28/1956	8 18	19 25	3 17	14 47	21 39	8 38	9 31	3sc 6	36 41	25 41	5 35	2 46	26 58	14 3	3 42	27 18	20 55	5 59	21 25	0cn49	1 11
4/ 7/1956	11 39	24 40	3 32	19 6	26 14	8 36	19 47	19 26	49 34	27 58	3 59	2 56	29 12	18 3	3 39	29 37	18 42	5 38	23 34	3 30	1 2
4/17/1956	14 56	29 56	3 45	23 17	0aq24	8 31	0sc 3	3sa10	55 20	0sc13	2 4	3 4	1 28	22 10	3 37	1 59	17 3	5 13	25 42	6 27	0 53
4/27/1956	18 9	5 9	3 58	27 22	4 4	8 24	10 9	12 23	57 6	2 28	0 4	3 9	3 45	26 21	3 35	4 23	16 6	4 46	27 49	9 36	0 45
5/ 7/1956	21 18	10 20	4 10	1cp21	7 9	8 13	19 53	17 21	56 47	4 41	28 13	3 10	6 4	0gm38	3 34	6 49	15 55	4 18	29 54	12 56	0 38
5/17/1956	24 24	15 30	4 21	5 14	9 33	7 58	29 8	19 12	55 7	6 54	26 44	3 9	8 25	4 58	3 33	9 17	16 27	3 51	1 58	16 24	0 32
5/27/1956	27 25	20 36	4 32	9 3	11 7	7 36	7 49	19 16R	52 18	9 5	25 45	3 5	10 47	9 21	3 33	11 48	17 39	3 26	4 0	19 58	0 26
6/ 6/1956	0cn24	25 39	4 42	12 48	11 43	7 7	15 52	18 35	48 31	11 15	25 21	3 0	13 10	13 46	3 32	14 21	19 26	3 2	6 1	23 37	0 20
6/16/1956	3 18	0cn39	4 53	16 30	11 11	6 26	23 19	17 58	43 59	13 23	25 32	2 54	15 36	18 13	3 32	16 57	21 45	2 40	8 0	27 21	0 15
6/26/1956	6 10	5 36	5 3	20 9	9 26	5 32	0cp11	17 53D	39 3	15 30	26 16	2 47	18 3	22 41	3 33	19 35	24 30	2 20	9 58	1le 6	0 10
7/ 6/1956	8 58	10 29	5 13	23 47	6 29	4 21	6 29	18 29	34 6	17 37	27 31	2 40	20 32	27 10	3 34	22 16	27 38	2 2	11 55	4 54	0 5
7/16/1956	11 43	15 18	5 23	27 23	2 36	2 56	12 17	19 47	29 25	19 41	29 13	2 34	23 2	1cn40	3 35	25 0	1li 7	1 45	13 50	8 43	0 1
7/26/1956	14 25	20 2	5 34	0aq58	28cp20	1 21	17 37	21 42	25 12	21 45	1 18	2 28	25 34	6 8	3 36	27 47	4 52	1 29	15 44	12 32	0 4
8/ 5/1956	17 4	24 41	5 45	4 33	24 18	0n14	22 33	24 7	21 30	23 47	3 43	2 22	28 8	10 36	3 38	0sc37	8 53	1 14	17 36	16 20	0 8
8/15/1956	19 40	29 15	5 56	8 9	21 5	1 41	27 6	26 57	18 20	25 48	6 25	2 17	0cn43	15 2	3 39	3 30	13 7	1 0	19 28	20 8	0 12
8/25/1956	22 14	3 43	6 9	11 45	19 2	2 55	1aq21	0cp 8	15 39	27 48	9 21	2 13	3 20	19 25	3 42	6 26	17 33	0 47	21 18	23 53	0 17
9/ 4/1956	24 45	8 4	6 22	15 23	18 12	3 56	5 18	3 33	13 23	29 47	12 29	2 8	5 59	23 44	3 44	9 26	22 10	0 35	23 6	27 35	0 21
9/14/1956	27 14	12 18	6 36	19 3	18 32	4 44	9 0	7 11	11 28	1 44	15 47	2 5	8 40	27 59	3 47	12 28	26 57	0 22	24 54	1vi13	0 26
9/24/1956	29 40	16 23	6 51	22 45	19 52	5 22	12 28	10 59	9 51	3 40	19 13	2 1	11 22	2 7	3 50	15 35	1sc52	0 11	26 40	4 47	0 30
10/ 4/1956	2 4	20 18	7 7	26 31	22 3	5 52	15 44	14 54	8 29	5 35	22 45	1 58	14 5	6 7	3 53	18 45	6 56	0n 1	28 25	8 15	0 35
10/14/1956	4 26	24 2	7 25	0pi21	24 57	6 16	18 50	18 56	7 18	7 28	26 23	1 56	16 50	9 56	3 57	21 58	12 7	0 12	0vi 9	11 36	0 40
10/24/1956	6 46	27 32	7 45	4 15	28 27	6 36	21 46	23 3	6 18	9 21	0sa 5	1 54	19 37	13 33	4 0	25 15	17 25	0 23	1 52	14 48	0 45
11/ 3/1956	9 4	0vi46	8 7	8 15	2 27	6 51	24 33	27 14	5 25	11 12	3 51	1 52	22 25	16 54	4 4	28 35	22 49	0 34	3 34	17 50	0 51
11/13/1956	11 20	3 41	8 31	12 20	6 52	7 4	27 13	1aq27	4 40	13 2	7 38	1 50	25 14	19 56	4 8	1 59	28 19	0 45	5 14	20 40	0 57
11/23/1956	13 34	6 14	8 58	16 33	11 41	7 15	29 45	5 43	4 0	14 51	11 26	1 49	28 5	22 34	4 12	5 27	3 54	0 56	6 54	23 16	1 4
12/ 3/1956	15 47	8 21	9 27	20 54	16 49	7 23	2 12	10 1	3 25	16 39	15 15	1 48	0le57	24 43	4 15	8 58	9 34	1 7	8 32	25 35	1 11
12/13/1956	17 58	9 57	9 58	25 23	22 15	7 29	4 33	14 20	2 54	18 26	19 2	1 47	3 51	26 17	4 17	12 33	15 18	1 18	10 10	27 34	1 19
12/23/1956	20 8	10 58	10 30	0ar 3	27 58	7 34	6 49	18 39	2 26	20 11	22 47	1 47	6 45	27 11	4 18	16 10	21 4	1 29	11 46	29 11	1 28
1/ 2/1957	22 16	11 18	11 3	4 53	3 57	7 36	9 0	22 58	2 1	21 56	26 30	1 47	9 40	27 21R	4 16	19 52	26 54	1 41	13 22	0li22	1 37
1/12/1957	24 23	10 55	11 35	9 55	10 11	7 37	11 7	27 17	1 38	23 39	0cp 8	1 47	12 37	26 44	4 10	23 36	2cp46	1 52	14 56	1 4	1 47
1/22/1957	26 28	9 47	12 2	15 10	16 40	7 36	13 11	1pi35	1 16	25 22	3 41	1 48	15 34	25 22	4 0	27 23	8 38	2 4	16 30	1 14	1 57
2/ 1/1957	28 33	7 58	12 22	20 39	23 23	7 32	15 11	5 51	0 56	27 3	7 8	1 49	18 32	23 25	3 44	1cp13	14 32	2 16	18 3	0 52	2 7
2/11/1957	0vi36	5 37	12 32	26 24	0ar19	7 26	17 8	10 6	0 38	28 43	10 27	1 50	21 30	21 8	3 22	5 5	20 25	2 28	19 34	29vi58	2 17
2/21/1957	2 38	2 57	12 27	2ta25	7 30	7 16	19 2	14 18	0 20	0cp23	13 36	1 52	24 29	18 51	2 55	8 59	26 17	2 40	21 6	28 35	2 27

Ephemeris	80			433			1221			1489			114			3018			2041		
0 Hours ET	Sappho			Eros			Amor			Attila			Kassandra			Godiva			Lancelot		
Mo Dy Year	h long	g long	g lat	h long	g long	g lat	h long	g long	g lat	h long	g long	g lat	h long	g long	g lat	h long	g long	g lat	h long	g long	g lat
3/ 3/1957	4 40	0 18	12 9	8 43	14 54	7 3	20 54	18 27	0 3	2 1	16 35	1 54	27 29	16 55	2 26	12 56	2aq 8	2 53	22 36	26 50	2 35
3/13/1957	6 40	27 57	11 40	15 20	22 32	6 46	22 43	22 34	0 14	3 39	19 20	1 56	0vi29	15 34	1 56	16 53	7 55	3 6	24 5	24 53	2 42
3/23/1957	8 40	26 8	11 1	22 16	0ta23	6 24	24 30	26 36	0 30	5 16	21 51	1 59	3 29	14 59	1 26	20 53	13 38	3 19	25 34	22 54	2 46
4/ 2/1957	10 39	24 58	10 18	29 30	8 28	5 58	26 16	0ar35	0 46	6 51	24 5	2 2	6 29	15 10D	0 59	24 53	19 16	3 33	27 2	21 4	2 48
4/12/1957	12 38	24 30	9 33	7 4	16 44	5 26	27 59	4 28	1 3	8 26	26 0	2 5	9 29	16 6	0 35	28 53	24 48	3 46	28 29	19 34	2 48
4/22/1957	14 35	24 42	8 49	14 55	25 12	4 49	29 42	8 17	1 20	10 0	27 32	2 8	12 29	17 42	0 13	2 54	0pi12	4 0	29 55	18 30	2 46
5/ 2/1957	16 33	25 31	8 7	23 4	3gm51	4 8	1 23	11 59	1 37	11 34	28 40	2 12	15 29	19 52	0n 6	6 54	5 26	4 15	1 21	17 54	2 44
5/12/1957	18 29	26 53	7 28	1cn27	12 40	3 21	3 3	15 35	1 56	13 6	29 20	2 16	18 28	22 31	0 23	10 54	10 29	4 30	2 46	17 49D	2 40
5/22/1957	20 26	28 42	6 53	10 3	21 37	2 31	4 41	19 3	2 15	14 38	29 31	2 19	21 27	25 35	0 38	14 53	15 18	4 45	4 11	18 12	2 36
6/ 1/1957	22 22	0vi55	6 21	18 48	0cn40	1 37	6 20	22 22	2 36	16 9	29 12	2 22	24 25	28 59	0 52	18 50	19 51	5 1	5 35	19 2	2 33
6/11/1957	24 18	3 27	5 51	27 38	9 48	0 41	7 57	25 32	2 58	17 39	28 23	2 24	27 22	2 40	1 4	22 46	24 5	5 17	6 58	20 16	2 29
6/21/1957	26 13	6 17	5 25	6 30	18 59	0s16	9 34	28 29	3 22	19 9	27 7	2 25	0li19	6 34	1 15	26 40	27 56	5 34	8 20	21 50	2 25
7/ 1/1957	28 9	9 20	5 1	15 21	28 11	1 12	11 10	1ta13	3 48	20 37	25 29	2 24	3 15	10 40	1 25	0pi32	1ar21	5 51	9 43	23 42	2 22
7/11/1957	0li 4	12 35	4 39	24 5	7 22	2 5	12 46	3 41	4 17	22 5	23 38	2 22	6 10	14 55	1 34	4 21	4 12	6 8	11 4	25 49	2 19
7/21/1957	2 0	16 0	4 18	2vi40	16 29	2 56	14 22	5 51	4 49	23 33	21 43	2 18	9 4	19 18	1 42	8 7	6 27	6 25	12 25	28 10	2 17
7/31/1957	3 55	19 34	4 0	11 3	25 33	3 41	15 57	7 38	5 25	25 0	19 55	2 12	11 56	23 48	1 50	11 50	7 57	6 41	13 46	0li41	2 15
8/10/1957	5 51	23 14	3 43	19 11	4 29	4 22	17 33	8 59	6 4	26 26	18 23	2 5	14 48	28 22	1 57	15 30	8 38	6 55	15 6	3 22	2 13
8/20/1957	7 47	27 1	3 27	27 3	13 18	4 56	19 9	9 49	6 48	27 52	17 14	1 57	17 38	3 1	2 4	19 7	8 25R	7 5	16 25	6 10	2 12
8/30/1957	9 43	0li53	3 12	4 37	21 58	5 24	20 45	10 3	7 35	29 17	16 33	1 49	20 27	7 43	2 10	22 40	7 19	7 9	17 45	9 5	2 11
9/ 9/1957	11 39	4 50	2 57	11 51	0li27	5 47	22 22	9 37	8 25	0aq41	16 21	1 41	23 14	12 27	2 17	26 10	5 28	7 4	19 3	12 4	2 10
9/19/1957	13 36	8 50	2 44	18 47	8 46	6 3	23 59	8 28	9 18	2 5	16 38	1 33	26 0	17 13	2 23	29 36	3 7	6 50	20 22	15 7	2 10
9/29/1957	15 34	12 53	2 31	25 25	16 54	6 15	25 37	6 34	10 9	3 29	17 22	1 26	28 45	22 0	2 29	2 58	0 37	6 25	21 40	18 13	2 10
10/ 9/1957	17 32	16 58	2 18	1sc44	24 50	6 21	27 15	4 2	10 55	4 52	18 31	1 19	1 27	26 48	2 35	6 17	28 22	5 52	22 57	21 20	2 11
10/19/1957	19 30	21 4	2 6	7 46	2sc35	6 23	28 55	1 1	11 33	6 14	20 2	1 12	4 9	1sc36	2 41	9 32	26 41	5 14	24 14	24 28	2 12
10/29/1957	21 29	25 12	1 54	13 32	10 9	6 21	0ta35	27 49	11 59	7 36	21 52	1 7	6 48	6 22	2 47	12 44	25 46	4 35	25 31	27 36	2 13
11/ 8/1957	23 30	29 20	1 42	19 2	17 32	6 17	2 17	24 45	12 10	8 58	24 0	1 1	9 26	11 8	2 53	15 52	25 41D	3 57	26 48	0sc41	2 15
11/18/1957	25 30	3 27	1 30	24 18	24 44	6 9	4 1	22 9	12 9	10 19	26 21	0 56	12 3	15 51	3 0	18 57	26 23	3 22	28 4	3 44	2 18
11/28/1957	27 32	7 32	1 18	29 21	1sa46	5 59	5 45	20 12	11 58	11 40	28 55	0 52	14 38	20 31	3 7	21 58	27 48	2 51	29 20	6 43	2 21
12/ 8/1957	29 35	11 35	1 5	4 12	8 39	5 48	7 32	19 3	11 39	13 0	1 39	0 48	17 11	25 7	3 14	24 55	29 49	2 23	0sc35	9 36	2 24
12/18/1957	1 39	15 35	0 52	8 53	15 22	5 35	9 20	18 40	11 17	14 20	4 31	0 44	19 42	29 38	3 21	27 49	2 21	1 58	1 51	12 21	2 28
12/28/1957	3 45	19 30	0 38	13 23	21 57	5 20	11 11	19 3	10 53	15 40	7 30	0 41	22 12	4 4	3 30	0ta40	5 17	1 36	3 6	14 58	2 33
1/ 7/1958	5 51	23 19	0 23	17 45	28 23	5 4	13 4	20 7	10 30	16 59	10 34	0 38	24 40	8 22	3 38	3 28	8 34	1 17	4 21	17 25	2 38
1/17/1958	7 59	27 0	0 6	21 58	4 41	4 48	14 59	21 46	10 8	18 18	13 41	0 35	27 7	12 33	3 48	6 13	12 8	1 0	5 35	19 38	2 44
1/27/1958	10 8	0sa32	0 11	26 5	10 51	4 30	16 58	23 56	9 48	19 37	16 51	0 32	29 31	16 34	3 58	8 55	15 54	0 44	6 50	21 36	2 51
2/ 6/1958	12 19	3 52	0 31	0cp 6	16 53	4 11	18 59	26 32	9 31	20 56	20 2	0 29	1 55	20 23	4 9	11 34	19 51	0 31	8 4	23 17	2 58
2/16/1958	14 32	6 58	0 53	4 1	22 49	3 51	21 4	29 30	9 15	22 14	23 12	0 27	4 16	24 0	4 21	14 10	23 56	0 19	9 18	24 37	3 6
2/26/1958	16 47	9 46	1 17	7 51	28 36	3 29	23 13	2 49	9 1	23 32	26 22	0 24	6 36	27 21	4 35	16 44	28 7	0 7	10 32	25 35	3 15
3/ 8/1958	19 3	12 14	1 45	11 37	4 17	3 7	25 25	6 24	8 49	24 49	29 29	0 22	8 55	0cp25	4 49	19 15	2 23	0s 3	11 46	26 7	3 24
3/18/1958	21 22	14 17	2 17	15 20	9 50	2 43	27 42	10 14	8 38	26 7	2 32	0 20	11 11	3 8	5 5	21 44	6 42	0 12	13 0	26 12R	3 33
3/28/1958	23 42	15 50	2 52	19 0	15 16	2 17	0gm 4	14 18	8 30	27 24	5 31	0 17	13 27	5 28	5 22	24 10	11 4	0 21	14 14	25 49	3 41

Ephemeris	80			433			1221			1489			114			3018			2041		
0 Hours ET	Sappho			Eros			Amor			Attila			Kassandra			Godiva			Lancelot		
Mo Dy Year	h long	g long	g lat	h long	g long	g lat	h long	g long	g lat	h long	g long	g lat	h long	g long	g lat	h long	g long	g lat	h long	g long	g lat
4/ 7/1958	26 5	16 48	3 32	22 38	20 34	1 49	2 32	18 34	8 22	28 41	8 24	0 15	15 41	7 20	5 40	26 34	15 27	0 30	15 27	24 58	3 49
4/17/1958	28 31	17 6	4 17	26 15	25 44	1 18	5 5	23 2	8 15	29 58	11 9	0 13	17 53	8 41	5 59	28 56	19 50	0 38	16 41	23 42	3 56
4/27/1958	0sa58	16 41	5 5	29 50	0pi45	0 44	7 45	27 40	8 9	1 15	13 46	0 10	20 4	9 28	6 19	1 16	24 13	0 46	17 54	22 6	4 0
5/ 7/1958	3 29	15 30	5 57	3 25	5 37	0 7	10 32	2 30	8 4	2 31	16 12	0 8	22 14	9 37R	6 38	3 34	28 36	0 53	19 7	20 17	4 2
5/17/1958	6 2	13 38	6 47	7 1	10 19	0 35	13 28	7 30	8 0	3 48	18 26	0 5	24 22	9 8	6 56	5 50	2 57	1 1	20 21	18 25	4 2
5/27/1958	8 39	11 15	7 35	10 37	14 49	1 21	16 32	12 40	7 55	5 4	20 25	0 2	26 29	8 0	7 11	8 5	7 16	1 8	21 34	16 38	3 58
6/ 6/1958	11 18	8 36	8 15	14 14	19 5	2 15	19 46	18 2	7 50	6 20	22 9	0s 1	28 35	6 20	7 21	10 18	11 33	1 16	22 47	15 6	3 53
6/16/1958	14 1	6 1	8 45	17 53	23 5	3 15	23 11	23 36	7 45	7 36	23 34	0 5	0cp39	4 16	7 24	12 29	15 48	1 24	24 1	13 54	3 45
6/26/1958	16 47	3 51	9 3	21 35	26 45	4 25	26 48	29 22	7 39	8 52	24 37	0 8	2 42	2 1	7 21	14 38	19 58	1 31	25 14	13 7	3 37
7/ 6/1958	19 36	2 20	9 12	25 19	0ar 0	5 46	0cn39	5 21	7 32	10 8	25 17	0 12	4 45	29sa52	7 10	16 46	24 5	1 39	26 27	12 48	3 27
7/16/1958	22 29	1 36	9 12	29 8	2 45	7 21	4 45	11 35	7 23	11 24	25 31	0 17	6 45	27 59	6 53	18 53	28 8	1 48	27 41	12 55	3 18
7/26/1958	25 26	1 42D	9 6	3 0	4 51	9 11	9 8	18 4	7 11	12 40	25 18	0 21	8 45	26 36	6 32	20 59	2 5	1 57	28 54	13 29	3 9
8/ 5/1958	28 27	2 36	8 57	6 58	6 8	11 18	13 49	24 50	6 57	13 56	24 37	0 26	10 44	25 46	6 8	23 3	5 56	2 6	0sa 8	14 26	3 0
8/15/1958	1cp32	4 13	8 45	11 2	6 25R	13 44	18 52	1le55	6 39	15 12	23 30	0 31	12 42	25 34	5 44	25 6	9 40	2 16	1 22	15 45	2 52
8/25/1958	4 42	6 29	8 33	15 13	5 27	16 23	24 18	9 19	6 16	16 28	22 1	0 36	14 38	25 56	5 20	27 8	13 15	2 27	2 35	17 23	2 44
9/ 4/1958	7 55	9 19	8 20	19 31	3 10	19 6	0le11	17 5	5 48	17 44	20 15	0 41	16 34	26 52	4 56	29 9	16 41	2 38	3 49	19 18	2 37
9/14/1958	11 13	12 38	8 7	23 57	29pi35	21 37	6 31	25 12	5 14	19 0	18 22	0 45	18 29	28 17	4 35	1 10	19 56	2 51	5 4	21 27	2 31
9/24/1958	14 36	16 23	7 54	28 33	25 10	23 31	13 23	3vi42	4 33	20 16	16 31	0 49	20 23	0cp 8	4 15	3 9	22 56	3 5	6 18	23 49	2 25
10/ 4/1958	18 3	20 31	7 42	3 20	20 40	24 34	20 47	12 35	3 44	21 32	14 52	0 52	22 16	2 22	3 57	5 7	25 41	3 20	7 32	26 22	2 20
10/14/1958	21 34	24 57	7 30	8 18	16 56	24 42	28 46	21 50	2 48	22 48	13 32	0 55	24 8	4 54	3 41	7 5	28 6	3 37	8 47	29 5	2 16
10/24/1958	25 10	29 41	7 18	13 29	14 34	24 6	7 19	1li26	1 45	24 5	12 37	0 57	25 59	7 42	3 26	9 3	0le 9	3 55	10 2	1 54	2 12
11/ 3/1958	28 51	4 39	7 6	18 54	13 49	23 1	16 26	11 19	0 36	25 21	12 10	0 58	27 50	10 44	3 12	10 59	1 45	4 15	11 17	4 51	2 8
11/13/1958	2 36	9 52	6 54	24 33	14 37	21 42	26 2	21 25	0n36	26 38	12 12D	0 59	29 40	13 57	3 0	12 55	2 50	4 37	12 32	7 52	2 5
11/23/1958	6 26	15 15	6 42	0ta29	16 49	20 15	6 1	1sc40	1 49	27 54	12 43	1 0	1 29	17 20	2 49	14 51	3 19	5 0	13 48	10 56	2 3
12/ 3/1958	10 20	20 50	6 31	6 42	20 10	18 47	16 15	11 56	3 0	29 11	13 39	1 0	3 18	20 50	2 39	16 46	3 9R	5 24	15 4	14 4	2 1
12/13/1958	14 18	26 33	6 18	13 13	24 31	17 17	26 33	22 8	4 6	0ar28	14 59	1 1	5 6	24 27	2 30	18 41	2 17	5 47	16 20	17 12	1 59
12/23/1958	18 20	2aq25	6 6	20 2	29 42	15 46	6 43	2sa 8	5 4	1 46	16 40	1 1	6 54	28 8	2 21	20 35	0 46	6 9	17 37	20 21	1 58
1/ 2/1959	22 26	8 23	5 53	27 11	5 34	14 14	16 36	11 53	5 55	3 3	18 39	1 2	8 41	1 54	2 13	22 29	28 41	6 26	18 53	23 29	1 57
1/12/1959	26 34	14 27	5 40	4 39	12 4	12 40	26 2	21 16	6 36	4 21	20 53	1 2	10 27	5 41	2 6	24 23	26 12	6 38	20 11	26 35	1 56
1/22/1959	0pi46	20 37	5 26	12 25	19 6	11 2	4 55	0cp15	7 8	5 39	23 21	1 3	12 13	9 31	2 0	26 17	23 38	6 43	21 28	29 36	1 56
2/ 1/1959	5 0	26 49	5 11	20 28	26 36	9 20	13 12	8 48	7 32	6 57	26 0	1 4	13 58	13 20	1 53	28 11	21 13	6 41	22 46	2 33	1 56
2/11/1959	9 17	3pi 6	4 56	28 47	4 31	7 34	20 52	16 53	7 50	8 16	28 48	1 4	15 44	17 9	1 48	0le 5	19 13	6 33	24 4	5 23	1 57
2/21/1959	13 35	9 24	4 40	7 19	12 47	5 44	27 56	24 31	8 1	9 34	1 43	1 5	17 28	20 57	1 42	1 59	17 49	6 20	25 23	8 5	1 58
3/ 3/1959	17 54	15 43	4 23	16 2	21 22	3 51	4 26	1aq41	8 8	10 54	4 44	1 7	19 13	24 42	1 37	3 53	17 5	6 5	26 42	10 37	1 59
3/13/1959	22 14	22 2	4 6	24 51	0gm13	1 57	10 24	8 24	8 12	12 13	7 50	1 8	20 57	28 24	1 32	5 47	17 2D	5 48	28 2	12 58	2 0
3/23/1959	26 34	28 21	3 47	3le43	9 17	0 4	15 53	14 41	8 13	13 33	10 58	1 10	22 41	2 1	1 27	7 41	17 37	5 31	29 22	15 4	2 2
4/ 2/1959	0ar54	4 38	3 28	12 34	18 30	1 45	20 57	20 32	8 12	14 53	14 9	1 11	24 24	5 33	1 22	9 35	18 46	5 15	0 43	16 53	2 4
4/12/1959	5 12	10 54	3 9	21 21	27 50	3 27	25 38	25 58	8 10	16 13	17 21	1 13	26 8	8 58	1 17	11 30	20 25	5 0	2 4	18 24	2 6
4/22/1959	9 30	17 7	2 48	29 59	7 14	4 59	29 59	0pi57	8 7	17 34	20 32	1 15	27 51	12 15	1 12	13 25	22 29	4 46	3 26	19 32	2 8
5/ 2/1959	13 46	23 16	2 27	8 26	16 37	6 20	4 2	5 30	8 4	18 56	23 42	1 18	29 34	15 23	1 6	15 21	24 54	4 33	4 48	20 16	2 10

Ephemeris	80			433			1221			1489			114			3018			2041		
0 Hours ET	Sappho			Eros			Amor			Attila			Kassandra			Godiva			Lancelot		
Mo Dy Year	h long	g long	g lat	h long	g long	g lat	h long	g long	g lat	h long	g long	g lat	h long	g long	g lat	h long	g long	g lat	h long	g long	g lat
5/12/1959	18 0	29 21	2 5	16 39	25 57	7 28	7 49	9 35	8 0	20 17	26 50	1 21	1 17	18 20	1 1	17 17	27 39	4 22	6 11	20 33	2 12
5/22/1959	22 11	5 22	1 42	24 36	5 12	8 22	11 22	13 10	7 55	21 39	29 55	1 24	2 59	21 4	0 55	19 14	0le38	4 12	7 34	20 22	2 14
6/ 1/1959	26 20	11 17	1 18	2li15	14 18	9 1	14 42	16 13	7 51	23 2	2 55	1 27	4 42	23 33	0 49	21 11	3 50	4 2	8 58	19 42	2 15
6/11/1959	0ta25	17 6	0 54	9 36	23 13	9 28	17 51	18 40	7 45	24 25	5 50	1 31	6 25	25 45	0 42	23 9	7 14	3 54	10 23	18 34	2 14
6/21/1959	4 27	22 48	0 29	16 38	1vi56	9 41	20 50	20 27	7 38	25 49	8 38	1 35	8 8	27 36	0 35	25 7	10 46	3 47	11 48	17 2	2 12
7/ 1/1959	8 25	28 23	0 3	23 21	10 25	9 45	23 40	21 28	7 29	27 13	11 18	1 40	9 50	29 4	0 27	27 7	14 27	3 40	13 15	15 14	2 9
7/11/1959	12 20	3 50	0 25	29 46	18 40	9 39	26 23	21 39R	7 18	28 38	13 48	1 45	11 33	0ar 4	0 18	29 7	18 15	3 34	14 41	13 19	2 4
7/21/1959	16 11	9 8	0 53	5 53	26 41	9 25	28 58	20 55	7 1	0ta 3	16 5	1 51	13 16	0 35	0 8	1 8	22 8	3 29	16 9	11 26	1 57
7/31/1959	19 57	14 15	1 23	11 44	4 28	9 6	1 26	19 16	6 38	1 29	18 8	1 58	14 59	0 33R	0s 4	3 10	26 6	3 24	17 37	9 47	1 49
8/10/1959	23 40	19 11	1 54	17 19	12 2	8 43	3 49	16 46	6 8	2 55	19 55	2 5	16 42	29pi56	0 16	5 13	0vi 9	3 20	19 6	8 29	1 40
8/20/1959	27 19	23 54	2 27	22 39	19 24	8 16	6 7	13 37	5 29	4 22	21 21	2 12	18 26	28 45	0 29	7 17	4 16	3 16	20 36	7 38	1 31
8/30/1959	0gm53	28 22	3 2	27 46	26 34	7 48	8 20	10 10	4 43	5 50	22 25	2 21	20 10	27 4	0 43	9 22	8 25	3 13	22 7	7 16	1 22
9/ 9/1959	4 23	2 32	3 40	2 41	3 34	7 18	10 28	6 46	3 53	7 18	23 4	2 29	21 53	25 1	0 57	11 29	12 37	3 10	23 38	7 26	1 13
9/19/1959	7 49	6 23	4 21	7 25	10 25	6 47	12 33	3 49	3 2	8 47	23 13	2 38	23 38	22 47	1 11	13 37	16 50	3 7	25 11	8 5	1 5
9/29/1959	11 11	9 50	5 5	11 58	17 7	6 16	14 34	1 33	2 13	10 16	22 53	2 48	25 22	20 36	1 23	15 46	21 5	3 5	26 44	9 12	0 57
10/ 9/1959	14 29	12 49	5 53	16 23	23 42	5 45	16 32	0 6	1 28	11 47	22 1	2 56	27 7	18 41	1 34	17 57	25 21	3 3	28 19	10 44	0 49
10/19/1959	17 44	15 14	6 45	20 39	0sa10	5 15	18 27	29 28	0 49	13 18	20 42	3 4	28 52	17 12	1 43	20 10	29 37	3 1	29 54	12 39	0 43
10/29/1959	20 54	17 2	7 41	24 48	6 33	4 45	20 20	29 36D	0 15	14 50	18 59	3 10	0ar38	16 16	1 50	22 24	3 51	3 0	1 30	14 53	0 36
11/ 8/1959	24 0	18 4	8 41	28 50	12 51	4 16	22 10	0pi24	0s14	16 22	17 1	3 14	2 24	15 57	1 56	24 39	8 5	2 59	3 8	17 25	0 30
11/18/1959	27 3	18 17R	9 43	2 47	19 5	3 47	23 58	1 46	0 39	17 56	15 0	3 15	4 11	16 14	2 0	26 57	12 15	2 58	4 46	20 12	0 25
11/28/1959	0cn 2	17 36	10 45	6 39	25 14	3 19	25 44	3 37	1 0	19 30	13 8	3 13	5 58	17 5	2 4	29 16	16 22	2 57	6 25	23 12	0 20
12/ 8/1959	2 58	16 3	11 42	10 26	1cp21	2 52	27 28	5 52	1 19	21 5	11 35	3 9	7 46	18 27	2 7	1 38	20 23	2 57	8 6	26 23	0 15
12/18/1959	5 50	13 46	12 29	14 10	7 25	2 25	29 11	8 27	1 36	22 41	10 29	3 4	9 34	20 16	2 9	4 1	24 18	2 56	9 48	29 43	0 10
12/28/1959	8 39	11 2	13 1	17 51	13 26	1 58	0 52	11 18	1 51	24 18	9 56	2 57	11 23	22 28	2 11	6 27	28 4	2 56	11 30	3 11	0 6
1/ 7/1960	11 25	8 13	13 14	21 29	19 25	1 32	2 33	14 22	2 4	25 55	9 56D	2 50	13 12	25 1	2 13	8 55	1sc39	2 56	13 14	6 47	0 2
1/17/1960	14 8	5 43	13 10	25 6	25 23	1 6	4 12	17 37	2 17	27 34	10 28	2 43	15 3	27 51	2 15	11 25	5 1	2 55	15 0	10 27	0 2
1/27/1960	16 48	3 49	12 51	28 42	1aq19	0 40	5 50	21 0	2 29	29 13	11 31	2 35	16 54	0ar56	2 18	13 58	8 6	2 54	16 46	14 12	0 7
2/ 6/1960	19 25	2 42	12 22	2 17	7 14	0 13	7 28	24 31	2 41	0 54	13 1	2 29	18 45	4 13	2 20	16 33	10 51	2 53	18 34	18 0	0 11
2/16/1960	21 59	2 24	11 47	5 52	13 8	0 14	9 5	28 6	2 52	2 35	14 56	2 22	20 38	7 40	2 22	19 11	13 11	2 50	20 23	21 50	0 15
2/26/1960	24 31	2 51	11 11	9 28	19 2	0 41	10 41	1ar47	3 3	4 18	17 11	2 17	22 31	11 15	2 25	21 52	15 1	2 47	22 13	25 41	0 19
3/ 7/1960	27 0	4 0	10 35	13 5	24 55	1 9	12 17	5 30	3 15	6 1	19 44	2 11	24 26	14 58	2 28	24 35	16 15	2 42	24 5	29 33	0 24
3/17/1960	29 27	5 43	10 1	16 43	0pi49	1 37	13 53	9 16	3 26	7 46	22 32	2 7	26 21	18 46	2 31	27 22	16 48	2 34	25 58	3 23	0 29
3/27/1960	1 52	7 56	9 30	20 24	6 43	2 7	15 29	13 3	3 38	9 32	25 33	2 2	28 17	22 39	2 34	0sc11	16 36	2 23	27 52	7 13	0 34
4/ 6/1960	4 14	10 32	9 1	24 7	12 38	2 38	17 4	16 51	3 51	11 18	28 45	1 59	0ta14	26 36	2 38	3 4	15 37	2 9	29 48	10 59	0 39
4/16/1960	6 35	13 28	8 35	27 54	18 35	3 10	18 40	20 38	4 4	13 6	2 6	1 55	2 12	0ta35	2 42	6 0	13 54	1 50	1 46	14 42	0 44
4/26/1960	8 53	16 39	8 12	1pi45	24 34	3 44	20 16	24 25	4 18	14 55	5 34	1 52	4 12	4 36	2 47	8 59	11 38	1 26	3 44	18 21	0 50
5/ 6/1960	11 10	20 3	7 50	5 42	0ar35	4 19	21 53	28 11	4 34	16 45	9 9	1 50	6 12	8 39	2 52	12 1	9 6	1 0	5 45	21 53	0 57
5/16/1960	13 25	23 37	7 31	9 43	6 40	4 56	23 30	1 55	4 50	18 37	12 49	1 48	8 13	12 41	2 57	15 7	6 41	0 31	7 46	25 18	1 4
5/26/1960	15 38	27 19	7 14	13 52	12 49	5 35	25 8	5 36	5 8	20 29	16 34	1 46	10 16	16 44	3 3	18 17	4 41	0 2	9 50	28 34	1 12
6/ 5/1960	17 49	1le 8	6 59	18 7	19 2	6 15	26 46	9 14	5 27	22 23	20 21	1 44	12 20	20 46	3 9	21 30	3 23	0 25	11 54	1 39	1 20

Ephemeris	80			433			1221			1489			114			3018			2041		
0 Hours ET	Sappho			Eros			Amor			Attila			Kassandra			Godiva			Lancelot		
Mo Dy Year	h long	g long	g lat	h long	g long	g lat	h long	g long	g lat	h long	g long	g lat	h long	g long	g lat	h long	g long	g lat	h long	g long	g lat
6/15/1960	19 59	5 2	6 45	22 31	25 22	6 58	28 25	12 47	5 48	24 17	24 11	1 42	14 25	24 45	3 16	24 46	2 55	0 49	14 1	4 31	1 29
6/25/1960	22 8	9 1	6 33	27 4	1ta49	7 42	0ta 6	16 16	6 11	26 13	28 3	1 41	16 31	28 43	3 24	28 7	3 18	1 11	16 8	7 7	1 39
7/ 5/1960	24 15	13 2	6 21	1ar47	8 23	8 28	1 47	19 38	6 37	28 11	1cn56	1 40	18 39	2 37	3 32	1sa30	4 29	1 30	18 18	9 25	1 51
7/15/1960	26 21	17 6	6 11	6 41	15 7	9 15	3 30	22 53	7 6	0cn 9	5 49	1 39	20 48	6 27	3 42	4 58	6 25	1 46	20 29	11 20	2 3
7/25/1960	28 26	21 12	6 2	11 47	22 2	10 3	5 14	25 59	7 38	2 9	9 42	1 39	22 59	10 12	3 52	8 29	8 58	2 0	22 41	12 50	2 16
8/ 4/1960	0vi29	25 18	5 54	17 7	29 7	10 51	7 0	28 53	8 14	4 10	13 33	1 38	25 11	13 50	4 3	12 3	12 5	2 11	24 55	13 50	2 31
8/14/1960	2 32	29 25	5 47	22 42	6 27	11 38	8 48	1 33	8 54	6 12	17 22	1 38	27 24	17 20	4 16	15 41	15 40	2 22	27 10	14 17	2 46
8/24/1960	4 34	3 32	5 41	28 32	14 0	12 22	10 38	3 57	9 40	8 15	21 9	1 38	29 39	20 39	4 29	19 22	19 39	2 31	29 27	14 7	3 2
9/ 3/1960	6 34	7 39	5 36	4 39	21 49	13 2	12 30	5 59	10 31	10 20	24 50	1 38	1 55	23 46	4 44	23 6	23 59	2 38	1 46	13 21	3 18
9/13/1960	8 34	11 44	5 31	11 5	29 55	13 34	14 25	7 35	11 28	12 26	28 26	1 39	4 13	26 37	5 1	26 53	28 37	2 45	4 6	12 2	3 32
9/23/1960	10 33	15 47	5 27	17 48	8 18	13 56	16 22	8 39	12 32	14 33	1 55	1 39	6 32	29 10	5 19	0cp43	3 30	2 51	6 27	10 16	3 44
10/ 3/1960	12 32	19 47	5 24	24 51	16 57	14 4	18 23	9 5	13 43	16 41	5 15	1 40	8 54	1 19	5 39	4 35	8 36	2 56	8 50	8 16	3 53
10/13/1960	14 30	23 44	5 22	2gm12	25 52	13 56	20 26	8 45	14 58	18 51	8 23	1 40	11 16	3 2	6 1	8 30	13 54	3 0	11 14	6 16	3 57
10/23/1960	16 27	27 37	5 20	9 53	5 1	13 26	22 34	7 34	16 15	21 2	11 18	1 41	13 41	4 11	6 23	12 26	19 22	3 3	13 39	4 32	3 57
11/ 2/1960	18 24	1li24	5 19	17 51	14 21	12 32	24 45	5 29	17 29	23 14	13 55	1 42	16 7	4 44	6 47	16 24	24 58	3 6	16 6	3 17	3 54
11/12/1960	20 21	5 4	5 19	26 5	23 49	11 10	27 1	2 35	18 31	25 27	16 12	1 43	18 34	4 35R	7 10	20 23	0cp41	3 9	18 34	2 38	3 48
11/22/1960	22 17	8 36	5 19	4 33	3vi18	9 21	29 21	29ta 5	19 14	27 41	18 5	1 43	21 4	3 43	7 30	24 24	6 30	3 10	21 3	2 40D	3 41
12/ 2/1960	24 13	11 57	5 20	13 13	12 45	7 3	1 47	25 23	19 30	29 56	19 28	1 44	23 35	2 11	7 46	28 24	12 24	3 12	23 33	3 21	3 32
12/12/1960	26 8	15 7	5 22	22 1	22 3	4 18	4 18	21 57	19 20	2 13	20 19	1 44	26 7	0 6	7 54	2 25	18 21	3 12	26 4	4 40	3 23
12/22/1960	28 4	18 2	5 24	0le52	1li 6	1 12	6 56	19 12	18 45	4 30	20 32	1 43	28 42	27 44	7 53	6 26	24 21	3 12	28 36	6 32	3 14
1/ 1/1961	29 59	20 39	5 27	9 45	9 48	2 12	9 41	17 25	17 54	6 49	20 7	1 41	1cn18	25 23	7 43	10 26	0aq23	3 12	1ta 9	8 53	3 6
1/11/1961	1 55	22 55	5 30	18 33	18 3	5 49	12 34	16 40	16 55	9 8	19 4	1 37	3 55	23 23	7 24	14 26	6 25	3 12	3 42	11 40	2 58
1/21/1961	3 50	24 46	5 33	27 15	25 44	9 32	15 35	16 57	15 52	11 29	17 29	1 32	6 35	21 58	6 58	18 23	12 28	3 11	6 17	14 48	2 51
1/31/1961	5 46	26 7	5 36	5 46	2sc43	13 18	18 46	18 11	14 49	13 50	15 31	1 25	9 16	21 16	6 29	22 20	18 29	3 9	8 52	18 13	2 44
2/10/1961	7 42	26 55	5 38	14 4	8 53	17 4	22 7	20 14	13 50	16 12	13 26	1 16	11 58	21 20D	5 59	26 14	24 29	3 7	11 27	21 54	2 38
2/20/1961	9 38	27 4R	5 38	22 7	14 1	20 47	25 40	23 1	12 53	18 35	11 30	1 6	14 42	22 9	5 29	0pi 6	0pi27	3 5	14 3	25 47	2 32
3/ 2/1961	11 34	26 33	5 36	29 52	17 55	24 24	29 27	26 26	12 0	20 59	9 57	0 56	17 28	23 38	5 1	3 55	6 21	3 3	16 39	29 50	2 27
3/12/1961	13 31	25 20	5 29	7 19	20 23	27 50	3 27	0gm25	11 9	23 23	8 58	0 45	20 15	25 42	4 34	7 42	12 11	3 0	19 15	4 1	2 22
3/22/1961	15 28	23 29	5 18	14 27	21 7	30 56	7 45	4 54	10 21	25 48	8 36	0 35	23 4	28 17	4 9	11 26	17 57	2 57	21 52	8 19	2 18
4/ 1/1961	17 26	21 9	5 0	21 16	19 59	33 25	12 20	9 51	9 33	28 14	8 55	0 25	25 54	1cn17	3 46	15 6	23 37	2 54	24 28	12 42	2 14
4/11/1961	19 25	18 33	4 36	27 47	17 8	34 58	17 16	15 13	8 46	0vi40	9 50	0 16	28 45	4 39	3 26	18 43	29 12	2 51	27 4	17 8	2 10
4/21/1961	21 24	15 58	4 6	4 0	13 4	35 13	22 34	21 1	7 58	3 6	11 20	0 8	1 37	8 19	3 6	22 17	4 41	2 47	29 41	21 38	2 7
5/ 1/1961	23 24	13 41	3 34	9 56	8 45	34 4	28 18	27 13	7 9	5 32	13 19	0 1	4 31	12 13	2 49	25 47	10 2	2 43	2 16	26 10	2 4
5/11/1961	25 25	11 55	2 59	15 36	5 9	31 44	4 29	3 48	6 18	7 59	15 44	0 6	7 26	16 20	2 32	29 14	15 16	2 39	4 52	0gm43	2 1
5/21/1961	27 26	10 48	2 26	21 1	2 49	28 41	11 10	10 48	5 23	10 26	18 30	0 11	10 21	20 38	2 17	2 37	20 22	2 34	7 27	5 16	1 58
5/31/1961	29 29	10 22	1 53	26 12	1 56	25 22	18 23	18 13	4 24	12 54	21 35	0 17	13 18	25 4	2 2	5 56	25 18	2 30	10 1	9 49	1 55
6/10/1961	1 33	10 37	1 24	1sa11	2 21	22 7	26 10	26 2	3 21	15 21	24 55	0 22	16 15	29 38	1 49	9 11	0ta 4	2 25	12 34	14 21	1 53
6/20/1961	3 38	11 30	0 56	5 58	3 51	19 6	4 32	4 16	2 13	17 48	28 27	0 26	19 13	4 18	1 36	12 23	4 39	2 20	15 7	18 51	1 50
6/30/1961	5 44	12 56	0 32	10 34	6 11	16 24	13 27	12 54	1 2	20 15	2 11	0 31	22 12	9 4	1 23	15 32	9 1	2 14	17 39	23 18	1 48
7/10/1961	7 52	14 52	0 9	15 1	9 10	14 2	22 54	21 55	0n13	22 42	6 3	0 35	25 11	13 53	1 11	18 36	13 9	2 8	20 10	27 43	1 46

Ephemeris	80			433			1221			1489			114			3018			2041		
0 Hours ET	Sappho			Eros			Amor			Attila			Kassandra			Godiva			Lancelot		
Mo Dy Year	h long	g long	g lat	h long	g long	g lat	h long	g long	g lat	h long	g long	g lat	h long	g long	g lat	h long	g long	g lat	h long	g long	g lat
7/20/1961	10 2	17 13	0 11	19 20	12 39	11 58	2li45	1vi18	1 29	25 8	10 3	0 38	28 11	18 47	1 0	21 38	17 1	2 2	22 40	2cn 4	1 44
7/30/1961	12 12	19 57	0 30	23 31	16 32	10 10	12 55	10 59	2 45	27 35	14 10	0 42	1vi10	23 43	0 48	24 36	20 34	1 55	25 9	6 19	1 42
8/ 9/1961	14 25	22 59	0 47	27 35	20 42	8 36	23 12	20 55	3 57	0li 0	18 21	0 45	4 10	28 42	0 37	27 30	23 46	1 47	27 36	10 30	1 41
8/19/1961	16 39	26 18	1 3	1cp34	25 7	7 13	3sc25	1li 2	5 3	2 26	22 36	0 49	7 10	3 42	0 26	0ta22	26 34	1 38	0cn 3	14 33	1 39
8/29/1961	18 56	29 51	1 18	5 27	29 44	6 1	13 24	11 12	6 0	4 51	26 55	0 52	10 10	8 43	0 16	3 10	28 52	1 29	2 28	18 29	1 37
9/ 8/1961	21 14	3 38	1 32	9 16	4 30	4 57	22 59	21 21	6 45	7 15	1li16	0 56	13 10	13 45	0 5	5 55	0gm37	1 18	4 52	22 16	1 35
9/18/1961	23 34	7 35	1 46	13 1	9 25	3 59	2sa 3	1sc22	7 19	9 38	5 38	0 59	16 9	18 47	0 6	8 37	1 43	1 5	7 14	25 51	1 34
9/28/1961	25 57	11 43	1 59	16 43	14 26	3 8	10 32	11 8	7 41	12 1	10 1	1 2	19 8	23 48	0 17	11 16	2 6	0 51	9 35	29 13	1 32
10/ 8/1961	28 22	15 59	2 12	20 22	19 34	2 22	18 24	20 36	7 52	14 23	14 25	1 6	22 6	28 48	0 28	13 53	1 42	0 35	11 55	2 20	1 30
10/18/1961	0sa50	20 24	2 25	23 59	24 48	1 40	25 39	29 42	7 53	16 44	18 47	1 9	25 4	3 46	0 39	16 27	0 30	0 16	14 13	5 8	1 28
10/28/1961	3 20	24 56	2 37	27 36	0cp 7	1 2	2cp19	8 24	7 47	19 5	23 8	1 13	28 1	8 41	0 51	18 58	28 37	0s 3	16 30	7 36	1 26
11/ 7/1961	5 54	29 35	2 50	1aq11	5 31	0 27	8 27	16 42	7 35	21 24	27 27	1 17	0li58	13 32	1 3	21 27	26 12	0 24	18 45	9 38	1 24
11/17/1961	8 30	4 20	3 3	4 46	11 0	0n 6	14 6	24 37	7 18	23 42	1sc42	1 21	3 53	18 19	1 15	23 53	23 33	0 44	20 59	11 11	1 21
11/27/1961	11 9	9 10	3 16	8 21	16 34	0 37	19 18	2cp 9	6 59	26 0	5 52	1 25	6 47	23 0	1 28	26 18	20 59	1 3	23 11	12 11	1 18
12/ 7/1961	13 52	14 6	3 30	11 58	22 12	1 5	24 6	9 20	6 38	28 16	9 57	1 30	9 40	27 33	1 42	28 40	18 50	1 19	25 21	12 35	1 14
12/17/1961	16 37	19 6	3 44	15 35	27 54	1 32	28 33	16 12	6 17	0sc31	13 55	1 35	12 32	1sc58	1 57	1gm 0	17 18	1 32	27 31	12 20	1 9
12/27/1961	19 27	24 10	3 59	19 15	3 41	1 58	2 42	22 45	5 55	2 45	17 45	1 40	15 23	6 12	2 12	3 18	16 30	1 43	29 38	11 27	1 3
1/ 6/1962	22 20	29 17	4 14	22 58	9 33	2 23	6 34	29 2	5 34	4 58	21 25	1 46	18 13	10 13	2 29	5 34	16 27D	1 51	1 44	10 0	0 56
1/16/1962	25 17	4 28	4 29	26 44	15 29	2 47	10 11	5 4	5 13	7 10	24 52	1 53	21 1	14 0	2 48	7 49	17 5	1 58	3 49	8 7	0 47
1/26/1962	28 18	9 40	4 46	0pi34	21 30	3 10	13 36	10 52	4 52	9 20	28 6	2 0	23 48	17 29	3 8	10 2	18 21	2 3	5 52	6 1	0 38
2/ 5/1962	1cp23	14 56	5 3	4 28	27 36	3 33	16 48	16 27	4 32	11 30	1sa 3	2 7	26 33	20 37	3 29	12 13	20 8	2 7	7 53	3 57	0 28
2/15/1962	4 32	20 12	5 22	8 28	3 48	3 55	19 51	21 49	4 13	13 38	3 41	2 16	29 17	23 21	3 53	14 22	22 21	2 10	9 53	2 8	0 19
2/25/1962	7 45	25 30	5 41	12 34	10 5	4 16	22 44	26 59	3 55	15 45	5 56	2 25	1 59	25 35	4 19	16 31	24 57	2 13	11 52	0 46	0 9
3/ 7/1962	11 3	0aq48	6 1	16 47	16 28	4 37	25 29	1pi59	3 37	17 50	7 46	2 34	4 40	27 17	4 46	18 38	27 50	2 16	13 49	29cn57	0 1
3/17/1962	14 26	6 6	6 22	21 8	22 58	4 57	28 6	6 46	3 20	19 55	9 6	2 44	7 19	28 21	5 16	20 43	0gm59	2 18	15 45	29 44	0 7
3/27/1962	17 53	11 23	6 45	25 38	29 35	5 16	0pi37	11 23	3 3	21 58	9 55	2 55	9 56	28 44	5 46	22 47	4 19	2 20	17 39	0le 5	0 14
4/ 6/1962	21 24	16 38	7 8	0ar18	6 20	5 34	3 2	15 49	2 47	24 0	10 9	3 6	12 32	28 23	6 17	24 51	7 49	2 22	19 32	0 58	0 20
4/16/1962	25 1	21 51	7 33	5 8	13 14	5 51	5 21	20 2	2 30	26 1	9 47	3 16	15 6	27 20	6 45	26 53	11 26	2 24	21 23	2 20	0 25
4/26/1962	28 42	27 0	7 59	10 10	20 16	6 6	7 36	24 4	2 14	28 0	8 50	3 25	17 38	25 40	7 9	28 54	15 10	2 27	23 14	4 6	0 30
5/ 6/1962	2 27	2pi 3	8 26	15 26	27 28	6 19	9 46	27 53	1 57	29 58	7 24	3 32	20 9	23 34	7 26	0cn54	18 59	2 29	25 2	6 13	0 34
5/16/1962	6 17	7 0	8 54	20 56	4 51	6 29	11 52	1ar28	1 39	1 55	5 36	3 36	22 38	21 18	7 35	2 54	22 51	2 32	26 50	8 37	0 38
5/26/1962	10 11	11 47	9 23	26 41	12 25	6 36	13 54	4 47	1 21	3 51	3 38	3 38	25 6	19 9	7 35	4 53	26 46	2 34	28 36	11 15	0 42
6/ 5/1962	14 9	16 23	9 53	2ta42	20 10	6 40	15 53	7 50	1 2	5 45	1 42	3 36	27 32	17 21	7 27	6 50	0cn43	2 38	0vi21	14 5	0 45
6/15/1962	18 11	20 44	10 24	9 1	28 7	6 39	17 49	10 34	0 42	7 39	0 0	3 31	29 56	16 7	7 14	8 48	4 41	2 41	2 5	17 6	0 48
6/25/1962	22 17	24 47	10 56	15 39	6 16	6 32	19 42	12 56	0 20	9 31	28 42	3 24	2 19	15 32	6 57	10 44	8 39	2 44	3 48	20 13	0 51
7/ 5/1962	26 26	28 27	11 28	22 35	14 37	6 20	21 33	14 53	0s 4	11 22	27 54	3 15	4 40	15 36D	6 37	12 40	12 38	2 48	5 29	23 28	0 54
7/15/1962	0pi39	1 37	11 59	29 50	23 10	6 1	23 22	16 21	0 30	13 12	27 39	3 6	6 59	16 18	6 17	14 36	16 36	2 53	7 10	26 47	0 57
7/25/1962	4 53	4 12	12 28	7 24	1cn53	5 35	25 9	17 16	0 59	15 1	27 56	2 56	9 17	17 33	5 58	16 31	20 32	2 58	8 49	0vi10	0 59
8/ 4/1962	9 10	6 3	12 53	15 17	10 46	5 2	26 53	17 33	1 31	16 48	28 43	2 47	11 34	19 18	5 39	18 26	24 26	3 3	10 27	3 37	1 2
8/14/1962	13 28	7 4	13 11	23 26	19 47	4 21	28 37	17 9	2 6	18 35	29 57	2 37	13 48	21 29	5 21	20 21	28 18	3 9	12 4	7 4	1 5

Ephemeris	80			433			1221			1489			114			3018			2041		
0 Hours ET	Sappho			Eros			Amor			Attila			Kassandra			Godiva			Lancelot		
Mo Dy Year	h long	g long	g lat	h long	g long	g lat	h long	g long	g lat	h long	g long	g lat	h long	g long	g lat	h long	g long	g lat	h long	g long	g lat
8/24/1962	17 48	7 8R	13 18	1cn49	28 56	3 34	0ar19	16 0	2 43	20 20	1 36	2 29	16 2	24 1	5 5	22 15	2 7	3 15	13 41	10 33	1 8
9/ 3/1962	22 8	6 16	13 9	10 25	8 9	2 42	1 59	14 8	3 22	22 5	3 36	2 21	18 14	26 52	4 50	24 9	5 51	3 22	15 16	14 2	1 11
9/13/1962	26 28	4 35	12 38	19 11	17 24	1 44	3 39	11 38	4 1	23 48	5 54	2 14	20 25	29 58	4 37	26 3	9 29	3 30	16 50	17 30	1 14
9/23/1962	0ar48	2 24	11 44	28 1	26 41	0 43	5 17	8 40	4 37	25 30	8 27	2 7	22 34	3 16	4 25	27 57	13 1	3 39	18 23	20 57	1 18
10/ 3/1962	5 7	0 7	10 28	6 54	5 56	0s20	6 55	5 31	5 8	27 12	11 14	2 2	24 42	6 46	4 14	29 51	16 25	3 49	19 56	24 21	1 22
10/13/1962	9 25	28 13	8 57	15 44	15 8	1 23	8 32	2 31	5 33	28 52	14 11	1 56	26 49	10 24	4 4	1 44	19 38	4 0	21 27	27 41	1 26
10/23/1962	13 42	27 3	7 21	24 28	24 14	2 25	10 9	29pi57	5 50	0cp31	17 18	1 52	28 54	14 9	3 55	3 38	22 40	4 12	22 58	0li56	1 30
11/ 2/1962	17 56	26 48	5 46	3vi 3	3li12	3 24	11 45	27 59	6 2	2 10	20 32	1 47	0cp58	18 1	3 47	5 32	25 27	4 26	24 27	4 6	1 35
11/12/1962	22 7	27 30	4 19	11 26	12 0	4 20	13 20	26 46	6 8	3 47	23 52	1 44	3 1	21 57	3 40	7 27	27 57	4 41	25 56	7 7	1 40
11/22/1962	26 16	29 4	3 1	19 34	20 37	5 11	14 56	26 16	6 10	5 24	27 16	1 40	5 3	25 56	3 33	9 21	0vi 5	4 58	27 24	10 0	1 45
12/ 2/1962	0ta22	1 23	1 53	27 25	29 2	5 58	16 32	26 29	6 10	7 0	0cp44	1 37	7 3	29 58	3 27	11 16	1 49	5 16	28 52	12 42	1 52
12/12/1962	4 24	4 20	0 55	4 58	7 13	6 39	18 7	27 19	6 8	8 35	4 14	1 35	9 3	4 2	3 22	13 11	3 2	5 36	0li18	15 10	1 58
12/22/1962	8 22	7 48	0 5	12 12	15 8	7 16	19 43	28 42	6 6	10 9	7 46	1 33	11 1	8 7	3 18	15 7	3 42	5 57	1 44	17 23	2 6
1/ 1/1963	12 17	11 41	0 38	19 8	22 49	7 48	21 20	0ar33	6 4	11 42	11 17	1 31	12 59	12 11	3 14	17 3	3 42R	6 18	3 10	19 18	2 14
1/11/1963	16 8	15 53	1 14	25 45	0sa12	8 16	22 56	2 49	6 3	13 14	14 47	1 29	14 55	16 14	3 10	18 59	3 2	6 39	4 34	20 53	2 23
1/21/1963	19 55	20 21	1 45	2sc 3	7 19	8 40	24 33	5 24	6 2	14 46	18 15	1 28	16 51	20 15	3 8	20 56	1 41	6 57	5 58	22 3	2 33
1/31/1963	23 38	25 1	2 12	8 5	14 9	9 1	26 11	8 16	6 1	16 17	21 39	1 27	18 46	24 13	3 5	22 54	29le43	7 11	7 21	22 48	2 43
2/10/1963	27 16	29 49	2 36	13 50	20 39	9 19	27 50	11 22	6 2	17 47	24 59	1 26	20 39	28 6	3 3	24 53	27 19	7 18	8 44	23 3	2 53
2/20/1963	0gm51	4 44	2 56	19 20	26 52	9 35	29 30	14 40	6 3	19 17	28 13	1 25	22 32	1 55	3 2	26 52	24 44	7 18	10 6	22 48	3 4
3/ 2/1963	4 21	9 45	3 14	24 35	2cp44	9 48	1 11	18 8	6 5	20 46	1aq20	1 25	24 24	5 38	3 1	28 52	22 15	7 10	11 28	22 3	3 14
3/12/1963	7 47	14 49	3 30	29 38	8 14	10 0	2 53	21 44	6 8	22 14	4 19	1 25	26 16	9 14	3 0	0vi53	20 8	6 55	12 49	20 49	3 24
3/22/1963	11 9	19 54	3 45	4 29	13 22	10 10	4 37	25 28	6 12	23 42	7 7	1 25	28 6	12 41	3 0	2 55	18 36	6 35	14 9	19 13	3 31
4/ 1/1963	14 27	25 1	3 57	9 9	18 4	10 19	6 22	29 18	6 17	25 9	9 44	1 25	29 56	15 58	3 0	4 58	17 44	6 12	15 29	17 22	3 37
4/11/1963	17 41	0gm 9	4 9	13 39	22 16	10 25	8 9	3 13	6 24	26 35	12 7	1 26	1 46	19 3	3 0	7 2	17 34D	5 48	16 49	15 26	3 39
4/21/1963	20 51	5 15	4 20	18 0	25 56	10 30	9 58	7 13	6 31	28 1	14 15	1 26	3 34	21 54	3 1	9 7	18 4	5 25	18 8	13 34	3 40
5/ 1/1963	23 58	10 21	4 30	22 14	28 57	10 33	11 49	11 17	6 39	29 26	16 4	1 27	5 22	24 30	3 2	11 14	19 10	5 2	19 26	11 57	3 37
5/11/1963	27 0	15 25	4 39	26 20	1 12	10 32	13 43	15 24	6 48	0 50	17 33	1 28	7 10	26 47	3 3	13 22	20 49	4 41	20 44	10 42	3 33
5/21/1963	29 59	20 28	4 48	0cp21	2 32	10 25	15 39	19 35	6 59	2 15	18 40	1 28	8 57	28 43	3 4	15 31	22 56	4 22	22 2	9 53	3 28
5/31/1963	2 55	25 28	4 56	4 16	2 48R	10 10	17 38	23 48	7 11	3 38	19 20	1 29	10 43	0pi15	3 5	17 41	25 27	4 4	23 19	9 32	3 21
6/10/1963	5 47	0cn25	5 5	8 6	1 52	9 43	19 40	28 5	7 24	5 1	19 34	1 29	12 29	1 20	3 5	19 54	28 18	3 48	24 36	9 39	3 14
6/20/1963	8 36	5 20	5 13	11 52	29cp41	9 0	21 46	2 24	7 39	6 24	19 19	1 29	14 15	1 54	3 6	22 7	1vi26	3 33	25 53	10 12	3 8
6/30/1963	11 22	10 11	5 21	15 35	26 20	7 57	23 55	6 45	7 55	7 46	18 35	1 29	16 0	1 54R	3 5	24 23	4 50	3 19	27 9	11 10	3 1
7/10/1963	14 5	14 59	5 29	19 15	22 13	6 34	26 9	11 9	8 13	9 8	17 24	1 28	17 45	1 20	3 3	26 40	8 27	3 6	28 25	12 29	2 55
7/20/1963	16 44	19 43	5 38	22 53	17 56	4 57	28 28	15 35	8 33	10 29	15 52	1 25	19 29	0 12	2 58	28 59	12 15	2 55	29 41	14 7	2 49
7/30/1963	19 22	24 23	5 47	26 29	14 9	3 14	0gm51	20 4	8 54	11 50	14 5	1 22	21 13	28 34	2 52	1 21	16 13	2 44	0 57	16 2	2 44
8/ 9/1963	21 56	28 58	5 56	0aq 5	11 21	1 36	3 20	24 35	9 19	13 11	12 11	1 18	22 57	26 34	2 43	3 44	20 20	2 33	2 12	18 10	2 39
8/19/1963	24 28	3 29	6 6	3 40	9 45	0 10	5 56	29 8	9 45	14 31	10 22	1 13	24 41	24 22	2 31	6 9	24 35	2 23	3 27	20 31	2 35
8/29/1963	26 57	7 53	6 17	7 15	9 21	1 4	8 38	3 43	10 15	15 51	8 45	1 7	26 24	22 12	2 17	8 37	28 57	2 14	4 41	23 2	2 32
9/ 8/1963	29 24	12 12	6 28	10 51	10 3	2 6	11 27	8 21	10 48	17 10	7 30	1 1	28 8	20 16	2 1	11 7	3 26	2 5	5 56	25 42	2 29
9/18/1963	1 48	16 23	6 41	14 28	11 39	2 56	14 25	13 0	11 24	18 30	6 40	0 55	29 51	18 46	1 45	13 39	8 0	1 56	7 10	28 29	2 26

Ephemeris	80			433			1221			1489			114			3018			2041		
0 Hours ET	Sappho			Eros			Amor			Attila			Kassandra			Godiva			Lancelot		
Mo Dy Year	h long	g long	g lat	h long	g long	g lat	h long	g long	g lat	h long	g long	g lat	h long	g long	g lat	h long	g long	g lat	h long	g long	g lat
9/28/1963	4 11	20 25	6 54	18 8	14 3	3 38	17 32	17 41	12 3	19 48	6 18	0 49	1 34	17 47	1 29	16 14	12 40	1 48	8 24	1sc22	2 24
10/ 8/1963	6 31	24 18	7 9	21 49	17 4	4 13	20 49	22 24	12 47	21 7	6 25D	0 43	3 17	17 23	1 14	18 51	17 25	1 39	9 38	4 19	2 23
10/18/1963	8 50	28 0	7 26	25 34	20 38	4 42	24 18	27 9	13 36	22 25	6 59	0 37	5 0	17 34	1 0	21 32	22 14	1 31	10 52	7 19	2 22
10/28/1963	11 6	1vi28	7 44	29 23	24 39	5 6	27 59	1le54	14 29	23 44	7 58	0 32	6 43	18 18	0 46	24 15	27 7	1 22	12 5	10 22	2 21
11/ 7/1963	13 21	4 41	8 4	3 16	29 3	5 27	1cn54	6 41	15 28	25 1	9 21	0 28	8 26	19 32	0 34	27 1	2sc 3	1 14	13 19	13 26	2 21
11/17/1963	15 34	7 36	8 27	7 14	3 47	5 46	6 5	11 28	16 31	26 19	11 3	0 24	10 9	21 13	0 24	29 50	7 2	1 5	14 32	16 29	2 22
11/27/1963	17 46	10 8	8 51	11 18	8 49	6 1	10 34	16 15	17 40	27 37	13 3	0 20	11 52	23 17	0 14	2 42	12 4	0 56	15 45	19 32	2 22
12/ 7/1963	19 56	12 15	9 18	15 29	14 7	6 15	15 22	21 2	18 52	28 54	15 18	0 16	13 35	25 41	0 5	5 37	17 7	0 46	16 59	22 31	2 24
12/17/1963	22 4	13 52	9 46	19 47	19 41	6 27	20 32	25 49	20 7	0pi11	17 45	0 13	15 18	28 23	0s 4	8 36	22 12	0 37	18 12	25 28	2 26
12/27/1963	24 11	14 54	10 17	24 14	25 28	6 37	26 7	0vi36	21 20	1 28	20 24	0 10	17 2	1pi18	0 11	11 38	27 17	0 26	19 25	28 18	2 28
1/ 6/1964	26 17	15 17	10 47	28 51	1pi28	6 45	2le 8	5 28	22 25	2 45	23 10	0 7	18 45	4 26	0 19	14 43	2sa22	0 15	20 38	1sa 2	2 31
1/16/1964	28 22	14 56	11 16	3 38	7 42	6 51	8 39	10 26	23 13	4 1	26 4	0 4	20 29	7 44	0 25	17 52	7 26	0 3	21 51	3 38	2 35
1/26/1964	0vi26	13 51	11 42	8 36	14 8	6 56	15 41	15 41	23 27	5 18	29 4	0 1	22 13	11 10	0 32	21 4	12 28	0 10	23 4	6 3	2 39
2/ 5/1964	2 28	12 5	12 0	13 47	20 46	6 58	23 18	21 29	22 38	6 34	2 7	0 1	23 58	14 43	0 38	24 20	17 27	0 24	24 17	8 15	2 44
2/15/1964	4 30	9 46	12 8	19 13	27 38	6 58	1vi29	28 15	19 59	7 51	5 13	0 4	25 43	18 21	0 45	27 39	22 21	0 40	25 30	10 13	2 49
2/25/1964	6 31	7 8	12 3	24 53	4 41	6 55	10 14	6 40	14 2	9 7	8 20	0 6	27 28	22 4	0 51	1sa 2	27 10	0 57	26 43	11 54	2 55
3/ 6/1964	8 30	4 29	11 45	0ta49	11 57	6 50	19 33	17 39	2 45	10 23	11 28	0 9	29 14	25 49	0 57	4 28	1cp52	1 16	27 57	13 15	3 1
3/16/1964	10 30	2 7	11 15	7 3	19 26	6 40	29 19	1sc44	14 28	11 40	14 34	0 11	1 0	29 37	1 3	7 58	6 24	1 37	29 10	14 14	3 8
3/26/1964	12 28	0 14	10 37	13 34	27 8	6 27	9 26	18 5	32 10	12 56	17 39	0 14	2 46	3 25	1 10	11 32	10 44	2 1	0sa23	14 48	3 16
4/ 5/1964	14 26	29 1	9 54	20 25	5 2	6 9	19 45	3sa59	43 53	14 12	20 40	0 17	4 33	7 14	1 16	15 9	14 50	2 27	1 37	14 56	3 23
4/15/1964	16 23	28 28	9 9	27 34	13 8	5 47	0sc 4	16 29	49 49	15 28	23 37	0 20	6 20	11 2	1 23	18 49	18 37	2 56	2 50	14 36	3 30
4/25/1964	18 20	28 37D	8 26	5 2	21 26	5 19	10 12	24 41	52 22	16 44	26 28	0 23	8 8	14 48	1 31	22 32	22 2	3 29	4 4	13 48	3 36
5/ 5/1964	20 17	29 22	7 44	12 49	29 55	4 47	19 59	29 3	53 5	18 1	29 13	0 26	9 57	18 32	1 38	26 18	25 0	4 6	5 18	12 35	3 40
5/15/1964	22 13	0vi40	7 6	20 53	8 35	4 9	29 16	0 22	52 33	19 17	1 49	0 29	11 46	22 13	1 47	0cp 7	27 25	4 46	6 32	11 2	3 43
5/25/1964	24 9	2 27	6 31	29 12	17 24	3 26	7 59	29sa35	50 52	20 33	4 15	0 33	13 36	25 50	1 55	3 58	29 9	5 31	7 46	9 14	3 44
6/ 4/1964	26 4	4 37	5 58	7 45	26 21	2 38	16 5	27 43	48 5	21 50	6 30	0 37	15 27	29 22	2 5	7 52	0aq 8	6 18	9 0	7 22	3 42
6/14/1964	28 0	7 7	5 29	16 27	5 24	1 48	23 33	25 40	44 18	23 6	8 31	0 41	17 18	2 47	2 15	11 48	0 17R	7 6	10 15	5 34	3 37
6/24/1964	29 55	9 55	5 3	25 16	14 32	0 54	0cp26	24 8	39 50	24 23	10 16	0 46	19 11	6 5	2 27	15 45	29cp34	7 52	11 30	3 59	3 31
7/ 4/1964	1 51	12 57	4 39	4le 8	23 42	0s 0	6 45	23 27	35 4	25 40	11 43	0 51	21 3	9 13	2 39	19 43	28 6	8 32	12 45	2 44	3 22
7/14/1964	3 46	16 11	4 17	13 0	2le54	0 55	12 33	23 39D	30 24	26 57	12 49	0 56	22 57	12 10	2 52	23 43	26 5	9 1	14 0	1 54	3 13
7/24/1964	5 42	19 35	3 57	21 46	12 4	1 47	17 54	24 42	26 5	28 14	13 31	1 2	24 52	14 54	3 7	27 43	23 56	9 15	15 16	1 32	3 3
8/ 3/1964	7 38	23 8	3 38	0vi24	21 11	2 37	22 50	26 26	22 15	29 31	13 49	1 9	26 48	17 21	3 24	1aq44	22 1	9 15	16 32	1 37D	2 53
8/13/1964	9 34	26 49	3 21	8 51	0vi14	3 23	27 24	28 44	18 57	0 48	13 38	1 15	28 44	19 30	3 42	5 44	20 41	9 1	17 48	2 9	2 44
8/23/1964	11 30	0li36	3 4	17 3	9 10	4 4	1aq38	1 29	16 8	2 6	12 59	1 22	0ta42	21 16	4 2	9 44	20 10	8 37	19 5	3 6	2 34
9/ 2/1964	13 27	4 28	2 49	24 59	17 58	4 40	5 36	4 35	13 45	3 24	11 54	1 28	2 40	22 35	4 23	13 43	20 31	8 8	20 22	4 25	2 26
9/12/1964	15 25	8 26	2 35	2li38	26 38	5 10	9 17	7 57	11 45	4 42	10 25	1 34	4 40	23 23	4 47	17 41	21 43	7 36	21 39	6 5	2 18
9/22/1964	17 22	12 27	2 21	9 58	5 7	5 34	12 46	11 32	10 4	6 0	8 39	1 40	6 41	23 37	5 11	21 37	23 40	7 3	22 57	8 2	2 10
10/ 2/1964	19 21	16 31	2 8	16 59	13 26	5 53	16 2	15 18	8 38	7 19	6 45	1 44	8 42	23 13	5 36	25 31	26 17	6 31	24 15	10 15	2 4
10/12/1964	21 20	20 37	1 55	23 42	21 33	6 7	19 7	19 11	7 25	8 38	4 51	1 47	10 45	22 10	6 1	29 23	29 28	6 1	25 33	12 41	1 57
10/22/1964	23 20	24 46	1 42	0sc 6	29 29	6 16	22 3	23 11	6 23	9 57	3 10	1 49	12 50	20 32	6 22	3 13	3 5	5 32	26 52	15 18	1 52

Ephemeris	80			433			1221			1489			114			3018			2041		
0 Hours ET	Sappho			Eros			Amor			Attila			Kassandra			Godiva			Lancelot		
Mo Dy Year	h long	g long	g lat	h long	g long	g lat	h long	g long	g lat	h long	g long	g lat	h long	g long	g lat	h long	g long	g lat	h long	g long	g lat
11/ 1/1964	25 21	28 56	1 29	6 13	7 13	6 21	24 51	27 16	5 29	11 17	1 48	1 49	14 55	18 27	6 39	7 0	7 4	5 6	28 12	18 6	1 47
11/11/1964	27 23	3 6	1 17	12 3	14 46	6 22	27 30	1aq25	4 42	12 36	0 53	1 49	17 2	16 7	6 50	10 44	11 21	4 41	29 32	21 2	1 42
11/21/1964	29 25	7 15	1 4	17 38	22 8	6 20	0pi 3	5 37	4 1	13 57	0 26	1 48	19 10	13 49	6 53	14 25	15 53	4 19	0 52	24 5	1 38
12/ 1/1964	1 29	11 24	0 51	22 58	29 18	6 15	2 29	9 52	3 25	15 17	0 30D	1 46	21 19	11 48	6 49	18 2	20 36	3 58	2 13	27 13	1 35
12/11/1964	3 34	15 30	0 37	28 4	6 19	6 8	4 50	14 8	2 54	16 38	1 2	1 44	23 30	10 18	6 40	21 36	25 28	3 39	3 35	0cp26	1 31
12/21/1964	5 41	19 34	0 23	2 59	13 8	5 59	7 6	18 25	2 25	17 59	2 1	1 42	25 42	9 27	6 26	25 7	0pi26	3 21	4 57	3 41	1 28
12/31/1964	7 49	23 33	0 8	7 42	19 49	5 48	9 17	22 42	2 0	19 21	3 25	1 40	27 55	9 18D	6 10	28 34	5 30	3 4	6 19	6 59	1 25
1/10/1965	9 58	27 27	0 9	12 15	26 19	5 36	11 24	26 59	1 36	20 43	5 10	1 39	0gm10	9 50	5 53	1 58	10 37	2 49	7 43	10 17	1 23
1/20/1965	12 9	1sa13	0 26	16 39	2cp40	5 22	13 28	1pi16	1 15	22 6	7 14	1 37	2 27	11 0	5 37	5 18	15 47	2 34	9 6	13 35	1 21
1/30/1965	14 21	4 51	0 46	20 55	8 53	5 7	15 27	5 31	0 55	23 29	9 34	1 36	4 45	12 44	5 20	8 34	20 58	2 21	10 31	16 52	1 19
2/ 9/1965	16 36	8 18	1 7	25 4	14 56	4 50	17 24	9 45	0 36	24 52	12 7	1 35	7 4	14 57	5 5	11 46	26 10	2 8	11 56	20 5	1 17
2/19/1965	18 52	11 32	1 31	29 6	20 50	4 32	19 18	13 57	0 18	26 17	14 51	1 34	9 26	17 36	4 51	14 56	1ar21	1 56	13 22	23 15	1 15
3/ 1/1965	21 10	14 29	1 58	3 2	26 36	4 13	21 10	18 7	0 1	27 41	17 45	1 34	11 48	20 36	4 38	18 1	6 31	1 45	14 49	26 19	1 13
3/11/1965	23 30	17 7	2 28	6 54	2aq13	3 53	22 59	22 13	0 15	29 6	20 47	1 34	14 12	23 54	4 26	21 3	11 39	1 34	16 16	29 16	1 12
3/21/1965	25 53	19 21	3 2	10 41	7 41	3 31	24 46	26 16	0 32	0ta32	23 55	1 34	16 38	27 28	4 16	24 2	16 45	1 23	17 44	2 4	1 10
3/31/1965	28 18	21 8	3 41	14 25	12 59	3 6	26 32	0ar15	0 48	1 58	27 7	1 34	19 6	1gm15	4 6	26 57	21 48	1 13	19 13	4 42	1 9
4/10/1965	0sa46	22 21	4 24	18 6	18 8	2 40	28 15	4 9	1 5	3 25	0ta24	1 35	21 35	5 12	3 57	29 49	26 48	1 4	20 43	7 8	1 7
4/20/1965	3 16	22 55	5 13	21 45	23 6	2 10	29 58	7 59	1 22	4 52	3 42	1 36	24 6	9 19	3 49	2 38	1ta45	0 54	22 13	9 19	1 6
4/30/1965	5 49	22 46R	6 5	25 22	27 52	1 38	1 39	11 43	1 39	6 21	7 3	1 37	26 38	13 34	3 41	5 24	6 37	0 45	23 44	11 13	1 4
5/10/1965	8 25	21 52	7 1	28 58	2 26	1 1	3 18	15 20	1 57	7 49	10 24	1 39	29 12	17 55	3 34	8 6	11 25	0 36	25 17	12 47	1 2
5/20/1965	11 4	20 14	7 56	2 33	6 45	0 20	4 57	18 50	2 16	9 19	13 44	1 41	1 48	22 22	3 27	10 46	16 8	0 27	26 50	13 59	1 0
5/30/1965	13 46	18 0	8 48	6 8	10 46	0 27	6 35	22 12	2 36	10 49	17 3	1 43	4 25	26 54	3 21	13 24	20 46	0 18	28 24	14 44	0 57
6/ 9/1965	16 32	15 23	9 32	9 44	14 27	1 22	8 13	25 24	2 58	12 20	20 20	1 45	7 4	1cn29	3 15	15 58	25 18	0 9	29 59	15 1	0 54
6/19/1965	19 21	12 46	10 4	13 20	17 44	2 25	9 49	28 24	3 22	13 51	23 34	1 48	9 44	6 8	3 10	18 30	29 44	0 0	1 35	14 48	0 50
6/29/1965	22 14	10 27	10 23	16 59	20 30	3 39	11 25	1ta12	3 48	15 24	26 43	1 52	12 26	10 50	3 4	21 0	4 3	0 9	3 13	14 4	0 46
7/ 9/1965	25 11	8 45	10 29	20 40	22 39	5 5	13 1	3 45	4 16	16 57	29 47	1 55	15 9	15 35	2 59	23 27	8 14	0 19	4 51	12 51	0 40
7/19/1965	28 11	7 50	10 26	24 24	24 1	6 46	14 37	6 0	4 48	18 30	2 43	2 0	17 54	20 20	2 54	25 52	12 17	0 29	6 30	11 14	0 34
7/29/1965	1cp16	7 45D	10 15	28 11	24 25	8 43	16 13	7 54	5 23	20 5	5 32	2 4	20 40	25 7	2 49	28 14	16 10	0 39	8 11	9 20	0 27
8/ 8/1965	4 25	8 31	10 0	2 2	23 39	10 54	17 49	9 22	6 1	21 41	8 9	2 10	23 28	29 55	2 44	0gm35	19 52	0 50	9 52	7 22	0 19
8/18/1965	7 38	10 2	9 43	5 59	21 38	13 16	19 24	10 21	6 44	23 17	10 34	2 16	26 16	4 43	2 39	2 54	23 21	1 2	11 35	5 29	0 11
8/28/1965	10 55	12 14	9 24	10 1	18 21	15 35	21 1	10 46	7 31	24 54	12 45	2 22	29 7	9 30	2 34	5 10	26 36	1 15	13 19	3 54	0 3
9/ 7/1965	14 17	15 2	9 6	14 10	14 9	17 36	22 37	10 31	8 21	26 33	14 37	2 30	1 58	14 16	2 28	7 25	29 33	1 29	15 4	2 44	0 5
9/17/1965	17 44	18 21	8 47	18 26	9 40	19 1	24 15	9 34	9 13	28 12	16 8	2 37	4 50	19 0	2 23	9 39	2 11	1 44	16 50	2 6	0 12
9/27/1965	21 15	22 7	8 29	22 50	5 40	19 43	25 52	7 52	10 6	29 52	17 14	2 46	7 44	23 41	2 17	11 50	4 25	2 1	18 38	2 2D	0 18
10/ 7/1965	24 51	26 16	8 11	27 23	2 49	19 47	27 31	5 29	10 54	1 33	17 53	2 55	10 38	28 18	2 11	14 0	6 11	2 19	20 27	2 31	0 24
10/17/1965	28 31	0cp46	7 54	2ar 6	1 25	19 23	29 11	2 34	11 35	3 15	18 0R	3 4	13 33	2 49	2 5	16 9	7 26	2 39	22 17	3 31	0 29
10/27/1965	2 16	5 33	7 37	7 1	1 28D	18 42	0 51	29ar22	12 4	4 58	17 35	3 12	16 29	7 15	1 58	18 16	8 3	3 0	24 9	5 1	0 34
11/ 6/1965	6 5	10 36	7 21	12 8	2 53	17 52	2 33	26 14	12 19	6 42	16 36	3 20	19 26	11 31	1 51	20 22	8 1R	3 23	26 2	6 57	0 38
11/16/1965	9 59	15 52	7 4	17 29	5 27	16 59	4 17	23 28	12 21	8 27	15 8	3 26	22 23	15 37	1 43	22 27	7 15	3 46	27 56	9 16	0 42
11/26/1965	13 56	21 21	6 48	23 4	8 59	16 4	6 2	21 20	12 11	10 13	13 16	3 29	25 20	19 30	1 34	24 31	5 48	4 8	29 52	11 54	0 46

Ephemeris	80			433			1221			1489			114			3018			2041		
0 Hours ET	Sappho			Eros			Amor			Attila			Kassandra			Godiva			Lancelot		
Mo Dy Year	h long	g long	g lat	h long	g long	g lat	h long	g long	g lat	h long	g long	g lat	h long	g long	g lat	h long	g long	g lat	h long	g long	g lat
12/ 6/1965	17 58	26 59	6 32	28 55	13 20	15 9	7 48	19 57	11 53	12 0	11 13	3 30	28 18	23 7	1 24	26 33	3 44	4 29	1 49	14 51	0 49
12/16/1965	22 3	2aq47	6 15	5 3	18 24	14 12	9 37	19 22	11 31	13 48	9 11	3 27	1vi17	26 25	1 13	28 35	1 16	4 45	3 48	18 2	0 52
12/26/1965	26 11	8 42	5 59	11 29	24 4	13 15	11 28	19 34	11 7	15 37	7 23	3 22	4 15	29 19	1 0	0cn35	28 39	4 56	5 48	21 27	0 55
1/ 5/1966	0pi23	14 44	5 42	18 14	0ar17	12 15	13 21	20 27	10 43	17 28	6 1	3 14	7 13	1 44	0 45	2 35	26 12	5 2	7 50	25 2	0 59
1/15/1966	4 37	20 52	5 24	25 17	6 58	11 13	15 17	21 58	10 20	19 19	5 10	3 5	10 12	3 36	0 28	4 34	24 10	5 3	9 53	28 47	1 2
1/25/1966	8 53	27 3	5 7	2gm40	14 5	10 7	17 15	24 0	9 59	21 12	4 56	2 55	13 10	4 49	0 9	6 32	22 43	4 59	11 58	2 40	1 5
2/ 4/1966	13 10	3 18	4 48	10 21	21 36	8 57	19 17	26 31	9 40	23 6	5 17	2 45	16 8	5 18	0 13	8 29	21 58	4 52	14 5	6 39	1 8
2/14/1966	17 29	9 36	4 30	18 19	29 28	7 41	21 22	29 25	9 23	25 1	6 12	2 35	19 5	5 0	0 38	10 26	21 54D	4 44	16 13	10 44	1 12
2/24/1966	21 49	15 54	4 10	26 34	7 39	6 22	23 31	2 39	9 8	26 57	7 37	2 25	22 2	3 57	1 5	12 22	22 29	4 35	18 22	14 53	1 15
3/ 6/1966	26 9	22 13	3 51	5 3	16 7	4 57	25 44	6 11	8 56	28 54	9 30	2 16	24 58	2 14	1 33	14 18	23 38	4 26	20 33	19 5	1 19
3/16/1966	0ar28	28 32	3 30	13 43	24 50	3 29	28 2	9 59	8 44	0cn53	11 46	2 8	27 54	0 5	2 0	16 13	25 17	4 17	22 46	23 20	1 22
3/26/1966	4 47	4 49	3 10	22 30	3gm45	1 59	0gm25	14 1	8 35	2 53	14 22	2 0	0li49	27 48	2 25	18 8	27 22	4 9	25 0	27 37	1 26
4/ 5/1966	9 5	11 5	2 48	1le22	12 50	0 28	2 53	18 15	8 26	4 54	17 15	1 54	3 43	25 43	2 47	20 3	29 48	4 2	27 15	1ar54	1 31
4/15/1966	13 21	17 18	2 26	10 14	22 3	1 1	5 27	22 42	8 19	6 56	20 23	1 47	6 36	24 6	3 3	21 57	2 31	3 55	29 32	6 11	1 35
4/25/1966	17 35	23 28	2 4	19 3	1cn21	2 27	8 7	27 20	8 12	8 59	23 43	1 42	9 28	23 10	3 16	23 51	5 29	3 50	1 51	10 28	1 40
5/ 5/1966	21 46	29 34	1 41	27 44	10 41	3 46	10 56	2gm 8	8 6	11 4	27 14	1 36	12 19	22 57	3 25	25 45	8 40	3 45	4 11	14 43	1 45
5/15/1966	25 55	5 35	1 18	6 14	20 0	4 57	13 52	7 8	8 1	13 9	0cn53	1 31	15 8	23 28	3 30	27 38	12 1	3 40	6 33	18 55	1 50
5/25/1966	0ta 0	11 31	0 54	14 32	29 17	5 59	16 57	12 19	7 56	15 16	4 40	1 27	17 57	24 39	3 34	29 32	15 30	3 37	8 56	23 4	1 56
6/ 4/1966	4 3	17 22	0 30	22 33	8 28	6 49	20 12	17 41	7 50	17 24	8 33	1 22	20 44	26 25	3 36	1 26	19 6	3 34	11 20	27 9	2 2
6/14/1966	8 1	23 7	0 5	0li18	17 31	7 29	23 39	23 14	7 44	19 34	12 31	1 18	23 30	28 40	3 36	3 20	22 47	3 31	13 46	1ta 9	2 9
6/24/1966	11 56	28 45	0 20	7 44	26 25	7 58	27 17	29 1	7 37	21 44	16 33	1 15	26 14	1li21	3 36	5 13	26 33	3 29	16 13	5 2	2 17
7/ 4/1966	15 47	4 15	0 46	14 51	5 7	8 17	1cn10	5 1	7 29	23 55	20 39	1 11	28 57	4 23	3 36	7 7	0le24	3 28	18 41	8 46	2 25
7/14/1966	19 34	9 38	1 13	21 39	13 37	8 26	5 18	11 15	7 20	26 8	24 48	1 8	1 38	7 43	3 35	9 2	4 17	3 27	21 10	12 21	2 33
7/24/1966	23 17	14 51	1 41	28 9	21 55	8 27	9 43	17 46	7 7	28 21	28 59	1 4	4 18	11 17	3 34	10 56	8 12	3 27	23 41	15 43	2 43
8/ 3/1966	26 56	19 54	2 10	4 21	29 59	8 21	14 27	24 33	6 52	0le36	3 10	1 1	6 57	15 3	3 33	12 51	12 9	3 27	26 12	18 51	2 53
8/13/1966	0gm31	24 47	2 41	10 17	7 50	8 9	19 33	1le38	6 33	2 52	7 23	0 58	9 34	18 58	3 32	14 46	16 7	3 28	28 45	21 41	3 4
8/23/1966	4 1	29 27	3 13	15 56	15 28	7 53	25 2	9 4	6 10	5 8	11 36	0 55	12 9	23 2	3 31	16 42	20 5	3 29	1 18	24 11	3 17
9/ 2/1966	7 28	3 52	3 47	21 20	22 55	7 34	0le58	16 50	5 41	7 26	15 47	0 52	14 42	27 13	3 30	18 38	24 4	3 31	3 52	26 15	3 30
9/12/1966	10 50	8 1	4 24	26 31	0sc10	7 12	7 23	24 59	5 6	9 44	19 57	0 48	17 15	1sc29	3 29	20 35	28 1	3 33	6 27	27 50	3 43
9/22/1966	14 9	11 50	5 3	1sa29	7 16	6 48	14 18	3vi30	4 24	12 4	24 4	0 45	19 45	5 50	3 29	22 32	1vi56	3 36	9 3	28 50	3 58
10/ 2/1966	17 23	15 17	5 45	6 16	14 11	6 23	21 47	12 24	3 35	14 24	28 8	0 42	22 14	10 13	3 29	24 30	5 49	3 39	11 39	29 13	4 12
10/12/1966	20 34	18 18	6 31	10 52	20 59	5 57	29 51	21 40	2 38	16 45	2 6	0 39	24 41	14 40	3 29	26 29	9 38	3 43	14 15	28 56	4 25
10/22/1966	23 41	20 47	7 21	15 19	27 39	5 31	8 29	1li17	1 35	19 6	5 58	0 35	27 7	19 8	3 29	28 28	13 22	3 48	16 52	27 59	4 36
11/ 1/1966	26 44	22 39	8 15	19 37	4 11	5 5	17 41	11 10	0 27	21 28	9 42	0 31	29 31	23 37	3 29	0vi29	17 0	3 54	19 28	26 29	4 44
11/11/1966	29 43	23 49	9 12	23 48	10 38	4 39	27 21	21 16	0 45	23 51	13 16	0 27	1 54	28 7	3 30	2 30	20 31	4 0	22 5	24 34	4 47
11/21/1966	2 39	24 10	10 12	27 52	17 0	4 13	7 23	1sc31	1 57	26 14	16 37	0 23	4 15	2 36	3 31	4 32	23 51	4 7	24 42	22 30	4 44
12/ 1/1966	5 32	23 40	11 12	1cp50	23 16	3 47	17 40	11 46	3 6	28 38	19 42	0 18	6 34	7 3	3 33	6 36	26 59	4 16	27 19	20 35	4 37
12/11/1966	8 21	22 17	12 7	5 43	29 28	3 21	27 58	21 57	4 10	1vi 2	22 30	0 12	8 52	11 29	3 35	8 40	29 53	4 25	29 56	19 4	4 25
12/21/1966	11 8	20 9	12 53	9 32	5 37	2 55	8 8	1sa57	5 7	3 27	24 55	0 6	11 9	15 52	3 37	10 46	2 28	4 35	2 32	18 8	4 9
12/31/1966	13 51	17 30	13 25	13 17	11 41	2 30	17 59	11 41	5 55	5 52	26 53	0n 0	13 24	20 11	3 40	12 53	4 41	4 46	5 8	17 52	3 53

Ephemeris 0 Hours ET	80 Sappho			433 Eros			1221 Amor			1489 Attila			114 Kassandra			3018 Godiva			2041 Lancelot		
Mo Dy Year	h long	g long	g lat	h long	g long	g lat	h long	g long	g lat	h long	g long	g lat	h long	g long	g lat	h long	g long	g lat	h long	g long	g lat
1/10/1967	16 31	14 41	13 38	16 58	17 43	2 4	27 22	21 3	6 35	8 17	28 22	0 8	15 37	24 25	3 44	15 2	6 27	4 59	7 44	18 18	3 36
1/20/1967	19 8	12 4	13 34	20 38	23 42	1 39	6 11	0cp 2	7 5	10 42	29 15	0 16	17 50	28 34	3 48	17 11	7 42	5 11	10 19	19 22	3 19
1/30/1967	21 43	9 59	13 14	24 15	29 39	1 13	14 24	8 34	7 28	13 7	29 30	0 26	20 0	2 35	3 52	19 23	8 20	5 24	12 53	21 1	3 4
2/ 9/1967	24 15	8 38	12 43	27 51	5 33	0 47	22 0	16 39	7 44	15 32	29 6	0 36	22 10	6 29	3 58	21 36	8 17R	5 36	15 26	23 10	2 49
2/19/1967	26 45	8 4	12 7	1aq26	11 26	0 20	28 59	24 16	7 54	17 57	28 4	0 46	24 18	10 14	4 3	23 50	7 31	5 46	17 58	25 44	2 36
3/ 1/1967	29 12	8 18	11 28	5 1	17 16	0n 7	5 25	1aq27	8 0	20 22	26 29	0 57	26 25	13 47	4 10	26 7	6 2	5 52	20 30	28 40	2 23
3/11/1967	1 37	9 13	10 49	8 37	23 6	0 35	11 19	8 11	8 3	22 47	24 34	1 7	28 31	17 7	4 18	28 25	3 57	5 53	23 0	1 53	2 12
3/21/1967	4 0	10 44	10 13	12 13	28 55	1 4	16 45	14 29	8 3	25 11	22 32	1 16	0cp35	20 13	4 26	0li45	1 26	5 47	25 30	5 20	2 2
3/31/1967	6 21	12 46	9 39	15 51	4 43	1 35	21 46	20 22	8 1	27 35	20 40	1 24	2 38	23 1	4 35	3 7	28vi50	5 34	27 58	8 59	1 53
4/10/1967	8 39	15 12	9 8	19 31	10 31	2 7	26 24	25 50	7 58	29 59	19 11	1 30	4 41	25 30	4 45	5 31	26 25	5 14	0cn25	12 48	1 45
4/20/1967	10 56	18 0	8 40	23 14	16 18	2 40	0aq42	0pi52	7 54	2 22	18 14	1 34	6 42	27 35	4 56	7 58	24 29	4 50	2 50	16 44	1 37
4/30/1967	13 11	21 3	8 14	27 0	22 7	3 16	4 43	5 29	7 49	4 45	17 54	1 37	8 42	29 14	5 7	10 27	23 14	4 22	5 15	20 45	1 30
5/10/1967	15 25	24 21	7 51	0pi50	27 57	3 54	8 28	9 39	7 45	7 7	18 12	1 39	10 41	0aq24	5 19	12 58	22 43	3 54	7 38	24 51	1 23
5/20/1967	17 36	27 49	7 31	4 45	3 48	4 34	11 59	13 20	7 39	9 28	19 5	1 40	12 38	1 1	5 31	15 31	22 59	3 27	9 59	29 0	1 17
5/30/1967	19 46	1le26	7 12	8 45	9 41	5 17	15 18	16 30	7 34	11 49	20 31	1 40	14 35	1 2R	5 42	18 8	23 56	3 1	12 19	3 11	1 11
6/ 9/1967	21 55	5 10	6 55	12 52	15 37	6 3	18 26	19 6	7 27	14 9	22 25	1 40	16 31	0 26	5 51	20 47	25 32	2 37	14 38	7 23	1 5
6/19/1967	24 3	9 0	6 40	17 5	21 37	6 52	21 23	21 4	7 20	16 28	24 43	1 40	18 26	29 15	5 57	23 28	27 42	2 14	16 55	11 36	1 0
6/29/1967	26 9	12 55	6 27	21 27	27 41	7 44	24 12	22 18	7 11	18 46	27 21	1 40	20 21	27 34	6 0	26 13	0li21	1 54	19 11	15 48	0 55
7/ 9/1967	28 14	16 53	6 15	25 57	3 51	8 41	26 54	22 44	6 59	21 3	0li17	1 39	22 14	25 30	5 57	29 1	3 25	1 35	21 25	20 0	0 50
7/19/1967	0vi17	20 54	6 4	0ar37	10 6	9 41	29 28	22 17	6 43	23 20	3 27	1 39	24 7	23 17	5 48	1 51	6 50	1 18	23 38	24 10	0 46
7/29/1967	2 20	24 57	5 54	5 28	16 28	10 44	1 56	20 56	6 21	25 35	6 50	1 39	25 59	21 8	5 35	4 45	10 35	1 1	25 49	28 18	0 41
8/ 8/1967	4 22	29 1	5 45	10 31	22 59	11 51	4 18	18 41	5 53	27 50	10 22	1 38	27 50	19 17	5 17	7 43	14 35	0 46	27 58	2 23	0 37
8/18/1967	6 22	3 6	5 37	15 47	29 39	13 2	6 35	15 43	5 16	0sc 3	14 3	1 38	29 40	17 52	4 56	10 43	18 51	0 32	0le 6	6 24	0 32
8/28/1967	8 22	7 11	5 30	21 18	6 30	14 14	8 47	12 20	4 33	2 15	17 51	1 38	1 30	17 1	4 34	13 47	23 19	0 19	2 12	10 20	0 28
9/ 7/1967	10 22	11 16	5 24	27 4	13 33	15 27	10 55	8 53	3 45	4 26	21 44	1 38	3 19	16 45	4 11	16 54	27 59	0 6	4 17	14 12	0 23
9/17/1967	12 20	15 20	5 18	3 6	20 49	16 40	12 59	5 46	2 55	6 36	25 42	1 39	5 8	17 4	3 50	20 5	2 50	0n 6	6 20	17 56	0 19
9/27/1967	14 18	19 22	5 14	9 26	28 19	17 48	15 0	3 17	2 7	8 45	29 43	1 39	6 56	17 56	3 29	23 20	7 50	0 18	8 22	21 33	0 14
10/ 7/1967	16 15	23 22	5 9	16 5	6 5	18 49	16 57	1 33	1 23	10 53	3 47	1 40	8 43	19 17	3 10	26 38	12 58	0 29	10 22	25 1	0 9
10/17/1967	18 12	27 18	5 6	23 2	14 7	19 38	18 52	0 40	0 43	13 0	7 52	1 40	10 30	21 3	2 53	29 60	18 15	0 40	12 21	28 18	0 4
10/27/1967	20 9	1li10	5 3	0gm18	22 25	20 8	20 44	0 33D	0 9	15 5	11 58	1 41	12 16	23 12	2 37	3 25	23 38	0 51	14 19	1vi23	0n 1
11/ 6/1967	22 5	4 57	5 1	7 53	0le57	20 13	22 34	1 8	0 20	17 10	16 4	1 42	14 3	25 40	2 23	6 54	29 9	1 2	16 14	4 12	0 6
11/16/1967	24 1	8 37	5 0	15 47	9 40	19 44	24 21	2 20	0 45	19 13	20 9	1 44	15 48	28 25	2 10	10 26	4 45	1 12	18 9	6 44	0 13
11/26/1967	25 56	12 9	4 59	23 56	18 32	18 31	26 7	4 2	1 6	21 15	24 12	1 45	17 33	1aq23	1 58	14 2	10 26	1 23	20 2	8 56	0 19
12/ 6/1967	27 52	15 32	4 58	2cn21	27 25	16 27	27 51	6 9	1 25	23 15	28 13	1 47	19 18	4 32	1 47	17 42	16 12	1 33	21 54	10 44	0 26
12/16/1967	29 47	18 42	4 59	10 57	6 10	13 22	29 34	8 37	1 41	25 15	2 9	1 49	21 3	7 52	1 37	21 24	22 2	1 43	23 44	12 4	0 34
12/26/1967	1 42	21 38	4 59	19 43	14 37	9 11	1 15	11 22	1 56	27 13	6 1	1 52	22 47	11 19	1 28	25 10	27 56	1 54	25 33	12 54	0 42
1/ 5/1968	3 38	24 16	5 0	28 34	22 34	3 52	2 55	14 22	2 10	29 10	9 46	1 54	24 32	14 52	1 20	28 58	3 52	2 4	27 21	13 10	0 51
1/15/1968	5 33	26 34	5 1	7 26	29 46	2s27	4 34	17 33	2 22	1 6	13 23	1 58	26 16	18 30	1 12	2 49	9 50	2 14	29 7	12 51	1 1
1/25/1968	7 29	28 27	5 2	16 16	5 58	9 32	6 12	20 54	2 34	3 1	16 52	2 1	27 59	22 12	1 4	6 43	15 49	2 24	0vi52	11 57	1 11
2/ 4/1968	9 25	29 51	5 2	25 0	10 51	16 57	7 50	24 22	2 45	4 55	20 10	2 5	29 43	25 57	0 57	10 38	21 49	2 35	2 36	10 33	1 20

Ephemeris 0 Hours ET Mo Dy Year	80 Sappho h long	g long	g lat	433 Eros h long	g long	g lat	1221 Amor h long	g long	g lat	1489 Attila h long	g long	g lat	114 Kassandra h long	g long	g lat	3018 Godiva h long	g long	g lat	2041 Lancelot h long	g long	g lat
2/14/1968	11 21	0 41	5 1	3vi34	14 6	24 9	9 27	27 56	2 56	6 47	23 16	2 10	1 26	29 42	0 50	14 36	27 48	2 45	4 19	8 44	1 29
2/24/1968	13 18	0 54	4 58	11 56	15 29	30 34	11 3	1ar34	3 7	8 39	26 7	2 15	3 10	3 29	0 44	18 35	3 46	2 55	6 0	6 42	1 37
3/ 5/1968	15 15	0 26	4 53	20 3	14 53	35 40	12 39	5 17	3 18	10 29	28 41	2 20	4 53	7 15	0 37	22 35	9 41	3 5	7 41	4 40	1 43
3/15/1968	17 12	29 17	4 43	27 53	12 38	38 57	14 15	9 2	3 30	12 18	0cp56	2 26	6 36	10 59	0 31	26 36	15 34	3 15	9 20	2 49	1 48
3/25/1968	19 10	27 29	4 29	5 25	9 31	40 17	15 50	12 48	3 41	14 6	2 48	2 32	8 20	14 41	0 24	0aq37	21 22	3 25	10 58	1 20	1 51
4/ 4/1968	21 9	25 10	4 9	12 39	6 34	39 46	17 26	16 36	3 54	15 54	4 16	2 39	10 3	18 20	0 18	4 38	27 6	3 36	12 35	0 19	1 53
4/14/1968	23 9	22 35	3 44	19 33	4 38	37 50	19 2	20 24	4 7	17 40	5 15	2 46	11 47	21 55	0 11	8 39	2pi42	3 46	14 11	29le51	1 54
4/24/1968	25 10	19 59	3 15	26 9	4 3	35 2	20 38	24 11	4 21	19 25	5 43	2 54	13 30	25 24	0 4	12 40	8 12	3 56	15 47	29 54D	1 53
5/ 4/1968	27 11	17 40	2 42	2sc27	4 46	31 48	22 15	27 57	4 35	21 8	5 40R	3 0	15 14	28 47	0s 3	16 39	13 32	4 6	17 21	0 27	1 53
5/14/1968	29 13	15 52	2 9	8 28	6 38	28 30	23 52	1ta42	4 51	22 51	5 3	3 6	16 58	2 1	0 11	20 37	18 41	4 16	18 54	1 27	1 52
5/24/1968	1 17	14 42	1 37	14 12	9 21	25 18	25 29	5 24	5 9	24 33	3 57	3 11	18 42	5 7	0 19	24 33	23 38	4 27	20 26	2 51	1 51
6/ 3/1968	3 22	14 13	1 6	19 41	12 43	22 20	27 8	9 4	5 28	26 14	2 24	3 14	20 26	8 2	0 28	28 26	28 21	4 37	21 57	4 36	1 50
6/13/1968	5 28	14 26	0 38	24 56	16 35	19 39	28 47	12 39	5 48	27 54	0 35	3 14	22 10	10 44	0 38	2 18	2 46	4 47	23 28	6 39	1 49
6/23/1968	7 35	15 17	0 13	29 58	20 49	17 14	0ta28	16 10	6 11	29 34	28 38	3 12	23 55	13 11	0 48	6 7	6 51	4 58	24 57	8 57	1 49
7/ 3/1968	9 44	16 43	0n10	4 48	25 18	15 4	2 9	19 36	6 36	1 12	26 46	3 8	25 40	15 20	1 0	9 52	10 33	5 9	26 26	11 27	1 48
7/13/1968	11 54	18 38	0 31	9 28	0sc 1	13 10	3 52	22 54	7 4	2 49	25 8	3 1	27 26	17 8	1 12	13 35	13 46	5 19	27 54	14 8	1 48
7/23/1968	14 6	21 0	0 50	13 58	4 52	11 28	5 37	26 3	7 35	4 26	23 54	2 53	29 11	18 33	1 26	17 15	16 27	5 30	29 21	16 58	1 48
8/ 2/1968	16 20	23 44	1 7	18 19	9 51	9 57	7 23	29 3	8 10	6 1	23 8	2 43	0 58	19 29	1 41	20 51	18 28	5 39	0 48	19 54	1 48
8/12/1968	18 36	26 48	1 23	22 32	14 55	8 36	9 11	1 49	8 49	7 36	22 52	2 34	2 44	19 54	1 58	24 23	19 46	5 48	2 13	22 57	1 48
8/22/1968	20 54	0sc 8	1 37	26 38	20 4	7 24	11 2	4 19	9 33	9 10	23 6	2 24	4 31	19 45	2 15	27 52	20 14	5 53	3 38	26 3	1 49
9/ 1/1968	23 13	3 44	1 51	0cp38	25 18	6 19	12 54	6 30	10 23	10 43	23 49	2 15	6 19	19 1	2 34	1ar18	19 49	5 55	5 3	29 14	1 50
9/11/1968	25 36	7 33	2 4	4 33	0sa35	5 21	14 49	8 16	11 19	12 16	24 58	2 6	8 7	17 42	2 52	4 39	18 32	5 51	6 26	2 26	1 51
9/21/1968	28 0	11 33	2 17	8 23	5 55	4 28	16 47	9 33	12 22	13 47	26 30	1 58	9 56	15 54	3 10	7 57	16 32	5 40	7 50	5 40	1 53
10/ 1/1968	0sa27	15 44	2 29	12 9	11 18	3 41	18 48	10 13	13 31	15 18	28 22	1 51	11 45	13 45	3 25	11 12	14 4	5 20	9 12	8 55	1 54
10/11/1968	2 57	20 5	2 42	15 52	16 44	2 57	20 52	10 10R	14 46	16 49	0cp32	1 44	13 36	11 28	3 38	14 22	11 31	4 54	10 34	12 8	1 57
10/21/1968	5 29	24 33	2 53	19 32	22 13	2 16	23 0	9 17	16 5	18 18	2 57	1 38	15 26	9 18	3 47	17 29	9 15	4 22	11 55	15 20	1 59
10/31/1968	8 5	29 10	3 5	23 10	27 45	1 39	25 12	7 30	17 21	19 47	5 35	1 33	17 18	7 28	3 52	20 33	7 34	3 47	13 16	18 29	2 2
11/10/1968	10 43	3 54	3 17	26 46	3 19	1 5	27 28	4 50	18 29	21 16	8 23	1 28	19 10	6 9	3 54	23 33	6 39	3 13	14 36	21 34	2 5
11/20/1968	13 25	8 44	3 30	0aq22	8 57	0 32	29 50	1 28	19 20	22 43	11 20	1 23	21 3	5 25	3 54	26 30	6 33D	2 40	15 56	24 33	2 9
11/30/1968	16 10	13 40	3 42	3 57	14 37	0 2	2 16	27 45	19 45	24 10	14 25	1 19	22 57	5 20D	3 51	29 23	7 13	2 10	17 15	27 26	2 14
12/10/1968	18 58	18 41	3 55	7 33	20 20	0 27	4 49	24 10	19 42	25 37	17 35	1 15	24 51	5 52	3 48	2 13	8 35	1 44	18 33	0sc10	2 19
12/20/1968	21 50	23 47	4 8	11 9	26 7	0 55	7 28	21 8	19 14	27 3	20 49	1 12	26 47	6 59	3 44	5 0	10 33	1 21	19 52	2 44	2 24
12/30/1968	24 46	28 57	4 22	14 46	1aq56	1 22	10 14	19 0	18 25	28 28	24 6	1 9	28 43	8 36	3 40	7 44	13 0	1 1	21 9	5 6	2 31
1/ 9/1969	27 46	4 11	4 36	18 26	7 49	1 47	13 8	17 55	17 26	29 53	27 25	1 6	0ta41	10 40	3 36	10 25	15 51	0 43	22 27	7 13	2 38
1/19/1969	0cp50	9 29	4 50	22 8	13 45	2 12	16 11	17 54D	16 21	1 17	0aq44	1 4	2 39	13 7	3 33	13 3	19 3	0 28	23 44	9 3	2 45
1/29/1969	3 58	14 50	5 6	25 53	19 45	2 37	19 23	18 52	15 16	2 41	4 3	1 2	4 39	15 54	3 30	15 38	22 30	0 14	25 1	10 33	2 54
2/ 8/1969	7 11	20 13	5 21	29 42	25 49	3 1	22 47	20 43	14 13	4 5	7 20	1 0	6 39	18 58	3 27	18 11	26 11	0 2	26 17	11 42	3 3
2/18/1969	10 28	25 38	5 38	3 35	1pi58	3 24	26 22	23 20	13 13	5 28	10 34	0 58	8 41	22 16	3 25	20 41	0ta 1	0 9	27 33	12 25	3 12
2/28/1969	13 49	1aq 4	5 55	7 34	8 11	3 48	0cn11	26 37	12 17	6 50	13 43	0 56	10 44	25 45	3 23	23 9	4 0	0 19	28 48	12 40	3 22
3/10/1969	17 15	6 32	6 13	11 38	14 29	4 11	4 15	0gm30	11 24	8 12	16 48	0 55	12 48	29 25	3 22	25 35	8 4	0 29	0sc 4	12 28	3 32

Ephemeris	80			433			1221			1489			114			3018			2041		
0 Hours ET	Sappho			Eros			Amor			Attila			Kassandra			Godiva			Lancelot		
Mo Dy Year	h long	g long	g lat	h long	g long	g lat	h long	g long	g lat	h long	g long	g lat	h long	g long	g lat	h long	g long	g lat	h long	g long	g lat
3/20/1969	20 46	12 0	6 31	15 50	20 53	4 33	8 35	4 54	10 32	9 34	19 45	0 53	14 53	3 13	3 21	27 58	12 13	0 37	1 19	11 46	3 42
3/30/1969	24 21	17 27	6 50	20 9	27 23	4 55	13 14	9 47	9 42	10 55	22 35	0 52	16 59	7 8	3 21	0gm19	16 26	0 45	2 33	10 38	3 50
4/ 9/1969	28 1	22 53	7 10	24 36	4 0	5 17	18 14	15 8	8 53	12 16	25 15	0 51	19 7	11 9	3 21	2 38	20 41	0 53	3 48	9 7	3 56
4/19/1969	1aq46	28 18	7 30	29 13	10 45	5 37	23 38	20 54	8 2	13 36	27 44	0 49	21 16	15 15	3 22	4 56	24 57	1 0	5 2	7 20	4 0
4/29/1969	5 34	3 39	7 51	4 1	17 37	5 57	29 26	27 5	7 11	14 57	29 60	0 48	23 27	19 24	3 23	7 11	29 14	1 7	6 16	5 27	4 2
5/ 9/1969	9 28	8 55	8 12	9 1	24 39	6 15	5 44	3 41	6 17	16 16	2 1	0 47	25 39	23 37	3 24	9 25	3 30	1 14	7 30	3 36	4 1
5/19/1969	13 25	14 6	8 33	14 13	1ta51	6 31	12 31	10 42	5 19	17 36	3 44	0 46	27 52	27 51	3 26	11 37	7 47	1 21	8 44	1 57	3 57
5/29/1969	17 27	19 9	8 55	19 39	9 13	6 45	19 52	18 8	4 18	18 55	5 8	0 44	0gm 6	2gm 8	3 28	13 47	12 1	1 28	9 57	0 37	3 51
6/ 8/1969	21 32	24 2	9 16	25 20	16 46	6 55	27 47	25 59	3 13	20 14	6 9	0 43	2 23	6 26	3 30	15 56	16 14	1 34	11 10	29li42	3 44
6/18/1969	25 40	28 43	9 38	1ta18	24 32	7 2	6 17	4 15	2 2	21 33	6 46	0 41	4 40	10 44	3 33	18 4	20 25	1 41	12 24	29 13	3 35
6/28/1969	29 52	3 7	9 59	7 32	2gm29	7 4	15 21	12 56	0 48	22 51	6 57	0 39	6 59	15 2	3 37	20 10	24 33	1 49	13 37	29 12D	3 27
7/ 8/1969	4 6	7 10	10 19	14 5	10 39	7 0	24 55	22 0	0n29	24 10	6 39	0 36	9 20	19 20	3 41	22 15	28 38	1 56	14 50	29 37	3 18
7/18/1969	8 23	10 49	10 37	20 56	19 1	6 51	4li54	1vi26	1 47	25 28	5 54	0 33	11 42	23 36	3 45	24 18	2 39	2 4	16 2	0sc26	3 10
7/28/1969	12 41	13 55	10 52	28 7	27 35	6 34	15 8	11 11	3 4	26 45	4 43	0 30	14 6	27 50	3 50	26 21	6 34	2 12	17 15	1 38	3 2
8/ 7/1969	17 1	16 23	11 4	5 36	6 21	6 9	25 28	21 11	4 18	28 3	3 11	0 26	16 32	2cn 1	3 55	28 23	10 25	2 21	18 28	3 9	2 55
8/17/1969	21 21	18 5	11 10	13 24	15 16	5 36	5 42	1li21	5 23	29 21	1 24	0 22	18 59	6 8	4 1	0cn23	14 9	2 31	19 41	4 57	2 48
8/27/1969	25 42	18 54	11 6	21 29	24 20	4 55	15 39	11 34	6 20	0pi38	29aq31	0 18	21 27	10 10	4 8	2 23	17 45	2 41	20 53	7 0	2 42
9/ 6/1969	0ar 3	18 45R	10 50	29 49	3le30	4 7	25 10	21 45	7 4	1 55	27 42	0 13	23 58	14 5	4 15	4 22	21 12	2 52	22 6	9 17	2 37
9/16/1969	4 22	17 38	10 17	8 22	12 46	3 11	4sa 9	1sc47	7 36	3 12	26 7	0 9	26 30	17 52	4 23	6 20	24 29	3 4	23 18	11 44	2 32
9/26/1969	8 41	15 45	9 26	17 5	22 4	2 10	12 31	11 33	7 56	4 29	24 51	0 4	29 3	21 29	4 31	8 18	27 33	3 18	24 31	14 20	2 28
10/ 6/1969	12 58	13 27	8 15	25 55	1vi23	1 4	20 16	21 1	8 4	5 46	24 1	0s 0	1 38	24 53	4 41	10 14	0le22	3 32	25 44	17 5	2 25
10/16/1969	17 13	11 9	6 51	4 47	10 40	0s 5	27 25	0sa 5	8 3	7 3	23 40	0 4	4 15	28 1	4 51	12 11	2 53	3 48	26 56	19 56	2 22
10/26/1969	21 26	9 19	5 20	13 38	19 52	1 15	3 58	8 46	7 54	8 19	23 47D	0 7	6 53	0le49	5 2	14 6	5 3	4 6	28 9	22 51	2 19
11/ 5/1969	25 35	8 17	3 51	22 24	28 58	2 24	10 0	17 1	7 39	9 36	24 21	0 11	9 33	3 13	5 14	16 1	6 47	4 26	29 22	25 51	2 17
11/15/1969	29 42	8 10D	2 29	1vi 1	7 55	3 31	15 33	24 53	7 21	10 53	25 20	0 14	12 15	5 8	5 27	17 56	8 2	4 47	0sa35	28 53	2 16
11/25/1969	3 45	8 59	1 17	9 27	16 41	4 34	20 40	2cp23	7 0	12 9	26 43	0 16	14 58	6 29	5 39	19 51	8 43	5 9	1 48	1 57	2 15
12/ 5/1969	7 45	10 38	0 14	17 38	25 15	5 33	25 23	9 31	6 38	13 26	28 25	0 19	17 42	7 11	5 50	21 45	8 45R	5 33	3 1	5 1	2 14
12/15/1969	11 41	13 1	0 38	25 34	3sc35	6 28	29 46	16 20	6 15	14 42	0pi25	0 21	20 28	7 8R	6 0	23 39	8 7	5 56	4 14	8 5	2 14
12/25/1969	15 33	15 59	1 22	3li11	11 40	7 18	3 51	22 51	5 52	15 59	2 40	0 23	23 15	6 20	6 5	25 33	6 48	6 18	5 28	11 6	2 14
1/ 4/1970	19 21	19 26	1 58	10 30	19 27	8 3	7 40	29 6	5 30	17 16	5 8	0 25	26 3	4 49	6 6	27 26	4 53	6 37	6 41	14 3	2 15
1/14/1970	23 5	23 16	2 29	17 31	26 57	8 44	11 14	5 6	5 8	18 32	7 47	0 27	28 53	2 45	5 58	29 20	2 31	6 50	7 55	16 56	2 16
1/24/1970	26 45	27 24	2 55	24 12	4 9	9 20	14 36	10 52	4 47	19 49	10 34	0 29	1 44	0 25	5 43	1 13	29cn57	6 56	9 9	19 42	2 18
2/ 3/1970	0gm21	1ta46	3 16	0sc35	11 0	9 53	17 46	16 26	4 27	21 6	13 28	0 31	4 36	28 8	5 21	3 7	27 27	6 55	10 23	22 20	2 20
2/13/1970	3 52	6 20	3 35	6 41	17 31	10 23	20 46	21 47	4 8	22 23	16 28	0 33	7 29	26 14	4 52	5 1	25 18	6 47	11 38	24 48	2 22
2/23/1970	7 20	11 2	3 50	12 30	23 39	10 50	23 37	26 57	3 49	23 40	19 31	0 35	10 24	24 57	4 21	6 55	23 41	6 34	12 52	27 4	2 26
3/ 5/1970	10 43	15 49	4 4	18 4	29 24	11 15	26 20	1pi56	3 31	24 57	22 38	0 38	13 18	24 26	3 48	8 49	22 44	6 18	14 7	29 6	2 29
3/15/1970	14 2	20 42	4 16	23 23	4 43	11 39	28 56	6 44	3 13	26 14	25 46	0 40	16 14	24 42	3 16	10 43	22 27	6 0	15 23	0 52	2 33
3/25/1970	17 17	25 37	4 26	28 29	9 33	12 1	1 25	11 21	2 56	27 32	28 54	0 42	19 11	25 42	2 47	12 38	22 50	5 42	16 38	2 19	2 37
4/ 4/1970	20 28	0gm35	4 36	3 23	13 51	12 22	3 49	15 47	2 39	28 49	2 1	0 45	22 8	27 22	2 19	14 33	23 48	5 24	17 54	3 24	2 42
4/14/1970	23 36	5 33	4 44	8 5	17 32	12 42	6 7	20 2	2 23	0ar 7	5 7	0 48	25 5	29 36	1 55	16 29	25 18	5 7	19 10	4 5	2 47

Ephemeris 0 Hours ET	80 Sappho			433 Eros			1221 Amor			1489 Attila			114 Kassandra			3018 Godiva			2041 Lancelot		
Mo Dy Year	h long	g long	g lat	h long	g long	g lat	h long	g long	g lat	h long	g long	g lat	h long	g long	g lat	h long	g long	g lat	h long	g long	g lat
4/24/1970	26 39	10 32	4 52	12 38	20 29	13 0	8 20	24 5	2 6	1 25	8 10	0 50	28 3	2 19	1 32	18 25	27 14	4 52	20 27	4 20	2 51
5/ 4/1970	29 39	15 30	4 59	17 1	22 35	13 15	10 29	27 56	1 49	2 43	11 8	0 54	1vi 1	5 27	1 12	20 21	29 34	4 38	21 44	4 7	2 56
5/14/1970	2 36	20 27	5 5	21 17	23 40	13 24	12 34	1ar34	1 32	4 2	14 2	0 57	4 0	8 54	0 53	22 18	2 13	4 25	23 1	3 25	2 59
5/24/1970	5 29	25 22	5 12	25 25	23 36R	13 25	14 35	4 57	1 13	5 20	16 49	1 1	6 58	12 39	0 36	24 16	5 8	4 13	24 19	2 18	3 2
6/ 3/1970	8 19	0cn16	5 18	29 27	22 14	13 12	16 34	8 3	0 54	6 39	19 28	1 5	9 56	16 37	0 21	26 15	8 17	4 3	25 37	0 47	3 3
6/13/1970	11 5	5 8	5 24	3 23	19 34	12 39	18 29	10 52	0 34	7 58	21 58	1 9	12 54	20 47	0 7	28 14	11 39	3 54	26 56	29 1	3 1
6/23/1970	13 49	9 57	5 30	7 15	15 50	11 39	20 22	13 20	0 12	9 18	24 16	1 14	15 52	25 6	0n 6	0vi14	15 10	3 45	28 15	27 7	2 58
7/ 3/1970	16 29	14 43	5 37	11 2	11 33	10 14	22 12	15 24	0s11	10 37	26 21	1 20	18 50	29 34	0 18	2 16	18 50	3 37	29 35	25 16	2 53
7/13/1970	19 7	19 26	5 43	14 46	7 22	8 28	24 0	17 1	0 38	11 58	28 11	1 26	21 47	4 7	0 30	4 18	22 38	3 31	0 55	23 36	2 45
7/23/1970	21 42	24 6	5 50	18 26	3 55	6 34	25 46	18 6	1 6	13 18	29 43	1 32	24 44	8 47	0 41	6 21	26 32	3 24	2 15	22 17	2 37
8/ 2/1970	24 14	28 42	5 57	22 5	1 35	4 44	27 31	18 36	1 38	14 39	0 54	1 39	27 39	13 30	0 51	8 26	0vi31	3 19	3 36	21 23	2 27
8/12/1970	26 44	3 14	6 5	25 42	0 28	3 4	29 14	18 25	2 12	16 0	1 42	1 46	0li35	18 17	1 1	10 31	4 36	3 13	4 58	20 57	2 17
8/22/1970	29 11	7 40	6 13	29 18	0 31D	1 37	0 55	17 30	2 50	17 21	2 3	1 54	3 29	23 7	1 10	12 38	8 44	3 9	6 21	21 0D	2 8
9/ 1/1970	1 36	12 2	6 22	2 53	1 35	0 23	2 35	15 51	3 29	18 43	1 56R	2 2	6 22	27 59	1 19	14 46	12 56	3 4	7 44	21 31	1 58
9/11/1970	3 59	16 17	6 32	6 28	3 30	0 39	4 15	13 32	4 8	20 6	1 21	2 10	9 14	2 53	1 28	16 56	17 11	3 0	9 7	22 28	1 49
9/21/1970	6 20	20 25	6 43	10 4	6 7	1 31	5 53	10 41	4 46	21 28	0 17	2 18	12 6	7 47	1 37	19 7	21 29	2 57	10 32	23 50	1 41
10/ 1/1970	8 38	24 25	6 55	13 41	9 18	2 15	7 31	7 34	5 18	22 52	28 48	2 25	14 56	12 42	1 46	21 20	25 48	2 53	11 57	25 34	1 33
10/11/1970	10 55	28 16	7 9	17 20	12 58	2 53	9 8	4 30	5 45	24 15	27 0	2 30	17 44	17 36	1 54	23 34	0li 9	2 50	13 22	27 36	1 26
10/21/1970	13 10	1 56	7 24	21 1	17 2	3 25	10 44	1 46	6 4	25 40	25 3	2 34	20 32	22 29	2 3	25 50	4 30	2 48	14 49	29 55	1 19
10/31/1970	15 24	5 23	7 40	24 45	21 26	3 53	12 20	29pi35	6 16	27 4	23 8	2 36	23 18	27 20	2 12	28 8	8 51	2 45	16 16	2 29	1 13
11/10/1970	17 35	8 34	7 59	28 33	26 9	4 18	13 56	28 8	6 22	28 29	21 24	2 36	26 3	2sc 9	2 21	0li28	13 12	2 42	17 44	5 16	1 8
11/20/1970	19 46	11 28	8 19	2 25	1aq 8	4 40	15 32	27 25	6 24	29 55	20 2	2 34	28 46	6 55	2 31	2 50	17 30	2 40	19 13	8 13	1 2
11/30/1970	21 54	14 0	8 41	6 22	6 20	5 0	17 7	27 24D	6 24	1 22	19 7	2 31	1 28	11 36	2 40	5 15	21 46	2 37	20 42	11 20	0 57
12/10/1970	24 2	16 7	9 6	10 24	11 46	5 17	18 43	28 3	6 22	2 48	18 42	2 28	4 8	16 13	2 51	7 41	25 58	2 35	22 13	14 34	0 53
12/20/1970	26 8	17 45	9 32	14 34	17 24	5 33	20 19	29 17	6 19	4 16	18 49D	2 23	6 47	20 43	3 1	10 9	0sc 4	2 33	23 44	17 55	0 49
12/30/1970	28 12	18 48	10 0	18 50	23 15	5 47	21 56	1 0	6 16	5 44	19 26	2 19	9 24	25 5	3 13	12 40	4 4	2 30	25 17	21 20	0 45
1/ 9/1971	0vi16	19 12	10 28	23 15	29 16	5 59	23 33	3 8	6 14	7 13	20 31	2 14	12 0	29 19	3 25	15 14	7 55	2 27	26 50	24 50	0 41
1/19/1971	2 19	18 54	10 55	27 49	5 28	6 10	25 10	5 38	6 12	8 42	22 1	2 10	14 34	3 22	3 39	17 50	11 34	2 24	28 24	28 22	0 37
1/29/1971	4 20	17 52	11 18	2 34	11 52	6 19	26 48	8 25	6 10	10 12	23 53	2 6	17 7	7 13	3 53	20 29	15 0	2 20	29 59	1 56	0 33
2/ 8/1971	6 21	16 8	11 34	7 29	18 27	6 27	28 28	11 28	6 10	11 43	26 5	2 3	19 37	10 50	4 8	23 11	18 8	2 15	1 36	5 31	0 30
2/18/1971	8 21	13 52	11 40	12 38	25 12	6 32	0ta 8	14 43	6 11	13 14	28 32	2 0	22 7	14 10	4 25	25 55	20 55	2 9	3 13	9 5	0 26
2/28/1971	10 20	11 15	11 34	17 59	2ar 9	6 35	1 49	18 9	6 12	14 47	1ta14	1 57	24 34	17 10	4 43	28 43	23 17	2 1	4 51	12 38	0 22
3/10/1971	12 19	8 36	11 16	23 36	9 18	6 35	3 32	21 44	6 15	16 20	4 7	1 55	27 0	19 47	5 3	1 33	25 7	1 52	6 31	16 8	0 19
3/20/1971	14 17	6 12	10 47	29 29	16 38	6 32	5 16	25 26	6 18	17 53	7 10	1 53	29 25	21 58	5 25	4 27	26 21	1 40	8 12	19 34	0 15
3/30/1971	16 14	4 17	10 9	5 38	24 10	6 26	7 2	29 16	6 23	19 28	10 21	1 52	1 48	23 38	5 48	7 24	26 53	1 25	9 54	22 56	0 11
4/ 9/1971	18 11	3 0	9 26	12 5	1ta54	6 16	8 49	3 11	6 28	21 3	13 38	1 50	4 9	24 44	6 11	10 24	26 38	1 7	11 37	26 11	0 6
4/19/1971	20 7	2 24	8 43	18 51	9 50	6 1	10 39	7 10	6 35	22 39	17 1	1 50	6 29	25 13	6 36	13 28	25 36	0 44	13 21	29 18	0 2
4/29/1971	22 3	2 29D	7 59	25 56	17 59	5 42	12 31	11 15	6 42	24 16	20 28	1 49	8 47	25 0	6 59	16 35	23 51	0 18	15 7	2 16	0 3
5/ 9/1971	23 59	3 11	7 19	3gm20	26 18	5 17	14 26	15 23	6 51	25 54	23 58	1 49	11 4	24 8	7 20	19 46	21 34	0n11	16 53	5 2	0 9
5/19/1971	25 54	4 25	6 41	11 2	4 48	4 47	16 23	19 34	7 1	27 33	27 30	1 49	13 20	22 39	7 37	23 1	19 5	0 42	18 42	7 35	0 15

Ephemeris	80			433			1221			1489			114			3018			2041		
0 Hours ET	Sappho			Eros			Amor			Attila			Kassandra			Godiva			Lancelot		
Mo Dy Year	h long	g long	g lat	h long	g long	g lat	h long	g long	g lat	h long	g long	g lat	h long	g long	g lat	h long	g long	g lat	h long	g long	g lat
5/29/1971	27 50	6 9	6 6	19 2	13 29	4 12	18 23	23 49	7 12	29 12	1gm 4	1 50	15 33	20 42	7 48	26 19	16 45	1 11	20 31	9 52	0 21
6/ 8/1971	29 45	8 17	5 34	27 18	22 19	3 31	20 26	28 7	7 24	0 53	4 38	1 50	17 46	18 29	7 51	29 40	14 56	1 39	22 22	11 50	0 28
6/18/1971	1 40	10 45	5 5	5 48	1cn16	2 46	22 33	2 29	7 38	2 35	8 12	1 51	19 57	16 16	7 46	3 5	13 52	2 2	24 15	13 25	0 36
6/28/1971	3 36	13 31	4 39	14 28	10 20	1 57	24 44	6 52	7 54	4 17	11 44	1 53	22 7	14 17	7 34	6 34	13 39D	2 22	26 9	14 35	0 45
7/ 8/1971	5 31	16 32	4 15	23 16	19 28	1 6	27 0	11 19	8 10	6 1	15 14	1 54	24 15	12 44	7 16	10 7	14 18	2 38	28 4	15 16	0 55
7/18/1971	7 27	19 45	3 53	2le 8	28 38	0 12	29 20	15 49	8 29	7 45	18 42	1 56	26 22	11 46	6 53	13 42	15 46	2 52	0pi 1	15 25	1 6
7/28/1971	9 23	23 8	3 33	11 0	7 50	0 41	1 45	20 21	8 50	9 30	22 5	1 59	28 28	11 25	6 30	17 21	17 57	3 2	1 59	15 0	1 17
8/ 7/1971	11 19	26 41	3 15	19 47	17 0	1 34	4 16	24 57	9 12	11 17	25 22	2 1	0cp33	11 42	6 5	21 4	20 45	3 10	3 59	14 1	1 29
8/17/1971	13 15	0li22	2 57	28 28	26 7	2 24	6 53	29 35	9 37	13 5	28 33	2 5	2 36	12 32	5 42	24 49	24 6	3 17	6 0	12 33	1 41
8/27/1971	15 13	4 9	2 41	6 57	5 10	3 11	9 38	4 17	10 5	14 53	1 36	2 8	4 38	13 53	5 19	28 38	27 54	3 22	8 3	10 43	1 52
9/ 6/1971	17 10	8 2	2 26	15 14	14 5	3 53	12 30	9 1	10 35	16 43	4 28	2 12	6 39	15 42	4 59	2 29	2sa 6	3 26	10 8	8 42	2 1
9/16/1971	19 8	12 0	2 11	23 14	22 53	4 30	15 30	13 49	11 8	18 34	7 7	2 17	8 39	17 54	4 40	6 22	6 37	3 28	12 14	6 44	2 9
9/26/1971	21 7	16 2	1 57	0li57	1li32	5 2	18 40	18 41	11 44	20 26	9 31	2 22	10 38	20 25	4 22	10 17	11 26	3 30	14 22	5 2	2 14
10/ 6/1971	23 7	20 8	1 44	8 21	10 1	5 28	22 1	23 36	12 24	22 19	11 36	2 27	12 36	23 14	4 7	14 15	16 28	3 32	16 31	3 47	2 18
10/16/1971	25 7	24 16	1 30	15 27	18 18	5 50	25 33	28 35	13 8	24 13	13 19	2 33	14 33	26 16	3 52	18 14	21 43	3 32	18 42	3 7	2 20
10/26/1971	27 9	28 27	1 17	22 14	26 25	6 6	29 19	3 38	13 55	26 8	14 36	2 39	16 29	29 31	3 39	22 14	27 8	3 32	20 55	3 5D	2 20
11/ 5/1971	29 11	2 38	1 4	28 43	4 19	6 17	3 19	8 46	14 46	28 5	15 23	2 46	18 25	2 55	3 28	26 15	2cp42	3 31	23 9	3 40	2 19
11/15/1971	1 15	6 51	0 51	4 54	12 2	6 25	7 35	13 59	15 40	0cn 3	15 37	2 52	20 19	6 28	3 17	0aq16	8 22	3 30	25 25	4 51	2 18
11/25/1971	3 20	11 3	0 37	10 48	19 33	6 29	12 9	19 19	16 37	2 1	15 15	2 58	22 12	10 7	3 8	4 17	14 9	3 28	27 42	6 34	2 16
12/ 5/1971	5 26	15 15	0 23	16 26	26 52	6 29	17 4	24 46	17 35	4 1	14 17	3 2	24 5	13 52	2 59	8 18	20 0	3 26	0ar 1	8 46	2 15
12/15/1971	7 33	19 25	0 9	21 49	3 59	6 27	22 21	0vi25	18 32	6 2	12 48	3 4	25 57	17 40	2 51	12 18	25 55	3 24	2 22	11 23	2 13
12/25/1971	9 42	23 32	0n 6	26 59	10 56	6 22	28 3	6 18	19 23	8 5	10 54	3 4	27 48	21 32	2 44	16 17	1aq52	3 21	4 44	14 22	2 11
1/ 4/1972	11 52	27 36	0 23	1sa56	17 42	6 15	4 13	12 33	20 2	10 8	8 49	3 0	29 38	25 25	2 38	20 15	7 50	3 18	7 8	17 39	2 10
1/14/1972	14 4	1sa34	0 40	6 42	24 17	6 7	10 53	19 18	20 17	12 13	6 47	2 54	1 28	29 20	2 32	24 11	13 48	3 14	9 33	21 12	2 9
1/24/1972	16 18	5 26	0 59	11 18	0cp41	5 56	18 5	26 48	19 49	14 19	5 2	2 45	3 17	3 14	2 26	28 5	19 47	3 10	12 0	24 58	2 8
2/ 3/1972	18 34	9 10	1 20	15 44	6 55	5 44	25 51	5 25	18 12	16 26	3 46	2 35	5 5	7 7	2 21	1 56	25 44	3 6	14 28	28 56	2 7
2/13/1972	20 52	12 44	1 43	20 2	12 59	5 31	4vi12	15 32	14 39	18 34	3 5	2 23	6 53	10 58	2 16	5 45	1pi38	3 2	16 57	3 2	2 6
2/23/1972	23 12	16 6	2 9	24 12	18 52	5 16	13 7	27 33	8 18	20 43	3 2D	2 11	8 40	14 46	2 12	9 31	7 31	2 57	19 28	7 17	2 6
3/ 4/1972	25 34	19 13	2 38	28 16	24 35	5 0	22 33	11 27	1n23	22 53	3 37	2 0	10 27	18 30	2 8	13 14	13 19	2 53	22 0	11 38	2 5
3/14/1972	27 59	22 1	3 10	2 14	0aq 8	4 42	2li26	26 28	13 7	25 5	4 46	1 48	12 14	22 10	2 4	16 53	19 4	2 48	24 33	16 5	2 5
3/24/1972	0sa26	24 28	3 47	6 6	5 30	4 22	12 36	11 0	23 58	27 17	6 27	1 38	13 59	25 43	2 1	20 29	24 44	2 43	27 7	20 35	2 5
4/ 3/1972	2 56	26 28	4 28	9 55	10 39	4 0	22 55	23 35	31 58	29 31	8 34	1 29	15 45	29 8	1 57	24 2	0ar19	2 38	29 42	25 8	2 6
4/13/1972	5 28	27 56	5 15	13 39	15 37	3 36	3sc11	3cp27	37 16	1 45	11 5	1 20	17 30	2 25	1 54	27 31	5 49	2 32	2 18	29 44	2 6
4/23/1972	8 4	28 47	6 6	17 21	20 21	3 8	13 13	10 28	40 40	4 1	13 55	1 12	19 15	5 32	1 50	0ar56	11 12	2 27	4 54	4 21	2 7
5/ 3/1972	10 42	28 56R	7 3	21 0	24 50	2 37	22 51	14 48	42 48	6 17	17 3	1 5	20 59	8 27	1 47	4 18	16 28	2 21	7 32	8 59	2 7
5/13/1972	13 24	28 20	8 3	24 38	29 1	2 2	1sa58	16 39	43 59	8 35	20 24	0 58	22 44	11 8	1 44	7 36	21 37	2 15	10 10	13 37	2 8
5/23/1972	16 9	26 58	9 3	28 14	2 52	1 22	10 30	16 15R	44 16	10 53	23 57	0 51	24 28	13 33	1 40	10 51	26 38	2 9	12 48	18 13	2 9
6/ 2/1972	18 58	24 56	9 59	1 49	6 20	0 35	18 24	14 4	43 31	13 12	27 41	0 45	26 11	15 40	1 36	14 2	1ta31	2 3	15 27	22 49	2 10
6/12/1972	21 50	22 25	10 47	5 24	9 18	0n20	25 41	10 48	41 38	15 32	1le33	0 40	27 55	17 24	1 31	17 9	6 14	1 56	18 6	27 22	2 12
6/22/1972	24 46	19 46	11 22	9 0	11 41	1 24	2cp24	7 21	38 40	17 52	5 32	0 35	29 38	18 44	1 26	20 12	10 46	1 50	20 46	1gm52	2 13

Ephemeris	80			433			1221			1489			114			30 8			2041		
0 Hours ET	Sappho			Eros			Amor			Attila			Kassandra			Godiva			Lancelot		
Mo Dy Year	h long	g long	g lat	h long	g long	g lat	h long	g long	g lat	h long	g long	g lat	h long	g long	g lat	h ong	g long	g lat	h long	g long	g lat
7/ 2/1972	27 46	17 21	11 42	12 36	13 21	2 41	8 33	4 29	34 54	20 13	9 37	0 30	1 22	19 36	1 20	23 13	15 7	1 42	23 25	6 18	2 15
7/12/1972	0cp50	15 27	11 47	16 14	14 8	4 10	14 13	2 37	30 44	22 35	13 47	0 25	3 5	19 56	1 14	26 9	19 15	1 35	26 4	10 40	2 17
7/22/1972	3 58	14 20	11 40	19 55	13 52R	5 54	19 27	1 52	26 37	24 58	18 1	0 20	4 48	19 43	1 6	29 3	23 8	1 27	28 44	14 56	2 19
8/ 1/1972	7 11	14 3	11 24	23 38	12 23	7 51	24 16	2 9	22 48	27 21	22 19	0 16	6 31	18 55	0 56	1 53	26 44	1 18	1 22	19 5	2 21
8/11/1972	10 28	14 39	11 3	27 24	9 39	9 55	28 44	3 17	19 25	29 44	26 39	0 11	8 14	17 35	0 46	4 40	0gm 0	1 9	4 1	23 5	2 23
8/21/1972	13 49	16 3	10 39	1pi15	5 51	11 56	2 53	5 7	16 30	2 8	1vi 0	0 7	9 57	15 46	0 34	7 24	2 54	0 59	6 39	26 55	2 26
8/31/1972	17 15	18 10	10 14	5 10	1 29	13 38	6 46	7 30	14 2	4 32	5 23	0 2	11 41	13 40	0 21	10 5	5 22	0 47	9 16	0cn33	2 29
9/10/1972	20 46	20 55	9 49	9 11	27 13	14 51	10 24	10 18	11 57	6 56	9 46	0n 2	13 24	11 26	0 8	12 44	7 18	0 35	11 53	3 57	2 32
9/25/1972	26 10	26 3	9 12	15 24	22 31	15 40	15 26	15 6	9 25	10 33	16 20	0 9	15 59	8 22	0 12	16 36	9 5	0 13	15 46	8 30	2 37
10/ 5/1972	29 52	0cp 2	8 48	19 42	21 1	15 41	18 34	18 36	8 3	12 57	20 41	0 14	17 43	6 48	0 24	19 8	9 25	0s 4	18 21	11 5	2 41
10/15/1972	3 39	4 24	8 25	24 8	20 54D	15 26	21 32	22 17	6 53	15 22	24 59	0 18	19 27	5 46	0 35	21 37	8 58	0 22	20 54	13 13	2 44
10/25/1972	7 30	9 6	8 2	28 44	22 3	15 3	24 21	26 7	5 54	17 46	29 14	0 23	21 11	5 18	0 45	24 4	7 45	0 42	23 26	14 52	2 48
11/ 4/1972	11 25	14 4	7 40	3 31	24 17	14 34	27 2	0aq 3	5 2	20 11	3 24	0 29	22 55	5 27D	0 53	26 28	5 51	1 3	25 57	15 55	2 51
11/14/1972	15 25	19 18	7 19	8 29	27 27	14 3	29 37	4 5	4 18	22 35	7 28	0 34	24 40	6 9	1 1	28 51	3 26	1 23	28 27	16 20	2 54
11/24/1972	19 28	24 44	6 58	13 39	1pi24	13 31	2 4	8 11	3 39	24 59	11 24	0 40	26 25	7 23	1 7	1 11	0 47	1 42	0cn56	16 4	2 55
12/ 4/1972	23 34	0aq21	6 37	19 4	6 0	12 58	4 26	12 19	3 5	27 22	15 11	0 46	28 11	9 4	1 13	3 29	28 13	1 58	3 23	15 7	2 54
12/14/1972	27 44	6 8	6 16	24 43	11 11	12 24	6 43	16 31	2 34	29 46	18 47	0 53	29 56	11 9	1 19	5 46	26 4	2 11	5 49	13 35	2 51
12/24/1972	1pi57	12 3	5 55	0ta39	16 53	11 49	8 56	20 44	2 7	2 8	22 9	1 0	1 43	13 35	1 23	8 0	24 31	2 21	8 13	11 38	2 44
1/ 3/1973	6pi12	18aq 5	5n34	6ta52	23pi 1	11n11	11pi 4	24aq57	1n42	4li31	25li15	1n 9	3ar29	16pi19	1s28	10 m13	23ta40R	2s28	10cn36	9cn31R	2s35
1/13/1973	10 29	24 12	5 13	13 23	29 35	10 31	13 8	29 11	1 20	6 52	28 1	1 17	5 17	19 17	1 32	12 25	23 34D	2 32	12 57	7 30	2 23
1/23/1973	14 47	0pi23	4 52	20 12	6 31	9 48	15 9	3 25	0 59	9 13	0sc25	1 27	7 4	22 28	1 37	14 34	24 9	2 35	15 17	5 50	2 9
2/ 2/1973	19 7	6 38	4 31	27 21	13 49	9 0	17 6	7 38	0 40	11 34	2 22	1 37	8 53	25 49	1 41	16 43	25 21	2 37	17 35	4 41	1 55
2/12/1973	23 27	12 55	4 9	4 48	21 26	8 8	19 1	11 49	0 22	13 53	3 48	1 49	10 42	29 19	1 45	18 50	27 4	2 38	19 52	4 11	1 40
2/22/1973	27 47	19 13	3 47	12 34	29 22	7 12	20 53	15 59	0 4	16 12	4 40	2 1	12 31	2 57	1 49	20 55	29 14	2 39	22 7	4 18D	1 27
3/ 4/1973	2ar 7	25 31	3 25	20 37	7 34	6 10	22 43	20 6	0 13	18 30	4 55	2 13	14 21	6 40	1 54	23 0	1 46	2 40	24 20	5 3	1 14
3/14/1973	6 26	1ar49	3 3	28 56	16 2	5 3	24 31	24 10	0 29	20 47	4 31	2 26	16 12	10 27	1 58	25 3	4 36	2 40	26 32	6 20	1 3
3/24/1973	10 43	8 5	2 40	7 28	24 43	3 52	26 16	28 11	0 45	23 3	3 31	2 38	18 4	14 18	2 3	27 5	7 41	2 40	28 42	8 5	0 52
4/ 3/1973	14 59	14 19	2 17	16 11	3gm35	2 37	28 1	2 8	1 2	25 19	2 0	2 48	19 56	18 12	2 8	29 7	10 57	2 41	0le50	10 15	0 43
4/13/1973	19 12	20 31	1 54	25 0	12 37	1 20	29 43	6 1	1 18	27 33	0 8	2 57	21 50	22 7	2 14	1 7	14 24	2 42	2 57	12 45	0 34
4/23/1973	23 23	26 38	1 31	3le52	21 46	0 3	1 24	9 49	1 35	29 46	28 8	3 2	23 44	26 3	2 19	3 6	17 59	2 43	5 2	15 32	0 27
5/ 3/1973	27 31	2ta43	1 7	12 44	1cn 0	1 13	3 5	13 31	1 52	1 58	26 14	3 5	25 39	29 59	2 26	5 5	21 39	2 44	7 6	18 32	0 20
5/13/1973	1ta35	8 42	0 43	21 31	10 17	2 26	4 44	17 6	2 11	4 9	24 40	3 4	27 35	3 54	2 32	7 3	25 25	2 45	9 8	21 43	0 13
5/23/1973	5 36	14 37	0 19	0vi10	19 34	3 33	6 22	20 34	2 30	6 19	23 35	3 1	29 32	7 48	2 40	9 0	29 15	2 47	11 8	25 4	0 8
6/ 2/1973	9 34	20 26	0s 5	8 37	28 49	4 34	7 59	23 54	2 51	8 28	23 4	2 57	1 30	11 40	2 47	10 57	3 7	2 49	13 7	28 31	0 2
6/12/1973	13 28	26 9	0 29	16 50	8 0	5 26	9 36	27 4	3 13	10 36	23 8D	2 51	3 29	15 29	2 56	12 53	7 2	2 51	15 5	2 4	0 3
6/22/1973	17 17	1gm46	0 54	24 48	17 5	6 10	11 13	0ta 2	3 38	12 43	23 46	2 45	5 29	19 14	3 5	14 48	10 58	2 54	17 0	5 42	0 7
7/ 2/1973	21 3	7 15	1 20	2li27	26 1	6 44	12 49	2 48	4 5	14 48	24 55	2 39	7 30	22 54	3 15	16 43	14 55	2 57	18 55	9 22	0 12
7/12/1973	24 45	12 37	1 46	9 49	4 47	7 10	14 25	5 18	4 34	16 53	26 31	2 33	9 32	26 28	3 26	18 38	18 52	3 0	20 48	13 5	0 16
7/22/1973	28 22	17 51	2 12	16 51	13 22	7 26	16 1	7 29	5 6	18 56	28 32	2 27	11 35	29 54	3 39	20 32	22 48	3 4	22 40	16 49	0 20
8/ 1/1973	1 56	22 55	2 40	23 34	21 46	7 35	17 36	9 19	5 42	20 58	0sc53	2 22	13 40	3 12	3 52	22 27	26 44	3 8	24 30	20 34	0 25
8/11/1973	5 25	27 49	3 9	29 60	29 57	7 37	19 12	10 43	6 22	22 58	3 31	2 17	15 46	6 17	4 7	24 20	0le37	3 13	26 19	24 18	0 29
8/21/1973	8 50	2 32	3 39	6 7	7 55	7 33	20 49	11 37	7 7	24 58	6 24	2 12	17 53	9 10	4 23	26 14	4 28	3 19	28 6	28 2	0 33

Ephemeris	80			433			1221			1489			114			30 8			2041		
0 Hours ET	Sappho			Eros			Amor			Attila			Kassandra			Godiva			Lancelot		
Mo Dy Year	h long	g long	g lat	h long	g long	g lat	h long	g long	g lat	h long	g long	g lat	h long	g long	g lat	h ong	g long	g lat	h long	g long	g lat
8/31/1973	12 11	7 2	4 11	11 58	15 42	7 24	22 25	11 55	7 55	26 57	9 29	2 8	20 2	11 45	4 41	28 8	8 15	3 25	29 52	1vi43	0 37
9/10/1973	15 28	11 17	4 45	17 33	23 17	7 12	24 2	11 33	8 46	28 54	12 45	2 5	22 11	14 1	5 1	0 e 2	11 59	3 31	1 37	5 22	0 41
9/20/1973	18 41	15 15	5 21	22 54	0sc40	6 56	25 40	10 28	9 40	0sa50	16 9	2 2	24 23	15 52	5 23	1 55	15 36	3 39	3 21	8 57	0 46
9/30/1973	21 50	18 53	6 0	28 1	7 53	6 37	27 19	8 38	10 32	2 45	19 40	1 59	26 35	17 15	5 46	3 49	19 7	3 48	5 3	12 28	0 50
10/10/1973	24 56	22 7	6 42	2 56	14 57	6 17	28 58	6 8	11 21	4 39	23 17	1 56	28 49	18 5	6 11	5 43	22 30	3 57	6 44	15 54	0 55
10/20/1973	27 57	24 54	7 28	7 40	21 51	5 56	0ta39	3 8	12 0	6 32	26 58	1 54	1gm 5	18 18	6 37	7 37	25 43	4 8	8 24	19 12	1 01
10/25/1973	29 27	26 6	7 52	9 58	25 15	5 45	1 30														
11/ 4/1973	2 23	28 2	8 43	14 27	1sa57	5 22	3 12	28 20	12 34	9 19	2 37	1 52	4 31	17 19	7 14	10 28	0vi 9	4 26	10 52	23 54	1 8
11/14/1973	5 16	29 18	9 38	18 47	8 32	4 59	4 56	25 24	12 39	11 9	6 24	1 50	6 51	15 51	7 34	12 23	2 46	4 40	12 30	26 49	1 14
11/24/1973	8 6	29 47	10 35	22 59	15 1	4 35	6 42	23 2	12 32	12 58	10 13	1 49	9 12	13 50	7 49	14 18	5 3	4 55	14 6	29 31	1 21
12/ 4/1973	10 53	29 25	11 32	27 5	21 24	4 12	8 29	21 24	12 15	14 45	14 1	1 49	11 34	11 30	7 56	16 13	6 57	5 13	15 41	1 58	1 28
12/14/1973	13 36	28 12	12 25	1cp 4	27 42	3 48	10 19	20 35	11 53	16 32	17 48	1 48	13 59	9 8	7 53	18 9	8 23	5 31	17 15	4 7	1 36
12/24/1973	16 17	26 13	13 10	4 59	3 54	3 24	12 10	20 32D	11 28	18 18	21 33	1 48	16 24	7 2	7 42	20 6	9 16	5 51	18 48	5 56	1 44
1/ 3/1974	18 55	23 39	13 41	8 48	10 3	3 0	14 5	21 14	11 2	20 3	25 15	1 48	18 52	5 27	7 23	22 3	9 32	6 11	20 21	7 21	1 54
1/ 8/1974	20 12	22 15	13 50	10 42	13 5	2 48	15 3	21 50	10 50	20 55	27 4	1 49	20 6	4 55	7 12	23 2	9 25R	6 21	21 6	7 54	1 58
1/18/1974	22 46	19 28	13 55	14 26	19 7	2 23	17 1	23 28	10 26	22 38	0cp39	1 49	22 36	4 23	6 48	25 0	8 39	6 41	22 37	8 38	2 9
1/23/1974	24 2	18 9	13 51	16 16	22 7	2 11	18 1	24 29	10 15	23 29	2 24	1 50	23 52	4 24D	6 35	25 59	8 1	6 49	23 22	8 49	2 14
2/ 2/1974	26 32	15 55	13 30	19 56	28 4	1 46	20 4	26 52	9 55	25 11	5 49	1 51	26 24	4 58	6 9	27 59	6 15	7 4	24 52	8 46R	2 24
2/ 7/1974	27cn46	15cn 3R	13s16	21cp45	1aq 1	1s33	21ta 7	28ar14	9s45	26sa 1	7cp29	1n52	27gm41	5gm31	5s57	28 e59	5vi11R	7s 9	25vi36	8li33R	2n30
2/17/1974	0le12	13 53	12 40	25 22	6 53	1 7	23 15	1ta13	9 28	27 42	10 41	1 54	0cn16	7 4	5 32	0 i59	2 44	7 14	27 5	7 43	2 40
2/27/1974	2 36	13 30	12 0	28 58	12 43	0 41	25 27	4 33	9 13	29 21	13 42	1 56	2 52	9 11	5 9	3 1	0 8	7 11	28 32	6 25	2 50
3/ 9/1974	4 58	13 52	11 19	2 33	18 30	0 13	27 44	8 11	8 59	0 59	16 31	1 58	5 31	11 46	4 47	5 3	27 41	7 0	29 59	4 43	2 58
3/19/1974	7 18	14 54	10 39	6 9	24 14	0 16	0gm 6	12 4	8 48	2 37	19 6	2 1	8 10	14 46	4 28	7 7	25 39	6 43	1 25	2 47	3 4
3/29/1974	9 36	16 31	10 2	9 44	29 57	0 46	2 33	16 11	8 37	4 14	21 24	2 4	10 52	18 6	4 9	9 12	24 12	6 21	2 50	0 48	3 8
4/ 8/1974	11 52	18 36	9 27	13 21	5 38	1 18	5 6	20 31	8 28	5 50	23 24	2 7	13 35	21 43	3 53	11 18	23 26	5 57	4 15	28 56	3 9
4/18/1974	14 7	21 5	8 55	17 0	11 17	1 52	7 46	25 3	8 20	7 25	25 1	2 11	16 19	25 35	3 37	13 25	23 23D	5 32	5 39	27 22	3 8
4/28/1974	16 19	23 53	8 27	20 41	16 55	2 28	10 33	29 47	8 13	8 59	26 15	2 15	19 5	29 39	3 23	15 34	24 0	5 8	7 2	26 12	3 6
5/ 8/1974	18 30	26 57	8 1	24 24	22 33	3 6	13 28	4 42	8 6	10 33	27 1	2 19	21 52	3 53	3 9	17 44	25 14	4 45	8 24	25 30	3 2
5/18/1974	20 40	0le15	7 38	28 12	28 9	3 48	16 32	9 48	8 0	12 6	27 19	2 23	24 41	8 16	2 57	19 56	26 59	4 24	9 46	25 18	2 57
5/28/1974	22 48	3 42	7 17	2 3	3 46	4 33	19 45	15 6	7 53	13 38	27 6	2 26	27 31	12 46	2 45	22 9	29 12	4 4	11 8	25 34	2 52
6/ 7/1974	24 55	7 18	6 59	5 59	9 22	5 22	23 10	20 35	7 46	15 9	26 23	2 28	0le22	17 22	2 33	24 24	1 48	3 46	12 29	26 16	2 47
6/17/1974	27 0	11 1	6 42	10 1	14 59	6 15	26 47	26 18	7 38	16 40	25 12	2 30	3 14	22 4	2 23	26 41	4 45	3 29	13 49	27 23	2 42
6/27/1974	29 5	14 50	6 27	14 10	20 37	7 13	0cn37	2cn14	7 30	18 9	23 38	2 30	6 7	26 50	2 12	29 0	7 59	3 14	15 9	28 50	2 38
7/ 7/1974	1 8	18 43	6 13	18 25	26 15	8 17	4 42	8 24	7 19	19 39	21 48	2 28	9 2	1le40	2 2	1 20	11 29	2 59	16 28	0li36	2 34
7/17/1974	3 10	22 40	6 0	22 49	1ta55	9 26	9 5	14 49	7 7	21 7	19 54	2 24	11 57	6 33	1 52	3 43	15 11	2 46	17 47	2 38	2 30
7/27/1974	5 12	26 40	5 49	27 22	7 36	10 43	13 46	21 31	6 51	22 35	18 4	2 19	14 53	11 28	1 42	6 8	19 5	2 34	19 5	4 53	2 27
8/ 6/1974	7 12	0vi41	5 39	2ar 5	13 18	12 8	18 48	28 32	6 32	24 3	16 29	2 12	17 50	16 25	1 32	8 35	23 10	2 23	20 23	7 20	2 24
8/16/1974	9 12	4 44	5 30	7 0	19 2	13 41	24 14	5 51	6 9	25 30	15 15	2 4	20 48	21 24	1 23	11 4	27 23	2 12	21 40	9 56	2 21
8/26/1974	11 11	8 48	5 21	12 6	24 47	15 22	0le 5	13 32	5 41	26 56	14 29	1 56	23 46	26 23	1 13	13 36	1li45	2 1	22 57	12 40	2 19
9/ 5/1974	13 9	12 51	5 14	17 26	0gm33	17 13	6 25	21 34	5 7	28 22	14 13	1 48	26 44	1vi23	1 3	16 10	6 14	1 51	24 14	15 31	2 18
9/15/1974	15 7	16 55	5 7	23 1	6 20	19 14	13 16	29 58	4 26	29 47	14 25	1 40	29 43	6 22	0 53	18 47	10 50	1 42	25 30	18 27	2 17
9/25/1974	17 4	20 57	5 1	28 51	12 5	21 23	20 40	8 46	3 39	1 11	15 4	1 32	2 42	11 20	0 43	21 26	15 32	1 32	26 46	21 27	2 16

Ephemeris	80			433			1221			1489			114			3018			2041		
0 Hours ET	Sappho			Eros			Amor			Attila			Kassandra			Godiva			Lancelot		
Mo Dy Year	h long	g long	g lat	h long	g long	g lat	h long	g long	g lat	h long	g long	g lat	h long	g long	g lat	h long	g long	g lat	h long	g long	g lat
10/ 5/1974	19 1	24 57	4 56	4 59	17 48	23 39	28 39	17 55	2 44	2 36	16 10	1 25	5 41	16 16	0 33	24 9	20 19	1 23	28 2	24 30	2 16
10/15/1974	20 57	28 55	4 51	11 24	23 26	25 59	7 12	27 25	1 43	3 59	17 38	1 18	8 40	21 10	0 22	26 54	25 12	1 13	29 17	27 35	2 16
10/25/1974	22 53	2 49	4 47	18 7	28 54	28 20	16 19	7 13	0 37	5 22	19 25	1 12	11 39	26 1	0 11	29 43	0sc10	1 4	0sc32	0sc40	2 17
11/ 4/1974	24 49	6 39	4 44	25 10	4 10	30 34	25 55	17 15	0n33	6 45	21 31	1 6	14 38	0li46	0n 1	2 34	5 11	0 55	1 46	3 45	2 18
11/14/1974	26 44	10 23	4 41	2gm31	9 5	32 32	5 55	27 26	1 43	8 8	23 51	1 1	17 36	5 26	0 13	5 29	10 16	0 45	3 1	6 49	2 20
11/24/1974	28 40	14 0	4 39	10 11	13 29	33 59	16 11	7 41	2 52	9 30	26 23	0 57	20 34	9 59	0 26	8 27	15 25	0 35	4 15	9 49	2 22
11/29/1974	29 37	15 46	4 37	14 8	15 27	34 25	21 21	12 84	3 25	10 10	27 44	0 54	22 3	12 12	0 33	9 57	18 0	0 30	4 52	11 18	2 23
12/ 9/1974	1 33	19 9	4 36	22 14	18 47	34 26	1sc38	22 56	4 25	11 32	0aq31	0 50	25 0	16 31	0 48	13 0	23 13	0 20	6 6	14 12	2 26
12/19/1974	3 28	22 20	4 34	0cn35	21 8	32 49	11 43	2sa52	5 18	12 53	3 27	0 47	27 56	20 38	1 3	16 6	28 27	0 9	7 19	16 59	2 29
12/29/1974	5 24	25 18	4 33	9 9	22 29	28 44	21 25	12 31	6 2	14 14	6 28	0 43	0li52	24 31	1 20	19 16	3 43	0n 2	8 33	19 39	2 33
1/ 3/1975	6 21	26 40	4 32	13 30	22 50	25 31	26 6	17 13	6 21	14 54	8 1	0 42	2 19	26 21	1 29	20 52	6 21	0 8	9 10	20 56	2 36
1/13/1975	8 17	29 11	4 31	22 17	23 4	16 24	5 1	26 18	6 52	16 14	11 9	0 39	5 13	29 47	1 49	24 8	11 37	0 21	10 23	23 20	2 40
1/23/1975	10 13	1 20	4 30	1le 9	23 16	4 22	13 19	4 59	7 15	17 34	14 19	0 36	8 6	2 50	2 10	27 26	16 53	0 34	11 36	25 32	2 46
2/ 2/1975	12 10	3 2	4 28	10 1	24 5	8 1	21 1	13 13	7 31	18 53	17 32	0 33	10 58	5 27	2 34	0sa49	22 7	0 49	12 49	27 28	2 52
2/12/1975	14 7	4 14	4 25	18 49	25 54	18 4	28 6	21 0	7 41	20 13	20 44	0 31	13 49	7 33	3 0	4 15	27 20	1 4	14 1	29 8	2 59
2/22/1975	16 4	4 50	4 20	27 31	28 55	24 51	4 36	28 20	7 47	21 32	23 55	0 28	16 39	9 2	3 28	7 45	2cp28	1 21	15 14	0sa27	3 7
3/ 4/1975	18 2	4 47R	4 13	6 2	2 58	28 49	10 35	5 14	7 49	22 50	27 4	0 26	19 27	9 51	3 59	11 18	7 32	1 40	16 27	1 23	3 15
3/14/1975	20li 0	4sc 3R	4s 3	14vi20	7le49	30s41	16cp 5	11aq44	7n48	24aq 9	0pi10	0n23	22li14	9sc56R	4n30	14sa54	12cp29	2n 0	17sc39	1sa55	3n23
3/24/1975	22 0	2 37	3 48	22 23	13 15	31 6	21 10	17 48	7 45	25 27	3 12	0 21	24 59	9 15	5 2	18 34	17 19	2 22	18 52	1 59R	3 32
4/ 3/1975	24 0	0 36	3 29	0li 8	19 4	30 32	25 52	23 28	7 42	26 46	6 7	0 19	27 43	7 53	5 32	22 17	21 58	2 46	20 4	1 36	3 40
4/13/1975	26 1	28 8	3 5	7 35	25 6	29 19	0aq13	28 44	7 37	28 3	8 56	0 16	0sc26	5 58	5 57	26 3	26 24	3 12	21 16	0 45	3 48
4/23/1975	28 2	25 30	2 36	14 44	1vi15	27 40	4 16	3 35	7 31	29 21	11 36	0 14	3 7	3 44	6 17	29 52	0aq35	3 42	22 29	29sc29	3 54
5/ 3/1975	0sc 5	23 0	2 5	21 33	7 27	25 48	8 3	8 2	7 25	0 39	14 7	0 12	5 46	1 30	6 28	3 44	4 26	4 14	23 41	27 53	3 58
5/13/1975	2 9	20 52	1 33	28 4	13 38	23 48	11 36	12 2	7 19	1 56	16 25	0 9	8 24	29li32	6 32	7 37	7 53	4 49	24 53	26 5	3 59
5/23/1975	4 15	19 18	1 1	4 17	19 48	21 47	14 57	15 34	7 12	3 14	18 30	0 6	11 1	28 6	6 29	11 33	10 50	5 28	26 6	24 12	3 58
6/ 2/1975	6 21	18 26	0 31	10 13	25 55	19 49	18 6	18 36	7 5	4 31	20 19	0 3	13 35	27 18	6 21	15 31	13 12	6 9	27 18	22 25	3 54
6/12/1975	8 29	18 15D	0 4	15 54	1li58	17 56	21 5	21 3	6 57	5 48	21 50	0s 0	16 8	27 11D	6 9	19 30	14 51	6 54	28 31	20 52	3 48
6/22/1975	10 39	18 46	0 21	21 19	7 58	16 9	23 56	22 52	6 48	7 5	23 0	0 4	18 40	27 44	5 56	23 30	15 43	7 39	29 43	19 40	3 41
7/ 2/1975	12 50	19 53	0 43	26 30	13 54	14 30	26 38	23 58	6 37	8 22	23 48	0 7	21 9	28 54	5 42	27 31	15 43R	8 23	0 56	18 52	3 32
7/12/1975	15 3	21 33	1 3	1sa29	19 47	12 58	29 13	24 16	6 23	9 39	24 10	0 11	23 38	0sc36	5 28	1aq32	14 50	9 2	2 9	18 32	3 22
7/22/1975	17 17	23 42	1 21	6 16	25 37	11 33	1 42	23 41	6 5	10 56	24 5R	0 16	26 4	2 45	5 15	5 33	13 13	9 32	3 22	18 38D	3 13
8/ 1/1975	19 34	26 16	1 37	10 53	1sc25	10 16	4 5	22 12	5 41	12 13	23 32	0 21	28 29	5 17	5 3	9 34	11 9	9 47	4 35	19 11	3 3
8/11/1975	21 52	29 12	1 52	15 20	7 11	9 5	6 23	19 52	5 11	13 30	22 32	0 25	0sa52	8 9	4 51	13 34	8 59	9 47	5 48	20 7	2 54
8/21/1975	24 13	2 26	2 6	19 39	12 55	8 0	8 35	16 52	4 34	14 47	21 9	0 30	3 14	11 18	4 40	17 33	7 9	9 31	7 2	21 26	2 46
8/31/1975	26 36	5 57	2 19	23 50	18 37	7 0	10 44	13 28	3 51	16 4	19 27	0 35	5 34	14 40	4 30	21 30	5 57	9 3	8 15	23 3	2 38
9/10/1975	29 2	9 43	2 32	27 55	24 19	6 5	12 49	10 4	3 4	17 21	17 36	0 40	7 53	18 13	4 21	25 25	5 34	8 28	9 29	24 58	2 31
9/20/1975	1 30	13 41	2 43	1cp53	0sa 0	5 15	14 50	7 2	2 16	18 38	15 43	0 44	10 10	21 56	4 13	29 18	6 4	7 49	10 43	27 7	2 25
9/30/1975	4 1	17 50	2 55	5 47	5 41	4 29	16 48	4 38	1 31	19 55	14 0	0 48	12 26	25 47	4 6	3 9	7 22	7 9	11 58	29 30	2 19
10/10/1975	6 34	22 10	3 6	9 36	11 22	3 46	18 43	3 1	0 50	21 12	12 34	0 50	14 40	29 44	3 59	6 57	9 25	6 31	13 12	2 3	2 14
10/20/1975	9 11	26 39	3 17	13 21	17 3	3 6	20 36	2 13	0 14	22 29	11 33	0 53	16 53	3 46	3 53	10 42	12 4	5 55	14 27	4 46	2 9

Ephemeris 0 Hours ET Mo Dy Year	80 Sappho h long	 g long	 g lat	433 Eros h long	 g long	 g lat	1221 Amor h long	 g long	 g lat	1489 Attila h long	 g long	 g lat	114 Kassandra h long	 g long	 g lat	3018 Godiva h long	 g long	 g lat	2041 Lancelot h long	 g long	 g lat
10/30/1975	11 51	1sa17	3 28	17 3	22 45	2 29	22 26	2 11D	0 17	23 46	10 59	0 54	19 4	7 52	3 48	14 24	15 14	5 22	15 42	7 36	2 5
11/ 9/1975	14 34	6 2	3 39	20 42	28 28	1 55	24 14	2 50	0 44	25 4	10 53D	0 56	21 14	12 1	3 43	18 3	18 49	4 51	16 58	10 33	2 1
11/19/1975	17 20	10 55	3 50	24 19	4 12	1 22	26 0	4 4	1 7	26 21	11 17	0 57	23 23	16 13	3 39	21 38	22 45	4 23	18 13	13 35	1 58
11/29/1975	20 10	15 54	4 1	27 56	9 58	0 51	27 45	5 48	1 26	27 39	12 7	0 58	25 30	20 25	3 36	25 10	26 57	3 58	19 30	16 41	1 55
12/ 9/1975	23 3	20 59	4 13	1aq31	15 45	0 21	29 27	7 57	1 44	28 57	13 21	0 58	27 36	24 38	3 33	28 38	1pi23	3 35	20 46	19 49	1 53
12/19/1975	26 1	26 9	4 25	5 6	21 33	0n 7	1 9	10 27	1 59	0ar15	14 57	0 59	29 41	28 50	3 31	2 3	5 58	3 14	22 3	22 59	1 51
12/29/1975	29 2	1cp24	4 37	8 42	27 24	0 34	2 49	13 13	2 12	1 33	16 52	0 59	1 45	3 1	3 29	5 23	10 41	2 55	23 20	26 9	1 49
1/ 8/1976	2 8	6 44	4 49	12 18	3 17	1 1	4 28	16 14	2 25	2 52	19 3	1 0	3 47	7 10	3 28	8 41	15 31	2 37	24 38	29 18	1 48
1/18/1976	5 17	12 7	5 2	15 56	9 12	1 27	6 7	19 26	2 36	4 10	21 28	1 1	5 48	11 16	3 27	11 54	20 25	2 21	25 56	2 25	1 47
1/28/1976	8 32	17 34	5 15	19 36	15 9	1 53	7 44	22 47	2 47	5 29	24 5	1 1	7 49	15 17	3 27	15 4	25 22	2 6	27 15	5 29	1 47
2/ 7/1976	11 50	23 4	5 29	23 19	21 9	2 18	9 21	26 15	2 58	6 48	26 51	1 2	9 48	19 13	3 27	18 10	0ar21	1 52	28 34	8 27	1 46
2/17/1976	15 13	28 37	5 43	27 5	27 13	2 43	10 58	29 50	3 9	8 8	29 45	1 3	11 46	23 4	3 27	21 13	5 21	1 39	29 53	11 20	1 46
2/27/1976	18 41	4 12	5 57	0pi55	3 20	3 8	12 34	3 29	3 19	9 28	2 45	1 5	13 43	26 46	3 29	24 12	10 21	1 27	1 13	14 4	1 47
3/ 8/1976	22 13	9 48	6 12	4 50	9 31	3 32	14 10	7 11	3 30	10 48	5 51	1 6	15 39	0aq21	3 30	27 8	15 20	1 16	2 34	16 38	1 47
3/18/1976	25 50	15 25	6 27	8 50	15 46	3 57	15 46	10 56	3 42	12 8	8 59	1 8	17 34	3 44	3 33	0ta 1	20 18	1 5	3 55	19 1	1 48
3/28/1976	29 31	21 3	6 42	12 56	22 7	4 22	17 22	14 43	3 53	13 29	12 10	1 9	19 29	6 57	3 36	2 51	25 14	0 55	5 17	21 10	1 49
4/ 7/1976	3 17	26 40	6 57	17 9	28 33	4 46	18 58	18 31	4 5	14 50	15 22	1 11	21 22	9 55	3 39	5 37	0ta 8	0 45	6 39	23 3	1 50
4/17/1976	7aq 7	2pi16	7n12	21pi30	5ar 5	5n11	20ar34	22ar19	4s18	16ar11	18ar34	1s13	23cp15	12aq37	3n43	8ta20	4ta59	0n36	8cp 2	24cp37	1n51
4/27/1976	11 2	7 50	7 27	26 1	11 44	5 34	22 10	26 7	4 32	17 33	21 46	1 16	25 7	15 1	3 47	11 1	9 46	0 26	9 26	25 50	1 52
5/ 7/1976	15 1	13 20	7 42	0ar40	18 31	5 58	23 47	29 54	4 47	18 55	24 55	1 19	26 58	17 4	3 51	13 38	14 30	0 17	10 50	26 38	1 53
5/17/1976	19 3	18 46	7 56	5 31	25 27	6 20	25 25	3 39	5 3	20 17	28 1	1 22	28 49	18 43	3 56	16 13	19 10	0 8	12 15	26 59	1 54
5/27/1976	23 9	24 6	8 10	10 34	2ta32	6 41	27 4	7 21	5 21	21 40	1ta 4	1 25	0aq38	19 54	4 1	18 45	23 46	0s 1	13 41	26 52	1 54
6/ 6/1976	27 19	29 18	8 23	15 50	9 47	6 59	28 43	11 1	5 40	23 4	4 1	1 29	2 27	20 34	4 5	21 15	28 16	0 9	15 8	26 14	1 54
6/16/1976	1pi31	4 20	8 35	21 20	17 14	7 16	0ta23	14 38	6 1	24 28	6 51	1 33	4 16	20 41R	4 9	23 43	2 41	0 19	16 35	25 9	1 52
6/26/1976	5 46	9 9	8 46	27 5	24 53	7 28	2 5	18 9	6 24	25 52	9 34	1 37	6 4	20 12	4 11	26 8	7 0	0 28	18 3	23 40	1 50
7/ 6/1976	10 3	13 41	8 55	3 7	2gm44	7 37	3 48	21 35	6 50	27 17	12 7	1 43	7 51	19 9	4 11	28 30	11 13	0 37	19 32	21 52	1 45
7/16/1976	14 21	17 54	9 3	9 27	10 48	7 40	5 32	24 54	7 18	28 42	14 28	1 48	9 38	17 34	4 8	0gm51	15 18	0 47	21 2	19 56	1 40
7/26/1976	18 41	21 41	9 7	16 4	19 5	7 36	7 19	28 4	7 50	0ta 8	16 36	1 54	11 24	15 36	4 2	3 10	19 14	0 57	22 32	18 1	1 33
8/ 5/1976	23 1	24 58	9 8	23 1	27 35	7 26	9 7	1gm 4	8 26	1 34	18 28	2 1	13 10	13 25	3 52	5 27	23 1	1 8	24 4	16 19	1 25
8/15/1976	27 21	27 36	9 4	0gm17	6 17	7 7	10 57	3 50	9 6	3 1	20 1	2 9	14 56	11 14	3 38	7 42	26 38	1 20	25 36	14 59	1 16
8/25/1976	1ar41	29 30	8 53	7 51	15 10	6 39	12 49	6 21	9 52	4 29	21 13	2 17	16 41	9 16	3 22	9 55	0cn 1	1 32	27 10	14 5	1 7
9/ 4/1976	6 0	0 31	8 33	15 44	24 13	6 2	14 44	8 32	10 43	5 57	21 59	2 25	18 26	7 42	3 4	12 7	3 10	1 45	28 44	13 42	0 59
9/14/1976	10 18	0 33R	8 3	23 53	3le24	5 15	16 41	10 18	11 40	7 26	22 19	2 34	20 10	6 40	2 45	14 17	6 2	2 0	0aq20	13 51	0 50
9/24/1976	14 34	29ar38	7 18	2cn17	12 41	4 20	18 42	11 34	12 45	8 55	22 8	2 43	21 55	6 12	2 27	16 25	8 33	2 16	1 56	14 30	0 42
10/ 4/1976	18 48	27 51	6 19	10 53	22 1	3 16	20 45	12 13	13 56	10 25	21 26	2 52	23 39	6 19D	2 9	18 33	10 41	2 33	3 34	15 38	0 35
10/14/1976	22 59	25 31	5 6	19 38	1vi22	2 6	22 53	12 7R	15 13	11 56	20 15	3 0	25 22	6 59	1 53	20 38	12 21	2 52	5 13	17 12	0 28
10/24/1976	27 7	23 5	3 45	28 28	10 42	0 50	25 4	11 11	16 33	13 27	18 40	3 7	27 6	8 9	1 38	22 43	13 28	3 13	6 52	19 9	0 22
11/ 3/1976	1ta13	21 1	2 22	7 21	19 58	0s28	27 20	9 19	17 50	14 59	16 46	3 11	28 49	9 47	1 24	24 46	13 59	3 35	8 33	21 27	0 16
11/13/1976	5 14	19 39	1 3	16 11	29 7	1 47	29 41	6 35	18 58	16 32	14 46	3 13	0pi32	11 48	1 11	26 49	13 49R	3 58	10 16	24 2	0 11
11/23/1976	9 13	19 11	0s 6	24 55	8 8	3 6	2 6	3 8	19 47	18 5	12 51	3 13	2 16	14 9	1 0	28 50	12 57	4 22	11 59	26 53	0 6

Ephemeris	80			433			1221			1489			114			3018			2041		
0 Hours ET	Sappho			Eros			Amor			Attila			Kassandra			Godiva			Lancelot		
Mo Dy Year	h long	g long	g lat	h long	g long	g lat	h long	g long	g lat	h long	g long	g lat	h long	g long	g lat	h long	g long	g lat	h long	g long	g lat
12/ 3/1976	13 7	19 40	1 5	3vi29	16 57	4 22	4 38	29ta23	20 10	19 40	11 12	3 10	3 59	16 48	0 49	0cn50	11 24	4 44	13 44	29 58	0 1
12/13/1976	16 58	20 59	1 54	11 51	25 33	5 35	7 16	25 47	20 3	21 15	9 58	3 5	5 42	19 41	0 40	2 50	9 16	5 3	15 30	3 14	0 4
12/23/1976	20 44	23 3	2 34	19 59	3sc54	6 43	10 1	22 49	19 29	22 51	9 14	2 58	7 25	22 46	0 31	4 49	6 46	5 18	17 17	6 40	0 8
1/ 2/1977	24 27	25 45	3 6	27 49	11 58	7 47	12 53	20 47	18 37	24 27	9 3	2 51	9 8	26 2	0 23	6 46	4 10	5 28	19 6	10 14	0 12
1/12/1977	28 5	28 57	3 32	5 22	19 43	8 46	15 55	19 49	17 33	26 5	9 25	2 44	10 51	29 26	0 15	8 44	1 46	5 31	20 56	13 55	0 17
1/22/1977	1gm39	2 33	3 53	12 36	27 8	9 40	19 5	19 56D	16 24	27 43	10 19	2 37	12 34	2 57	0 8	10 40	29gm47	5 28	22 47	17 42	0 21
2/ 1/1977	5 9	6 28	4 11	19 31	4 11	10 30	22 27	21 3	15 16	29 22	11 40	2 30	14 18	6 33	0 1	12 36	28 26	5 22	24 40	21 33	0 25
2/11/1977	8 34	10 38	4 25	26 8	10 51	11 17	26 0	23 3	14 10	1 2	13 26	2 24	16 1	10 14	0 6	14 32	27 45	5 13	26 34	25 28	0 29
2/21/1977	11 56	15 0	4 36	2sc26	17 5	12 1	29 46	25 49	13 8	2 43	15 34	2 18	17 45	13 57	0 13	16 27	27 45D	5 2	28 30	29 26	0 33
3/ 3/1977	15 14	19 31	4 46	8 27	22 53	12 43	3 47	29 16	12 9	4 25	18 0	2 13	19 29	17 43	0 20	18 21	28 24	4 51	0pi27	3 26	0 38
3/13/1977	18 27	24 8	4 54	14 12	28 9	13 24	8 4	3 18	11 12	6 7	20 42	2 8	21 13	21 29	0 26	20 16	29 36	4 40	2 26	7 26	0 43
3/23/1977	21 37	28 51	5 1	19 41	2 52	14 3	12 39	7 53	10 18	7 51	23 38	2 4	22 57	25 15	0 33	22 9	1 17	4 30	4 27	11 26	0 47
4/ 2/1977	24 43	3 37	5 7	24 57	6 56	14 41	17 34	12 57	9 24	9 36	26 44	2 0	24 42	29 1	0 40	24 3	3 23	4 21	6 29	15 25	0 53
4/12/1977	27 45	8 25	5 13	29 59	10 15	15 19	22 52	18 28	8 30	11 21	0gm 1	1 57	26 27	2 45	0 47	25 57	5 50	4 12	8 32	19 22	0 58
4/22/1977	0cn44	13 14	5 17	4 50	12 40	15 54	28 35	24 26	7 36	13 8	3 25	1 54	28 13	6 26	0 55	27 50	8 35	4 5	10 37	23 16	1 4
5/ 2/1977	3 39	18 4	5 22	9 30	14 2	16 26	4 46	0cn50	6 39	14 55	6 55	1 51	29 58	10 4	1 2	29 44	11 34	3 58	12 44	27 7	1 10
5/12/1977	6 31	22 54	5 26	14 0	14 12R	16 50	11 26	7 39	5 40	16 44	10 31	1 49	1 45	13 37	1 11	1 37	14 45	3 52	14 53	0ar52	1 17
5/22/1977	9cn20	27gm44	5s30	18sa21	13cp 0R	17s 0	18le39	14cn54	4s37	18gm33	14gm12	1s47	3ar32	17ar 5	1s20	3le31	18cn 6	3s47	17pi 3	4ar32	1s24
6/ 1/1977	12 5	2 32	5 34	22 35	10 27	16 47	26 25	22 34	3 30	20 24	17 55	1 45	5 19	20 25	1 29	5 24	21 36	3 42	19 15	8 3	1 32
6/11/1977	14 48	7 18	5 38	26 41	6 46	16 3	4 46	0le40	2 19	22 16	21 41	1 44	7 7	23 38	1 39	7 18	25 13	3 39	21 28	11 25	1 41
6/21/1977	17 27	12 3	5 42	0cp41	2 26	14 44	13 41	9 11	1 4	24 8	25 29	1 43	8 55	26 41	1 51	9 12	28 55	3 35	23 43	14 36	1 51
7/ 1/1977	20 4	16 46	5 47	4 36	28sa11	12 56	23 7	18 6	0n15	26 2	29 18	1 42	10 44	29 32	2 3	11 6	2 42	3 33	26 0	17 34	2 1
7/11/1977	22 38	21 25	5 51	8 26	24 40	10 50	2li58	27 24	1 35	27 57	3 7	1 41	12 34	2 10	2 16	13 1	6 33	3 31	28 18	20 15	2 13
7/21/1977	25 9	26 2	5 56	12 12	22 16	8 42	13 7	7 3	2 55	29 53	6 55	1 40	14 24	4 31	2 31	14 56	10 27	3 29	0ar38	22 36	2 26
7/31/1977	27 38	0le36	6 2	15 55	21 8	6 42	23 24	16 58	4 11	1 49	10 42	1 40	16 15	6 33	2 48	16 51	14 24	3 28	3 0	24 34	2 39
8/10/1977	0le 5	5 6	6 7	19 35	21 10D	4 56	3sc37	27 5	5 20	3 47	14 27	1 40	18 7	8 11	3 5	18 47	18 23	3 28	5 23	26 5	2 55
8/20/1977	2 29	9 31	6 14	23 13	22 12	3 24	13 37	7 18	6 20	5 46	18 8	1 40	20 0	9 24	3 25	20 44	22 22	3 28	7 47	27 4	3 11
8/30/1977	4 51	13 52	6 21	26 50	24 5	2 6	23 12	17 30	7 8	7 46	21 45	1 41	21 53	10 5	3 46	22 41	26 23	3 28	10 14	27 27	3 27
9/ 9/1977	7 11	18 7	6 29	0aq25	26 40	1 0	2sa17	27 36	7 43	9 48	25 16	1 41	23 48	10 12R	4 9	24 39	0vi23	3 30	12 41	27 13	3 44
9/19/1977	9 29	22 17	6 38	4 1	29 48	0 3	10 46	7 28	8 5	11 50	28 40	1 42	25 43	9 43	4 32	26 37	4 23	3 31	15 10	26 21	4 0
9/29/1977	11 45	26 18	6 48	7 36	3 24	0 46	18 38	17 2	8 16	13 53	1 54	1 43	27 39	8 36	4 54	28 37	8 20	3 33	17 41	24 55	4 14
10/ 9/1977	14 0	0vi12	7 0	11 12	7 23	1 28	25 54	26 14	8 16	15 57	4 58	1 44	29 36	6 56	5 15	0vi37	12 16	3 36	20 12	23 4	4 24
10/19/1977	16 13	3 55	7 12	14 49	11 42	2 5	2cp35	5 2	8 7	18 3	7 47	1 45	1 34	4 50	5 32	2 38	16 8	3 39	22 45	21 2	4 30
10/29/1977	18 24	7 27	7 26	18 29	16 18	2 37	8 44	13 26	7 53	20 9	10 19	1 46	3 33	2 32	5 44	4 40	19 55	3 43	25 19	19 5	4 30
11/ 8/1977	20 33	10 45	7 42	22 11	21 8	3 6	14 23	21 25	7 34	22 16	12 31	1 47	5 34	0 15	5 50	6 43	23 37	3 48	27 54	17 29	4 26
11/18/1977	22 42	13 47	8 0	25 56	26 11	3 32	19 35	29 1	7 12	24 25	14 19	1 49	7 35	28 16	5 51	8 48	27 10	3 53	0ta30	16 26	4 18
11/28/1977	24 48	16 30	8 19	29 44	1aq26	3 55	24 24	6 15	6 49	26 34	15 38	1 50	9 38	26 47	5 47	10 53	0li34	3 59	3 7	16 2	4 8
12/ 8/1977	26 54	18 49	8 41	3 37	6 53	4 17	28 52	13 10	6 25	28 44	16 25	1 50	11 41	25 55	5 39	13 0	3 46	4 6	5 45	16 19	3 56
12/18/1977	28 58	20 42	9 4	7 36	12 30	4 36	3 1	19 46	6 1	0le55	16 35	1 50	13 46	25 44	5 28	15 9	6 44	4 14	8 23	17 17	3 44

Ephemeris	80			433			1221			1489			114			3018			2041		
0 Hours ET	Sappho			Eros			Amor			Attila			Kassandra			Godiva			Lancelot		
Mo Dy Year	h long	g long	g lat	h long	g long	g lat	h long	g long	g lat	h long	g long	g lat	h long	g long	g lat	h long	g long	g lat	h long	g long	g lat
12/28/1977	1 2	22 3	9 29	11 40	18 17	4 54	6 53	26 6	5 37	3 8	16 8	1 49	15 53	26 12	5 17	17 18	9 23	4 23	11 2	18 52	3 32
1/ 7/1978	3 4	22 49	9 55	15 51	24 13	5 11	10 31	2aq10	5 14	5 21	15 4	1 47	18 0	27 17	5 5	19 30	11 40	4 32	13 41	20 59	3 20
1/17/1978	5 5	22 53R	10 20	20 10	0pi20	5 26	13 55	8 1	4 52	7 34	13 28	1 43	20 9	28 55	4 54	21 43	13 30	4 42	16 21	23 34	3 9
1/27/1978	7 6	22 15	10 42	24 37	6 35	5 39	17 8	13 39	4 31	9 49	11 31	1 37	22 19	1ta 3	4 43	23 57	14 50	4 52	19 1	26 32	2 59
2/ 6/1978	9 6	20 55	11 0	29 14	13 1	5 51	20 11	19 5	4 11	12 4	9 27	1 29	24 31	3 35	4 33	26 13	15 32	5 3	21 41	29 51	2 50
2/16/1978	11 4	18 55	11 10	4 1	19 36	6 2	23 4	24 20	3 51	14 20	7 31	1 19	26 44	6 28	4 24	28 32	15 34R	5 12	24 21	3 25	2 41
2/26/1978	13 3	16 29	11 9	9 0	26 21	6 10	25 49	29 23	3 33	16 37	5 57	1 9	28 59	9 39	4 16	0li52	14 51	5 18	27 2	7 14	2 33
3/ 8/1978	15 0	13 49	10 57	14 11	3ar17	6 17	28 27	4 17	3 15	18 54	4 55	0 58	1 15	13 5	4 8	3 14	13 25	5 22	29 41	11 13	2 26
3/18/1978	16 57	11 15	10 32	19 37	10 23	6 21	0pi58	8 59	2 57	21 12	4 30	0 48	3 32	16 45	4 1	5 38	11 20	5 19	2 21	15 21	2 19
3/28/1978	18 54	9 3	9 58	25 17	17 40	6 23	3 23	13 32	2 40	23 31	4 45	0 38	5 51	20 35	3 56	8 5	8 51	5 10	5 0	19 36	2 13
4/ 7/1978	20 50	7 24	9 19	1ta14	25 9	6 21	5 43	17 53	2 23	25 50	5 35	0 29	8 12	24 34	3 50	10 34	6 14	4 55	7 39	23 57	2 7
4/17/1978	22 46	6 24	8 36	7 28	2ta49	6 15	7 57	22 4	2 6	28 9	7 0	0 21	10 34	28 41	3 46	13 5	3 49	4 33	10 17	28 22	2 2
4/27/1978	24 42	6 6	7 53	14 0	10 41	6 5	10 7	26 3	1 49	0vi29	8 53	0 14	12 58	2 54	3 41	15 39	1 53	4 7	12 54	2 49	1 57
5/ 7/1978	26 38	6 27	7 11	20 51	18 45	5 51	12 14	29 50	1 32	2 49	11 12	0 7	15 24	7 13	3 38	18 16	0 38	3 39	15 30	7 20	1 52
5/17/1978	28 33	7 23	6 32	28 0	27 0	5 31	14 16	3 23	1 14	5 9	13 53	0 1	17 51	11 36	3 34	20 55	0 10	3 11	18 6	11 51	1 48
5/27/1978	0li28	8 50	5 57	5 29	5 26	5 6	16 15	6 42	0 55	7 30	16 51	0 5	20 20	16 4	3 31	23 37	0 28	2 43	20 40	16 23	1 44
6/ 6/1978	2 23	10 44	5 24	13 16	14 2	4 36	18 12	9 44	0 36	9 50	20 5	0 10	22 50	20 34	3 29	26 22	1 30	2 18	23 13	20 54	1 40
6/16/1978	4 19	13 2	4 54	21 20	22 48	4 0	20 5	12 29	0 15	12 11	23 32	0 15	25 23	25 8	3 26	29 10	3 11	1 54	25 45	25 24	1 36
6/26/1978	6li14	15vi38	4s27	29gm40	1cn43	3n19	21pi56	14ar52	0s 8	14vi32	27le10	0n19	27gm56	29gm43	3s24	2sc 1	5li26	1s32	28gm16	29gm53	1s32
7/ 6/1978	8 10	18 30	4 2	8 13	10 44	2 33	23 45	16 50	0 33	16 53	0vi57	0 24	0cn32	4 20	3 22	4 55	8 11	1 12	0cn46	4 20	1 29
7/16/1978	10 6	21 37	3 40	16 56	19 51	1 43	25 32	18 21	1 0	19 13	4 51	0 28	3 9	8 57	3 21	7 53	11 22	0 54	3 14	8 43	1 25
7/26/1978	12 2	24 55	3 19	25 45	29 1	0 51	27 17	19 20	1 30	21 34	8 52	0 31	5 48	13 36	3 19	10 54	14 55	0 37	5 40	13 3	1 22
8/ 5/1978	13 59	28 24	3 0	4 37	8 13	0s 2	29 1	19 42	2 3	23 54	12 58	0 35	8 28	18 14	3 18	13 58	18 47	0 21	8 6	17 19	1 19
8/15/1978	15 56	2 1	2 42	13 28	17 25	0 56	0ar43	19 23	2 39	26 14	17 8	0 39	11 10	22 51	3 16	17 6	22 57	0 7	10 29	21 29	1 16
8/25/1978	17 54	5 46	2 25	22 15	26 34	1 48	2 24	18 21	3 17	28 34	21 21	0 42	13 53	27 27	3 15	20 18	27 21	0 7	12 51	25 33	1 12
9/ 4/1978	19 53	9 37	2 9	0vi53	5 39	2 38	4 3	16 35	3 57	0li54	25 37	0 46	16 38	2le 0	3 14	23 33	1sc58	0 20	15 12	29 29	1 9
9/14/1978	21 52	13 34	1 54	9 19	14 39	3 24	5 42	14 9	4 37	3 12	29 54	0 49	19 25	6 31	3 13	26 52	6 48	0 32	17 31	3 17	1 6
9/24/1978	23 52	17 36	1 40	17 32	23 31	4 6	7 20	11 14	5 14	5 31	4 12	0 53	22 12	10 56	3 12	0sa15	11 48	0 44	19 48	6 55	1 3
10/ 4/1978	25 53	21 41	1 26	25 28	2li14	4 42	8 57	8 6	5 45	7 49	8 30	0 57	25 2	15 17	3 11	3 41	16 57	0 55	22 4	10 21	0 59
10/14/1978	27 54	25 50	1 12	3li 6	10 47	5 14	10 34	5 4	6 9	10 6	12 48	1 1	27 52	19 29	3 9	7 11	22 16	1 6	24 18	13 33	0 56
10/24/1978	29 57	0sc 2	0 59	10 26	19 9	5 40	12 11	2 26	6 26	12 23	17 4	1 4	0le44	23 33	3 8	10 44	27 42	1 16	26 30	16 28	0 52
11/ 3/1978	2 1	4 15	0 45	17 27	27 20	6 1	13 47	0 23	6 36	14 39	21 18	1 9	3 36	27 25	3 6	14 21	3 16	1 26	28 41	19 4	0 48
11/13/1978	4 6	8 31	0 32	24 9	5 19	6 18	15 23	29 3	6 40	16 54	25 28	1 13	6 30	1vi 2	3 4	18 1	8 55	1 36	0le50	21 16	0 43
11/23/1978	6 13	12 46	0 18	0sc33	13 5	6 30	16 59	28 28	6 41	19 9	29 34	1 18	9 25	4 22	3 1	21 45	14 41	1 45	2 57	23 3	0 38
12/ 3/1978	8 21	17 2	0 3	6 40	20 39	6 39	18 35	28 36D	6 39	21 22	3 35	1 22	12 21	7 21	2 58	25 31	20 31	1 55	5 3	24 19	0 33
12/13/1978	10 30	21 17	0 12	12 29	28 1	6 43	20 11	29 22	6 35	23 35	7 28	1 28	15 17	9 53	2 54	29 21	26 26	2 4	7 7	25 1	0 27
12/23/1978	12 41	25 30	0 27	18 4	5 10	6 45	21 47	0ar41	6 31	25 47	11 12	1 33	18 14	11 53	2 48	3 13	2cp24	2 13	9 10	25 7R	0 20
1/ 2/1979	14 54	29 40	0 44	23 23	12 7	6 44	23 24	2 30	6 28	27 58	14 46	1 40	21 12	13 16	2 40	7 7	8 25	2 21	11 11	24 34	0 12
1/12/1979	17 9	3 46	1 2	28 30	18 52	6 41	25 2	4 43	6 25	0sc 9	18 8	1 46	24 10	13 57	2 30	11 4	14 28	2 30	13 10	23 26	0 4
1/22/1979	19 25	7 48	1 21	3 24	25 26	6 36	26 40	7 16	6 22	2 18	21 15	1 54	27 9	13 52R	2 17	15 3	20 32	2 38	15 8	21 47	0 5

Ephemeris	80			433			1221			1489			114			3018			2041		
0 Hours ET	Sappho			Eros			Amor			Attila			Kassandra			Godiva			Lancelot		
Mo Dy Year	h long	g long	g lat	h long	g long	g lat	h long	g long	g lat	h long	g long	g lat	h long	g long	g lat	h long	g long	g lat	h long	g long	g lat
2/ 1/1979	21 44	11 43	1 42	8 7	1cp47	6 29	28 20	10 7	6 20	4 26	24 5	2 2	0vi 8	12 59	2 0	19 3	26 37	2 46	17 4	19 48	0 14
2/11/1979	24 5	15 29	2 6	12 40	7 57	6 20	29 60	13 13	6 20	6 33	26 35	2 10	3 7	11 25	1 39	23 4	2aq41	2 54	18 59	17 41	0 23
2/21/1979	26 28	19 6	2 32	17 4	13 55	6 9	1 41	16 31	6 20	8 40	28 42	2 20	6 6	9 20	1 15	27 6	8 44	3 2	20 53	15 41	0 31
3/ 3/1979	28 53	22 30	3 1	21 19	19 41	5 57	3 24	20 0	6 21	10 45	0sa22	2 30	9 5	7 1	0 48	1aq 8	14 44	3 9	22 44	14 0	0 39
3/13/1979	1 21	25 39	3 33	25 28	25 14	5 43	5 8	23 37	6 23	12 49	1 32	2 41	12 4	4 51	0 21	5 10	20 42	3 16	24 35	12 47	0 45
3/23/1979	3 52	28 29	4 10	29 30	0aq35	5 28	6 54	27 22	6 27	14 52	2 8	2 52	15 2	3 7	0n 5	9 12	26 36	3 23	26 24	12 8	0 50
4/ 2/1979	6 26	0cp57	4 51	3 26	5 42	5 10	8 41	1ta14	6 31	16 55	2 10R	3 3	18 0	2 2	0 29	13 13	2pi24	3 30	28 12	12 3D	0 55
4/12/1979	9 2	2 58	5 37	7 18	10 34	4 50	10 31	5 11	6 36	18 56	1 35	3 13	20 58	1 42	0 51	17 13	8 7	3 36	29 58	12 31	0 58
4/22/1979	11 42	4 27	6 29	11 5	15 10	4 27	12 22	9 14	6 43	20 56	0 28	3 23	23 55	2 7	1 9	21 11	13 42	3 43	1 43	13 29	1 2
5/ 2/1979	14 25	5 18	7 26	14 49	19 27	4 1	14 17	13 20	6 50	22 55	28 53	3 30	26 51	3 15	1 25	25 8	19 10	3 49	3 27	14 53	1 4
5/12/1979	17 11	5 25R	8 28	18 30	23 22	3 30	16 13	17 31	6 59	24 52	27 0	3 34	29 47	5 0	1 38	29 2	24 28	3 55	5 9	16 40	1 6
5/22/1979	20 1	4 47	9 33	22 8	26 54	2 55	18 13	21 46	7 8	26 49	25 2	3 35	2 41	7 17	1 49	2 54	29 35	4 0	6 51	18 46	1 8
6/ 1/1979	22 55	3 22	10 37	25 45	29 55	2 13	20 16	26 4	7 19	28 45	23 11	3 33	5 35	10 0	1 59	6 43	4 30	4 6	8 31	21 9	1 10
6/11/1979	25 52	1 17	11 35	29 21	2 20	1 23	22 23	0gm25	7 32	0sa40	21 38	3 28	8 28	13 6	2 7	10 29	9 10	4 11	10 10	23 44	1 12
6/21/1979	28 54	28sa46	12 21	2 56	4 2	0 24	24 33	4 50	7 45	2 33	20 32	3 21	11 19	16 30	2 14	14 12	13 33	4 16	11 47	26 31	1 13
7/ 1/1979	2 0	26 8	12 52	6 31	4 51	0 47	26 48	9 18	8 0	4 26	19 58	3 13	14 10	20 10	2 20	17 51	17 37	4 22	13 24	29 27	1 15
7/11/1979	5 10	23 48	13 5	10 7	4 37R	2 11	29 7	13 49	8 17	6 18	19 56D	3 4	16 59	24 2	2 26	21 27	21 19	4 26	15 0	2 31	1 17
7/21/1979	8 24	22 2	13 2	13 44	3 12	3 48	1 32	18 24	8 35	8 8	20 26	2 55	19 46	28 5	2 30	25 0	24 34	4 31	16 35	5 40	1 19
7/31/1979	11cp42	21sa 5R	12n46	17aq23	0pi33R	5n35	4gm 2	23gm 2	8s55	9sa58	21sc26	2n46	22li33	2li16	2n35	28pi29	27ar18	4n35	18vi 8	8vi54	1n20
8/10/1979	15 6	21 1D	12 22	21 4	26 52	7 25	6 38	27 44	9 17	11 46	22 51	2 37	25 18	6 34	2 39	1 54	29 26	4 38	19 41	12 12	1 22
8/20/1979	18 34	21 50	11 52	24 48	22 34	9 6	9 21	2 30	9 42	13 34	24 40	2 30	28 1	10 58	2 43	5 15	0 52	4 40	21 13	15 32	1 24
8/30/1979	22 6	23 27	11 20	28 35	18 20	10 28	12 12	7 19	10 8	15 20	26 49	2 22	0sc43	15 27	2 47	8 33	1 30	4 40	22 43	18 54	1 27
9/ 9/1979	25 43	25 48	10 47	2 27	14 50	11 26	15 11	12 13	10 37	17 6	29 15	2 16	3 23	19 59	2 50	11 47	1 18R	4 37	24 13	22 17	1 29
9/19/1979	29 25	28 47	10 14	6 24	12 29	11 59	18 19	17 12	11 8	18 50	1 56	2 10	6 2	24 34	2 54	14 57	0 14	4 29	25 42	25 39	1 32
9/29/1979	3 11	2 18	9 43	10 26	11 27	12 15	21 38	22 16	11 42	20 34	4 49	2 4	8 39	29 11	2 57	18 4	28 24	4 16	27 10	29 0	1 35
10/ 9/1979	7 2	6 17	9 12	14 35	11 43D	12 18	25 8	27 26	12 19	22 17	7 52	1 59	11 15	3 50	3 1	21 7	26 1	3 57	28 38	2 19	1 38
10/19/1979	10 57	10 39	8 43	18 51	13 7	12 13	28 51	2 42	12 59	23 58	11 4	1 55	13 49	8 29	3 4	24 7	23 24	3 32	0li 4	5 35	1 42
10/29/1979	14 56	15 23	8 14	23 15	15 29	12 3	2 49	8 5	13 41	25 39	14 23	1 51	16 22	13 8	3 8	27 3	20 56	3 4	1 30	8 46	1 45
11/ 8/1979	18 59	20 24	7 46	27 49	18 41	11 50	7 2	13 38	14 25	27 19	17 47	1 48	18 52	17 47	3 12	29 56	18 56	2 34	2 55	11 52	1 50
11/18/1979	23 6	25 40	7 20	2 33	22 34	11 35	11 33	19 21	15 10	28 59	21 16	1 45	21 22	22 24	3 16	2 46	17 39	2 4	4 20	14 50	1 55
11/28/1979	27 15	1aq 9	6 53	7 28	27 4	11 19	16 23	25 17	15 55	0cp37	24 47	1 42	23 49	26 58	3 21	5 32	17 9	1 37	5 43	17 40	2 0
12/ 8/1979	1pi28	6 49	6 28	12 35	2pi 3	11 1	21 36	1vi29	16 36	2 15	28 21	1 40	26 15	1sa30	3 25	8 16	17 26	1 12	7 6	20 19	2 6
12/18/1979	5 43	12 38	6 2	17 56	7 30	10 42	27 13	8 3	17 11	3 51	1 56	1 38	28 39	5 58	3 31	10 56	18 25	0 50	8 29	22 46	2 13
12/28/1979	10 1	18 35	5 37	23 32	13 22	10 21	3 17	15 6	17 32	5 27	5 30	1 36	1sa 2	10 22	3 37	13 34	20 3	0 31	9 50	24 59	2 20
1/ 7/1980	14 19	24 38	5 12	29 24	19 37	9 59	9 51	22 46	17 30	7 2	9 3	1 35	3 23	14 39	3 43	16 9	22 12	0 14	11 12	26 53	2 28
1/17/1980	18 39	0pi46	4 47	5 32	26 12	9 33	16 57	1li17	16 50	8 37	12 33	1 34	5 43	18 50	3 50	18 41	24 48	0s 0	12 32	28 28	2 37
1/27/1980	23 0	6 58	4 23	11 58	3ar 7	9 4	24 36	10 52	15 11	10 10	16 0	1 33	8 1	22 52	3 57	21 11	27 45	0 13	13 52	29 41	2 46
2/ 6/1980	27 21	13 13	3 58	18 43	10 21	8 32	2vi49	21 46	12 6	11 43	19 23	1 33	10 18	26 45	4 6	23 38	0ta59	0 24	15 12	0 27	2 56
2/16/1980	1ar41	19 29	3 34	25 47	17 52	7 55	11 37	4sc 4	7 10	13 16	22 39	1 33	12 33	0cp27	4 15	26 3	4 27	0 34	16 31	0 46	3 6
2/26/1980	6 0	25 46	3 10	3gm10	25 40	7 14	20 57	17 28	0 19	14 47	25 48	1 33	14 47	3 56	4 25	28 26	8 7	0 43	17 49	0 35	3 17

Ephemeris	80			433			1221			1489			114			3018			2041		
0 Hours ET	Sappho			Eros			Amor			Attila			Kassandra			Godiva			Lancelot		
Mo Dy Year	h long	g long	g lat	h long	g long	g lat	h long	g long	g lat	h long	g long	g lat	h long	g long	g lat	h long	g long	g lat	h long	g long	g lat
3/ 7/1980	10 18	2ar 2	2 45	10 51	3ta44	6 27	0li43	1sa20	7 43	16 18	28 48	1 33	16 59	7 10	4 37	0gm47	11 55	0 51	19 7	29li55	3 27
3/17/1980	14 34	8 17	2 21	18 50	12 2	5 36	10 50	14 39	15 35	17 48	1 38	1 34	19 10	10 6	4 49	3 6	15 50	0 59	20 25	28 47	3 36
3/27/1980	18 48	14 30	1 57	27 5	20 34	4 39	21 7	26 32	22 11	19 18	4 16	1 35	21 20	12 42	5 2	5 22	19 51	1 6	21 42	27 15	3 43
4/ 6/1980	23 0	20 41	1 33	5 34	29 17	3 39	1sc23	6 34	27 15	20 47	6 39	1 36	23 28	14 55	5 17	7 37	23 56	1 12	22 59	25 27	3 48
4/16/1980	27 8	26 48	1 9	14 14	8 11	2 34	11 27	14 34	30 59	22 16	8 47	1 37	25 35	16 41	5 32	9 51	28 3	1 19	24 15	23 32	3 51
4/26/1980	1ta14	2ta52	0 45	23 1	17 13	1 27	21 9	20 32	33 45	23 43	10 36	1 38	27 41	17 57	5 48	12 2	2 12	1 25	25 31	21 39	3 51
5/ 6/1980	5 16	8 51	0 21	1le53	26 22	0 18	0sa21	24 29	35 48	25 11	12 4	1 39	29 45	18 39	6 4	14 12	6 23	1 31	26 47	19 59	3 48
5/16/1980	9 14	14 46	0s 2	10 45	5 34	0 50	8 59	26 24	37 16	26 37	13 9	1 41	1 48	18 44R	6 20	16 21	10 33	1 37	28 2	18 39	3 44
5/26/1980	13 8	20 35	0 26	19 33	14 49	1 55	17 0	26 18R	38 4	28 4	13 47	1 42	3 50	18 11	6 34	18 28	14 44	1 43	29 17	17 43	3 38
6/ 5/1980	16 59	26 19	0 50	28 14	24 4	2 57	24 24	24 23	38 5	29 29	13 58	1 43	5 51	17 2	6 45	20 33	18 53	1 49	0sc32	17 15	3 30
6/15/1980	20 45	1gm57	1 14	6 45	3le16	3 54	1cp12	21 5	37 7	0 55	13 40	1 44	7 51	15 21	6 52	22 38	23 1	1 55	1 46	17 14D	3 23
6/25/1980	24 27	7 29	1 38	15 2	12 24	4 43	7 28	17 9	35 4	2 19	12 52	1 44	9 50	13 18	6 53	24 41	27 8	2 2	3 0	17 40	3 15
7/ 5/1980	28 6	12 53	2 3	23 3	21 26	5 26	13 13	13 26	32 6	3 44	11 39	1 43	11 48	11 4	6 47	26 44	1cn11	2 8	4 14	18 30	3 8
7/15/1980	1gm39	18 11	2 28	0li47	0vi20	6 0	18 31	10 33	28 34	5 8	10 4	1 41	13 45	8 55	6 35	28 45	5 12	2 16	5 28	19 43	3 1
7/25/1980	5 9	23 20	2 53	8 13	9 5	6 26	23 25	8 48	24 53	6 31	8 15	1 38	15 41	7 3	6 17	0cn45	9 9	2 23	6 41	21 14	2 55
8/ 4/1980	8 35	28 20	3 20	15 20	17 39	6 45	27 57	8 13	21 22	7 54	6 22	1 33	17 36	5 39	5 56	2 45	13 1	2 31	7 54	23 3	2 49
8/14/1980	11 56	3 10	3 47	22 8	26 2	6 57	2 10	8 37	18 13	9 17	4 34	1 28	19 30	4 48	5 33	4 43	16 48	2 40	9 7	25 7	2 43
8/24/1980	15 14	7 49	4 16	28 37	4 13	7 2	6 6	9 52	15 28	10 39	3 1	1 22	21 24	4 33	5 9	6 41	20 29	2 49	10 20	27 23	2 39
9/ 3/1980	18gm27	12cn15	4s46	4sc49	12li12	7s 2	9aq47	11cp46	13n 7	12aq 1	1aq50R	1n15	23cp16	4cp54	4n45	8cn38	24cn 2	2s59	11sc33	29li50	2n35
9/13/1980	21 37	16 28	5 18	10 44	19 59	6 57	13 14	14 11	11 9	13 23	1 5	1 8	25 8	5 47	4 23	10 35	27 26	3 10	12 45	2 26	2 31
9/23/1980	24 43	20 23	5 52	16 23	27 35	6 49	16 30	17 0	9 28	14 44	0 50	1 2	26 59	7 10	4 2	12 30	0le39	3 22	13 58	5 10	2 28
10/ 3/1980	27 45	24 0	6 29	21 47	5 0	6 36	19 34	20 9	8 4	16 5	1 2	0 55	28 49	8 59	3 43	14 26	3 40	3 36	15 10	8 0	2 26
10/13/1980	0cn43	27 15	7 10	26 57	12 15	6 22	22 30	23 32	6 52	17 25	1 42	0 49	0aq39	11 9	3 26	16 21	6 25	3 50	16 23	10 55	2 24
10/23/1980	3 38	0le 2	7 53	1sa55	19 20	6 5	25 17	27 7	5 51	18 46	2 47	0 44	2 28	13 39	3 10	18 15	8 52	4 7	17 35	13 53	2 22
11/ 2/1980	6 30	2 19	8 40	6 42	26 16	5 47	27 56	0aq52	4 58	20 6	4 15	0 39	4 16	16 25	2 56	20 10	10 57	4 25	18 47	16 54	2 21
11/12/1980	9 19	4 0	9 30	11 18	3sa 4	5 28	0pi28	4 44	4 13	21 26	6 2	0 34	6 4	19 25	2 43	22 4	12 36	4 44	19 59	19 56	2 21
11/22/1980	12 4	4 58	10 24	15 44	9 43	5 8	2 54	8 41	3 33	22 45	8 6	0 30	7 51	22 36	2 31	23 57	13 46	5 6	21 11	22 59	2 21
12/ 2/1980	14 46	5 10R	11 18	20 3	16 16	4 47	5 15	12 43	2 59	24 5	10 25	0 26	9 38	25 56	2 21	25 51	14 21	5 28	22 23	26 0	2 21
12/12/1980	17 26	4 31	12 11	24 13	22 42	4 25	7 31	16 49	2 28	25 24	12 56	0 22	11 25	29 25	2 11	27 44	14 17R	5 52	23 35	28 59	2 22
12/22/1980	20 2	3 2	12 59	28 17	29 2	4 3	9 42	20 57	2 1	26 43	15 38	0 19	13 10	3 0	2 2	29 38	13 32	6 14	24 47	1 55	2 24
1/ 1/1981	22 36	0 51	13 37	2 15	5 16	3 41	11 49	25 7	1 36	28 1	18 28	0 16	14 56	6 39	1 54	1 31	12 7	6 35	26 0	4 45	2 26
1/11/1981	25 7	28 10	14 0	6 8	11 25	3 18	13 52	29 18	1 13	29 20	21 25	0 13	16 41	10 23	1 46	3 24	10 7	6 52	27 12	7 29	2 28
1/21/1981	27 36	25 22	14 5	9 57	17 29	2 55	15 52	3 29	0 53	0pi38	24 27	0 10	18 26	14 9	1 39	5 18	7 42	7 3	28 24	10 4	2 31
1/31/1981	0le 3	22 48	13 52	13 42	23 28	2 32	17 49	7 39	0 33	1 57	27 33	0 7	20 10	17 56	1 32	7 12	5 7	7 7	29 36	12 30	2 35
2/10/1981	2 27	20 44	13 25	17 23	29 23	2 7	19 43	11 49	0 15	3 15	0pi42	0 4	21 55	21 44	1 26	9 6	2 39	7 3	0 49	14 43	2 39
2/20/1981	4 49	19 23	12 49	21 3	5 14	1 42	21 35	15 57	0s 2	4 33	3 51	0 2	23 39	25 31	1 19	11 0	0 35	6 53	2 2	16 42	2 44
3/ 2/1981	7 9	18 48	12 8	24 40	11 1	1 16	23 24	20 4	0 19	5 51	7 1	0s 1	25 22	29 17	1 13	12 55	29 4	6 38	3 14	18 25	2 49
3/12/1981	9 27	18 58D	11 25	28 16	16 44	0 49	25 11	24 8	0 35	7 9	10 10	0 3	27 6	3 0	1 8	14 50	28 13	6 20	4 27	19 48	2 55
3/22/1981	11 43	19 49	10 43	1 51	22 24	0 20	26 57	28 8	0 51	8 27	13 16	0 6	28 49	6 40	1 2	16 45	28 3D	6 0	5 40	20 49	3 1
4/ 1/1981	13 57	21 15	10 4	5 27	28 0	0n10	28 41	2 6	1 7	9 45	16 19	0 9	0pi33	10 16	0 56	18 41	28 32	5 41	6 53	21 26	3 8

Ephemeris 0 Hours ET	80 Sappho			433 Eros			1221 Amor			1489 Attila			114 Kassandra			3018 Godiva			2041 Lancelot		
Mo Dy Year	h long	g long	g lat	h long	g long	g lat	h long	g long	g lat	h long	g long	g lat	h long	g long	g lat	h long	g long	g lat	h long	g long	g lat
4/11/1981	16 10	23 11	9 28	9 2	3 34	0 42	0ar23	5 59	1 24	11 2	19 18	0 11	2 16	13 46	0 50	20 37	29 36	5 22	8 7	21 37	3 14
4/21/1981	18 21	25 32	8 54	12 39	9 4	1 17	2 4	9 48	1 40	12 20	22 11	0 14	3 59	17 10	0 44	22 34	1 10	5 5	9 20	21 20	3 20
5/ 1/1981	20 31	28 13	8 25	16 17	14 31	1 54	3 44	13 31	1 58	13 38	24 57	0 17	5 42	20 25	0 38	24 32	3 12	4 48	10 34	20 35	3 26
5/11/1981	22 39	1le11	7 58	19 57	19 55	2 34	5 23	17 8	2 16	14 55	27 34	0 21	7 25	23 32	0 32	26 31	5 36	4 34	11 48	19 25	3 30
5/21/1981	24 46	4 23	7 33	23 40	25 16	3 18	7 1	20 38	2 35	16 13	0ar 1	0 24	9 9	26 27	0 25	28 30	8 19	4 20	13 3	17 53	3 32
5/31/1981	26 52	7 45	7 11	27 26	0ar34	4 6	8 39	24 1	2 56	17 31	2 17	0 28	10 52	29 10	0 17	0vi31	11 18	4 8	14 17	16 6	3 32
6/10/1981	28 56	11 17	6 52	1pi17	5 49	5 0	10 16	27 14	3 18	18 49	4 19	0 32	12 35	1 37	0 9	2 32	14 31	3 57	15 32	14 13	3 30
6/20/1981	0vi59	14 56	6 34	5 12	11 0	5 58	11 52	0ta16	3 42	20 7	6 4	0 37	14 19	3 46	0 1	4 34	17 56	3 47	16 48	12 23	3 25
6/30/1981	3 2	18 41	6 18	9 12	16 8	7 4	13 28	3 6	4 8	21 24	7 32	0 41	16 2	5 36	0 9	6 37	21 31	3 38	18 3	10 46	3 18
7/10/1981	5 3	22 31	6 3	13 19	21 11	8 17	15 4	5 42	4 37	22 42	8 38	0 47	17 46	7 1	0 19	8 42	25 16	3 29	19 19	9 28	3 10
7/20/1981	7 4	26 26	5 50	17 33	26 8	9 40	16 40	8 0	5 8	24 1	9 21	0 52	19 30	7 58	0 31	10 47	29 7	3 22	20 36	8 35	3 1
7/30/1981	9 3	0vi23	5 38	21 55	0ta58	11 13	18 16	9 58	5 44	25 19	9 38	0 58	21 14	8 26	0 44	12 54	3 5	3 15	21 52	8 10	2 51
8/ 9/1981	11 2	4 22	5 27	26 26	5 39	12 59	19 52	11 31	6 23	26 37	9 27	1 4	22 59	8 19R	0 58	15 3	7 10	3 8	23 10	8 12D	2 41
8/19/1981	13 1	8 23	5 17	1ar 6	10 8	14 58	21 29	12 36	7 7	27 56	8 48	1 11	24 44	7 38	1 13	17 12	11 19	3 2	24 27	8 42	2 31
8/29/1981	14 58	12 26	5 8	5 58	14 20	17 14	23 6	13 7	7 54	29 14	7 43	1 17	26 29	6 23	1 29	19 24	15 32	2 57	25 45	9 37	2 22
9/ 8/1981	16 55	16 29	5 0	11 2	18 11	19 48	24 43	12 59R	8 46	0ar33	6 14	1 23	28 15	4 38	1 44	21 37	19 50	2 52	27 4	10 55	2 13
9/18/1981	18 52	20 31	4 53	16 18	21 32	22 40	26 21	12 9	9 39	1 52	4 28	1 28	0ar 1	2 32	1 59	23 51	24 11	2 47	28 23	12 34	2 5
9/28/1981	20 48	24 33	4 46	21 50	24 13	25 50	28 0	10 34	10 33	3 11	2 35	1 33	1 47	0 17	2 13	26 8	28 34	2 42	29 42	14 31	1 58
10/ 8/1981	22vi44	28vi33	4s40	27ar36	26ta 0	29n13	29ar40	8ta16R	11s24	4ar31	0ar42R	1s36	3ar34	28pi 7R	2s24	28vi26	3li 0	2s38	1cp 2	16sa45	1n51
10/18/1981	24 40	2 30	4 34	3 39	26 38	32 40	1 21	5 24	12 7	5 50	29 2	1 38	5 22	26 14	2 33	0li46	7 27	2 34	2 23	19 12	1 45
10/28/1981	26 36	6 25	4 29	10 0	25 52	35 51	3 4	2 12	12 38	7 10	27 42	1 39	7 10	24 50	2 39	3 8	11 56	2 29	3 44	21 52	1 39
11/ 7/1981	28 31	10 15	4 25	16 39	23 38	38 19	4 48	29ar 1	12 55	8 30	26 47	1 39	8 58	24 1	2 43	5 33	16 25	2 25	5 6	24 41	1 34
11/17/1981	0li27	14 0	4 20	23 37	20 15	39 31	6 33	26 10	12 56	9 51	26 21	1 38	10 48	23 49	2 46	7 59	20 53	2 21	6 28	27 40	1 29
11/27/1981	2 22	17 38	4 17	0gm54	16 32	39 1	8 20	23 54	12 46	11 11	26 25D	1 37	12 38	24 13	2 47	10 28	25 21	2 17	7 51	0cp45	1 25
12/ 7/1981	4 17	21 7	4 13	8 30	13 38	36 41	10 9	22 25	12 27	12 32	26 59	1 36	14 28	25 11	2 48	13 0	29 46	2 13	9 15	3 57	1 21
12/17/1981	6 13	24 26	4 10	16 23	12 24	32 45	12 0	21 44	12 3	13 53	27 58	1 35	16 19	26 40	2 48	15 34	4 8	2 8	10 39	7 13	1 17
12/27/1981	8 9	27 33	4 7	24 34	13 11	27 38	13 54	21 50D	11 36	15 15	29 22	1 33	18 11	28 36	2 48	18 10	8 25	2 4	12 5	10 33	1 14
1/ 6/1982	10 5	0sc25	4 4	2cn59	15 55	21 46	15 50	22 40	11 10	16 37	1 7	1 32	20 4	0ar56	2 48	20 49	12 36	1 58	13 30	13 55	1 11
1/16/1982	12 1	2 58	4 0	11 35	20 19	15 31	17 49	24 8	10 44	17 59	3 11	1 31	21 58	3 35	2 48	23 31	16 40	1 53	14 57	17 18	1 8
1/26/1982	13 58	5 10	3 57	20 21	26 6	9 14	19 52	26 9	10 21	19 22	5 30	1 30	23 52	6 31	2 48	26 16	20 34	1 46	16 25	20 41	1 5
2/ 5/1982	15 55	6 56	3 52	29 12	2gm58	3 13	21 57	28 39	10 0	20 45	8 3	1 30	25 48	9 42	2 48	29 4	24 15	1 38	17 53	24 3	1 3
2/15/1982	17 53	8 11	3 46	8 4	10 40	2s17	24 7	1 33	9 41	22 9	10 47	1 29	27 44	13 5	2 49	1 55	27 41	1 29	19 22	27 23	1 0
2/25/1982	19 52	8 52	3 39	16 53	18 59	7 5	26 21	4 49	9 24	23 33	13 40	1 29	29 41	16 38	2 50	4 50	0sa48	1 19	20 52	0aq39	0 58
3/ 7/1982	21 51	8 55R	3 28	25 37	27 44	11 2	28 39	8 23	9 9	24 57	16 41	1 29	1 40	20 19	2 51	7 47	3 32	1 7	22 23	3 51	0 55
3/17/1982	23 51	8 15	3 15	4vi11	6 45	14 7	1gm 2	12 13	8 56	26 22	19 48	1 29	3 39	24 7	2 53	10 48	5 48	0 52	23 55	6 56	0 53
3/27/1982	25 52	6 54	2 58	12 32	15 50	16 20	3 31	16 18	8 44	27 47	23 0	1 30	5 39	28 1	2 55	13 53	7 30	0 34	25 27	9 54	0 50
4/ 6/1982	27 54	4 56	2 36	20 38	24 55	17 47	6 6	20 37	8 34	29 13	26 15	1 31	7 41	1ta59	2 57	17 0	8 32	0 12	27 1	12 42	0 48
4/16/1982	29 57	2 31	2 10	28 28	3le51	18 33	8 47	25 8	8 25	0ta39	29 32	1 32	9 44	6 1	3 0	20 12	8 50	0n13	28 36	15 19	0 45
4/26/1982	2 0	29li54	1 41	5 59	12 35	18 46	11 36	29 52	8 16	2 6	2 51	1 34	11 48	10 6	3 3	23 27	8 19	0 42	0aq12	17 43	0 42
5/ 6/1982	4 6	27 21	1 9	13 12	21 4	18 32	14 33	4 47	8 8	3 33	6 10	1 35	13 53	14 13	3 7	26 45	7 1	1 14	1 49	19 50	0 39

Ephemeris	80			433			1221			1489			114			3018			2041		
0 Hours ET	Sappho			Eros			Amor			Attila			Kassandra			Godiva			Lancelot		
Mo Dy Year	h long	g long	g lat	h long	g long	g lat	h long	g long	g lat	h long	g long	g lat	h long	g long	g lat	h long	g long	g lat	h long	g long	g lat
5/16/1982	6 12	25 10	0 38	20 6	29 16	17 59	17 40	9 53	8 1	5 1	9 29	1 37	15 59	18 22	3 11	0sa 8	5 4	1 49	3 27	21 40	0 35
5/26/1982	8 20	23 33	0 7	26 41	7 12	17 11	20 56	15 12	7 53	6 30	12 46	1 40	18 7	22 31	3 15	3 34	2 43	2 23	5 6	23 7	0 31
6/ 5/1982	10 29	22 36	0 21	2sc58	14 51	16 14	24 24	20 43	7 45	7 59	16 0	1 42	20 16	26 40	3 20	7 3	0 20	2 55	6 46	24 10	0 27
6/15/1982	12 40	22 23	0 47	8 58	22 14	15 11	28 4	26 26	7 36	9 29	19 10	1 45	22 26	0gm49	3 25	10 36	28 18	3 22	8 28	24 45	0 22
6/25/1982	14 53	22 50	1 9	14 41	29 23	14 5	1cn57	2cn24	7 26	10 59	22 16	1 49	24 38	4 56	3 31	14 13	26 55	3 43	10 10	24 49R	0 16
7/ 5/1982	17 8	23 56	1 30	20 10	6 20	12 59	6 7	8 35	7 14	12 30	25 16	1 53	26 51	9 1	3 38	17 52	26 23	3 59	11 54	24 22	0 9
7/15/1982	19 24	25 36	1 47	25 24	13 5	11 53	10 33	15 3	7 0	14 2	28 9	1 57	29 6	13 4	3 45	21 35	26 44	4 10	13 40	23 24	0 2
7/25/1982	21 42	27 45	2 3	0sa26	19 40	10 50	15 19	21 47	6 43	15 34	0gm53	2 2	1 22	17 3	3 53	25 21	27 56	4 17	15 26	21 57	0 6
8/ 4/1982	24 3	0sc19	2 18	5 16	26 7	9 49	20 27	28 50	6 22	17 7	3 26	2 8	3 39	20 57	4 2	29 10	29 54	4 21	17 14	20 9	0 15
8/14/1982	26 26	3 16	2 31	9 55	2sc27	8 52	25 58	6 12	5 57	18 41	5 45	2 14	5 58	24 45	4 12	3 2	2 34	4 22	19 4	18 9	0 23
8/24/1982	28 51	6 33	2 43	14 24	8 40	7 57	1le56	13 54	5 28	20 15	7 50	2 20	8 19	28 26	4 22	6 56	5 48	4 22	20 55	16 11	0 31
9/ 3/1982	1 19	10 6	2 55	18 45	14 49	7 6	8 23	21 59	4 52	21 50	9 36	2 28	10 41	1 57	4 34	10 52	9 31	4 21	22 47	14 25	0 39
9/13/1982	3 49	13 55	3 6	22 58	20 52	6 19	15 21	0vi25	4 10	23 26	11 0	2 36	13 5	5 16	4 47	14 50	13 39	4 19	24 41	13 4	0 46
9/23/1982	6 23	17 57	3 16	27 4	26 53	5 34	22 52	9 14	3 21	25 3	12 0	2 44	15 31	8 21	5 1	18 49	18 8	4 16	26 36	12 13	0 52
10/ 3/1982	8 59	22 10	3 26	1cp 3	2sa51	4 51	0vi57	18 25	2 26	26 40	12 31	2 53	17 58	11 9	5 17	22 50	22 55	4 12	28 33	11 58	0 57
10/13/1982	11 38	26 34	3 36	4 58	8 46	4 12	9 38	27 56	1 24	28 19	12 32R	3 2	20 27	13 34	5 34	26 51	27 56	4 8	0pi31	12 17	1 2
10/23/1982	14 21	1sa 8	3 46	8 48	14 40	3 35	18 50	7 44	0 18	29 58	12 0	3 11	22 57	15 34	5 52	0aq53	3 9	4 3	2 31	13 11	1 6
11/ 2/1982	17 7	5 51	3 56	12 33	20 32	2 59	28 31	17 45	0 51	1 38	10 56	3 19	25 29	17 3	6 11	4 55	8 32	3 58	4 32	14 37	1 9
11/12/1982	19sa56	10sa42	4n 6	16cp16	26sa24	2s26	8li34	27li56	1n59	3gm19	9gm24R	3s25	28gm 3	17cn55	6s31	8aq56	14cp 4	3n53	6pi36	16aq31	1s12
11/22/1982	22 49	15 40	4 16	19 56	2cp16	1 54	18 50	8 8	3 6	5 1	7 32	3 28	0cn38	18 6	6 50	12 56	19 42	3 47	8 40	18 50	1 14
12/ 2/1982	25 46	20 45	4 26	23 34	8 7	1 24	29 7	18 18	4 7	6 44	5 30	3 29	3 15	17 33	7 7	16 56	25 25	3 41	10 47	21 31	1 16
12/12/1982	28 47	25 56	4 36	27 10	13 59	0 55	9 15	28 18	5 1	8 27	3 30	3 27	5 53	16 16	7 19	20 53	1aq12	3 35	12 55	24 31	1 19
12/22/1982	1 52	1cp13	4 46	0aq45	19 51	0 27	19 3	8 2	5 46	10 12	1 46	3 21	8 33	14 23	7 25	24 49	7 2	3 29	15 5	27 47	1 21
1/ 1/1983	5 1	6 36	4 57	4 20	25 44	0n 1	28 22	17 28	6 23	11 57	0 27	3 14	11 15	12 6	7 22	28 43	12 53	3 22	17 16	1pi17	1 23
1/11/1983	8 15	12 2	5 8	7 56	1aq38	0 27	7 8	26 30	6 51	13 44	29 41	3 6	13 58	9 43	7 10	2 35	18 45	3 16	19 30	5 0	1 25
1/21/1983	11 33	17 34	5 19	11 32	7 34	0 54	15 18	5 7	7 11	15 31	29 29	2 56	16 43	7 35	6 49	6 23	24 37	3 9	21 45	8 53	1 27
1/31/1983	14 55	23 8	5 30	15 9	13 30	1 20	22 50	13 18	7 25	17 20	29 52	2 46	19 29	5 58	6 21	10 9	0pi27	3 2	24 1	12 54	1 29
2/10/1983	18 22	28 46	5 41	18 49	19 29	1 46	29 47	21 3	7 33	19 9	0 47	2 37	22 16	5 3	5 50	13 52	6 16	2 56	26 19	17 3	1 32
2/20/1983	21 53	4 27	5 52	22 30	25 30	2 12	6 10	28 22	7 37	21 0	2 12	2 28	25 5	4 55D	5 18	17 31	12 3	2 49	28 39	21 18	1 34
3/ 2/1983	25 29	10 10	6 4	26 16	1pi33	2 38	12 2	5 15	7 38	22 51	4 4	2 20	27 55	5 33	4 46	21 7	17 46	2 42	1ar 1	25 37	1 37
3/12/1983	29 10	15 55	6 15	0pi 4	7 39	3 4	17 26	11 44	7 36	24 44	6 18	2 12	0le46	6 53	4 16	24 39	23 25	2 35	3 24	0ar 1	1 39
3/22/1983	2 55	21 41	6 26	3 58	13 49	3 31	22 25	17 48	7 32	26 37	8 51	2 5	3 38	8 49	3 48	28 8	29 0	2 28	5 49	4 27	1 42
4/ 1/1983	6 44	27 27	6 37	7 57	20 2	3 57	27 2	23 29	7 27	28 32	11 41	1 59	6 32	11 18	3 22	1ar33	4 30	2 21	8 16	8 56	1 45
4/11/1983	10 38	3 13	6 48	12 1	26 21	4 25	1aq19	28 47	7 21	0cn27	14 44	1 53	9 26	14 13	2 58	4 55	9 56	2 14	10 43	13 26	1 48
4/21/1983	14 36	8 58	6 58	16 13	2ar44	4 52	5 19	3 41	7 15	2 24	18 0	1 48	12 22	17 31	2 37	8 12	15 15	2 7	13 13	17 57	1 52
5/ 1/1983	18 38	14 41	7 7	20 32	9 13	5 19	9 3	8 11	7 8	4 22	21 25	1 43	15 18	21 7	2 17	11 26	20 28	2 0	15 44	22 27	1 56
5/11/1983	22 43	20 21	7 16	24 59	15 50	5 47	12 33	12 16	7 0	6 20	24 59	1 38	18 14	24 59	1 59	14 37	25 35	1 53	18 16	26 57	1 59
5/21/1983	26 52	25 57	7 23	29 37	22 34	6 14	15 51	15 54	6 53	8 20	28 40	1 34	21 12	29 4	1 42	17 43	0ta34	1 46	20 49	1ta25	2 4
5/31/1983	1pi 3	1ar26	7 30	4 25	29 26	6 40	18 58	19 3	6 45	10 21	2 27	1 31	24 9	3 20	1 26	20 47	5 26	1 39	23 24	5 51	2 8
6/10/1983	5 17	6 49	7 35	9 24	6 29	7 6	21 55	21 39	6 36	12 23	6 19	1 27	27 8	7 45	1 12	23 46	10 9	1 31	26 0	10 13	2 13

Ephemeris	80			433			1221			1489			114			3018			2041		
0 Hours ET	Sappho			Eros			Amor			Attila			Kassandra			Godiva			Lancelot		
Mo Dy Year	h long	g long	g lat	h long	g long	g lat	h long	g long	g lat	h long	g long	g lat	h long	g long	g lat	h long	g long	g lat	h long	g long	g lat
6/20/1983	9 34	12 2	7 38	14 37	13 42	7 30	24 44	23 39	6 26	14 26	10 14	1 24	0vi 6	12 17	0 58	26 43	14 43	1 23	28 36	14 31	2 18
6/30/1983	13 51	17 3	7 40	20 3	21 6	7 51	27 25	24 59	6 14	16 30	14 13	1 21	3 5	16 55	0 45	29 36	19 7	1 15	1ta14	18 44	2 23
7/10/1983	18 10	21 49	7 39	25 45	28 43	8 9	29 59	25 32	6 0	18 35	18 14	1 18	6 4	21 39	0 33	2 26	23 20	1 7	3 53	22 51	2 29
7/20/1983	22 30	26 17	7 36	1ta42	6 33	8 23	2 26	25 15	5 42	20 41	22 16	1 15	9 2	26 27	0 21	5 12	27 20	0 58	6 32	26 49	2 36
7/30/1983	26 50	0ta22	7 29	7 57	14 36	8 31	4 48	24 5	5 19	22 48	26 19	1 13	12 1	1vi19	0 9	7 56	1gm 5	0 48	9 12	0gm38	2 42
8/ 9/1983	1ar 9	4 0	7 18	14 31	22 54	8 33	7 5	22 2	4 51	24 56	0le23	1 10	14 59	6 13	0n 2	10 37	4 34	0 38	11 53	4 15	2 50
8/19/1983	5 28	7 3	7 3	21 22	1cn25	8 26	9 17	19 15	4 15	27 5	4 26	1 8	17 57	11 10	0 13	13 14	7 44	0 27	14 34	7 38	2 58
8/29/1983	9 46	9 26	6 42	28 33	10 9	8 11	11 25	15 59	3 34	29 15	8 28	1 5	20 54	16 8	0 23	15 50	10 31	0 15	17 15	10 43	3 7
9/ 8/1983	14 1	11 1	6 13	6 3	19 5	7 44	13 29	12 33	2 49	1 26	12 27	1 3	23 51	21 7	0 34	18 22	12 53	0 2	19 57	13 29	3 16
9/18/1983	18 15	11 40	5 35	13 52	28 11	7 7	15 30	9 22	2 3	3 38	16 23	1 1	26 47	26 6	0 44	20 52	14 44	0 13	22 38	15 49	3 26
9/28/1983	22 26	11 20	4 46	21 57	7 25	6 18	17 27	6 44	1 19	5 50	20 14	0 58	29 42	1li 5	0 55	23 20	16 1	0 29	25 20	17 42	3 36
10/ 8/1983	26 35	10 2	3 46	0cn17	16 46	5 18	19 22	4 50	0 38	8 4	24 0	0 56	2 37	6 3	1 5	25 45	16 38	0 47	28 1	19 0	3 46
10/18/1983	0ta40	7 58	2 36	8 51	26 10	4 7	21 14	3 45	0 2	10 18	27 38	0 53	5 30	11 0	1 16	28 8	16 31R	1 6	0gm42	19 41	3 57
10/28/1983	4 42	5 29	1 20	17 34	5 35	2 48	23 4	3 27	0 28	12 33	1vi 6	0 51	8 23	15 54	1 26	0gm29	15 39	1 27	3 23	19 41R	4 6
11/ 7/1983	8 41	3 1	0 4	26 24	14 57	1 21	24 52	3 51	0 55	14 48	4 23	0 48	11 14	20 46	1 37	2 48	14 4	1 48	6 3	18 59	4 13
11/17/1983	12 36	1 2	1 7	5 16	24 14	0s10	26 38	4 54	1 17	17 5	7 26	0 45	14 4	25 33	1 49	5 5	11 52	2 9	8 42	17 39	4 17
11/27/1983	16 26	29ar50	2 7	14 7	3li22	1 44	28 22	6 27	1 36	19 22	10 11	0 41	16 53	0sc15	2 1	7 20	9 18	2 27	11 21	15 50	4 17
12/ 7/1983	20 13	29 33	2 57	22 53	12 21	3 19	0ar 4	8 28	1 53	21 39	12 36	0 37	19 41	4 52	2 13	9 34	6 40	2 43	13 59	13 46	4 12
12/17/1983	23ta56	0ta10	3s37	1vi30	21li 5	4s52	1ar46	10pi50	2s 8	23le57	14vi36	0s33	22li27	9sc20	2n26	11gm46	4gm16R	2s55	16gm36	11gm44R	4s 2
12/27/1983	27 35	1 37	4 8	9 56	29 34	6 21	3 26	13 31	2 21	26 16	16 7	0 28	25 12	13 40	2 40	13 56	2 23	3 3	19 12	10 1	3 48
1/ 6/1984	1gm 9	3 45	4 31	18 7	7 44	7 47	5 5	16 26	2 33	28 35	17 5	0 22	27 56	17 50	2 55	16 5	1 9	3 8	21 46	8 50	3 32
1/16/1984	4 39	6 29	4 49	26 2	15 34	9 8	6 44	19 34	2 44	0vi54	17 27	0 16	0sc38	21 47	3 12	18 12	0 39	3 10	24 20	8 17	3 15
1/26/1984	8 6	9 40	5 2	3li39	23 1	10 25	8 21	22 52	2 55	3 14	17 9	0 8	3 18	25 29	3 29	20 18	0 51	3 10	26 52	8 26D	2 57
2/ 5/1984	11 28	13 14	5 13	10 58	0sa 2	11 38	9 58	26 18	3 5	5 34	16 14	0n 0	5 57	28 54	3 48	22 23	1 42	3 10	29 23	9 13	2 40
2/15/1984	14 46	17 6	5 21	17 58	6 36	12 48	11 35	29 50	3 16	7 54	14 45	0 9	8 34	1 59	4 9	24 26	3 7	3 8	1 53	10 37	2 24
2/25/1984	18 0	21 11	5 27	24 39	12 40	13 55	13 11	3 28	3 26	10 14	12 53	0 19	11 10	4 40	4 31	26 29	5 0	3 7	4 21	12 31	2 9
3/ 6/1984	21 11	25 28	5 31	1sc 1	18 9	15 0	14 47	7 9	3 36	12 35	10 50	0 28	13 44	6 53	4 56	28 30	7 18	3 5	6 47	14 51	1 56
3/16/1984	24 17	29 52	5 35	7 7	22 59	16 4	16 23	10 54	3 47	14 55	8 54	0 37	16 17	8 34	5 22	0cn31	9 56	3 3	9 12	17 33	1 44
3/26/1984	27 20	4 23	5 38	12 55	27 4	17 7	17 59	14 40	3 58	17 15	7 18	0 44	18 48	9 39	5 49	2 30	12 50	3 2	11 36	20 33	1 33
4/ 5/1984	0cn19	8 58	5 40	18 28	0cp17	18 10	19 35	18 28	4 10	19 36	6 12	0 51	21 17	10 4	6 18	4 29	15 58	3 1	13 57	23 48	1 24
4/15/1984	3 15	13 37	5 42	23 47	2 28	19 10	21 12	22 17	4 23	21 56	5 42	0 57	23 45	9 47	6 45	6 27	19 17	3 0	16 18	27 15	1 15
4/25/1984	6 8	18 18	5 44	28 53	3 26	20 6	22 49	26 5	4 36	24 16	5 51D	1 1	26 11	8 48	7 11	8 24	22 45	3 0	18 36	0cn51	1 6
5/ 5/1984	8 57	23 0	5 45	3 46	3 1	20 50	24 26	29 53	4 51	26 36	6 35	1 5	28 36	7 13	7 32	10 21	26 20	2 59	20 53	4 35	0 59
5/15/1984	11 43	27 43	5 47	8 28	1 8	21 14	26 4	3 39	5 6	28 55	7 54	1 8	0sa59	5 11	7 46	12 17	0cn 1	3 0	23 8	8 25	0 52
5/25/1984	14 26	2 25	5 48	13 1	27 54	21 5	27 43	7 24	5 24	1 14	9 41	1 10	3 20	2 55	7 52	14 13	3 47	3 0	25 21	12 19	0 46
6/ 4/1984	17 6	7 7	5 50	17 24	23 43	20 13	29 23	11 6	5 42	3 33	11 53	1 12	5 40	0 44	7 50	16 8	7 36	3 1	27 33	16 16	0 40
6/14/1984	19 43	11 48	5 52	21 39	19 17	18 37	1 4	14 44	6 3	5 51	14 27	1 14	7 58	28 51	7 40	18 3	11 28	3 3	29 43	20 16	0 34
6/24/1984	22 18	16 27	5 55	25 47	15 22	16 27	2 46	18 19	6 25	8 8	17 19	1 15	10 15	27 28	7 24	19 57	15 22	3 4	1 52	24 17	0 29
7/ 4/1984	24 50	21 5	5 57	29 49	12 29	14 2	4 29	21 48	6 50	10 25	20 27	1 17	12 31	26 42	7 4	21 51	19 17	3 7	3 58	28 19	0 24
7/14/1984	27 19	25 40	6 0	3 45	10 53	11 38	6 14	25 11	7 18	12 42	23 47	1 18	14 45	26 35D	6 42	23 45	23 13	3 9	6 3	2 20	0 19

Ephemeris	80			433			1221			1489			114			3018			2041		
0 Hours ET	Sappho			Eros			Amor			Attila			Kassandra			Godiva			Lancelot		
Mo Dy Year	h long	g long	g lat	h long	g long	g lat	h long	g long	g lat	h long	g long	g lat	h long	g long	g lat	h long	g long	g lat	h long	g long	g lat
7/24/1984	29 46	0le13	6 3	7 36	10 31	9 24	8 1	28 27	7 48	14 58	27 18	.1 19	16 57	27 5	6 20	25 39	27 9	3 12	8 7	6 20	0 14
8/ 3/1984	2 11	4 43	6 7	11 23	11 14	7 26	9 50	1 33	8 23	17 13	0li59	1 21	19 8	28 10	5 58	27 32	1le 4	3 16	10 8	10 19	0 9
8/13/1984	4 33	9 9	6 12	15 6	12 52	5 44	11 41	4 27	9 2	19 27	4 46	1 22	21 18	29 44	5 37	29 26	4 58	3 20	12 9	14 16	0 5
8/23/1984	6 54	13 31	6 17	18 47	15 13	4 17	13 34	7 6	9 46	21 40	8 40	1 23	23 27	1 45	5 18	1 19	8 50	3 24	14 7	18 10	0n 0
9/ 2/1984	9 13	17 49	6 23	22 25	18 11	3 2	15 30	9 28	10 36	23 53	12 38	1 24	25 34	4 8	5 0	3 13	12 39	3 29	16 4	22 0	0 5
9/12/1984	11 29	22 2	6 29	26 2	21 37	1 57	17 28	11 27	11 31	26 5	16 41	1 26	27 40	6 50	4 44	5 6	16 24	3 35	18 0	25 45	0 9
9/22/1984	13 44	26 9	6 37	29 38	25 26	1 2	19 30	12 58	12 34	28 16	20 46	1 27	29 44	9 47	4 29	7 0	20 5	3 42	19 54	29 25	0 14
10/ 2/1984	15 57	0vi 9	6 46	3 13	29 36	0 13	21 35	13 55	13 44	0sc26	24 53	1 29	1 48	12 59	4 16	8 54	23 39	3 50	21 46	2 57	0 19
10/12/1984	18 9	4 0	6 56	6 48	4 2	0 29	23 44	14 10	15 0	2 35	29 2	1 30	3 50	16 21	4 4	10 48	27 7	3 58	23 37	6 22	0 24
10/22/1984	20 19	7 42	7 7	10 24	8 41	1 7	25 57	13 36	16 21	4 43	3 11	1 32	5 52	19 53	3 53	12 43	0vi25	4 8	25 27	9 38	0 30
11/ 1/1984	22 28	11 13	7 19	14 1	13 34	1 41	28 14	12 8	17 42	6 50	7 20	1 34	7 52	23 33	3 43	14 38	3 32	4 18	27 15	12 42	0 36
11/11/1984	24 35	14 30	7 34	17 39	18 38	2 12	0gm36	9 43	18 57	8 57	11 27	1 36	9 51	27 18	3 34	16 33	6 26	4 31	29 2	15 32	0 42
11/21/1984	26 41	17 31	7 49	21 20	23 52	2 39	3 4	6 30	19 56	11 2	15 32	1 39	11 49	1cp 9	3 26	18 29	9 3	4 44	0vi47	18 7	0 48
12/ 1/1984	28 46	20 13	8 7	25 4	29 15	3 5	5 38	2 47	20 30	13 6	19 34	1 42	13 46	5 4	3 19	20 26	11 20	4 59	2 32	20 23	0 56
12/11/1984	0vi49	22 33	8 27	28 52	4 47	3 28	8 18	29ta 2	20 33	15 9	23 32	1 45	15 42	9 1	3 12	22 23	13 12	5 15	4 15	22 19	1 4
12/21/1984	2 52	24 26	8 48	2 44	10 28	3 50	11 6	25 43	20 8	17 11	27 24	1 48	17 37	13 0	3 7	24 21	14 37	5 33	5 56	23 49	1 12
12/31/1984	4 54	25 49	9 10	6 40	16 17	4 11	14 1	23 16	19 19	19 12	1sa10	1 52	19 32	17 0	3 1	26 19	15 28	5 52	7 37	24 52	1 21
1/10/1985	6 55	26 36	9 34	10 43	22 14	4 30	17 6	21 53	18 15	21 12	4 47	1 56	21 25	20 59	2 57	28 19	15 42	6 11	9 16	25 24	1 31
1/20/1985	8vi55	26vi43R	9s56	14pi52	28aq19	4n48	20gm20	21ta37	17s 4	23sc11	8sa14	2n 0	23cp18	24cp57	2n53	0vi19	15vi14R	6s30	10vi54	25vi23R	1n41
1/30/1985	10 54	26 7	10 17	19 8	4 33	5 4	23 45	22 24	15 52	25 9	11 30	2 6	25 10	28 54	2 49	2 20	14 5	6 46	12 31	24 47	1 52
2/ 9/1985	12 52	24 49	10 32	23 33	10 54	5 20	27 23	24 9	14 41	27 6	14 33	2 11	27 1	2 47	2 46	4 22	12 17	6 58	14 7	23 40	2 2
2/19/1985	14 50	22 52	10 40	28 7	17 25	5 34	1cn14	26 42	13 34	29 2	17 19	2 17	28 51	6 36	2 43	6 25	9 58	7 4	15 42	22 6	2 11
3/ 1/1985	16 48	20 27	10 38	2 51	24 4	5 46	5 20	29 60	12 30	0sa56	19 47	2 24	0aq41	10 21	2 40	8 30	7 23	7 3	17 16	20 12	2 19
3/11/1985	18 45	17 49	10 25	7 47	0ar52	5 57	9 43	3 56	11 29	2 50	21 54	2 31	2 30	13 59	2 38	10 35	4 50	6 54	18 49	18 10	2 25
3/21/1985	20 41	15 14	10 0	12 55	7 51	6 6	14 25	8 25	10 30	4 43	23 37	2 39	4 19	17 29	2 36	12 42	2 36	6 37	20 21	16 11	2 29
3/31/1985	22 37	13 0	9 27	18 17	14 59	6 13	19 28	13 26	9 32	6 34	24 52	2 47	6 7	20 51	2 35	14 50	0 54	6 16	21 52	14 28	2 31
4/10/1985	24 33	11 19	8 47	23 53	22 18	6 17	24 54	18 56	8 34	8 25	25 38	2 56	7 54	24 3	2 33	17 0	29le52	5 51	23 22	13 8	2 31
4/20/1985	26 29	10 16	8 5	29 45	29 48	6 18	0le47	24 54	7 36	10 14	25 51	3 4	9 41	27 3	2 32	19 11	29 33	5 26	24 52	12 16	2 30
4/30/1985	28 24	9 55	7 23	5 54	7 29	6 15	7 7	1cn19	6 35	12 3	25 30	3 12	11 27	29 49	2 31	21 23	29 57	5 0	26 20	11 55	2 28
5/10/1985	0li20	10 13	6 42	12 21	15 22	6 7	13 59	8 10	5 31	13 50	24 37	3 19	13 13	2 19	2 30	23 38	0 58	4 36	27 48	12 4	2 25
5/20/1985	2 15	11 6	6 4	19 7	23 27	5 56	21 23	15 27	4 24	15 37	23 15	3 24	14 59	4 31	2 29	25 54	2 34	4 13	29 14	12 41	2 23
5/30/1985	4 11	12 32	5 29	26 12	1gm43	5 38	29 21	23 11	3 13	17 23	21 31	3 26	16 44	6 21	2 28	28 12	4 40	3 52	0li40	13 44	2 20
6/ 9/1985	6 6	14 24	4 57	3gm35	10 10	5 16	7 54	1le21	1 57	19 7	19 36	3 26	18 29	7 46	2 26	0li32	7 11	3 33	2 6	15 9	2 17
6/19/1985	8 2	16 40	4 27	11 18	18 47	4 47	17 0	9 57	0 38	20 51	17 40	3 24	20 13	8 44	2 24	2 53	10 4	3 15	3 30	16 54	2 14
6/29/1985	9 58	19 15	4 1	19 18	27 34	4 13	26 36	18 57	0 45	22 34	15 56	3 18	21 57	9 10	2 22	5 17	13 16	2 58	4 54	18 56	2 12
7/ 9/1985	11 55	22 7	3 36	27 33	6 29	3 33	6 34	28 20	2 8	24 16	14 33	3 10	23 41	9 4R	2 18	7 44	16 45	2 43	6 17	21 12	2 10
7/19/1985	13 51	25 13	3 14	6 3	15 32	2 48	16 47	8 3	3 29	25 57	13 37	3 2	25 25	8 22	2 13	10 12	20 28	2 29	7 40	23 39	2 8
7/29/1985	15 49	28 32	2 53	14 44	24 39	1 58	27 4	18 3	4 46	27 37	13 11	2 52	27 8	7 7	2 5	12 43	24 23	2 16	9 1	26 17	2 6
8/ 8/1985	17 47	2 0	2 34	23 32	3le51	1 6	7 14	28 13	5 55	29 17	13 16D	2 42	28 51	5 24	1 56	15 16	28 30	2 4	10 23	29 3	2 5
8/18/1985	19 45	5 38	2 17	2le24	13 3	0 12	17 6	8 29	6 53	0 55	13 51	2 32	0pi35	3 20	1 45	17 53	2 46	1 52	11 43	1 56	2 5

Ephemeris	80			433			1221			1489			114			3018			2041		
0 Hours ET	Sappho			Eros			Amor			Attila			Kassandra			Godiva			Lancelot		
Mo Dy Year	h long	g long	g lat	h long	g long	g lat	h long	g long	g lat	h long	g long	g lat	h long	g long	g lat	h long	g long	g lat	h long	g long	g lat
8/28/1985	21 44	9 24	2 0	11 16	22 16	0 43	26 31	18 43	7 39	2 33	14 54	2 23	2 18	1 8	1 33	20 31	7 12	1 41	13 3	4 54	2 4
9/ 7/1985	23 44	13 16	1 44	20 4	1vi26	1 37	5 24	28 48	8 10	4 10	16 21	2 15	4 1	29aq 0	1 19	23 13	11 46	1 30	14 23	7 57	2 4
9/17/1985	25 45	17 15	1 29	28 45	10 32	2 29	13 40	8 39	8 28	5 46	18 9	2 7	5 44	27 8	1 4	25 57	16 27	1 19	15 42	11 2	2 5
9/27/1985	27 47	21 18	1 15	7 15	19 32	3 17	21 20	18 9	8 34	7 21	20 16	1 59	7 26	25 43	0 49	28 45	21 15	1 9	17 1	14 10	2 5
10/ 7/1985	29 49	25 25	1 0	15 31	28 24	4 2	28 23	27 17	8 30	8 56	22 40	1 53	9 9	24 52	0 36	1 35	26 9	0 59	18 19	17 19	2 6
10/17/1985	1 53	29 37	0 46	23 32	7 7	4 41	4 53	5 59	8 18	10 29	25 17	1 47	10 52	24 35	0 23	4 29	1sc 8	0 49	19 36	20 27	2 8
10/27/1985	3 58	3 51	0 33	1li15	15 39	5 16	10 51	14 17	8 0	12 3	28 5	1 42	12 35	24 53	0 11	7 26	6 14	0 38	20 53	23 35	2 10
11/ 6/1985	6 4	8 7	0 19	8 40	24 0	5 46	16 20	22 10	7 38	13 35	1cp 3	1 37	14 18	25 44	0 0	10 26	11 23	0 28	22 10	26 40	2 12
11/16/1985	8 12	12 25	0 5	15 46	2sc 9	6 10	21 24	29 40	7 14	15 7	4 9	1 32	16 2	27 5	0 9	13 30	16 38	0 18	23 27	29 42	2 15
11/26/1985	10 21	16 45	0 10	22 34	10 5	6 30	26 5	6 49	6 48	16 38	7 21	1 29	17 45	28 52	0 18	16 37	21 56	0 7	24 43	2 39	2 18
12/ 6/1985	12 32	21 4	0 25	29 2	17 49	6 46	0aq26	13 38	6 23	18 8	10 38	1 25	19 29	1pi 1	0 26	19 48	27 17	0n 4	25 58	5 30	2 22
12/16/1985	14 45	25 23	0 40	5 13	25 19	6 57	4 30	20 10	5 58	19 38	13 59	1 22	21 13	3 30	0 33	23 2	2sa41	0 15	27 14	8 13	2 27
12/26/1985	16 59	29 41	0 57	11 7	2sa36	7 6	8 17	26 25	5 33	21 7	17 22	1 19	22 57	6 15	0 40	26 20	8 8	0 27	28 28	10 46	2 32
1/ 5/1986	19 15	3 56	1 14	16 45	9 39	7 11	11 50	2aq26	5 9	22 36	20 46	1 17	24 41	9 15	0 46	29 42	13 36	0 40	29 43	13 8	2 37
1/15/1986	21 33	8 8	1 33	22 9	16 30	7 13	15 10	8 14	4 47	24 4	24 10	1 14	26 26	12 26	0 52	3 7	19 5	0 53	0 57	15 16	2 44
1/25/1986	23 54	12 16	1 54	27 18	23 7	7 13	18 19	13 49	4 25	25 32	27 34	1 12	28 11	15 47	0 58	6 36	24 35	1 6	2 12	17 8	2 51
2/ 4/1986	26 16	16 18	2 16	2sa16	29 31	7 11	21 19	19 13	4 4	26 59	0aq54	1 11	29 57	19 16	1 4	10 8	0cp 4	1 21	3 25	18 42	2 59
2/14/1986	28 41	20 12	2 41	7 1	5 41	7 7	24 9	24 26	3 44	28 25	4 12	1 9	1 43	22 52	1 9	13 44	5 31	1 37	4 39	19 54	3 7
2/24/1986	1sa 9	23sa58	3n 8	11sa37	11cp37	7s 1	26aq52	29aq28	3n25	29cp51	7aq25	1n 8	3ar29	26pi32	1s15	17sa24	10cp56	1n53	5sc52	20sc42	3n16
3/ 6/1986	3 39	27 32	3 38	16 3	17 20	6 53	29 27	4 20	3 7	1 16	10 32	1 7	5 16	0ar17	1 21	21 6	16 18	2 11	7 5	21 4	3 26
3/16/1986	6 12	0cp52	4 12	20 21	22 48	6 44	1 56	9 2	2 49	2 41	13 31	1 6	7 3	4 4	1 26	24 52	21 35	2 30	8 18	20 58R	3 35
3/26/1986	8 48	3 56	4 50	24 31	28 0	6 32	4 19	13 35	2 32	4 6	16 23	1 5	8 51	7 54	1 32	28 41	26 46	2 51	9 31	20 24	3 44
4/ 5/1986	11 27	6 39	5 33	28 34	2 55	6 19	6 37	17 56	2 15	5 30	19 4	1 4	10 40	11 44	1 39	2 32	1aq49	3 13	10 44	19 23	3 52
4/15/1986	14 9	8 58	6 21	2 32	7 32	6 3	8 50	22 8	1 58	6 54	21 33	1 3	12 29	15 35	1 45	6 26	6 41	3 37	11 56	17 58	3 58
4/25/1986	16 55	10 46	7 15	6 25	11 48	5 45	10 59	26 8	1 41	8 17	23 48	1 2	14 19	19 25	1 52	10 22	11 22	4 2	13 9	16 16	4 2
5/ 5/1986	19 44	11 59	8 15	10 13	15 40	5 23	13 4	29 57	1 24	9 40	25 48	1 2	16 10	23 14	1 59	14 19	15 47	4 30	14 21	14 24	4 4
5/15/1986	22 37	12 30	9 20	13 58	19 4	4 56	15 5	3 33	1 6	11 2	27 29	1 1	18 1	27 1	2 7	18 19	19 52	5 0	15 33	12 32	4 3
5/25/1986	25 33	12 16	10 28	17 39	21 55	4 25	17 3	6 55	0 47	12 25	28 49	1 1	19 53	0ta44	2 15	22 19	23 35	5 33	16 45	10 49	3 59
6/ 4/1986	28 34	11 15	11 36	21 18	24 5	3 46	18 58	10 1	0 27	13 46	29 47	1 0	21 46	4 24	2 25	26 21	26 49	6 7	17 57	9 23	3 53
6/14/1986	1 39	9 28	12 39	24 55	25 28	2 58	20 51	12 50	0 6	15 8	0 19	0 59	23 40	7 58	2 34	0aq23	29 29	6 44	19 9	8 20	3 45
6/24/1986	4 47	7 8	13 31	28 31	25 53	2 0	22 41	15 19	0 16	16 29	0 23R	0 57	25 34	11 26	2 45	4 25	1 29	7 22	20 21	7 43	3 37
7/ 4/1986	8 1	4 32	14 7	2 6	25 11	0 50	24 29	17 25	0 41	17 50	29aq59	0 56	27 30	14 47	2 57	8 26	2 42	8 1	21 33	7 32	3 28
7/14/1986	11 18	2 4	14 23	5 41	23 16	0n34	26 16	19 5	1 8	19 10	29 8	0 53	29 26	17 59	3 10	12 27	3 2	8 38	22 45	7 49	3 19
7/24/1986	14 40	0 4	14 19	9 17	20 11	2 9	28 0	20 14	1 38	20 31	27 51	0 50	1 24	20 59	3 24	16 27	2 29	9 9	23 57	8 30	3 10
8/ 3/1986	18 7	28 50	14 0	12 53	16 13	3 48	29 43	20 48	2 10	21 51	26 14	0 47	3 23	23 47	3 40	20 25	1 6	9 32	25 9	9 34	3 1
8/13/1986	21 38	28 29	13 29	16 31	11 55	5 24	1 25	20 43R	2 46	23 11	24 25	0 43	5 22	26 18	3 57	24 22	29aq 6	9 41	26 21	10 59	2 54
8/23/1986	25 14	29 2	12 52	20 12	7 56	6 45	3 5	19 54	3 24	24 30	22 33	0 38	7 23	28 30	4 16	28 16	26 51	9 35	27 33	12 42	2 46
9/ 2/1986	28 55	0cp27	12 11	23 55	4 50	7 49	4 45	18 22	4 5	25 50	20 47	0 33	9 25	0gm20	4 37	2 8	24 45	9 15	28 46	14 41	2 40
9/12/1986	2 40	2 38	11 30	27 41	2 57	8 34	6 23	16 8	4 45	27 9	19 16	0 28	11 28	1 42	5 0	5 58	23 10	8 42	29 58	16 53	2 34
9/22/1986	6 29	5 29	10 49	1pi31	2 21	9 4	8 1	13 22	5 23	28 28	18 8	0 23	13 32	2 33	5 24	9 44	22 20	8 1	1 10	19 17	2 29

Ephemeris	80			433			1221			1489			114			3018			2041		
0 Hours ET	Sappho			Eros			Amor			Attila			Kassandra			Godiva			Lancelot		
Mo Dy Year	h long	g long	g lat	h long	g long	g lat	h long	g long	g lat	h long	g long	g lat	h long	g long	g lat	h long	g long	g lat	h long	g long	g lat
10/ 2/1986	10 23	8 56	10 10	5 26	2 56	9 22	9 38	10 17	5 56	29 47	17 27	0 18	15 37	2 49	5 50	13 28	22 21D	7 18	2 23	21 51	2 24
10/12/1986	14 21	12 51	9 32	9 27	4 33	9 33	11 14	7 11	6 23	1 5	17 14	0 13	17 44	2 26	6 15	17 8	23 11	6 34	3 35	24 34	2 20
10/22/1986	18 23	17 13	8 55	13 34	7 3	9 39	12 51	4 23	6 41	2 24	17 29	0 9	19 52	1 23	6 40	20 44	24 48	5 53	4 48	27 24	2 17
11/ 1/1986	22 28	21 55	8 20	17 48	10 16	9 41	14 27	2 8	6 52	3 42	18 11	0 5	22 2	29ta45	7 1	24 17	27 3	5 14	6 1	0sa19	2 14
11/11/1986	26 37	26 56	7 46	22 10	14 8	9 40	16 3	0 34	6 57	5 1	19 19	0 1	24 13	27 38	7 17	27 47	29 51	4 40	7 14	3 19	2 11
11/21/1986	0pi49	2aq12	7 14	26 41	18 31	9 37	17 38	29pi44	6 57	6 19	20 48	0 2	26 25	25 16	7 25	1ar13	3 7	4 8	8 28	6 22	2 9
12/ 1/1986	5 3	7 41	6 42	1ar22	23 21	9 33	19 14	29 38D	6 54	7 37	22 36	0 5	28 39	22 57	7 26	4 35	6 43	3 40	9 41	9 26	2 8
12/11/1986	9 19	13 22	6 12	6 14	28 37	9 27	20 51	0ar12	6 50	8 55	24 42	0 8	0gm54	20 57	7 18	7 53	10 38	3 15	10 55	12 32	2 6
12/21/1986	13 38	19 11	5 42	11 18	4 14	9 19	22 27	1 22	6 45	10 13	27 2	0 11	3 11	19 29	7 5	11 8	14 47	2 52	12 9	15 37	2 6
12/31/1986	17 57	25 8	5 13	16 35	10 12	9 9	24 4	3 1	6 41	11 31	29 34	0 13	5 29	18 42	6 47	14 19	19 6	2 31	13 24	18 40	2 5
1/10/1987	22 17	1pi11	4 45	22 6	16 29	8 57	25 42	5 7	6 37	12 49	2 16	0 16	7 49	18 37D	6 27	17 27	23 34	2 13	14 38	21 41	2 5
1/20/1987	26 37	7 18	4 17	27 54	23 3	8 43	27 21	7 35	6 33	14 7	5 6	0 18	10 11	19 14	6 7	20 31	28 9	1 56	15 53	24 37	2 6
1/30/1987	0ar57	13 28	3 49	3 57	29 55	8 25	29 0	10 21	6 31	15 25	8 3	0 20	12 34	20 30	5 47	23 31	2 49	1 41	17 9	27 27	2 7
2/ 9/1987	5 17	19 40	3 22	10 18	7 4	8 4	0ta40	13 23	6 29	16 43	11 5	0 23	14 59	22 20	5 27	26 28	7 32	1 27	18 24	0cp10	2 8
2/19/1987	9 35	25 53	2 56	16 58	14 28	7 40	2 22	16 38	6 28	18 1	14 11	0 25	17 25	24 40	5 9	29 21	12 18	1 14	19 40	2 44	2 10
3/ 1/1987	13 51	2ar 7	2 30	23 56	22 8	7 11	4 5	20 4	6 29	19 19	17 18	0 27	19 53	27 25	4 53	2 12	17 5	1 2	20 57	5 7	2 12
3/11/1987	18 5	8 19	2 4	1gm14	0ta 3	6 37	5 50	23 39	6 30	20 37	20 27	0 30	22 23	0gm32	4 37	4 59	21 52	0 50	22 14	7 16	2 14
3/21/1987	22 17	14 30	1 39	8 50	8 11	5 59	7 36	27 23	6 33	21 55	23 36	0 32	24 54	3 56	4 23	7 43	26 39	0 40	23 31	9 10	2 17
3/31/1987	26ar26	20ar39	1n14	16gm44	16ta33	5n15	9ta24	1ta14	6s36	23pi13	26pi44	0s35	27gm27	7gm36	4s10	10ta24	1ta25	0n30	24sa49	10cp46	2n20
4/10/1987	0ta32	26 45	0 50	24 55	25 7	4 26	11 14	5 10	6 41	24 31	29 50	0 38	0cn 2	11 29	3 58	13 2	6 9	0 20	26 7	12 2	2 24
4/20/1987	4 35	2ta47	0 26	3cn20	3gm52	3 33	13 6	9 13	6 47	25 50	2 52	0 41	2 39	15 33	3 48	15 38	10 52	0 11	27 25	12 54	2 27
4/30/1987	8 33	8 45	0 2	11 57	12 47	2 36	15 1	13 19	6 54	27 8	5 50	0 44	5 17	19 45	3 37	18 10	15 32	0 2	28 44	13 20	2 30
5/10/1987	12 29	14 40	0 22	20 43	21 49	1 35	16 58	17 31	7 2	28 27	8 43	0 47	7 56	24 6	3 28	20 41	20 8	0 7	0cp 4	13 19R	2 34
5/20/1987	16 20	20 29	0 45	29 33	0cn56	0 33	18 59	21 46	7 11	29 46	11 28	0 51	10 38	28 34	3 19	23 9	24 42	0 16	1 24	12 49	2 36
5/30/1987	20 7	26 13	1 8	8 25	10 7	0s29	21 3	26 5	7 21	1 4	14 5	0 55	13 21	3 7	3 10	25 34	29 11	0 24	2 45	11 52	2 37
6/ 9/1987	23 50	1gm51	1 31	17 15	19 21	1 30	23 10	0gm28	7 32	2 24	16 32	0 59	16 5	7 45	3 2	27 57	3 37	0 33	4 6	10 29	2 37
6/19/1987	27 29	7 24	1 54	25 58	28 33	2 28	25 22	4 55	7 45	3 43	18 47	1 4	18 51	12 27	2 54	0gm19	7 57	0 42	5 28	8 47	2 36
6/29/1987	1gm 4	12 50	2 17	4 32	7 44	3 21	27 38	9 25	7 59	5 2	20 48	1 9	21 39	17 12	2 47	2 38	12 12	0 51	6 51	6 53	2 32
7/ 9/1987	4 35	18 10	2 41	12 53	16 50	4 9	29 58	13 59	8 15	6 22	22 33	1 15	24 28	22 1	2 39	4 55	16 22	1 0	8 14	4 59	2 27
7/19/1987	8 1	23 22	3 4	20 59	25 50	4 51	2 24	18 37	8 32	7 42	23 59	1 21	27 18	26 51	2 32	7 10	20 24	1 10	9 38	3 13	2 20
7/29/1987	11 23	28 26	3 29	28 48	4 43	5 26	4 56	23 18	8 51	9 2	25 4	1 28	0le 9	1le44	2 25	9 24	24 19	1 20	11 3	1 44	2 11
8/ 8/1987	14 42	3 22	3 54	6 18	13 26	5 53	7 34	28 4	9 11	10 22	25 45	1 35	3 2	6 37	2 18	11 36	28 5	1 31	12 28	0 40	2 2
8/18/1987	17 56	8 8	4 20	13 31	21 59	6 14	10 20	2 55	9 34	11 43	26 0	1 42	5 56	11 32	2 10	13 46	1cn41	1 42	13 54	0 4	1 53
8/28/1987	21 7	12 44	4 47	20 24	0li22	6 28	13 13	7 50	9 58	13 4	25 46	1 50	8 51	16 26	2 3	15 55	5 6	1 54	15 21	29 59D	1 43
9/ 7/1987	24 13	17 7	5 16	26 58	8 33	6 37	16 14	12 51	10 24	14 25	25 4	1 58	11 47	21 20	1 55	18 2	8 17	2 7	16 49	0 23	1 34
9/17/1987	27 16	21 16	5 46	3 15	16 32	6 40	19 26	17 58	10 53	15 47	23 55	2 5	14 44	26 13	1 48	20 8	11 13	2 22	18 18	1 15	1 25
9/27/1987	0cn16	25 10	6 19	9 14	24 20	6 38	22 48	23 11	11 24	17 9	22 23	2 12	17 41	1vi 4	1 40	22 13	13 49	2 38	19 47	2 32	1 17
10/ 7/1987	3 12	28 45	6 54	14 58	1sc57	6 33	26 21	28 32	11 56	18 31	20 34	2 17	20 39	5 51	1 31	24 17	16 4	2 55	21 18	4 13	1 9
10/17/1987	6 4	1 59	7 32	20 26	9 23	6 24	0cn 8	4 1	12 31	19 53	18 39	2 20	23 38	10 35	1 23	26 19	17 52	3 14	22 49	6 15	1 2
10/27/1987	8 53	4 47	8 13	25 40	16 39	6 13	4 10	9 40	13 7	21 16	16 47	2 22	26 37	15 14	1 13	28 21	19 11	3 34	24 21	8 35	0 56

Ephemeris	80			433			1221			1489			114			3018			2041		
0 Hours ET	Sappho			Eros			Amor			Attila			Kassandra			Godiva			Lancelot		
Mo Dy Year	h long	g long	g lat	h long	g long	g lat	h long	g long	g lat	h long	g long	g lat	h long	g long	g lat	h long	g long	g lat	h long	g long	g lat
11/ 6/1987	11 40	7 5	8 58	0sa41	23 44	5 59	8 28	15 31	13 44	22 39	15 9	2 22	29 37	19 47	1 3	0cn21	19 54	3 57	25 54	11 10	0 50
11/16/1987	14 23	8 48	9 46	5 31	0sa41	5 44	13 5	21 37	14 20	24 3	13 52	2 21	2 37	24 11	0 53	2 21	19 59R	4 20	27 28	14 0	0 44
11/26/1987	17 3	9 50	10 37	10 9	7 29	5 28	18 2	28 0	14 53	25 27	13 4	2 19	5 37	28 26	0 41	4 20	19 23	4 44	29 4	17 1	0 39
12/ 6/1987	19 40	10 7	11 29	14 38	14 8	5 10	23 22	4 45	15 21	26 51	12 45	2 16	8 37	2 27	0 28	6 18	18 4	5 6	0aq40	20 12	0 34
12/16/1987	22 14	9 35	12 19	18 59	20 40	4 51	29 8	11 59	15 39	28 16	12 57	2 12	11 37	6 14	0 14	8 15	16 9	5 27	2 17	23 32	0 29
12/26/1987	24 46	8 13	13 6	23 11	27 5	4 32	5 21	19 47	15 40	29 41	13 39	2 9	14 36	9 42	0n 1	10 12	13 46	5 43	3 56	26 59	0 25
1/ 5/1988	27 16	6 7	13 42	27 17	3 23	4 12	12 5	28 20	15 13	1 7	14 47	2 5	17 36	12 48	0 19	12 8	11 11	5 53	5 36	0aq32	0 21
1/15/1988	29 43	3 32	14 4	1cp17	9 35	3 51	19 21	7 48	14 6	2 33	16 19	2 2	20 35	15 28	0 38	14 3	8 40	5 57	7 16	4 10	0 16
1/25/1988	2 8	0 44	14 9	5 11	15 41	3 29	27 11	18 19	12 1	4 0	18 12	1 59	23 33	17 35	1 0	15 58	6 30	5 54	8 59	7 51	0 12
2/ 4/1988	4 30	28 6	13 57	9 1	21 41	3 7	5 37	29 57	8 43	5 27	20 24	1 56	26 31	19 6	1 24	17 53	4 54	5 47	10 42	11 35	0 8
2/14/1988	6 51	25 56	13 30	12 46	27 36	2 44	14 37	12 35	4 4	6 55	22 50	1 53	29 28	19 55	1 51	19 47	3 57	5 37	12 27	15 20	0 4
2/24/1988	9 10	24 26	12 53	16 29	3 25	2 20	24 7	25 44	1n41	8 23	25 30	1 51	2 24	19 59R	2 20	21 41	3 40	5 25	14 13	19 6	0s 0
3/ 5/1988	11 26	23 42	12 10	20 9	9 9	1 55	4li 3	8 46	7 51	9 52	28 21	1 50	5 19	19 16	2 51	23 35	4 4	5 12	16 0	22 51	0 4
3/15/1988	13 41	23 43D	11 26	23 47	14 48	1 28	14 16	21 2	13 39	11 21	1ta21	1 48	8 13	17 50	3 21	25 29	5 2	4 59	17 49	26 35	0 9
3/25/1988	15 54	24 25	10 43	27 23	20 22	1 0	24 36	2cp 2	18 35	12 51	4 29	1 47	11 6	15 52	3 50	27 22	6 32	4 47	19 39	0pi17	0 14
4/ 4/1988	18 6	25 43	10 3	0aq58	25 51	0 30	4sc52	11 32	22 31	14 22	7 42	1 47	13 58	13 37	4 14	29 16	8 28	4 36	21 30	3 55	0 19
4/14/1988	20 16	27 31	9 25	4 33	1pi16	0n 2	14 52	19 26	25 36	15 53	11 0	1 46	16 49	11 24	4 33	1 9	10 47	4 25	23 23	7 29	0 24
4/24/1988	22 25	29 46	8 51	8 9	6 35	0 37	24 28	25 44	28 2	17 25	14 21	1 46	19 38	9 32	4 45	3 3	13 24	4 16	25 18	10 56	0 30
5/ 4/1988	24le32	2le21	8s20	11aq45	11pi49	1n15	3sa31	0aq23	29n58	18ta57	17ta45	1s47	22li26	8li14R	4n52	4le56	16cn18	4s 8	27aq14	14pi16	0s36
5/14/1988	26 38	5 14	7 52	15 22	16 57	1 57	11 58	3 21	31 31	20 30	21 11	1 47	25 12	7 38	4 54	6 50	19 24	4 1	29 12	17 28	0 43
5/24/1988	28 43	8 21	7 26	19 2	22 0	2 43	19 48	4 32	32 41	22 4	24 37	1 48	27 57	7 45D	4 52	8 44	22 42	3 54	1 11	20 28	0 50
6/ 3/1988	0vi46	11 39	7 3	22 44	26 56	3 34	27 2	3 55	33 21	23 38	28 2	1 49	0sc41	8 33	4 48	10 38	26 9	3 48	3 12	23 16	0 58
6/13/1988	2 49	15 7	6 43	26 29	1ar44	4 31	3 40	1 38	33 21	25 13	1gm27	1 51	3 23	9 58	4 43	12 33	29 43	3 44	5 14	25 48	1 7
6/23/1988	4 51	18 43	6 24	0pi18	6 24	5 36	9 46	28 3	32 27	26 49	4 49	1 53	6 3	11 54	4 37	14 28	3 24	3 39	7 19	28 2	
7/ 3/1988	6 51	22 26	6 8	4 12	10 53	6 49	15 22	23 55	30 36	28 25	8 8	1 55	8 42	14 18	4 31	16 23	7 11	3 36	9 24	29 54	1 28
7/13/1988	8 51	26 14	5 52	8 11	15 9	8 12	20 32	19 59	27 56	0gm 2	11 24	1 58	11 19	17 5	4 25	18 19	11 2	3 33	11 32	1 20	1 40
7/23/1988	10 51	0vi 6	5 38	12 15	19 9	9 48	25 19	16 54	24 47	1 40	14 33	2 1	13 54	20 10	4 19	20 15	14 57	3 30	13 41	2 17	1 53
8/ 2/1988	12 49	4 1	5 26	16 27	22 48	11 39	29 44	14 58	21 31	3 19	17 36	2 4	16 28	23 32	4 13	22 12	18 54	3 28	15 52	2 42	2 7
8/12/1988	14 47	8 0	5 14	20 46	26 1	13 46	3 51	14 9	18 26	4 59	20 31	2 8	19 0	27 6	4 8	24 10	22 55	3 27	18 5	2 31	2 22
8/22/1988	16 44	12 0	5 4	25 14	28 38	16 13	7 42	14 23D	15 40	6 39	23 15	2 13	21 31	0sc51	4 3	26 8	26 57	3 26	20 19	1 43	2 37
9/ 1/1988	18 41	16 1	4 54	29 52	0ta29	18 59	11 18	15 27	13 16	8 20	25 48	2 18	23 59	4 46	3 58	28 8	1vi 0	3 25	22 35	0 23	2 51
9/11/1988	20 38	20 3	4 45	4 40	1 22	22 5	14 41	17 12	11 14	10 2	28 5	2 23	26 27	8 47	3 54	0vi 8	5 4	3 25	24 53	28 37	3 4
9/21/1988	22 34	24 5	4 37	9 40	0 59	25 22	17 53	19 30	9 30	11 45	0cn 5	2 29	28 52	12 55	3 51	2 9	9 7	3 26	27 13	26 37	3 14
10/ 1/1988	24 30	28 7	4 29	14 53	29ar11	28 37	20 54	22 12	8 3	13 28	1 43	2 36	1 16	17 7	3 47	4 11	13 10	3 27	29 34	24 36	3 21
10/11/1988	26 26	2 7	4 22	20 20	25 56	31 24	23 46	25 14	6 49	15 13	2 58	2 43	3 38	21 24	3 45	6 14	17 11	3 28	1 57	22 51	3 24
10/21/1988	28 21	6 5	4 16	26 2	21 39	33 13	26 30	28 32	5 46	16 58	3 44	2 50	5 59	25 43	3 42	8 18	21 10	3 30	4 21	21 34	3 24

Ephemeris	80			433			1221			1489			114			3018			2041		
0 Hours ET	Sappho			Eros			Amor			Attila			Kassandra			Godiva			Lancelot		
Mo Dy Year	h long	g long	g lat	h long	g long	g lat	h long	g long	g lat	h long	g long	g lat	h long	g long	g lat	h long	g long	g lat	h long	g long	g lat
10/31/1988	0li17	10 0	4 10	2ta 0	17 11	33 44	29 7	2 2	4 53	18 44	3 59	2 58	8 18	0sa 4	3 40	10 24	25 4	3 33	6 48	20 53	3 21
11/10/1988	2 12	13 51	4 4	8 16	13 34	32 55	1 37	5 42	4 6	20 31	3 40	3 5	10 36	4 26	3 39	12 31	28 54	3 36	9 15	20 51D	3 17
11/20/1988	4 8	17 36	3 59	14 50	11 33	31 3	4 1	9 30	3 26	22 19	2 47	3 11	12 52	8 49	3 38	14 39	2 38	3 39	11 44	21 30	3 12
11/30/1988	6 3	21 15	3 54	21 42	11 25D	28 30	6 20	13 24	2 51	24 8	1 23	3 15	15 6	13 12	3 37	16 48	6 13	3 44	14 15	22 47	3 6
12/10/1988	7 59	24 46	3 49	28 54	13 4	25 34	8 34	17 23	2 20	25 58	29gm35	3 17	17 20	17 33	3 37	18 59	9 38	3 48	16 47	24 37	3 0
12/20/1988	9 55	28 7	3 45	6 24	16 17	22 26	10 43	21 25	1 53	27 48	27 34	3 16	19 31	21 52	3 37	21 12	12 50	3 54	19 21	26 57	2 54
12/30/1988	11 52	1sc16	3 40	14 13	20 49	19 13	12 49	25 30	1 28	29 40	25 31	3 12	21 42	26 9	3 38	23 26	15 47	4 0	21 56	29 43	2 48
1/ 9/1989	13 48	4 10	3 35	22 19	26 24	15 55	14 51	29 36	1 5	1 33	23 42	3 5	23 50	0cp21	3 39	25 43	18 24	4 6	24 31	2 51	2 43
1/19/1989	15 46	6 47	3 29	0cn40	2ta54	12 36	16 50	3 44	0 45	3 26	22 17	2 56	25 58	4 30	3 41	28 1	20 37	4 13	27 9	6 18	2 38
1/29/1989	17 44	9 2	3 23	9 14	10 8	9 15	18 45	7 52	0 25	5 20	21 24	2 46	28 4	8 32	3 43	0li20	22 23	4 21	29 47	10 0	2 33
2/ 8/1989	19 42	10 52	3 16	17 57	17 57	5 56	20 38	11 59	0 7	7 16	21 6	2 35	0cp 9	12 28	3 46	2 42	23 35	4 28	2 26	13 56	2 29
2/18/1989	21 41	12 12	3 7	26 47	26 18	2 40	22 29	16 6	0 10	9 12	21 25	2 24	2 13	16 15	3 50	5 6	24 8	4 34	5 5	18 2	2 25
2/28/1989	23 41	12 59	2 56	5 39	5 2	0s28	24 18	20 11	0 26	11 9	22 17	2 14	4 16	19 53	3 54	7 33	23 59R	4 38	7 46	22 17	2 22
3/10/1989	25 42	13 6R	2 43	14 30	14 4	3 22	26 4	24 14	0 43	13 8	23 40	2 4	6 17	23 20	3 59	10 1	23 4	4 39	10 27	26 40	2 19
3/20/1989	27 44	12 33	2 26	23 16	23 21	5 59	27 49	28 14	0 59	15 7	25 31	1 55	8 17	26 34	4 4	12 32	21 26	4 36	13 9	1ta 8	2 16
3/30/1989	29 46	11 17	2 6	1vi53	2cn45	8 14	29 32	2 12	1 15	17 7	27 44	1 46	10 17	29 34	4 10	15 6	19 14	4 28	15 51	5 41	2 13
4/ 9/1989	1 50	9 23	1 42	10 18	12 12	10 5	1 14	6 5	1 31	19 8	0cn19	1 38	12 15	2 16	4 17	17 42	16 40	4 13	18 33	10 17	2 11
4/19/1989	3 55	7 0	1 14	18 29	21 38	11 31	2 55	9 54	1 47	21 10	3 10	1 31	14 12	4 39	4 25	20 21	14 5	3 52	21 16	14 56	2 9
4/29/1989	6 2	4 23	0 43	26 23	0le57	12 32	4 34	13 39	2 4	23 13	6 16	1 25	16 8	6 39	4 33	23 3	11 50	3 26	23 58	19 36	2 7
5/ 9/1989	8 9	1 48	0 11	4 0	10 5	13 9	6 13	17 17	2 22	25 16	9 34	1 18	18 3	8 13	4 41	25 47	10 9	2 58	26 40	24 18	2 5
5/19/1989	10 19	29li34	0 20	11 18	19 0	13 26	7 51	20 50	2 41	27 21	13 2	1 13	19 58	9 17	4 50	28 35	9 13	2 30	29 22	28 59	2 3
5/29/1989	12 29	27 53	0 49	18 17	27 41	13 26	9 28	24 14	3 1	29 26	16 40	1 8	21 51	9 49	4 58	1sc25	9 5D	2 2	2 4	3 39	2 1
6/ 8/1989	14sc42	26li52R	1n15	24li57	6vi 5	13s11	11ar 4	27ar30	3s23	1le33	20cn24	1s 3	23cp44	9aq46R	5n 6	4sc19	9li44	1s36	4gm45	8gm19	2s 0
6/18/1989	16 56	26 35	1 39	1sc19	14 12	12 46	12 41	0ta37	3 46	3 40	24 15	0 58	25 36	9 7	5 11	7 16	11 6	1 12	7 26	12 56	1 59
6/28/1989	19 12	27 0	1 59	7 24	22 4	12 12	14 17	3 31	4 12	5 48	28 10	0 54	27 27	7 54	5 14	10 17	13 7	0 50	10 6	17 30	1 57
7/ 8/1989	21 31	28 4	2 17	13 12	29 40	11 33	15 53	6 12	4 40	7 56	2 10	0 49	29 17	6 10	5 14	13 21	15 41	0 30	12 45	22 2	1 56
7/18/1989	23 51	29 42	2 33	18 45	7 2	10 51	17 28	8 37	5 12	10 6	6 14	0 45	1 6	4 6	5 8	16 28	18 45	0 12	15 23	26 28	1 55
7/28/1989	26 14	1 51	2 47	24 3	14 11	10 6	19 4	10 42	5 46	12 16	10 19	0 41	2 55	1 53	4 58	19 40	22 13	0n 5	18 0	0cn50	1 54
8/ 7/1989	28 39	4 27	2 59	29 8	21 9	9 21	20 41	12 25	6 25	14 27	14 27	0 38	4 44	29cp45	4 44	22 54	26 3	0 20	20 36	5 6	1 53
8/17/1989	1sa 6	7 25	3 11	4 1	27 56	8 37	22 17	13 40	7 8	16 38	18 36	0 34	6 31	27 55	4 26	26 12	0sc11	0 34	23 11	9 14	1 52
8/27/1989	3 36	10 44	3 21	8 43	4 34	7 53	23 54	14 24	7 55	18 50	22 45	0 30	8 19	26 31	4 6	29 34	4 36	0 47	25 45	13 15	1 51
9/ 6/1989	6 10	14 20	3 31	13 15	11 4	7 10	25 32	14 31R	8 46	21 3	26 54	0 26	10 5	25 40	3 45	3 0	9 16	0 59	28 17	17 5	1 50
9/16/1989	8 45	18 12	3 40	17 38	17 28	6 29	27 10	13 56	9 40	23 16	1vi 1	0 23	11 51	25 25	3 25	6 29	14 8	1 11	0cn48	20 42	1 50
9/26/1989	11 24	22 18	3 49	21 53	23 46	5 49	28 50	12 37	10 35	25 30	5 7	0 19	13 37	25 44	3 5	10 2	19 12	1 21	3 17	24 7	1 49
10/ 6/1989	14 7	26 35	3 58	26 1	29 59	5 12	0ta30	10 33	11 28	27 44	9 9	0 15	15 22	26 35	2 46	13 38	24 26	1 31	5 45	27 14	1 48
10/16/1989	16 52	1sa 4	4 6	0cp 2	6 7	4 36	2 12	7 51	12 15	29 59	13 7	0 10	17 7	27 56	2 29	17 18	29 49	1 41	8 11	0le 2	1 47
10/26/1989	19 41	5 43	4 15	3 58	12 13	4 1	3 54	4 44	12 51	2 14	17 0	0 6	18 52	29 42	2 13	21 1	5 20	1 50	10 36	2 26	1 47
11/ 5/1989	22 34	10 32	4 23	7 49	18 16	3 28	5 39	1 29	13 12	4 29	20 45	0 1	20 36	1 50	1 59	24 47	10 59	1 59	12 59	4 24	1 45
11/15/1989	25 31	15 28	4 31	11 36	24 16	2 57	7 25	28 27	13 18	6 45	24 22	0 3	22 20	4 18	1 46	28 36	16 44	2 7	15 20	5 51	1 44
11/25/1989	28 31	20 33	4 40	15 19	0cp15	2 26	9 13	25 57	13 11	9 1	27 47	0 9	24 4	7 1	1 34	2 28	22 35	2 15	17 40	6 42	1 41

Ephemeris	80			433			1221			1489			114			3018			2041		
0 Hours ET	Sappho			Eros			Amor			Attila			Kassandra			Godiva			Lancelot		
Mo Dy Year	h long	g long	g lat	h long	g long	g lat	h long	g long	g lat	h long	g long	g lat	h long	g long	g lat	h long	g long	g lat	h long	g long	g lat
12/ 5/1989	1 36	25 44	4 48	19 0	6 12	1 57	11 2	24 11	12 53	11 17	1li 0	0 15	25 47	9 59	1 23	6 22	28 30	2 22	19 58	6 55	1 38
12/15/1989	4 45	1cp 2	4 57	22 38	12 9	1 28	12 55	23 13	12 28	13 33	3 56	0 21	27 30	13 8	1 13	10 18	4 29	2 29	22 14	6 28	1 34
12/25/1989	7 58	6 26	5 5	26 15	18 4	1 1	14 49	23 4D	12 1	15 49	6 33	0 28	29 13	16 26	1 4	14 16	10 32	2 36	24 29	5 23	1 28
1/ 4/1990	11 15	11 56	5 14	29 50	24 0	0 33	16 46	23 40	11 32	18 5	8 47	0 36	0 56	19 52	0 56	18 16	16 36	2 42	26 42	3 44	1 21
1/14/1990	14 37	17 30	5 23	3 25	29 55	0 7	18 47	24 57	11 5	20 22	10 35	0 44	2 39	23 24	0 48	22 16	22 42	2 48	28 53	1 44	1 12
1/24/1990	18 4	23 9	5 32	7 1	5 51	0 20	20 50	26 49	10 40	22 38	11 52	0 53	4 22	27 1	0 40	26 18	28 49	2 54	1le 2	29cn36	1 2
2/ 3/1990	21 35	28 52	5 40	10 36	11 47	0 46	22 58	29 11	10 17	24 54	12 34	1 3	6 4	0pi42	0 33	0aq20	4 56	2 59	3 10	27 35	0 51
2/13/1990	25 10	4 38	5 49	14 13	17 43	1 13	25 9	1 59	9 56	27 10	12 38R	1 14	7 47	4 26	0 26	4 22	11 1	3 4	5 15	25 55	0 40
2/23/1990	28 51	10 26	5 57	17 52	23 41	1 40	27 24	5 10	9 37	29 26	12 5	1 25	9 30	8 10	0 19	8 24	17 5	3 9	7 20	24 47	0 30
3/ 5/1990	2 35	16 18	6 5	21 33	29 41	2 7	29 44	8 41	9 21	1 42	10 56	1 36	11 12	11 56	0 12	12 25	23 5	3 13	9 22	24 14	0 20
3/15/1990	6 24	22 11	6 13	25 17	5 42	2 34	2 10	12 29	9 6	3 57	9 18	1 47	12 55	15 40	0 6	16 25	29 2	3 17	11 23	24 17D	0 11
3/25/1990	10 18	28 5	6 20	29 5	11 46	3 2	4 41	16 33	8 53	6 12	7 22	1 56	14 38	19 24	0s 1	20 23	4 55	3 20	13 23	24 55	0 3
4/ 4/1990	14 15	3 59	6 26	2 57	17 52	3 31	7 19	20 51	8 41	8 26	5 22	2 3	16 21	23 5	0 8	24 20	10 42	3 24	15 20	26 5	0 4
4/14/1990	18 17	9 53	6 32	6 54	24 2	4 0	10 3	25 22	8 31	10 40	3 33	2 8	18 4	26 43	0 15	28 14	16 24	3 26	17 16	27 41	0 11
4/24/1990	22 22	15 46	6 37	10 57	0ar16	4 30	12 56	0gm 6	8 21	12 54	2 7	2 11	19 48	0ar16	0 22	2 6	21 58	3 29	19 11	29 40	0 16
5/ 4/1990	26 30	21 37	6 40	15 6	6 35	5 1	15 56	5 2	8 11	15 7	1 12	2 12	21 31	3 45	0 30	5 55	27 24	3 31	21 4	1 59	0 21
5/14/1990	0pi42	27 24	6 43	19 23	13 0	5 33	19 7	10 10	8 2	17 19	0 53	2 11	23 15	7 6	0 38	9 42	2 41	3 33	22 56	4 34	0 26
5/24/1990	4 55	3 8	6 44	23 48	19 31	6 5	22 28	15 30	7 53	19 31	1 10	2 10	24 59	10 21	0 47	13 25	7 47	3 35	24 46	7 22	0 30
6/ 3/1990	9 11	8 46	6 43	28 22	26 10	6 37	26 0	21 4	7 43	21 42	2 0	2 8	26 44	13 26	0 56	17 5	12 42	3 36	26 35	10 21	0 34
6/13/1990	13 29	14 16	6 41	3 7	2ta57	7 9	29 46	26 50	7 33	23 53	3 22	2 5	28 29	16 20	1 6	20 42	17 24	3 37	28 22	13 29	0 38
6/23/1990	17 48	19 38	6 36	8 3	9 54	7 41	3 46	2cn51	7 21	26 3	5 10	2 3	0ar14	19 1	1 17	24 14	21 51	3 38	0vi 8	16 43	0 41
7/ 3/1990	22 8	24 48	6 29	13 11	17 2	8 12	8 3	9 7	7 7	28 12	7 22	2 1	2 0	21 27	1 29	27 44	26 0	3 39	1 52	20 4	0 44
7/13/1990	26pi28	29ar44	6n20	18ar33	24ta21	8n41	12cn38	15cn39	6s51	0sc20	9li54	1n58	3ar46	23ar36	1s43	1ar10	29ar49	3n39	3vi36	23le29	0n48
7/23/1990	0ar47	4 24	6 8	24 10	1gm53	9 7	17 33	22 28	6 32	2 28	12 43	1 56	5 33	25 23	1 57	4 31	3 15	3 39	5 18	26 57	0 51
8/ 2/1990	5 6	8 42	5 52	0ta 3	9 39	9 29	22 51	29 36	6 9	4 35	15 46	1 54	7 20	26 45	2 13	7 50	6 14	3 38	6 59	0vi27	0 54
8/12/1990	9 23	12 36	5 32	6 13	17 40	9 46	28 34	7 4	5 41	6 41	19 1	1 52	9 8	27 40	2 31	11 4	8 41	3 37	8 38	3 59	0 57
8/22/1990	13 39	15 59	5 8	12 40	25 55	9 56	4 44	14 52	5 9	8 46	22 25	1 51	10 57	28 2	2 50	14 15	10 32	3 34	10 17	7 31	1 0
9/ 1/1990	17 53	18 46	4 37	19 27	4 25	9 57	11 24	23 3	4 30	10 50	25 59	1 50	12 46	27 50	3 10	17 22	11 40	3 30	11 54	11 3	1 4
9/11/1990	22 5	20 48	4 1	26 32	13 9	9 47	18 37	1vi36	3 45	12 53	29 39	1 49	14 36	27 2	3 30	20 26	12 2	3 23	13 31	14 34	1 7
9/21/1990	26 13	22 0	3 16	3 56	22 7	9 24	26 24	10 31	2 54	14 56	3 25	1 48	16 26	25 39	3 50	23 26	11 33	3 14	15 6	18 3	1 11
10/ 1/1990	0ta19	22 13R	2 23	11 39	1le15	8 48	4 46	19 47	1 57	16 57	7 15	1 47	18 17	23 47	4 8	26 23	10 15	3 1	16 40	21 28	1 15
10/11/1990	4 21	21 28	1 20	19 39	10 34	7 57	13 42	29 23	0 54	18 58	11 9	1 47	20 9	21 35	4 24	29 16	8 14	2 44	18 13	24 50	1 19
10/21/1990	8 20	19 47	0 11	27 56	19 59	6 52	23 9	9 15	0n12	20 57	15 5	1 46	22 2	19 17	4 35	2 6	5 44	2 23	19 45	28 6	1 23
10/31/1990	12 15	17 27	1 2	6 26	29 27	5 31	3li 2	19 19	1 20	22 56	19 3	1 47	23 56	17 8	4 41	4 53	3 5	1 59	21 16	1li16	1 28
11/10/1990	16 6	14 50	2 13	15 6	8 55	3 58	13 13	29 30	2 27	24 54	23 1	1 47	25 51	15 21	4 44	7 37	0 39	1 33	22 47	4 18	1 34
11/20/1990	19 53	12 24	3 16	23 55	18 19	2 13	23 32	9 42	3 30	26 51	26 59	1 47	27 46	14 6	4 42	10 18	28 44	1 8	24 16	7 11	1 39
11/30/1990	23 36	10 34	4 9	2le47	27 36	0 20	3sc48	19 49	4 27	28 47	0sa56	1 48	29 43	13 29	4 38	12 56	27 31	0 44	25 45	9 52	1 46
12/10/1990	27 15	9 35	4 50	11 38	6 43	1 38	13 49	29 44	5 16	0sa42	4 50	1 49	1 40	13 30D	4 33	15 32	27 5	0 23	27 12	12 19	1 53
12/20/1990	0gm50	9 30D	5 20	20 26	15 36	3 38	23 27	9 24	5 57	2 36	8 41	1 51	3 39	14 10	4 26	18 4	27 25	0 4	28 39	14 31	2 0
12/30/1990	4 21	10 17	5 41	29 7	24 12	5 39	2sa33	18 43	6 29	4 29	12 28	1 53	5 38	15 23	4 19	20 35	28 26	0 12	0li 5	16 24	2 9

Ephemeris 0 Hours ET Mo Dy Year	80 Sappho h long	 g long	 g lat	433 Eros h long	 g long	 g lat	1221 Amor h long	 g long	 g lat	1489 Attila h long	 g long	 g lat	114 Kassandra h long	 g long	 g lat	3018 Godiva h long	 g long	 g lat	2041 Lancelot h long	 g long	 g lat
1/ 9/1991	7 47	11 52	5 56	7 36	2sc27	7 38	11 4	27 39	6 52	6 21	16 9	1 55	7 39	17 8	4 12	23 2	0ta 4	0 26	1 30	17 56	2 17
1/19/1991	11 10	14 5	6 5	15 52	10 19	9 34	18 58	6 9	7 9	8 12	19 43	1 57	9 41	19 19	4 6	25 28	2 11	0 38	2 55	19 3	2 27
1/29/1991	14 28	16 51	6 10	23 52	17 43	11 26	26 14	14 14	7 19	10 2	23 9	2 0	11 44	21 53	4 0	27 51	4 44	0 48	4 19	19 44	2 37
2/ 8/1991	17 43	20 2	6 13	1li35	24 37	13 14	2cp56	21 53	7 24	11 52	26 25	2 3	13 48	24 46	3 54	0gm12	7 38	0 57	5 42	19 55	2 48
2/18/1991	20 53	23 34	6 14	8 59	0sa57	15 0	9 5	29 7	7 25	13 40	29 29	2 7	15 54	27 56	3 50	2 31	10 48	1 5	7 4	19 35	2 59
2/28/1991	24 0	27 23	6 15	16 4	6 36	16 43	14 44	5 56	7 24	15 27	2 20	2 11	18 1	1ta20	3 46	4 48	14 12	1 12	8 26	18 45	3 9
3/10/1991	27 3	1gm24	6 14	22 51	11 29	18 25	19 57	12 22	7 20	17 14	4 55	2 16	20 9	4 56	3 42	7 4	17 47	1 18	9 48	17 27	3 18
3/20/1991	0cn 3	5 36	6 13	29 19	15 28	20 6	24 45	18 24	7 14	19 0	7 11	2 21	22 19	8 41	3 39	9 17	21 30	1 24	11 8	15 47	3 25
3/30/1991	2 59	9 55	6 11	5 29	18 23	21 45	29 13	24 3	7 7	20 45	9 6	2 26	24 30	12 35	3 37	11 29	25 20	1 30	12 28	13 53	3 31
4/ 9/1991	5 52	14 19	6 10	11 23	20 3	23 20	3 22	29 20	7 0	22 29	10 38	2 32	26 42	16 36	3 35	13 39	29 16	1 35	13 48	11 56	3 33
4/19/1991	8 41	18 48	6 9	17 0	20 16R	24 45	7 14	4 15	6 52	24 12	11 42	2 38	28 56	20 43	3 34	15 48	3 15	1 41	15 7	10 5	3 34
4/29/1991	11 28	23 21	6 7	22 23	18 54	25 49	10 52	8 47	6 44	25 54	12 18	2 44	1 12	24 55	3 33	17 56	7 17	1 46	16 25	8 31	3 31
5/ 9/1991	14 11	27 55	6 6	27 33	16 0	26 17	14 16	12 55	6 35	27 35	12 22R	2 50	3 29	29 11	3 32	20 2	11 21	1 51	17 44	7 19	3 27
5/19/1991	16 51	2 30	6 5	2 29	11 55	25 54	17 29	16 38	6 26	29 16	11 54	2 55	5 48	3 30	3 32	22 7	15 26	1 56	19 1	6 34	3 22
5/29/1991	19 29	7 6	6 4	7 15	7 26	24 34	20 31	19 53	6 17	0 56	10 55	2 59	8 8	7 51	3 32	24 10	19 32	2 1	20 18	6 17	3 16
6/ 8/1991	22 4	11 41	6 4	11 50	3 20	22 24	23 25	22 39	6 7	2 35	9 30	3 2	10 30	12 15	3 33	26 13	23 38	2 6	21 35	6 28	3 9
6/18/1991	24 36	16 16	6 4	16 16	0 15	19 44	26 10	24 50	5 56	4 13	7 45	3 2	12 53	16 40	3 34	28 14	27 42	2 12	22 51	7 5	3 3
6/28/1991	27 5	20 50	6 4	20 33	28 31	16 56	28 47	26 23	5 43	5 51	5 51	3 0	15 18	21 6	3 35	0cn15	1cn46	2 18	24 7	8 6	2 57
7/ 8/1991	29 33	25 23	6 5	24 43	28 4	14 16	1 18	27 13	5 28	7 28	3 58	2 55	17 45	25 33	3 37	2 14	5 47	2 24	25 23	9 29	2 51
7/18/1991	1 58	29 54	6 7	28 47	28 45	11 51	3 43	27 14R	5 10	9 4	2 17	2 49	20 13	29 59	3 39	4 13	9 46	2 30	26 38	11 9	2 46
7/28/1991	4 21	4 23	6 8	2 44	0sa22	9 44	6 2	26 25	4 48	10 39	0 56	2 41	22 44	4 24	3 41	6 11	13 42	2 37	27 53	13 6	2 41
8/ 7/1991	6 41	8 49	6 11	6 37	2 44	7 54	8 17	24 43	4 21	12 14	0 2	2 32	25 15	8 48	3 44	8 9	17 34	2 44	29 7	15 16	2 37
8/17/1991	9le 0	13le11	6s14	10cp25	5sa42	6s19	10pi27	22pi13R	3n48	13cp48	29sa38R	2n22	27gm49	13cn 9	3s47	10cn 5	21cn21	2s52	0sc22	17li38	2n33
8/27/1991	11 17	17 31	6 18	14 9	9 9	4 58	12 33	19 8	3 9	15 21	29 44D	2 13	0cn24	17 27	3 50	12 1	25 3	3 1	1 36	20 10	2 30
9/ 6/1991	13 32	21 46	6 22	17 51	12 58	3 47	14 35	15 45	2 26	16 54	0cp18	2 4	3 1	21 40	3 54	13 57	28 37	3 10	2 49	22 51	2 27
9/16/1991	15 46	25 56	6 28	21 30	17 5	2 46	16 34	12 27	1 42	18 26	1 20	1 55	5 40	25 48	3 58	15 52	2 4	3 21	4 3	25 38	2 25
9/26/1991	17 58	0vi 0	6 34	25 7	21 29	1 52	18 30	9 35	0 59	19 57	2 46	1 48	8 20	29 48	4 2	17 47	5 21	3 32	5 16	28 30	2 23
10/ 6/1991	20 8	3 58	6 41	28 43	26 5	1 5	20 24	7 22	0 20	21 28	4 32	1 40	11 2	3 39	4 7	19 41	8 26	3 45	6 29	1sc27	2 22
10/16/1991	22 17	7 48	6 50	2 18	0cp53	0 22	22 15	5 57	0s15	22 58	6 38	1 34	13 46	7 18	4 12	21 35	11 16	3 59	7 42	4 27	2 21
10/26/1991	24 24	11 28	6 59	5 53	5 51	0n16	24 3	5 19	0 45	24 28	8 59	1 27	16 31	10 42	4 18	23 29	13 50	4 14	8 55	7 29	2 21
11/ 5/1991	26 30	14 58	7 10	9 28	10 58	0 50	25 50	5 26D	1 11	25 57	11 33	1 22	19 17	13 49	4 24	25 23	16 4	4 32	10 7	10 32	2 21
11/15/1991	28 35	18 14	7 23	13 5	16 14	1 22	27 35	6 13	1 33	27 25	14 19	1 17	22 5	16 33	4 30	27 16	17 54	4 50	11 20	13 34	2 22
11/25/1991	0vi39	21 14	7 37	16 43	21 37	1 51	29 19	7 33	1 51	28 53	17 14	1 12	24 55	18 51	4 36	29 10	19 15	5 11	12 32	16 34	2 23
12/ 5/1991	2 42	23 56	7 53	20 23	27 7	2 18	1 1	9 22	2 7	0aq21	20 17	1 8	27 46	20 36	4 41	1 3	20 3	5 33	13 44	19 32	2 25
12/15/1991	4 44	26 16	8 10	24 6	2aq44	2 43	2 41	11 35	2 21	1 48	23 26	1 5	0le38	21 43	4 45	2 57	20 14	5 56	14 56	22 26	2 27
12/25/1991	6 45	28 10	8 29	27 53	8 28	3 7	4 21	14 8	2 33	3 15	26 39	1 1	3 31	22 8	4 48	4 51	19 45	6 18	16 8	25 14	2 30
1/ 4/1992	8 45	29 33	8 49	1pi43	14 19	3 29	6 0	16 57	2 45	4 41	29 56	0 58	6 26	21 46	4 47	6 45	18 35	6 40	17 20	27 55	2 33
1/14/1992	10 44	0 21	9 10	5 39	20 16	3 50	7 37	19 59	2 55	6 6	3 15	0 55	9 21	20 38	4 41	8 39	16 47	6 58	18 32	0sa27	2 37
1/24/1992	12 43	0 30R	9 30	9 39	26 20	4 11	9 15	23 13	3 6	7 32	6 35	0 52	12 18	18 51	4 29	10 33	14 30	7 10	19 44	2 49	2 42
2/ 3/1992	14 41	29vi57	9 48	13 47	2pi31	4 30	10 51	26 35	3 15	8 56	9 55	0 50	15 15	16 37	4 11	12 28	11 57	7 15	20 55	4 57	2 47

Ephemeris	80			433			1221			1489			114			3018			2041		
0 Hours ET	Sappho			Eros			Amor			Attila			Kassandra			Godiva			Lancelot		
Mo Dy Year	h long	g long	g lat	h long	g long	g lat	h long	g long	g lat	h long	g long	g lat	h long	g long	g lat	h long	g long	g lat	h long	g long	g lat
2/13/1992	16 38	28 41	10 1	18 1	8 49	4 48	12 28	0ar 5	3 25	10 21	13 13	0 48	18 13	14 17	3 47	14 23	9 24	7 13	22 7	6 50	2 52
2/23/1992	18 35	26 47	10 7	22 23	15 14	5 5	14 4	3 40	3 35	11 45	16 29	0 45	21 12	12 12	3 19	16 18	7 10	7 3	23 19	8 26	2 59
3/ 4/1992	20 32	24 24	10 4	26 54	21 48	5 21	15 40	7 19	3 45	13 9	19 40	0 43	24 12	10 38	2 48	18 14	5 26	6 48	24 31	9 41	3 6
3/14/1992	22 28	21 47	9 50	1ar36	28 29	5 36	17 16	11 3	3 55	14 32	22 47	0 41	27 12	9 46	2 17	20 10	4 20	6 29	25 43	10 34	3 13
3/24/1992	24 24	19 12	9 25	6 28	5 20	5 49	18 51	14 48	4 6	15 55	25 48	0 40	0vi12	9 42D	1 47	22 8	3 55	6 9	26 55	11 1	3 21
4/ 3/1992	26 19	16 56	8 52	11 32	12 19	6 1	20 28	18 36	4 17	17 17	28 41	0 38	3 12	10 23	1 20	24 5	4 11	5 48	28 7	11 2R	3 28
4/13/1992	28 15	15 12	8 13	16 50	19 29	6 10	22 4	22 24	4 29	18 40	1 25	0 36	6 13	11 46	0 55	26 4	5 3	5 27	29 19	10 35	3 36
4/23/1992	0li10	14 7	7 32	22 22	26 49	6 17	23 41	26 13	4 42	20 2	3 59	0 34	9 13	13 46	0 33	28 3	6 28	5 8	0sa31	9 40	3 42
5/ 3/1992	2 6	13 43	6 50	28 9	4 19	6 20	25 19	0ta 2	4 56	21 24	6 20	0 32	12 13	16 17	0 13	0vi 4	8 21	4 50	1 44	8 22	3 47
5/13/1992	4 1	13 58	6 10	4 13	12 1	6 20	26 57	3 49	5 11	22 45	8 27	0 30	15 14	19 15	0n 5	2 5	10 39	4 34	2 56	6 44	3 50
5/23/1992	5 57	14 49	5 33	10 35	19 54	6 15	28 36	7 35	5 28	24 6	10 17	0 28	18 13	22 33	0 20	4 7	13 17	4 19	4 9	4 54	3 50
6/ 2/1992	7 52	16 12	4 59	17 15	27 59	6 6	0ta16	11 19	5 46	25 27	11 49	0 25	21 13	26 10	0 34	6 10	16 13	4 6	5 22	3 1	3 48
6/12/1992	9 48	18 3	4 27	24 14	6 16	5 51	1 58	15 0	6 6	26 48	12 59	0 23	24 11	0vi 2	0 47	8 14	19 24	3 53	6 35	1 15	3 43
6/22/1992	11 45	20 17	3 59	1gm32	14 43	5 31	3 40	18 38	6 27	28 9	13 45	0 20	27 9	4 6	0 59	10 20	22 48	3 42	7 48	29sc44	3 36
7/ 2/1992	13 41	22 51	3 32	9 9	23 22	5 4	5 24	22 11	6 51	29 29	14 5	0 17	0li 7	8 20	1 9	12 27	26 23	3 32	9 2	28 34	3 28
7/12/1992	15 38	25 42	3 9	17 4	2cn10	4 31	7 10	25 39	7 18	0 49	13 58	0 14	3 3	12 42	1 19	14 35	0vi 8	3 22	10 15	27 50	3 19
7/22/1992	17 36	28 48	2 47	25 15	11 6	3 52	8 57	29 0	7 48	2 9	13 22	0 10	5 58	17 11	1 28	16 44	4 0	3 14	11 29	27 33	3 9
8/ 1/1992	19 34	2 7	2 27	3cn41	20 10	3 7	10 47	2 12	8 22	3 29	12 20	0 6	8 53	21 46	1 37	18 55	8 1	3 6	12 44	27 44	2 59
8/11/1992	21 33	5 36	2 8	12 18	29 18	2 18	12 39	5 14	8 59	4 49	10 54	0 1	11 46	26 26	1 45	21 8	12 7	2 58	13 58	28 20	2 49
8/21/1992	23 33	9 15	1 50	21 4	8 31	1 24	14 33	8 2	9 41	6 8	9 11	0 3	14 38	1li 9	1 52	23 22	16 20	2 51	15 13	29 21	2 40
8/31/1992	25 33	13 1	1 34	29 56	17 45	0 28	16 30	10 35	10 29	7 28	7 19	0 8	17 28	5 54	2 0	25 38	20 37	2 44	16 28	0sa43	2 32
9/10/1992	27 35	16 55	1 18	8 48	26 58	0 29	18 30	12 47	11 23	8 47	5 28	0 12	20 17	10 42	2 7	27 56	24 59	2 38	17 44	2 25	2 24
9/20/1992	29li37	20li55	1s 3	17le37	6vi 9	1s25	20ta33	14gm34	12s24	10pi 6	3pi48R	0s17	23li 5	15li32	2n14	0li16	29vi25	2s32	19sa 0	4sa24	2n17
9/30/1992	1 41	25 0	0 48	26 20	15 16	2 20	22 39	15 50	13 33	11 26	2 26	0 21	25 52	20 22	2 20	2 38	3 54	2 26	20 16	6 38	2 10
10/10/1992	3 45	29 9	0 34	4 54	24 16	3 12	24 50	16 28	14 48	12 45	1 29	0 24	28 37	25 12	2 27	5 2	8 26	2 20	21 33	9 5	2 4
10/20/1992	5 51	3 23	0 20	13 14	3li 8	4 0	27 4	16 19R	16 9	14 4	0 59	0 27	1sc20	0sc 2	2 34	7 28	13 1	2 15	22 50	11 43	1 59
10/30/1992	7 59	7 40	0 6	21 20	11 50	4 44	29 24	15 18	17 33	15 23	0 59D	0 30	4 2	4 50	2 41	9 56	17 37	2 9	24 8	14 30	1 54
11/ 9/1992	10 8	11 59	0 8	29 8	20 22	5 23	1 48	13 18	18 54	16 42	1 27	0 32	6 42	9 37	2 48	12 27	22 15	2 4	25 26	17 26	1 50
11/19/1992	12 18	16 20	0 23	6 38	28 41	5 57	4 19	10 25	20 3	18 1	2 21	0 34	9 20	14 21	2 55	15 0	26 53	1 58	26 44	20 27	1 46
11/29/1992	14 30	20 43	0 38	13 50	6 48	6 26	6 55	6 50	20 50	19 20	3 39	0 36	11 57	19 2	3 3	17 36	1sc31	1 52	28 3	23 34	1 42
12/ 9/1992	16 44	25 6	0 53	20 42	14 41	6 50	9 38	2 58	21 7	20 39	5 19	0 38	14 32	23 38	3 11	20 14	6 8	1 46	29 23	26 45	1 39
12/19/1992	18 59	29 30	1 9	27 16	22 20	7 10	12 29	29ta21	20 52	21 58	7 16	0 39	17 6	28 9	3 19	22 56	10 43	1 39	0 43	29 59	1 37
12/29/1992	21 17	3 52	1 27	3 33	29 45	7 26	15 29	26 26	20 8	23 17	9 30	0 41	19 37	2 34	3 28	25 40	15 15	1 32	2 4	3 15	1 34
1/ 8/1993	23 37	8 12	1 45	9 31	6 55	7 38	18 38	24 34	19 6	24 36	11 57	0 42	22 8	6 52	3 38	28 27	19 43	1 25	3 25	6 30	1 32
1/18/1993	25 59	12 30	2 4	15 14	13 51	7 48	21 57	23 50	17 53	25 55	14 35	0 44	24 36	11 1	3 48	1sc17	24 5	1 16	4 47	9 45	1 30
1/28/1993	28 23	16 45	2 26	20 42	20 32	7 54	25 27	24 13	16 36	27 15	17 22	0 45	27 3	14 59	3 59	4 10	28 20	1 7	6 9	12 58	1 28
2/ 7/1993	0sa50	20 54	2 49	25 55	26 57	7 58	29 11	25 38	15 20	28 34	20 17	0 47	29 28	18 45	4 12	7 7	2 26	0 56	7 33	16 8	1 27
2/17/1993	3 19	24 56	3 15	0sa56	3 8	8 0	3 9	27 57	14 6	29 53	23 18	0 49	1 51	22 18	4 25	10 7	6 20	0 44	8 56	19 14	1 26
2/27/1993	5 52	28 51	3 43	5 45	9 2	8 0	7 23	1gm 3	12 56	1 13	26 23	0 50	4 13	25 34	4 40	13 10	9 59	0 30	10 21	22 13	1 25
3/ 9/1993	8 27	2 35	4 14	10 23	14 40	7 59	11 55	4 51	11 49	2 33	29 32	0 52	6 34	28 30	4 56	16 17	13 20	0 14	11 46	25 5	1 24

Ephemeris	80			433			1221			1489			114			3018			2041		
0 Hours ET	Sappho			Eros			Amor			Attila			Kassandra			Godiva			Lancelot		
Mo Dy Year	h long	g long	g lat	h long	g long	g lat	h long	g long	g lat	h long	g long	g lat	h long	g long	g lat	h long	g long	g lat	h long	g long	g lat
3/19/1993	11 5	6 8	4 49	14 52	20 0	7 55	16 46	9 16	10 44	3 53	2 42	0 54	8 52	1cp 5	5 13	19 27	16 19	0n 5	13 13	27 47	1 23
3/29/1993	13 46	9 25	5 29	19 12	25 1	7 50	22 0	14 14	9 40	5 13	5 53	0 57	11 9	3 15	5 31	22 41	18 51	0 27	14 39	0aq19	1 22
4/ 8/1993	16 31	12 23	6 13	23 25	29 41	7 42	27 39	19 43	8 37	6 33	9 4	0 59	13 25	4 55	5 51	25 58	20 50	0 53	16 7	2 37	1 22
4/18/1993	19 19	14 59	7 2	27 30	3 58	7 33	3 45	25 42	7 32	7 53	12 13	1 2	15 39	6 2	6 11	29 19	22 11	1 23	17 36	4 40	1 21
4/28/1993	22 11	17 7	7 57	1cp29	7 48	7 20	10 21	2cn 9	6 26	9 14	15 20	1 5	17 52	6 33	6 32	2 44	22 47	1 56	19 5	6 25	1 20
5/ 8/1993	25 7	18 43	8 59	5 23	11 7	7 4	17 28	9 4	5 16	10 35	18 24	1 8	20 3	6 25R	6 52	6 12	22 35R	2 34	20 36	7 48	1 19
5/18/1993	28 6	19 40	10 5	9 13	13 49	6 43	25 10	16 27	4 3	11 56	21 23	1 11	22 13	5 38	7 10	9 44	21 34	3 14	22 7	8 48	1 18
5/28/1993	1cp10	19 52	11 16	12 58	15 46	6 17	3vi27	24 17	2 45	13 17	24 16	1 15	24 21	4 14	7 24	13 19	19 51	3 55	23 39	9 20	1 16
6/ 7/1993	4 18	19 17	12 28	16 41	16 50	5 42	12 18	2le34	1 23	14 39	27 2	1 19	26 29	2 21	7 32	16 57	17 39	4 32	25 12	9 24R	1 14
6/17/1993	7 30	17 54	13 36	20 21	16 53R	4 58	21 41	11 18	0n 2	16 1	29 39	1 24	28 34	0 10	7 33	20 39	15 20	5 4	26 46	8 57	1 11
6/27/1993	10 46	15 50	14 34	23 58	15 44	4 0	1li31	20 27	1 30	17 23	2 6	1 29	0cp39	27 56	7 27	24 24	13 18	5 28	28 22	8 0	1 7
7/ 7/1993	14 7	13 22	15 14	27 34	13 22	2 49	11 40	29 59	2 57	18 45	4 21	1 34	2 42	25 53	7 13	28 12	11 53	5 43	29 58	6 36	1 2
7/17/1993	17 32	10 49	15 34	1aq10	9 55	1 24	22 0	9 50	4 22	20 8	6 22	1 40	4 45	24 13	6 54	2 2	11 16	5 50	1 36	4 51	0 57
7/27/1993	21 3	8 37	15 31	4 45	5 45	0n10	2sc18	19 58	5 40	21 31	8 5	1 47	6 46	23 6	6 32	5 55	11 33	5 51	3 14	2 53	0 50
8/ 6/1993	24 37	7 4	15 9	8 20	1 31	1 45	12 23	0li16	6 47	22 54	9 30	1 54	8 45	22 35	6 7	9 50	12 43	5 48	4 54	0 55	0 42
8/16/1993	28 16	6 24	14 33	11 56	27 50	3 11	22 5	10 37	7 42	24 18	10 32	2 2	10 44	22 40D	5 43	13 47	14 40	5 42	6 35	29 8	0 34
8/26/1993	2 0	6 39	13 48	15 34	25 10	4 23	1sa17	20 54	8 23	25 42	11 9	2 10	12 42	23 20	5 19	17 45	17 18	5 34	8 17	27 42	0 25
9/ 5/1993	5 49	7 48	12 59	19 13	23 44	5 20	9 54	1sc 0	8 48	27 7	11 18	2 18	14 39	24 31	4 56	21 45	20 32	5 24	10 1	26 43	0 17
9/15/1993	9 41	9 47	12 8	22 55	23 31D	6 5	17 53	10 48	8 59	28 32	10 58	2 27	16 35	26 10	4 35	25 46	24 16	5 14	11 45	26 17	0 10
9/25/1993	13 38	12 29	11 19	26 41	24 24	6 38	25 16	20 14	8 59	29 57	10 9	2 35	18 29	28 13	4 16	29 47	28 24	5 3	13 31	26 24D	0 2
10/ 5/1993	17 39	15 49	10 30	0pi30	26 13	7 3	2cp 3	29 15	8 48	1 23	8 52	2 43	20 23	0cp37	3 58	3 48	2 53	4 53	15 19	27 4	0 4
10/15/1993	21 43	19 40	9 44	4 24	28 50	7 22	8 16	7 49	8 30	2 49	7 13	2 49	22 17	3 18	3 42	7 49	7 38	4 42	17 8	28 15	0 10
10/25/1993	25aq51	23cp57	9n 0	8pi23	2aq 8	7n37	14cp 0	15sa58	8n 7	4ta16	5ta20R	2s53	24cp 9	6cp14	3n28	11aq49	12cp38	4n31	18aq58	29cp54	0s16
11/ 4/1993	0pi 2	28 38	8 19	12 28	5 58	7 48	19 16	23 42	7 41	5 43	3 22	2 55	26 0	9 22	3 15	15 49	17 48	4 21	20 50	1 57	0 21
11/14/1993	4 16	3 37	7 39	16 40	10 17	7 56	24 9	1cp 4	7 14	7 10	1 31	2 55	27 51	12 41	3 3	19 46	23 8	4 10	22 43	4 22	0 25
11/24/1993	8 32	8 52	7 0	20 59	15 1	8 2	28 39	8 4	6 46	8 39	29ar58	2 53	29 41	16 8	2 53	23 43	28 35	3 59	24 38	7 7	0 30
12/ 4/1993	12 50	14 20	6 24	25 28	20 7	8 6	2 51	14 45	6 18	10 7	28 50	2 50	1 30	19 43	2 43	27 37	4 7	3 49	26 34	10 8	0 34
12/14/1993	17 9	19 59	5 49	0ar 5	25 33	8 8	6 46	21 10	5 51	11 36	28 11	2 45	3 19	23 22	2 34	1pi28	9 44	3 39	28 32	13 23	0 38
12/24/1993	21 29	25 47	5 15	4 54	1pi17	8 8	10 25	27 19	5 25	13 6	28 5D	2 39	5 7	27 7	2 26	5 18	15 23	3 29	0pi31	16 50	0 42
1/ 3/1994	25 49	1pi43	4 42	9 54	7 18	8 7	13 51	3 14	5 0	14 36	28 29	2 33	6 55	0aq54	2 18	9 4	21 5	3 19	2 33	20 29	0 45
1/13/1994	0ar 9	7 43	4 11	15 8	13 35	8 4	17 6	8 57	4 37	16 7	29 23	2 28	8 42	4 43	2 12	12 47	26 47	3 10	4 35	24 16	0 49
1/23/1994	4 29	13 48	3 40	20 35	20 8	7 58	20 10	14 27	4 14	17 39	0 44	2 22	10 28	8 34	2 5	16 27	2pi28	3 1	6 40	28 11	0 53
2/ 2/1994	8 47	19 55	3 11	26 17	26 56	7 49	23 4	19 47	3 53	19 11	2 28	2 17	12 14	12 24	1 59	20 4	8 9	2 51	8 46	2 12	0 56
2/12/1994	13 4	26 5	2 42	2ta16	3 58	7 38	25 50	24 57	3 33	20 44	4 32	2 13	13 59	16 13	1 54	23 37	13 48	2 42	10 54	6 18	1 0
2/22/1994	17 19	2ar14	2 14	8 32	11 16	7 23	28 29	29 57	3 14	22 17	6 54	2 9	15 45	20 1	1 48	27 7	19 25	2 34	13 3	10 28	1 4
3/ 4/1994	21 31	8 23	1 47	15 6	18 47	7 4	1pi 1	4 47	2 55	23 51	9 31	2 5	17 29	23 45	1 43	0ar33	24 59	2 25	15 15	14 42	1 8
3/14/1994	25 41	14 32	1 21	21 59	26 32	6 41	3 26	9 27	2 37	25 26	12 21	2 2	19 14	27 26	1 38	3 55	0ar29	2 17	17 28	18 58	1 12
3/24/1994	29 48	20 38	0 55	29 11	4 31	6 13	5 46	13 59	2 20	27 1	15 21	1 59	20 58	1pi 2	1 34	7 14	5 55	2 8	19 42	23 16	1 17
4/ 3/1994	3 51	26 41	0 30	6 42	12 43	5 41	8 2	18 20	2 2	28 38	18 29	1 56	22 41	4 31	1 29	10 29	11 17	2 0	21 59	27 34	1 21
4/13/1994	7 51	2ta42	0 6	14 31	21 7	5 3	10 12	22 32	1 45	0gm14	21 45	1 55	24 25	7 54	1 24	13 41	16 34	1 52	24 17	1ar52	1 26

Ephemeris	80			433			1221			1489			114			3018			2041		
0 Hours ET	Sappho			Eros			Amor			Attila			Kassandra			Godiva			Lancelot		
Mo Dy Year	h long	g long	g lat	h long	g long	g lat	h long	g long	g lat	h long	g long	g lat	h long	g long	g lat	h long	g long	g lat	h long	g long	g lat
4/23/1994	11 48	8 39	0 18	22 38	29 43	4 20	12 19	26 33	1 28	1 52	25 7	1 53	26 8	11 8	1 20	16 48	21 46	1 44	26 37	6 10	1 32
5/ 3/1994	15 40	14 32	0 41	0cn59	8 29	3 32	14 22	0ar23	1 11	3 30	28 33	1 52	27 51	14 12	1 15	19 53	26 52	1 36	28 58	10 25	1 37
5/13/1994	19 28	20 20	1 4	9 33	17 23	2 40	16 21	4 1	0 53	5 10	2 3	1 51	29 34	17 4	1 10	22 53	1ta52	1 27	1 22	14 39	1 43
5/23/1994	23 13	26 4	1 26	18 16	26 25	1 45	18 18	7 26	0 34	6 49	5 35	1 50	1 17	19 42	1 4	25 51	6 45	1 19	3 47	18 48	1 49
6/ 2/1994	26 53	1gm43	1 48	27 6	5 32	0 48	20 11	10 37	0 14	8 30	9 10	1 50	3 0	22 4	0 59	28 45	11 31	1 11	6 13	22 54	1 56
6/12/1994	0gm29	7 16	2 10	5 58	14 42	0s11	22 3	13 31	0s 7	10 12	12 45	1 50	4 42	24 8	0 52	1 35	16 10	1 3	8 41	26 54	2 4
6/22/1994	4 1	12 43	2 32	14 49	23 54	1 8	23 52	16 6	0 29	11 54	16 20	1 50	6 25	25 49	0 45	4 23	20 40	0 54	11 11	0ta47	2 12
7/ 2/1994	7 29	18 5	2 54	23 35	3le 6	2 3	25 39	18 20	0 54	13 37	19 54	1 50	8 8	27 6	0 37	7 8	25 0	0 45	13 42	4 31	2 21
7/12/1994	10 52	23 19	3 16	2vi11	12 15	2 55	27 24	20 8	1 21	15 21	23 26	1 51	9 50	27 55	0 29	9 49	29 11	0 36	16 15	8 5	2 30
7/22/1994	14 12	28 27	3 38	10 36	21 20	3 42	29 8	21 28	1 50	17 6	26 56	1 52	11 33	28 12	0 19	12 28	3 9	0 26	18 48	11 26	2 40
8/ 1/1994	17 27	3 27	4 1	18 46	0vi19	4 24	0ar50	22 16	2 23	18 52	0cn22	1 54	13 16	27 55	0 8	15 4	6 55	0 16	21 24	14 32	2 52
8/11/1994	20 39	8 19	4 25	26 40	9 11	4 59	2 31	22 25R	2 59	20 38	3 44	1 55	14 59	27 4	0s 4	17 38	10 25	0 5	24 0	17 20	3 4
8/21/1994	23 47	13 1	4 49	4 16	17 54	5 28	4 10	21 53	3 37	22 26	6 59	1 58	16 42	25 40	0 17	20 8	13 38	0 7	26 37	19 46	3 17
8/31/1994	26 51	17 33	5 15	11 34	26 26	5 51	5 49	20 37	4 18	24 14	10 7	2 0	18 25	23 49	0 30	22 37	16 31	0 20	29 16	21 45	3 31
9/10/1994	29 51	21 53	5 42	18 33	4 49	6 8	7 27	18 38	4 59	26 4	13 5	2 3	20 9	21 40	0 44	25 3	19 0	0 34	1 55	23 14	3 47
9/20/1994	2 48	26 0	6 10	25 13	13 0	6 19	9 5	16 3	5 39	27 54	15 52	2 6	21 52	19 26	0 57	27 27	21 1	0 50	4 35	24 7	4 2
9/30/1994	5 42	29 52	6 41	1sc34	20 59	6 25	10 41	13 3	6 14	29 45	18 24	2 9	23 37	17 20	1 9	29 48	22 30	1 7	7 16	24 21	4 17
10/10/1994	8 32	3 26	7 14	7 38	28 48	6 26	12 18	9 55	6 42	1 37	20 40	2 13	25 21	15 35	1 20	2 8	23 23	1 26	9 58	23 54	4 32
10/20/1994	11 19	6 38	7 50	13 26	6 25	6 24	13 54	6 58	7 3	3 30	22 35	2 18	27 6	14 19	1 29	4 26	23 34	1 46	12 40	22 49	4 43
10/30/1994	14 3	9 27	8 30	18 59	13 51	6 18	15 30	4 29	7 15	5 24	24 6	2 22	28 51	13 38	1 37	6 42	23 2	2 7	15 22	21 10	4 51
11/ 9/1994	16 44	11 46	9 12	24 17	21 6	6 10	17 6	2 39	7 20	7 19	25 9	2 27	0ar36	13 33D	1 43	8 56	21 45	2 29	18 5	19 12	4 54
11/19/1994	19 22	13 31	9 57	29 21	28 12	5 59	18 41	1 33	7 20	9 15	25 41	2 31	2 22	14 3	1 48	11 8	19 48	2 50	20 47	17 10	4 51
11/29/1994	21cn57	14le38	10s45	4sa14	5sa 8	5s47	20ar18	1ar11R	7s17	11cn12	25cn38R	2s35	4ar 9	15pi 7	1s52	13gm19	17gm23R	3s 9	23ta30	15ta21R	4s43
12/ 9/1994	24 30	14 59	11 35	8 56	11 55	5 33	21 54	1 31	7 12	13 10	24 59	2 38	5 55	16 39	1 55	15 28	14 45	3 25	26 13	14 0	4 30
12/19/1994	27 0	14 32	12 23	13 27	18 33	5 17	23 31	2 28	7 6	15 8	23 46	2 40	7 43	18 37	1 58	17 36	12 13	3 37	28 55	13 17	4 14
12/29/1994	29 28	13 17	13 7	17 50	25 3	5 1	25 8	3 57	7 0	17 8	22 5	2 39	9 31	20 57	2 1	19 43	10 4	3 44	1 37	13 16D	3 57
1/ 8/1995	1 53	11 18	13 42	22 5	1cp25	4 43	26 47	5 54	6 54	19 8	20 6	2 36	11 20	23 36	2 4	21 48	8 30	3 47	4 18	13 57	3 40
1/18/1995	4 16	8 46	14 4	26 12	7 40	4 24	28 26	8 15	6 49	21 10	18 2	2 30	13 9	26 31	2 6	23 52	7 38	3 47	6 59	15 16	3 24
1/28/1995	6 37	5 59	14 9	0cp14	13 48	4 5	0ta 6	10 55	6 45	23 12	16 7	2 22	14 59	29 40	2 9	25 55	7 28D	3 45	9 39	17 8	3 8
2/ 7/1995	8 56	3 19	13 56	4 9	19 48	3 45	1 47	13 53	6 43	25 15	14 35	2 13	16 50	3 0	2 12	27 57	7 59	3 41	12 18	19 30	2 53
2/17/1995	11 14	1 4	13 29	8 0	25 42	3 23	3 29	17 4	6 41	27 19	13 34	2 2	18 41	6 30	2 15	29 58	9 6	3 38	14 56	22 16	2 40
2/27/1995	13 29	29cn27	12 52	11 47	1aq30	3 0	5 13	20 27	6 40	29 24	13 10	1 51	20 34	10 7	2 18	1 58	10 44	3 33	17 34	25 23	2 28
3/ 9/1995	15 43	28 34	12 9	15 30	7 10	2 37	6 59	24 1	6 41	1 30	13 23	1 40	22 27	13 51	2 21	3 57	12 48	3 30	20 10	28 46	2 17
3/19/1995	17 54	28 26D	11 24	19 11	12 45	2 11	8 46	27 43	6 42	3 37	14 11	1 29	24 21	17 40	2 25	5 55	15 14	3 26	22 45	2 23	2 7
3/29/1995	20 5	28 59	10 40	22 49	18 13	1 44	10 35	1ta33	6 45	5 44	15 32	1 20	26 16	21 33	2 28	7 53	17 58	3 23	25 18	6 10	1 58
4/ 8/1995	22 14	0le 9	9 58	26 26	23 34	1 15	12 26	5 30	6 49	7 53	17 21	1 11	28 12	25 29	2 33	9 50	20 58	3 20	27 51	10 6	1 49
4/18/1995	24 21	1 51	9 19	0aq 1	28 48	0 43	14 20	9 33	6 54	10 2	19 35	1 2	0ta 9	29 27	2 37	11 46	24 9	3 18	0cn22	14 9	1 41
4/28/1995	26 28	3 59	8 44	3 36	3 54	0 8	16 17	13 40	7 0	12 11	22 10	0 55	2 6	3 27	2 42	13 42	27 31	3 16	2 51	18 17	1 34
5/ 8/1995	28 33	6 28	8 12	7 12	8 53	0 30	18 16	17 53	7 6	14 22	25 3	0 48	4 5	7 28	2 48	15 38	1cn 1	3 14	5 19	22 29	1 27
5/18/1995	0vi37	9 16	7 43	10 47	13 43	1 13	20 19	22 10	7 15	16 33	28 12	0 41	6 5	11 29	2 53	17 33	4 38	3 13	7 45	26 44	1 21

Ephemeris	80			433			1221			1489			114			3018			2041		
0 Hours ET	Sappho			Eros			Amor			Attila			Kassandra			Godiva			Lancelot		
Mo Dy Year	h long	g long	g lat	h long	g long	g lat	h long	g long	g lat	h long	g long	g lat	h long	g long	g lat	h long	g long	g lat	h long	g long	g lat
5/28/1995	2 39	12 18	7 17	14 24	18 24	2 1	22 25	26 32	7 24	18 44	1le33	0 35	8 7	15 29	3 0	19 27	8 20	3 13	10 10	1cn 1	1 15
6/ 7/1995	4 41	15 33	6 53	18 3	22 53	2 54	24 35	0gm58	7 34	20 57	5 5	0 30	10 9	19 27	3 7	21 22	12 6	3 13	12 33	5 18	1 10
6/17/1995	6 42	18 58	6 32	21 44	27 8	3 55	26 49	5 28	7 46	23 10	8 46	0 25	12 13	23 24	3 14	23 16	15 56	3 13	14 54	9 36	1 4
6/27/1995	8 42	22 31	6 13	25 28	1ar 8	5 5	29 7	10 2	7 59	25 23	12 34	0 20	14 17	27 17	3 23	25 10	19 48	3 14	17 14	13 53	0 59
7/ 7/1995	10 42	26 11	5 55	29 16	4 48	6 26	1 31	14 40	8 13	27 37	16 30	0 15	16 23	1gm 7	3 32	27 3	23 42	3 15	19 31	18 9	0 54
7/17/1995	12 40	29 56	5 39	3 8	8 4	7 59	4 1	19 23	8 28	29 51	20 30	0 10	18 31	4 52	3 42	28 57	27 38	3 17	21 48	22 23	0 50
7/27/1995	14 38	3 47	5 25	7 6	10 48	9 47	6 36	24 11	8 45	2 6	24 35	0 6	20 40	8 30	3 53	0le51	1le33	3 19	24 2	26 35	0 45
8/ 6/1995	16 36	7 40	5 12	11 8	12 52	11 53	9 19	29 4	9 4	4 21	28 43	0 2	22 50	12 1	4 5	2 44	5 29	3 21	26 15	0le43	0 41
8/16/1995	18 33	11 37	4 59	15 18	14 5	14 17	12 8	4 3	9 23	6 37	2 54	0 3	25 1	15 22	4 18	4 38	9 24	3 25	28 26	4 47	0 36
8/26/1995	20 29	15 36	4 48	19 35	14 15R	17 0	15 7	9 7	9 45	8 52	7 7	0 7	27 14	18 32	4 33	6 32	13 18	3 28	0le35	8 47	0 32
9/ 5/1995	22 25	19 37	4 38	24 0	13 8	19 54	18 14	14 19	10 8	11 8	11 22	0 11	29 29	21 28	4 49	8 26	17 9	3 33	2 43	12 41	0 28
9/15/1995	24 21	23 38	4 28	28 35	10 38	22 47	21 31	19 38	10 32	13 24	15 36	0 15	1 45	24 6	5 7	10 20	20 57	3 38	4 49	16 29	0 23
9/25/1995	26 17	27 40	4 19	3 20	6 51	25 16	25 0	25 6	10 58	15 40	19 51	0 19	4 2	26 24	5 27	12 15	24 42	3 43	6 53	20 9	0 18
10/ 5/1995	28 12	1li41	4 11	8 16	2 20	26 58	28 42	0le44	11 25	17 56	24 4	0 24	6 21	28 16	5 48	14 10	28 21	3 50	8 55	23 39	0 14
10/15/1995	0li 8	5 41	4 3	13 25	27pi54	27 39	2 38	6 33	11 52	20 12	28 15	0 28	8 42	29 38	6 11	16 6	1 54	3 57	10 56	26 59	0 9
10/25/1995	2 3	9 40	3 56	18 48	24 27	27 19	6 49	12 36	12 19	22 29	2 23	0 33	11 4	0 25	6 35	18 2	5 18	4 6	12 55	0vi 6	0 4
11/ 4/1995	3 58	13 35	3 49	24 25	22 32	26 13	11 18	18 54	12 45	24 45	6 27	0 38	13 28	0 33R	7 0	19 58	8 33	4 15	14 53	2 58	0n 2
11/14/1995	5 54	17 27	3 43	0ta19	22 18D	24 40	16 6	25 32	13 7	27 1	10 26	0 43	15 54	29gm58	7 22	21 55	11 35	4 26	16 49	5 33	0 8
11/24/1995	7 49	21 14	3 36	6 29	23 41	22 53	21 17	2vi32	13 24	29 16	14 18	0 49	18 21	28 42	7 41	23 53	14 22	4 38	18 44	7 47	0 14
12/ 4/1995	9 45	24 54	3 30	12 57	26 26	21 1	26 52	10 0	13 32	1 32	18 2	0 55	20 50	26 49	7 55	25 51	16 51	4 51	20 37	9 38	0 21
12/14/1995	11 41	28 26	3 24	19 44	0ar22	19 6	2le53	18 2	13 25	3 47	21 35	1 1	23 21	24 32	8 0	27 51	18 58	5 6	22 28	11 1	0 29
12/24/1995	13 38	1 49	3 17	26 50	5 16	17 10	9 24	26 44	12 56	6 2	24 56	1 8	25 53	22 8	7 55	29 51	20 38	5 22	24 19	11 53	0 37
1/ 3/1996	15li35	5sc 0	3s11	4gm15	11ar 0	15n12	16le27	6li11	11s58	8li17	28li 2	1n16	28gm27	19gm57R	7s41	1vi52	21vi47	5s39	26le 7	12vi12	0n46
1/13/1996	17 33	7 57	3 4	11 59	17 25	13 12	24 3	16 30	10 20	10 31	0sc50	1 24	1cn 3	18 14	7 20	3 54	22 21	5 57	27 55	11 56	0 55
1/23/1996	19 31	10 36	2 56	20 0	24 26	11 9	2vi14	27 39	7 53	12 44	3 17	1 33	3 41	17 12	6 53	5 57	22 15R	6 14	29 41	11 4	1 5
2/ 2/1996	21 30	12 55	2 48	28 17	1ta59	9 3	10 59	9 31	4 34	14 57	5 19	1 43	6 20	16 55	6 24	8 1	21 26	6 30	1 25	9 41	1 15
2/12/1996	23 29	14 49	2 38	6 47	9 58	6 53	20 17	21 50	0 32	17 10	6 54	1 54	9 0	17 23	5 55	10 6	19 55	6 42	3 9	7 52	1 24
2/22/1996	25 30	16 14	2 26	15 28	18 20	4 41	0li 3	4sa 9	3 54	19 22	7 56	2 5	11 43	18 33	5 26	12 13	17 49	6 49	4 51	5 50	1 32
3/ 3/1996	27 31	17 5	2 12	24 17	27 2	2 28	10 10	16 3	8 19	21 33	8 23	2 17	14 27	20 21	4 59	14 20	15 20	6 50	6 31	3 46	1 38
3/13/1996	29 33	17 19	1 55	3le 9	5 59	0 18	20 28	27 10	12 19	23 44	8 13	2 30	17 12	22 42	4 34	16 30	12 43	6 42	8 11	1 53	1 43
3/23/1996	1 37	16 51	1 36	12 0	15 9	1 48	0sc46	7 14	15 45	25 54	7 26	2 41	19 59	25 30	4 11	18 40	10 18	6 27	9 50	0 22	1 47
4/ 2/1996	3 42	15 40	1 12	20 48	24 28	3 45	10 53	16 8	18 35	28 3	6 7	2 52	22 47	28 41	3 50	20 53	8 20	6 5	11 27	29 19	1 49
4/12/1996	5 47	13 51	0 45	29 28	3cn52	5 30	20 39	23 50	20 53	0sc12	4 23	3 1	25 37	2 11	3 30	23 7	7 0	5 41	13 3	28 48	1 50
4/22/1996	7 55	11 31	0 15	7 57	13 18	7 2	29 56	0aq17	22 47	2 20	2 25	3 7	28 28	5 58	3 12	25 23	6 23	5 14	14 38	28 49D	1 50
5/ 2/1996	10 3	8 54	0 17	16 13	22 42	8 18	8 38	5 25	24 23	4 27	0 29	3 9	1le20	9 59	2 55	27 40	6 30D	4 48	16 13	29 20	1 50
5/12/1996	12 14	6 18	0 49	24 12	2le 0	9 17	16 43	9 13	25 44	6 33	28 46	3 9	4 14	14 11	2 40	29 60	7 17	4 22	17 46	0vi19	1 49
5/22/1996	14 26	4 0	1 20	1li54	11 10	10 0	24 11	11 33	26 53	8 38	27 27	3 7	7 8	18 32	2 25	2 21	8 41	3 58	19 18	1 42	1 48
6/ 1/1996	16 39	2 15	1 48	9 18	20 10	10 27	1cp 3	12 18	27 47	10 43	26 40	3 2	10 4	23 2	2 11	4 44	10 38	3 36	20 49	3 25	1 48
6/11/1996	18 55	1 10	2 12	16 22	28 56	10 40	7 22	11 25	28 18	12 46	26 26	2 57	13 1	27 39	1 58	7 10	13 2	3 15	22 19	5 27	1 47
6/21/1996	21 13	0 49	2 33	23 8	7 28	10 41	13 10	9 0	28 17	14 49	26 47	2 50	15 58	2le21	1 46	9 38	15 51	2 56	23 49	7 44	1 46

Ephemeris	80			433			1221			1489			114			3018			2041		
0 Hours ET	Sappho			Eros			Amor			Attila			Kassandra			Godiva			Lancelot		
Mo Dy Year	h long	g long	g lat	h long	g long	g lat	h long	g long	g lat	h long	g long	g lat	h long	g long	g lat	h long	g long	g lat	h long	g long	g lat
7/ 1/1996	23 32	1 11	2 51	29 36	15 45	10 32	18 30	5 23	27 31	16 51	27 40	2 44	18 56	7 9	1 34	12 8	19 0	2 39	25 17	10 13	1 46
7/11/1996	25 54	2 12	3 6	5 45	23 48	10 15	23 26	1 12	25 57	18 52	29 2	2 37	21 55	12 0	1 22	14 41	22 28	2 23	26 45	12 53	1 46
7/21/1996	28 19	3 50	3 20	11 38	1li36	9 51	28 0	27 11	23 41	20 52	0sc49	2 31	24 54	16 55	1 11	17 17	26 11	2 8	28 12	15 42	1 46
7/31/1996	0sa46	5 59	3 31	17 15	9 10	9 23	2 15	23 56	21 0	22 51	2 58	2 25	27 54	21 52	1 0	19 55	0li 7	1 54	29 38	18 38	1 46
8/10/1996	3 15	8 35	3 41	22 38	16 31	8 52	6 12	21 45	18 13	24 49	5 26	2 20	0vi53	26 51	0 49	22 36	4 16	1 40	1 3	21 40	1 46
8/20/1996	5 47	11 35	3 50	27 47	23 41	8 19	9 54	20 42	15 35	26 46	8 10	2 15	3 54	1vi52	0 38	25 19	8 36	1 28	2 28	24 46	1 47
8/30/1996	8 23	14 56	3 59	2 43	0sc40	7 45	13 22	20 41D	13 13	28 42	11 8	2 10	6 54	6 54	0 27	28 6	13 5	1 16	3 52	27 56	1 48
9/ 9/1996	11 1	18 35	4 7	7 28	7 29	7 11	16 38	21 31	11 10	0sa38	14 16	2 6	9 54	11 56	0 16	0sc56	17 44	1 4	5 15	1li 8	1 49
9/19/1996	13 42	22 30	4 14	12 3	14 10	6 36	19 44	23 4	9 25	2 32	17 35	2 3	12 54	16 58	0 6	3 48	22 30	0 53	6 37	4 21	1 51
9/29/1996	16 27	26 39	4 21	16 28	20 43	6 3	22 40	25 11	7 56	4 25	21 1	2 0	15 53	21 59	0n 5	6 44	27 24	0 42	7 59	7 35	1 53
10/ 9/1996	19 15	1sa 1	4 28	20 45	27 10	5 29	25 27	27 44	6 41	6 18	24 33	1 57	18 53	26 59	0 16	9 44	2sc25	0 31	9 20	10 48	1 55
10/19/1996	22 7	5 35	4 35	24 55	3 31	4 57	28 7	0aq37	5 37	8 9	28 11	1 55	21 51	1li56	0 28	12 47	7 32	0 20	10 41	13 59	1 57
10/29/1996	25 3	10 19	4 42	28 58	9 47	4 26	0pi40	3 48	4 43	10 0	1 53	1 52	24 50	6 50	0 40	15 53	12 45	0 9	12 1	17 7	2 0
11/ 8/1996	28 2	15 13	4 49	2 56	15 59	3 55	3 6	7 12	3 56	11 50	5 37	1 51	27 47	11 40	0 52	19 3	18 3	0n 1	13 21	20 12	2 4
11/18/1996	1cp 6	20 15	4 55	6 48	22 8	3 25	5 27	10 46	3 15	13 38	9 24	1 49	0li44	16 25	1 5	22 16	23 26	0 12	14 40	23 11	2 8
11/28/1996	4 14	25 25	5 2	10 36	28 14	2 57	7 43	14 28	2 40	15 26	13 11	1 48	3 39	21 3	1 18	25 33	28 53	0 23	15 58	26 3	2 12
12/ 8/1996	7 26	0cp43	5 9	14 20	4 17	2 28	9 54	18 18	2 9	17 13	16 58	1 48	6 34	25 33	1 32	28 54	4 24	0 35	17 16	28 47	2 17
12/18/1996	10 43	6 8	5 15	18 2	10 18	2 1	12 2	22 12	1 41	18 59	20 44	1 47	9 28	29 54	1 47	2 18	9 58	0 47	18 34	1 21	2 23
12/28/1996	14 4	11 38	5 22	21 40	16 17	1 34	14 5	26 10	1 17	20 44	24 28	1 47	12 20	4 4	2 4	5 46	15 35	0 59	19 51	3 42	2 29
1/ 7/1997	17 29	17 14	5 28	25 18	22 15	1 7	16 5	0pi11	0 54	22 29	28 8	1 47	15 12	8 0	2 22	9 17	21 14	1 11	21 8	5 49	2 36
1/17/1997	20 59	22 56	5 35	28 53	28 11	0 41	18 2	4 14	0 34	24 12	1cp45	1 47	18 2	11 39	2 41	12 52	26 54	1 24	22 24	7 39	2 44
1/27/1997	24 34	28 42	5 41	2 29	4 7	0 14	19 56	8 18	0 15	25 55	5 15	1 48	20 50	14 59	3 2	16 30	2cp35	1 38	23 40	9 10	2 52
2/ 6/1997	28cp14	4aq32	5n47	6aq 4	10aq 2	0n12	21pi48	12pi22	0s 3	27sa37	8cp38	1n49	23li38	17sc57	3n25	20sa12	8cp17	1n52	24li55	10sc18	3n 1
2/16/1997	1 58	10 25	5 52	9 39	15 57	0 39	23 37	16 26	0 20	29 18	11 53	1 51	26 23	20 27	3 50	23 56	13 57	2 6	26 11	11 1	3 11
2/26/1997	5 46	16 21	5 57	13 16	21 52	1 6	25 25	20 29	0 36	0 58	14 59	1 52	29 8	22 27	4 17	27 44	19 35	2 22	27 25	11 18	3 21
3/ 8/1997	9 39	22 19	6 2	16 54	27 48	1 34	27 10	24 31	0 52	2 37	17 52	1 55	1 50	23 50	4 46	1cp35	25 11	2 38	28 40	11 5	3 31
3/18/1997	13 36	28 19	6 5	20 34	3 44	2 3	28 54	28 30	1 8	4 16	20 32	1 57	4 31	24 34	5 16	5 27	0aq42	2 55	29 54	10 24	3 40
3/28/1997	17 37	4 20	6 8	24 17	9 41	2 32	0ar36	2 27	1 24	5 54	22 56	2 0	7 11	24 34R	5 48	9 23	6 9	3 13	1 8	9 16	3 48
4/ 7/1997	21 41	10 21	6 10	28 4	15 41	3 2	2 18	6 20	1 40	7 31	25 3	2 3	9 49	23 50	6 18	13 20	11 28	3 32	2 22	7 46	3 55
4/17/1997	25 49	16 22	6 11	1pi54	21 42	3 34	3 58	10 10	1 56	9 7	26 48	2 6	12 25	22 25	6 45	17 18	16 39	3 52	3 35	5 59	3 59
4/27/1997	0pi 0	22 21	6 11	5 49	27 46	4 6	5 36	13 55	2 13	10 43	28 10	2 9	15 0	20 29	7 7	21 18	21 40	4 13	4 49	4 5	4 1
5/ 7/1997	4 14	28 18	6 9	9 50	3 54	4 40	7 15	17 35	2 30	12 18	29 7	2 13	17 33	18 16	7 21	25 19	26 28	4 36	6 2	2 14	4 0
5/17/1997	8 30	4 11	6 6	13 58	10 6	5 16	8 52	21 9	2 49	13 52	29 35	2 16	20 4	16 1	7 27	29 21	1pi 0	4 59	7 14	0 34	3 56
5/27/1997	12 48	10 0	6 1	18 12	16 24	5 53	10 29	24 36	3 9	15 26	29 33R	2 19	22 34	14 3	7 24	3 22	5 13	5 24	8 27	29 13	3 51
6/ 6/1997	17 7	15 43	5 54	22 34	22 47	6 31	12 5	27 55	3 30	16 59	29 0	2 22	25 2	12 34	7 15	7 24	9 3	5 51	9 40	28 17	3 44
6/16/1997	21 27	21 19	5 45	27 6	29 17	7 10	13 41	1ta 5	3 53	18 31	27 58	2 23	27 28	11 43	7 0	11 25	12 25	6 19	10 52	27 47	3 36
6/26/1997	25 47	26 46	5 34	1ar47	5 56	7 50	15 17	4 5	4 18	20 3	26 32	2 23	29 53	11 31	6 43	15 24	15 14	6 48	12 4	27 44D	3 27
7/ 6/1997	0ar 7	2ta 3	5 20	6 39	12 44	8 31	16 53	6 51	4 46	21 34	24 47	2 21	2 16	11 59	6 24	19 23	17 23	7 17	13 16	28 8	3 19
7/16/1997	4 26	7 6	5 4	11 44	19 42	9 12	18 29	9 23	5 17	23 5	22 54	2 18	4 37	13 2	6 5	23 19	18 46	7 46	14 28	28 56	3 11
7/26/1997	8 45	11 53	4 45	17 2	26 51	9 52	20 5	11 36	5 51	24 35	21 2	2 13	6 57	14 37	5 47	27 14	19 18	8 13	15 40	0sc 6	3 3

Ephemeris	80			433			1221			1489			114			3018			2041		
0 Hours ET	Sappho			Eros			Amor			Attila			Kassandra			Godiva			Lancelot		
Mo Dy Year	h long	g long	g lat	h long	g long	g lat	h long	g long	g lat	h long	g long	g lat	h long	g long	g lat	h long	g long	g lat	h long	g long	g lat
8/ 5/1997	13 1	16 21	4 22	22 34	4 14	10 30	21 42	13 29	6 28	26 4	19 22	2 6	9 16	16 40	5 29	1pi 7	18 55	8 35	16 52	1 36	2 56
8/15/1997	17 16	20 26	3 55	28 22	11 51	11 5	23 19	14 56	7 10	27 33	18 2	1 59	11 32	19 5	5 13	4 56	17 40	8 49	18 4	3 23	2 49
8/25/1997	21 28	24 4	3 24	4 26	19 43	11 35	24 56	15 53	7 57	29 2	17 8	1 51	13 48	21 51	4 58	8 43	15 44	8 52	19 16	5 25	2 44
9/ 4/1997	25 38	27 7	2 47	10 48	27 51	11 59	26 34	16 16	8 48	0aq29	16 43	1 42	16 1	24 52	4 45	12 27	13 25	8 41	20 28	7 40	2 38
9/14/1997	29 45	29 31	2 4	17 29	6 15	12 13	28 13	15 58	9 42	1 57	16 48D	1 34	18 14	28 8	4 33	16 8	11 6	8 16	21 39	10 6	2 34
9/24/1997	3 48	1 7	1 14	24 28	14 56	12 15	29 53	14 56	10 39	3 24	17 21	1 27	20 25	1sa35	4 21	19 45	9 12	7 42	22 51	12 41	2 30
10/ 4/1997	7 48	1 49	0 17	1gm47	23 51	12 2	1 34	13 9	11 34	4 50	18 20	1 19	22 34	5 11	4 11	23 19	7 58	7 0	24 3	15 25	2 26
10/14/1997	11 44	1 32	0 47	9 24	3le 0	11 31	3 16	10 40	12 25	6 16	19 43	1 13	24 43	8 56	4 2	26 49	7 34	6 17	25 15	18 14	2 23
10/24/1997	15 36	0 17	1 57	17 19	12 19	10 41	5 0	7 39	13 6	7 41	21 26	1 6	26 50	12 46	3 54	0ar16	8 0	5 34	26 27	21 9	2 21
11/ 3/1997	19 24	28 12	3 9	25 31	21 47	9 29	6 45	4 24	13 33	9 7	23 28	1 1	28 55	16 42	3 47	3 39	9 12	4 53	27 39	24 8	2 19
11/13/1997	23 8	25 37	4 17	3cn57	1vi17	7 55	8 32	1 14	13 44	10 31	25 46	0 56	1cp 0	20 43	3 40	6 58	11 6	4 16	28 51	27 9	2 17
11/23/1997	26 48	22 58	5 17	12 34	10 47	6 0	10 21	28 29	13 40	11 56	28 17	0 51	3 3	24 46	3 34	10 14	13 33	3 43	0sa 3	0sa12	2 16
12/ 3/1997	0gm24	20 41	6 4	21 21	20 12	3 46	12 12	26 24	13 24	13 20	0aq59	0 47	5 5	28 51	3 29	13 25	16 29	3 14	1 16	3 15	2 16
12/13/1997	3 56	19 7	6 37	0le12	29 26	1 15	14 6	25 8	12 59	14 43	3 51	0 43	7 6	2 57	3 25	16 34	19 48	2 47	2 28	6 17	2 16
12/23/1997	7 23	18 25	6 59	9 4	8 26	1 26	16 2	24 41	12 31	16 6	6 50	0 39	9 6	7 3	3 21	19 38	23 26	2 24	3 41	9 17	2 16
1/ 2/1998	10 47	18 37D	7 11	17 54	17 6	4 16	18 1	25 2	12 1	17 29	9 55	0 36	11 4	11 9	3 17	22 39	27 19	2 3	4 54	12 14	2 17
1/12/1998	14 6	19 39	7 16	26 37	25 21	7 10	20 3	26 6	11 31	18 52	13 4	0 32	13 2	15 13	3 14	25 37	1ar24	1 45	6 7	15 5	2 19
1/22/1998	17 21	21 24	7 16	5 10	3sc 6	10 4	22 9	27 47	11 4	20 14	16 17	0 29	14 59	19 15	3 12	28 32	5 38	1 28	7 20	17 51	2 20
2/ 1/1998	20 33	23 45	7 14	13 31	10 17	12 58	24 18	0ta 0	10 38	21 36	19 31	0 27	16 54	23 13	3 10	1ta23	9 59	1 13	8 33	20 28	2 23
2/11/1998	23 41	26 36	7 10	21 36	16 45	15 49	26 32	2 42	10 15	22 58	22 46	0 24	18 49	27 6	3 9	4 11	14 26	0 59	9 47	22 55	2 25
2/21/1998	26 44	29 50	7 5	29 24	22 25	18 38	28 50	5 48	9 54	24 19	26 0	0 21	20 43	0aq55	3 8	6 55	18 57	0 47	11 1	25 10	2 29
3/ 3/1998	29 45	3 22	6 59	6 54	27 6	21 25	1 14	9 16	9 35	25 40	29 13	0 19	22 36	4 36	3 7	9 37	23 30	0 35	12 15	27 11	2 33
3/13/1998	2cn42	7gm10	6s53	14li 5	0sa37	24s 7	3gm42	13ta 1	9s19	27aq 1	2pi23	0n16	24cp29	8aq10	3n 7	12ta16	28ar 6	0n24	13sa30	28sa56	2n37
3/23/1998	5 35	11 8	6 48	20 58	2 45	26 42	6 17	17 4	9 4	28 22	5 29	0 14	26 20	11 34	3 7	14 53	2 41	0 14	14 45	0cp22	2 41
4/ 2/1998	8 25	15 16	6 42	27 31	3 17	29 3	8 59	21 21	8 50	29 42	8 30	0 11	28 11	14 48	3 8	17 26	7 17	0 4	16 0	1 26	2 46
4/12/1998	11 12	19 31	6 37	3 47	2 2	30 54	11 48	25 53	8 38	1 3	11 24	0 9	0aq 1	17 49	3 9	19 57	11 53	0 5	17 15	2 5	2 52
4/22/1998	13 56	23 51	6 32	9 46	29sc 8	31 59	14 45	0gm37	8 26	2 23	14 11	0 6	1 50	20 35	3 10	22 26	16 27	0 14	18 31	2 19	2 57
5/ 2/1998	16 37	28 15	6 28	15 28	25 0	31 58	17 51	5 35	8 15	3 43	16 48	0 3	3 39	23 5	3 12	24 52	20 59	0 22	19 47	2 4	3 1
5/12/1998	19 15	2 42	6 24	20 55	20 33	30 42	21 8	10 46	8 4	5 3	19 14	0 0	5 27	25 15	3 14	27 16	25 29	0 30	21 4	1 22	3 5
5/22/1998	21 50	7 10	6 21	26 9	16 40	28 24	24 36	16 10	7 53	6 22	21 28	0 3	7 15	27 2	3 15	29 38	29 57	0 39	22 21	0 13	3 8
6/ 1/1998	24 23	11 40	6 18	1sa 9	13 58	25 26	28 16	21 47	7 41	7 42	23 27	0 6	9 2	28 24	3 17	1 58	4 21	0 47	23 39	28 41	3 9
6/11/1998	26 53	16 9	6 15	5 58	12 41	22 16	2 10	27 38	7 28	9 1	25 9	0 10	10 48	29 17	3 19	4 15	8 42	0 55	24 57	26 54	3 8
6/21/1998	29 21	20 39	6 14	10 36	12 43D	19 11	6 20	3 44	7 14	10 21	26 31	0 14	12 34	29 39	3 19	6 31	12 58	1 4	26 15	25 0	3 5
7/ 1/1998	1 46	25 9	6 12	15 5	13 51	16 22	10 47	10 5	6 57	11 40	27 31	0 18	14 20	29 26	3 19	8 46	17 10	1 12	27 34	23 9	2 59
7/11/1998	4 9	29 37	6 11	19 24	15 53	13 52	15 34	16 44	6 38	12 59	28 7	0 23	16 5	28 38	3 17	10 58	21 16	1 21	28 54	21 30	2 52
7/21/1998	6 30	4 4	6 11	23 37	18 36	11 42	20 42	23 40	6 16	14 18	28 17	0 28	17 50	27 17	3 12	13 9	25 17	1 31	0cp14	20 10	2 43
7/31/1998	8 49	8 28	6 11	27 42	21 52	9 48	26 14	0le56	5 50	15 37	27 59	0 33	19 34	25 29	3 5	15 18	29 11	1 40	1 34	19 17	2 34
8/10/1998	11 7	12 51	6 12	1cp41	25 33	8 10	2le13	8 32	5 19	16 56	27 13	0 38	21 19	23 23	2 55	17 26	2 57	1 51	2 56	18 51	2 24
8/20/1998	13 22	17 11	6 14	5 35	29 33	6 44	8 42	16 29	4 43	18 15	26 1	0 44	23 3	21 10	2 43	19 33	6 33	2 2	4 17	18 55D	2 14
8/30/1998	15 36	21 27	6 17	9 24	3 51	5 30	15 41	24 48	4 1	19 34	24 28	0 49	24 46	19 4	2 28	21 38	9 59	2 14	5 40	19 26	2 5

Ephemeris	80			433			1221			1489			114			3018			2041		
0 Hours ET	Sappho			Eros			Amor			Attila			Kassandra			Godiva			Lancelot		
Mo Dy Year	h long	g long	g lat	h long	g long	g lat	h long	g long	g lat	h long	g long	g lat	h long	g long	g lat	h long	g long	g lat	h long	g long	g lat
9/ 9/1998	17 48	25 40	6 20	13 10	8 21	4 25	23 15	3vi30	3 13	20 52	22 40	0 54	26 30	17 18	2 12	23 42	13 13	2 27	7 3	20 24	1 55
9/19/1998	19 58	29 48	6 24	16 52	13 2	3 27	1vi23	12 33	2 18	22 11	20 48	0 59	28 13	16 0	1 55	25 45	16 12	2 41	8 27	21 46	1 47
9/29/1998	22 7	3 50	6 29	20 32	17 53	2 35	10 6	21 57	1 19	23 30	18 59	1 2	29 56	15 14	1 38	27 47	18 53	2 57	9 51	23 30	1 39
10/ 9/1998	24 15	7 46	6 35	24 9	22 53	1 49	19 22	1li39	0 15	24 49	17 25	1 5	1 39	15 5	1 23	29 48	21 14	3 14	11 17	25 32	1 32
10/19/1998	26 21	11 34	6 42	27 46	27 59	1 8	29 6	11 35	0 52	26 8	16 12	1 7	3 22	15 29	1 8	1 48	23 11	3 32	12 43	27 52	1 25
10/29/1998	28 26	15 13	6 50	1aq21	3 12	0 30	9 12	21 42	1 58	27 27	15 25	1 9	5 5	16 25	0 55	3 47	24 40	3 53	14 10	0cp26	1 19
11/ 8/1998	0vi30	18 41	7 0	4 56	8 32	0n 5	19 30	1sc53	3 2	28 46	15 8	1 10	6 48	17 50	0 43	5 46	25 36	4 14	15 37	3 12	1 13
11/18/1998	2 33	21 57	7 10	8 31	13 57	0 37	29 50	12 3	4 0	0ar 5	15 19	1 10	8 31	19 41	0 32	7 43	25 55	4 38	17 6	6 9	1 8
11/28/1998	4 35	24 57	7 23	12 8	19 28	1 7	9 59	22 5	4 52	1 24	15 58	1 10	10 13	21 53	0 21	9 40	25 33	5 2	18 35	9 16	1 3
12/ 8/1998	6 36	27 39	7 37	15 45	25 4	1 35	19 48	1sa54	5 35	2 43	17 2	1 10	11 56	24 24	0 12	11 37	24 30	5 25	20 5	12 30	0 58
12/18/1998	8 36	29 59	7 52	19 25	0aq45	2 1	29 8	11 25	6 10	4 3	18 30	1 10	13 39	27 11	0 4	13 33	22 47	5 46	21 37	15 50	0 54
12/28/1998	10 35	1 53	8 9	23 7	6 32	2 26	7 54	20 35	6 36	5 22	20 17	1 10	15 23	0pi11	0 4	15 28	20 33	6 3	23 9	19 16	0 50
1/ 7/1999	12 34	3 17	8 27	26 52	12 24	2 50	16 3	29 22	6 54	6 42	22 22	1 11	17 6	3 23	0 11	17 23	18 1	6 15	24 42	22 46	0 46
1/17/1999	14 32	4 7	8 45	0pi42	18 21	3 13	23 34	7 43	7 6	8 2	24 42	1 11	18 49	6 44	0 18	19 18	15 27	6 19	26 16	26 18	0 43
1/27/1999	16 29	4 18	9 2	4 35	24 24	3 35	0cp30	15 39	7 12	9 21	27 14	1 11	20 33	10 12	0 25	21 12	13 8	6 18	27 51	29 52	0 39
2/ 6/1999	18 26	3 47	9 18	8 34	0pi32	3 56	6 51	23 10	7 13	10 41	29 57	1 12	22 17	13 47	0 32	23 6	11 18	6 10	29 28	3 27	0 35
2/16/1999	20 23	2 34	9 29	12 40	6 47	4 17	12 42	0aq16	7 11	12 2	2 48	1 12	24 1	17 26	0 38	25 0	10 6	5 59	1 5	7 2	0 32
2/26/1999	22 19	0 42	9 33	16 52	13 7	4 37	18 5	6 59	7 7	13 22	5 47	1 13	25 46	21 9	0 44	26 54	9 34	5 45	2 44	10 35	0 28
3/ 8/1999	24 15	28 21	9 28	21 12	19 35	4 56	23 3	13 19	7 1	14 43	8 50	1 14	27 31	24 55	0 51	28 47	9 43D	5 30	4 23	14 5	0 25
3/18/1999	26 11	25 44	9 13	25 40	26 10	5 14	27 38	19 17	6 54	16 4	11 58	1 15	29 16	28 42	0 57	0le41	10 28	5 16	6 4	17 32	0 21
3/28/1999	28 6	23 9	8 48	0ar18	2ar52	5 31	1aq54	24 54	6 45	17 25	15 8	1 17	1 2	2 29	1 4	2 35	11 47	5 2	7 46	20 54	0 17
4/ 7/1999	0li 2	20 52	8 15	5 7	9 43	5 47	5 53	0pi 9	6 36	18 46	18 20	1 18	2 48	6 17	1 10	4 29	13 33	4 49	9 30	24 10	0 13
4/17/1999	1li57	19vi 5R	7s37	10ar 8	16ar43	6n 1	9aq36	5pi 3	6n27	20ar 8	21ar32	1s20	4ar35	10ar 3	1s18	6le22	15cn44	4s37	11aq15	27aq18	0n 8
4/27/1999	3 52	17 57	6 57	15 21	23 53	6 13	13 6	9 36	6 17	21 29	24 44	1 23	6 22	13 47	1 25	8 17	18 15	4 26	13 1	0pi16	0 4
5/ 7/1999	5 48	17 30	6 16	20 49	1ta13	6 22	16 23	13 46	6 7	22 51	27 54	1 25	8 10	17 29	1 33	10 11	21 3	4 16	14 48	3 3	0s 2
5/17/1999	7 44	17 43	5 37	26 32	8 43	6 28	19 29	17 33	5 57	24 14	1ta 1	1 28	9 58	21 7	1 41	12 6	24 5	4 7	16 37	5 37	0 7
5/27/1999	9 40	18 31	5 1	2ta31	16 25	6 31	22 26	20 54	5 46	25 37	4 5	1 31	11 47	24 40	1 50	14 1	27 19	4 0	18 27	7 54	0 14
6/ 6/1999	11 36	19 52	4 27	8 47	24 19	6 29	25 14	23 47	5 35	27 0	7 4	1 35	13 36	28 7	2 0	15 56	0le44	3 53	20 19	9 52	0 21
6/16/1999	13 32	21 41	3 57	15 22	2gm24	6 23	27 54	26 8	5 23	28 23	9 58	1 38	15 26	1ta27	2 11	17 52	4 17	3 47	22 13	11 29	0 29
6/26/1999	15 29	23 54	3 29	22 15	10 41	6 11	0pi27	27 54	5 9	29 47	12 43	1 43	17 17	4 39	2 22	19 49	7 57	3 41	24 7	12 39	0 37
7/ 6/1999	17 27	26 28	3 3	29 28	19 10	5 52	2 54	29 0	4 54	1 11	15 20	1 48	19 9	7 40	2 35	21 46	11 43	3 37	26 4	13 20	0 47
7/16/1999	19 25	29 19	2 40	6 59	27 49	5 27	5 16	29 21	4 35	2 35	17 47	1 53	21 1	10 29	2 49	23 43	15 35	3 33	28 2	13 29	0 57
7/26/1999	21 24	2 25	2 19	14 48	6 38	4 55	7 32	28 52	4 13	4 0	20 0	1 59	22 55	13 4	3 4	25 42	19 30	3 29	0pi 1	13 4	1 9
8/ 5/1999	23 24	5 44	1 59	22 55	15 36	4 17	9 44	27 33	3 47	5 25	21 59	2 5	24 49	15 20	3 21	27 41	23 30	3 26	2 2	12 6	1 20
8/15/1999	25 24	9 14	1 40	1cn16	24 41	3 32	11 51	25 23	3 15	6 51	23 40	2 12	26 44	17 16	3 40	29 41	27 32	3 24	4 5	10 37	1 32
8/25/1999	27 25	12 53	1 23	9 50	3le52	2 41	13 55	22 33	2 39	8 17	25 1	2 20	28 40	18 48	4 0	1 42	1vi37	3 22	6 10	8 46	1 43
9/ 4/1999	29 28	16 41	1 7	18 34	13 6	1 46	15 55	19 17	1 59	9 43	25 59	2 28	0ta36	19 51	4 22	3 44	5 43	3 20	8 16	6 45	1 53
9/14/1999	1 31	20 37	0 51	27 24	22 21	0 48	17 53	15 56	1 17	11 10	26 30	2 37	2 34	20 21	4 45	5 47	9 51	3 19	10 24	4 46	2 1
9/24/1999	3 36	24 38	0 36	6 16	1vi36	0s13	19 47	12 52	0 35	12 38	26 32R	2 46	4 33	20 15R	5 10	7 51	13 59	3 18	12 34	3 4	2 7
10/ 4/1999	5 42	28 46	0 22	15 6	10 48	1 13	21 39	10 22	0s 3	14 6	26 5	2 55	6 33	19 31	5 34	9 57	18 7	3 18	14 45	1 50	2 11

Ephemeris	80			433			1221			1489			114			3018			2041		
0 Hours ET	Sappho			Eros			Amor			Attila			Kassandra			Godiva			Lancelot		
Mo Dy Year	h long	g long	g lat	h long	g long	g lat	h long	g long	g lat	h long	g long	g lat	h long	g long	g lat	h long	g long	g lat	h long	g long	g lat
10/14/1999	7 49	2 58	0 7	23 52	19 55	2 13	23 29	8 35	0 37	15 34	25 7	3 3	8 34	18 10	5 57	12 3	22 14	3 18	16 59	1 11	2 13
10/24/1999	9 57	7 14	0n 7	2vi28	28 56	3 9	25 16	7 37	1 6	17 3	23 43	3 10	10 36	16 18	6 16	14 11	26 19	3 19	19 14	1 10D	2 14
11/ 3/1999	12 8	11 34	0 21	10 53	7 47	4 3	27 2	7 24	1 31	18 33	21 57	3 15	12 39	14 4	6 31	16 20	0li22	3 20	21 30	1 46	2 14
11/13/1999	14 19	15 57	0 36	19 3	16 29	4 52	28 46	7 53	1 51	20 3	20 0	3 18	14 44	11 43	6 38	18 31	4 20	3 21	23 49	2 59	2 14
11/23/1999	16 33	20 22	0 51	26 56	24 59	5 36	0ar28	8 58	2 9	21 33	18 2	3 18	16 50	9 32	6 39	20 43	8 14	3 23	26 9	4 45	2 12
12/ 3/1999	18 48	24 49	1 6	4 32	3sc16	6 16	2 9	10 34	2 24	23 4	16 15	3 15	18 57	7 45	6 34	22 57	12 0	3 25	28 31	6 59	2 11
12/13/1999	21 6	29 17	1 22	11 49	11 18	6 50	3 50	12 36	2 37	24 36	14 48	3 11	21 5	6 32	6 24	25 13	15 39	3 28	0ar54	9 39	2 10
12/23/1999	23 25	3 45	1 38	18 47	19 7	7 20	5 29	15 0	2 49	26 9	13 50	3 5	23 15	6 0	6 10	27 31	19 6	3 30	3 20	12 40	2 9

About the Author

Nebraska-born J. Lee Lehman is an incongruous blend; an ultra-traditional astrologer with expertise in asteroids and heliocentric, her interests range from mysticism to statistics. Dr. Lehman has degrees from Rutgers University in botany, and uses her scientific training and discipline in her approach to astrology. A political activist in college, she has succeeded in integrating irreverence for authority with a profound respect for historical precedence.

Since 1980 she has published numerous asteroid ephemerides, an ephemeris of Halley's Comet, and articles on such diverse subjects as asteroids, heliocentric Mercury, research methods in astrology, essential dignities, archaeoastronomy and computers in astrology. Dr. Lehman is an expert in database programming, and is especially interested in the ways that astrology databases can be used to verify traditional and new astrological methods. An avid reader and trained debater, she is concerned about scientific ethics and politics during this period when massive changes are taking place in the underlying assumptions about the way science is practiced.

Dr. Lehman resides in the San Francisco Bay area, where in addition to her astrological interests, she enjoys hiking, marine mammal watching and traveling.